Supernovae and Nucleosynthesis

David Arnett

SUPERNOVAE AND NUCLEOSYNTHESIS

An Investigation of the History of Matter, from the Big Bang to the Present

PRINCETON UNIVERSITY PRESS

Princeton, New Jersey

Copyright © 1996 by Princeton University Press
Published by Princeton University Press, 41 William Street,
Princeton, New Jersey 08540
In the United Kingdom: Princeton University Press,
Chichester, West Sussex

All Rights Reserved

Library of Congress Cataloging-in-Publication Data

Arnett, David, 1940–
   Supernovae and nucleosynthesis : an investigation of the history
of matter, from the big bang to the present / David Arnett.
      p.   cm.—(Princeton series in astrophysics)
   Includes bibliographical references and index.
   ISBN 0-691-01148-6 (alk. paper).—ISBN 0-691-01147-8 (pbk. :
alk. paper)
      1. Cosmology.   2. Stars–Evolution.   3. Supernovae.
   4. Nucleosynthesis.   5. Nuclear Astrophysics.   I. Title.
II. Series.
QB981.A66   1996
523.1—dc20   95-41534
                                                                           CIP

This book has been composed in Times Roman

Princeton University Press books are printed on acid-free paper and meet the
guidelines for permanence and durability of the Committee on Production
Guidelines for Book Longevity of the Council on Library Resources

Printed in the United States of America
by Princeton Academic Press

1 3 5 7 9 10 8 6 4 2
1 3 5 7 9 10 8 6 4 2
(pbk.)

To Bette, Eric, Jennifer, and Delaney

More probably what astronomers are really
viewing is precisely what they have always
viewed—the edge of their own vision.

—*TIME*, April 23, 1973

Contents

List of Figures

List of Tables

Preface

This book is not intended to be a scientific history, a textbook, or a review, although it has some of these elements and could serve these purposes. It is intended to be what was well described by Prof. S. Chandrasekhar in reference to his own goals for scientific books: "...a certain viewpoint of the field, written by one who has been an active participant in its development..." The subject is the synthesis and evolution of atomic nuclei, by thermonuclear reactions, from the Big Bang to the present. What is the origin and history of the matter of which we are made? While much has been accomplished, much more remains to be done. The intent is to illustrate the intellectual basis and the interconnections among topics too seldom considered together. In such a broad and actively developing subject, newer results will soon outdate many of the details presented here, but the ideas should survive longer. Many areas of research bear on this broad topic, and in my opinion the weak communication between these areas is an impediment to progress. By attempting a broad synthesis of the topic, it is hoped that change will be assisted and catalyzed.

This book is intended to be a new synthesis; essentially all of the calculations and figures have been redone for this purpose. The gestation period has been painfully long. Some of these results have appeared in journals and conference proceedings, but many have not. Some have been independently discovered and published by others—to whom just credit is due—but in these cases I have to beg pardon for not re-writing the relevant chapters unless the work broke new ground. To keep the project finite, I have had to choose and to omit; other authors—or this author writing a few months later—would probably have produced a significantly different result. To guide the reader, there are extensive references to provide an introduction to the literature. Many equally good references were not quoted due to lack of space, to choice of emphasis, or simply oversight, and the reader should expect pleasure from their discovery. I am painfully aware of this limitation, and I appologize to any authors so slighted. My only defense is that, at over 700 citations, I may have lost control!

Special thanks are due to those who taught me so much about this subject, Al Cameron and Willy Fowler, and to Sir Fred Hoyle, who taught us all. The entry of Hans Bethe into the area of supernova theory has enriched it enormously, and given me special pleasure. I thank my colleagues and students, especially Don Clayton, Jim Truran, and

Stan Woosley, with whom this material was learned. Friedel Thielemann provided the reaction rate data set so extensively used for the text. Anurag Shankar helped translate this data to a convenient form. Bruce Fryxell and Ewald Müller provided the basic version of PROMETHEUS, the multidimensional hydrodynamics code which has influenced much of the discussion. Brian Schmidt and Nick Suntzeff provided machine-readable versions of their observational data which greatly improved Chapter 13. Many colleagues and students helped by reading the various drafts, including most recently, Roger Bell, Philip Pinto, Grant Bazan, and Jave Kane. Support from the National Science Foundation and the National Aeronautics and Space Administration is gratefully acknowledged.

Tucson, Arizona, March 15, 1995

Supernovae and Nucleosynthesis

1

Introduction

There is something fascinating about science.
One gets such wholesale returns of conjecture
out of such a trifling investment of fact.
—Mark Twain (1835–1910),
Life on the Mississippi

We live in a Universe of matter. At the present epoch, the matter that we see shining contains far more energy in its rest mass than all of the energy of the microwave radiation left from the Big Bang. This matter is composed of electrons, protons, and neutrons, and their combined forms of nuclei, atoms, and molecules. Examination of the dynamics of galaxies indicates the existence of gravitational forces which imply still more matter, by roughly a factor of ten, which is not shining in ways we have yet detected. This is the *dark matter*, and its nature is a deep puzzle.

Our understanding of the physical universe is beginning to allow us to trace the history of matter from the Big Bang to the present. With the discovery of the positron in 1932, and its identification as a consequence of Dirac's theory of relativistic quantum mechanics, it seemed that there was a strange sort of mirror image of everything: the positron was an antielectron, and there were antiprotons, antineutrons, and so on, perhaps to antiworlds and antipeople. But this does not seem to be so; while antiparticles are routinely made by experimental physicists and by cosmic rays, antimatter seems to be rare in our universe. Matter and antimatter do not coexist well; particles and their antiparticles combine (annihilate) to form pure radiation (photons). This is conversion of mass into energy with 100% efficiency! Particle physics, driven by experiments in terrestrial laboratories and by the search for underlying patterns, may explain why an ever so slight excess of matter over antimatter developed early in the Big Bang. This then resulted in extensive annihilation of matter and antimatter as the universe cooled, converting essentially all the antimatter—as well as almost all the matter—to energy, and generating far more photons than the few particles of matter left over. This gave a *high entropy* universe, a theoretical concept proved correct by the observation of a universal background of microwave radiation at 2.7 degrees kelvin. In this context, high entropy means many photons (particles of light) per particle of matter (electrons, protons, and neutrons), or by terrestrial and stellar standards, a very low density for any given temperature. As a consequence of this high entropy, the major products of cosmological nucleosynthesis were hydrogen and helium—fuels for stars, not ashes. This is the primordial matter. Thus stars could shine and produce elements such as carbon, oxygen, and iron, from which planets and people are made.

Nuclear physics is fundamental to our understanding of stellar phenomena. The nature and behavior of stars affects all of astronomy, and stars themselves are gravitationally contained thermonuclear reactors. Stars are seen forming from the gas in the interstellar medium. Many stars are observed to eject considerable amounts of matter back into space, which presumably will itself form new stars. The observed abundances of nuclei in stars and galaxies are thought to be the result of thermonuclear processing in previous generations of stars. The most spectacular example is the explosion of a star as a *supernova*. Other astronomical situations have been observed, or suggested, in which conditions may occur that are sufficiently extreme for nuclear processing—for example, the "Big Bang" itself; the violent events in galactic nuclei; Seyfert galaxies and quasi-stellar objects; x-ray sources and pulsars.

Nucleosynthesis is a term coined to refer to those processes by which atomic nuclei are transformed on the cosmic scale. It involves the study of how primordial matter is processed into the abundances observed in astronomical objects (e.g., the solar system, stars, interstellar gas, cosmic rays). The observed abundances of nuclei probably represent the ashes of previous stages of thermonuclear burning. Or, to change the metaphor, this book deals with an attempt to construct a sort of "nuclear paleontology." Instead of a fossil record, the cosmic abundances of atomic nuclei are used to infer the events by which the universe was formed and evolved to its present state.

In a given sample of matter the dominant influences in determining the composition include (1) atomic and molecular processes and (2) nuclear processes. To reflect this, one may divide compositional information into two classes: elemental and isotopic abundances. The abundance ratios of different elements will often be sensitive to the previous effects of atomic processes. For example, the composition of condensed bits of interstellar matter would favor easily condensed elements over the noble gases. The abundance ratios of different isotopes of the same element will be relatively less sensitive to atomic processes, and therefore should provide more reliable clues as to the last nuclear processes to which the matter was exposed.

The isotopic composition of matter represents the contributions of many sources of nucleosynthesis. The process of mixing appears here as a fundamental difficulty. For a complex ensemble one cannot infer the complete nature of the individual nucleosynthesis events from a knowledge only of the average result. Thus one must try to determine the nature of individual events as well as their accumulated effect.

The theory of nucleosynthesis will be powerful only to the extent that it is quantitative. This requirement places severe demands upon knowledge of cosmic abundances, nuclear reaction rates, and the conditions in and the nature of astronomical objects. A consideration of thermonuclear conditions involves us in a broad variety of astronomical problems. The initial conditions for stellar nucleosynthesis are thought to be the results of cosmological nucleosynthesis. Our ideas about the nature of the Big Bang are largely constrained by the degree to which we can identify some abundances being the relic of cosmological nucleosynthesis rather than the result of subsequent stellar processes.

The nature of nucleosynthesis is deeply connected with the nature of stellar evolution. It is in stars that conditions are found which are capable of providing the abundances of almost all of the nuclei we observe. Not only must the conditions be right for thermonuclear burning, but the nuclei so processed must be ejected into the interstellar medium for subsequent incorporation into the stars, meteorites, nebulae, and cosmic rays in which we observe these abundances. Because of this, and also because most of the nuclear processing occurs during the last, most extreme conditions, the late and final stages of stellar evolution are of particular importance. The stages of "ordinary" stellar evolution (hydrogen and helium burning) and the final states of stars (white dwarfs, neutron stars, and black holes?) must be connected by events (supernovae, novae, planetary nebula ejection). One deals with *events* more than *objects*; dynamics become all important.

Even with a thorough understanding of stellar evolution (which is far from available), we would have a serious difficulty. How do ensembles of stars and gas clouds interact as galaxies? As yet we have only the rude beginnings of an answer. It is toward these questions that this book is directed.

2

Abundances of Nuclei

We see how we may determine their forms, their distances,
their bulk, their motions, but we can never know anything of
their chemical or mineralogical structure. . . . We
must keep carefully apart the idea of the solar system
and that of the universe, and be always assured that
our only true interest is with the former. . . . The
stars serve us scientifically only as providing positions.
—Auguste Comte (1798–1857),
Course of Positive Philosophy

It is a capital mistake to theorize before one
has data. Insensibly one begins to twist facts to
suit theories, instead of theories to suit facts.
—Arthur Conan Doyle (1859–1930),
A Scandal in Bohemia

Before considering theories of how the atomic nuclei were formed in
their observed abundances, we must consider the abundance data itself, and how it is obtained. We encounter two fundamental problems:
sampling and accuracy. The most accurate methods of abundance determination involve the analysis of a sample in a laboratory, or a spacecraft.
Obviously such samples represent only the most minuscule fragment of
an astronomical object; to what extent is it representative? With matter
so measured we can only probe a restricted region of space and time.
To probe further we must use less accurate methods. One of the tri-
umphs of twentieth-century astronomy is the use of stellar spectra to
infer abundance information from a much larger region of the universe.
These data are fundamental to our study, but suffer from some quite
understandable flaws.

First, the information so obtained refers almost entirely to the abundance of atomic elements, not of isotopes; that is, it tells us only about
the number of protons in a nucleus, but not the number of neutrons.
While different isotopes of the same element will have almost identical

atomic properties, they will have drastically different nuclear properties. It is the nuclear properties that contain the information we seek concerning past history in thermonuclear environments. Only for some molecular lines do we extract subtle shifts of energy due to the different masses of the isotopes, and obtain isotopic information from stellar spectra.

Second, as will be discussed below from a relatively simple viewpoint, to analyze stellar abundances we must build a model of a complex physical system—a stellar atmosphere. The inferred abundances depend sensitively upon the way in which the model is built, and upon the choice of the parameters of the model, especially the temperature, the gravity, and the atomic, ionic, and molecular physics. Our ability to determine elemental abundances in this manner is limited by our knowledge of the physical system.

Third, the atmospheric abundances may not be representative of the information we seek. A stellar atmosphere is only a thin layer at the surface of the star, and may be affected in an unrepresentative way by the poorly understood processes of mass loss and mixing. The star itself may be subject to relatively superficial processes which obscure its original abundance pattern.

Meteorites provide another source of abundance data, and one of the most important. Unlike planetary and lunar material, some types of meteorite seem to be chemically representative samples of solar system matter. Other meteorites (or parts of meteorites) seem to have originated as solid bodies beyond the solar system; they are truly star dust. But a key point about meteorites is the extraordinary accuracy with which their abundances are measured. At present, the accuracy obtained from meteoritic analysis is roughly a thousand times better than can be obtained from stellar spectral analysis.

But how representative of astronomical objects are these tiny bits of alien matter? We are not sure. A further complication arises from atomic processes that cause elements to separate into different forms (water, ammonia, methane, and so on). This *chemical fractionation* masks the original nuclear abundance information. Fortunately, isotopic ratios, and elemental ratios for elements with similar chemical properties such as the rare earths, are insensitive to such effects. Historically the stellar and meteoritic data have formed a more or less complementary set. The most influential development in abundance determination has been the discovery of *isotopic anomalies* in meteorites and stars. By an *anomaly* we mean a variation from the average abundance which is beyond that attributable to the known sources solar system abundance variations, such as chemical fractionation or cosmic-ray irradiation. Many of the anomalies first found were associated with extinct radioactivities: when the solid was formed it contained a radioactive parent nucleus which subsequently decayed away, leaving variations in the abundance of a daughter nucleus. The origin of these anomalies is still being debated, but they seem to represent direct evidence for some particular nucleosynthesis events, the sum of which presumably add up to give as a total result the solar system abundance pattern. The evidence provided by radioactivities has an additional aspect which is of fundamental importance: it provides information concerning time-scales ("cosmic clocks") in the production of these abundances.

Cosmic rays represent a relatively fresh sample of matter from outside the solar system. The discovery of meteoritic abundance anomalies became possible because of improvement in experimental techniques, and cosmic-ray experiments are passing a similar threshold. The mere existence of a set of isotopic abundances for matter from outside the solar system will be enormously influential. As with meteorites, cosmic rays present a problem with regard to sampling. Their sources and their acceleration mechanisms are actively debated. Cosmic rays might be a sample of some freshly synthesized matter from supernovae, or perhaps they might be a sample of interstellar matter that has seen 4.55 billion years more galactic evolution than did the meteoritic material. Either way the cosmic rays represent an important possible source of abundance data.

A growing set of isotopic abundance data is being obtained from molecular lines, either in cool stellar atmospheres or interstellar clouds and nebulae. Most of the information obtained so far is related to hydrogen, oxygen, carbon, and nitrogen because of their large abundances. Such information can test ideas concerning the first stages of stellar evolution, hydrogen and helium burning, during which these elements are thought to be most affected.

Finally, we should note that spectroscopy of dilute gas—such as planetary nebulae, supernova remnants, novae, supernovae, and jets—provides some information directly concerning the events which eject and perhaps synthesize newly processed nuclei. With the recent nearby supernovae SN1987A and SN1993J, there is an unprecedented wealth of high-quality data now available.

Progress in these various lines of research has now given us hints as to the course of thermonuclear evolution in stars, galaxies, and the Big Bang. Most likely our ideas will need modification. We may hope that by considering the various data, with their strengths and weaknesses in mind, we may discover the essential clues to the mysteries of the cosmic abundance pattern.

2.1 WHAT ARE ABUNDANCES?

By the term *abundance* we mean the fraction of a given sample which is in a particular form. This implies that we consider the ratio of number (for example) of this particular species to some standard. It is desirable to consider this standard to be invariant with respect to compression, so that changes in abundance reflect only nuclear processing. Before discussing the *solar system abundances*, we will carefully define what we mean by an "abundance."

Consider number density N_j, the number of species j per unit volume. If expansion or contraction occurs—a common phenomenon for gases and plasmas—changes in N_j reflect these effects rather than nuclear transmutations. In order to avoid such inconvenient and perpetual bookkeeping, it is useful to consider the number of species j per unit of some conserved quantity, instead of per unit volume. Nucleon number is such a quantity, being conserved even in the relativistic limit.

Avogadro's number N_A is defined as the number of atoms of some species j which makes W_j grams, where W_j is the atomic weight of species j. For helium, W_{He} is 4.0026 for example. If m_j is the mass of a nucleus of species j as measured in the lab, then for a mixture consisting only of species j, the mass density is $\rho_m = N_j m_j$ and $N_A m_j = W_j$. For a mixture of species, $\rho_m = \Sigma_j N_j m_j$.

Atomic weights are expressed on a scale where $W(^{12}C) = A(^{12}C) = 12$. The mass unit M_u is $M_u = m_{^{12}C}/12 = 1/N_A$, where $m_{^{12}C}$ is the mass of a single neutral atom having a nucleus of ^{12}C. In general,

$$W_j = m_j/M_u \tag{2.1}$$

$$= (Z_j m_H + (A_j - Z_j)m_n - B_j/c^2)/M_u \tag{2.2}$$

where m_p, m_e, and m_n refer to the masses of protons, electrons, and neutrons respectively; and $m_H = m_p + m_e$. The *nuclear binding energy* of the nucleus j is B_j. The *mass excess* is $m_j - A_j M_u$. Using the binding energy for ^{12}C, we find

$$B(^{12}C)/(m(^{12}C)c^2) = (W_H + W_n)/2 - 1, \tag{2.3}$$

so that

$$W_j = A_j + A_j \frac{B(^{12}C)/12 - B_j/A_j}{M_u} + \left(Z_j - \frac{A_j}{2}\right)(W_H - W_n). \tag{2.4}$$

Since $W_H = 1.007825$ and $W_n = 1.008665$, and usually Z_j is roughly $A_j/2$, the last term is quite small. The second term can be more important, but is still much less than A_j. The mass density is

$$\rho_m = \frac{\sum_j N_j W_j}{N_A} \tag{2.5}$$

which becomes

$$\rho_m = \sum_j \left(\frac{N_j A_j}{N_A}\right)(1 + b_j)$$

where

$$b_j = \frac{B(^{12}C) - 12B_j/A_j}{m(^{12}C)c^2}.$$

This quantity ρ_m changes with composition and is not relativistically invariant. However, use of the related quantity

$$\rho \equiv \frac{\sum_j N_j A_j}{N_A} \tag{2.6}$$

avoids these difficulties. Note that ρ almost equals ρ_m. We now define the *nucleon fraction* X_j for species j to be

$$X_j \equiv \frac{N_j A_j}{\rho N_A}, \tag{2.7}$$

which is just the fraction of the nucleons in the sample which are tied up in the form of particles of species j. In the astrophysical literature this and a similar quantity involving ρ_m, not ρ, are both called the *mass fraction*. We will attempt to use the more exact term *nucleon fraction* for X_j in what follows. From 2.6 and 2.7,

$$\sum_j X_j = 1. \tag{2.8}$$

A related quantity is

$$Y_j = \frac{X_j}{A_j} = \frac{N_j}{\rho N_A}, \tag{2.9}$$

which is a ratio of the number of nuclei of species j to the total number of nucleons in the system. Now, the total number of nucleons is

$$n = \sum_j N_j A_j, \tag{2.10}$$

so

$$\sum_j Y_j = \frac{\sum_j N_j}{n}, \tag{2.11}$$

which is not equal to unity, in general. Notice that Y_j is the fraction of a mole of particles in the form of species j, so that Y_j is a *mole fraction*. In actual calculations the distinction between ρ and ρ_m is usually not important numerically. From this point on, the expression 2.6 will be used. The only problem that could arise would be if we were careless about defining rest mass for newtonian gravity,[1] and even then the error would be small.

Experimentally, the choice of normalizing to the total number of nucleons has a disadvantage: because this involves a sum over all species, errors in species of large abundance dominate. In fact, the estimated solar abundances of hydrogen plus helium have varied by 0.5% in the last decade; the corresponding mass fraction of heavier elements has varied by about 1/3 of its value. To minimize such problems, it is traditional to express elemental abundances by number with a normalization. In the meteoritics community the choice is to make the number of silicon atoms equal to 10^6. This is the "$\log Si = 6$" scale, on which the abundances y_i are related to the Y_i by

$$\log y_i = \log f_{Si} + \log Y_i. \tag{2.12}$$

Notice that

$$\sum_j y_j A_j = \sum_j f Y_j A_j, \tag{2.13}$$

$$= f, \tag{2.14}$$

for any normalizing constant f. The Anders-Grevesse [13] abundances are given on this scale; for atomic silicon (all isotopes), $Y_{Si} = 2.529 \times 10^{-5}$, and therefore

$$\log f_{Si} = 10.5970, \tag{2.15}$$

which is consistent with direct evaluation of $\sum_j y_j A_j$.

Astronomers sometimes use a scale on which the number of hydrogen atoms equals 10^{12}. If abundances y_i are expressed on this scale,

$$\log y_i = \log f_H + \log Y_i, \tag{2.16}$$

---

[1]In special and general relativity, we would consider such effects and include them in the internal energy.

where the Anders-Grevesse [13] abundances (rescaled) give

$$\log f_H = 12.1514. \tag{2.17}$$

A common alternative is to express abundance *ratios* relative to solar values:

$$\log[N_i/(N_i)_\odot] - \log[N_j/(N_j)_\odot] \equiv [i/j]. \tag{2.18}$$

Thus, a ratio of sodium to iron which is half solar would be [Na/Fe] = −0.3. Historically some abundances have been inferred by differential comparison with the sun, or a standard star, so that ratios of abundances might be expected to be more accurate than absolute values.

2.2 SOLAR SYSTEM ABUNDANCES

The fundamental set of abundance data is that for the solar system; this is our standard of reference. It is constructed from a variety of data sets, but primarily (1) solar photospheric elemental abundances, and (2) isotopic abundances from carbonaceous chondritic meteorites. The Sun comprises most of the mass of the solar system, and therefore is more representative than, for example, planets, which have undergone extensive chemical fractionation. The carbonaceous chondritic meteorites, though containing only a minuscule fraction of the nucleons in the solar system, are thought to be the most primitive of solar system material, and can be subjected to highly accurate laboratory analysis.

The solar system abundances show patterns that are directly related to the systematic behavior of nuclei. This will become clearer as the reader progresses through subsequent chapters; this section might be reviewed as they are read, especially chapter 3.

The twenty-five most abundant nuclei (by mass) in the solar system abundance table of Anders and Grevesse [13] are shown in table 2.1. The complete abundances are given in tabular form in Appendix A. Almost all the mass (about 98 percent) is contained in two nuclei, ^{1}H and ^{4}He, which are thought to have been produced in the Big Bang. These nuclei are not the most tightly bound, so that this abundance pattern is incompletely relaxed to equilibrium. The prevalence of low A (i.e., 1 and 4) rather than high A ($A \approx 56$) is a consequence of initial conditions, and of cosmological significance (see chapter 5).

Due to the history of the technology, the most easily identified of the elements heavier than helium exist in the metallic state under terrestrial conditions. Hence the amusingly inaccurate[2] term *metallicity* is used to

---

[2] Oxygen is the most abundant of these "metals."

TABLE 2.1

The 25 Most Abundant Nuclei

Rank	Z	Symbol	A	Nucleon Fraction	Source (process)
1	1	H	1	7.057e-01	Big Bang
2	2	He	4	2.752e-01	Big Bang, CNO, pp
3	8	O	16	9.592e-03	Helium
4	6	C	12	3.032e-03	Helium
5	10	Ne	20	1.548e-03	Carbon
6	26	Fe	56	1.169e-03	e-process
7	7	N	14	1.105e-03	CNO
8	14	Si	28	6.530e-04	Oxygen
9	12	Mg	24	5.130e-04	Carbon
10	16	S	32	3.958e-04	Oxygen
11	10	Ne	22	2.076e-04	Helium
12	12	Mg	26	7.892e-05	Carbon
13	18	Ar	36	7.740e-05	Silicon, Oxygen
14	26	Fe	54	7.158e-05	e-process, Silicon
15	12	Mg	25	6.893e-05	Carbon
16	20	Ca	40	5.990e-05	Silicon, Oxygen
17	13	Al	27	5.798e-05	Carbon
18	28	Ni	58	4.915e-05	Silicon, e-process
19	6	C	13	3.683e-05	CNO
20	2	He	3	3.453e-05	Big Bang, pp
21	14	Si	29	3.448e-05	Carbon, Neon
22	11	Na	23	3.339e-05	Carbon
23	26	Fe	57	2.840e-05	e-process
24	14	Si	30	2.345e-05	Carbon, Neon
25	1	H	2	2.317e-05	Big Bang

denote the net abundance of matter with $Z > 2$. To obtain the nucleon fractions X_j relative to these $Z > 2$ elements only, divide the entry in table 2.1 by 0.0191. Notice that about half of all such matter is in the form of ^{16}O.

Given the high abundance of ^{1}H and ^{4}He, how can the heavier yet more tightly bound nuclei have remained so rare? An obvious inhibitor is the coulomb repulsion between the positively charged nuclei. As will be discussed in chapter 3, the probability of penetrating this coulomb barrier has an exponential dependence on the product of the charge of reactants, Z_1Z_2. For example, the fusion of two oxygen nuclei implies a product 64 times larger than for hydrogen fusion, and this product occurs in an exponential function. This strong inhibition can be countered by higher relative velocities of reactants, that is, higher temperatures for the nuclei. To obtain such higher temperatures the matter must drop deeper into the gravitational potential well of the star. To be seen in material from which we define the solar system pattern, the matter must escape from this deep well. This difficulty of escape—which leads to containment in white dwarfs, neutron stars, and black holes—also acts to keep the more massive nuclei rare.

What of the nuclei which do escape the stellar furnace to be measured in solar system abundances? After ^{1}H the next most abundant odd A nucleus is ^{25}Mg, at fifteenth in the list. There is clearly a strong effect which favors *even-A* nuclei over *odd-A* ones. Further, of the intervening nuclei, only ^{14}N is not *even-even*, a term meaning a nucleus having both even-Z and even-N. As we shall see, this exception has to do with the nuclear peculiarities of the first (and easiest to attain) burning stage— hydrogen burning. Except for ^{56}Fe, many of the most abundant nuclei are *even-even* and have $Z = N$, and hence are called *alpha-particle* nuclei. The most abundant are ^{16}O and ^{12}C, followed by ^{20}Ne, ^{24}Mg, ^{28}Si, ^{32}S, ^{36}Ar, and ^{40}Ca; the latter are denoted *α-nuclei* in figure 2.1.

There are other favored configurations. Nuclei, like atoms, have a shell structure. Certain combinations of protons (or of neutrons) are particularly tightly bound; these are called *magic numbers*. Notice the high abundance of doubly magic nuclei: ^{4}He with $Z = N = 2$, ^{16}O with $Z = N = 8$, ^{40}Ca with $Z = N = 20$, and for Fe made as Ni, ^{56}Ni with $Z = N = 28$. It has long been suggested that ^{56}Fe was formed as an alpha-particle nucleus ^{56}Ni [597, 483]. The steep peak in the abundances at ^{56}Fe is due to the fact that the most tightly bound of nuclei having equal numbers of neutrons and protons is ^{56}Ni. The light curves of both Type I and Type II supernovae bear witness to the radioactive decay sequence Ni $\rightarrow$ Co $\rightarrow$ Fe for $A = 56$.

The complete set of solar system abundances is given in figure 2.1. The solid squares are data from the 1989 compilation of Anders and Grevesse [13], while the open squares are due to Cameron [148], from 1982. The overall agreement is excellent, suggesting that the values are relatively well agreed upon. Several of the effects just mentioned are

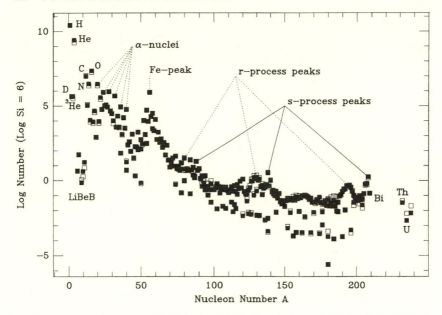

Fig. 2.1. Abundance Features

obvious in the figure. The striking decline in abundance with A is relatively smooth, with valleys at $A = 5$ to 11 and around $A = 45$, and the dramatic peak at iron ($A = 56$). There is a pronounced odd-even effect: even A nuclei are favored. This is especially true for nuclei with both even N and even Z, such as C, O, and the nuclei labeled "α-particle."

A detailed examination reveals more evidence of nuclear systematics reflected in the abundance pattern. In figure 2.2 are shown the abundances versus nucleon number A, with A ranging from 0 to 64. Isotopes of the same element are connected by lines. There are two deep valleys, at LiBeB and around Sc, which reflect unusually low binding energies for nuclei in those regions. Note that F (fluorine), which lies just beyond the magic proton number $Z = 8$ and is therefore weakly bound, has a correspondingly small abundance.

A pronounced odd-even effect in Z may be seen to modulate the abundance pattern, with odd Z elements being rarer (except for H). For a proton number $Z \le 20$, the most abundant isotopes are those which are least neutron-rich. Superimposed on this is a tendency for even-N to be favored over odd-N; sulfur ($Z = 16$) is a clear example.

For $30 > Z > 20$, the most abundant isotopes tend to be those which have $N = Z + 2$; this is the case for Ti, Cr, and Fe, and nearly so for Ni. For larger Z, the most abundant isotopes are more neutron rich.

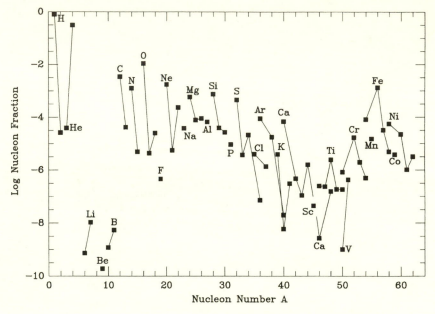

Fig. 2.2. Abundance ($A = 1$, 64)

The most abundant isotopes of Ti, Cr, and Fe are $A = 48$, 52, and 56 respectively. These are thought to be formed as ^{48}Cr, ^{52}Fe, and ^{56}Ni, all of which are unstable to positron emission and/or electron capture. In an explosive situation, these weak decays may not have time to occur, so nuclei with $Z = N$ are formed, to later decay (twice) to their even-even daughters with $N = Z + 2$. As we shall see, the resulting abundance patterns agree well with those shown here.

Hydrogen burning forms ^{4}He which corresponds to a change from matter that is proton-rich to matter that has $Z = N$. The abundance pattern, with its preference for $Z = N$, suggests that the *ejected* matter was not processed to extreme densities, at which electron capture would make it neutron-rich. This may be seen explicitly by examining the abundance ratios of ^{26}Mg/^{24}Mg, ^{30}Si/^{28}Si, ^{34}S/^{32}S, ^{38}Ar/^{36}Ar, and ^{42}Ca/^{40}Ca, for example.

Beyond the *iron-peak* at $A \approx 56$, the isotopic ratios change character. This may be seen in figure 2.3, which shows abundance versus nucleon number $50 < A < 100$. At Ge ($Z = 32$), the dominance of neutron-poor isotopes switches to favor the neutron-rich ones. Because of the increasingly large coulomb barrier for charged particle reactions, these heavier nuclei are produced by neutron capture processes. Notice that

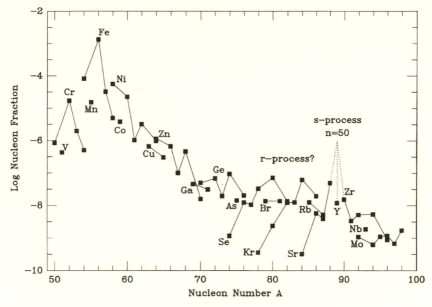

Fig. 2.3. Abundance ($A = 50, 100$)

the level of abundances by mass has dropped by a factor of more than 10^7 relative to hydrogen, so that these nuclei are very rare.

Although they do not represent major yields in the sense of total mass processed, the large number of isotopic ratios available for $A > 80$ contain a large amount of information about their production mechanism.

As we move toward higher A, we see evidence of the *s-process* and *r-process* of Burbidge, Burbidge, Fowler, and Hoyle [135]. This may be seen clearly in figure 2.4, which shows another broad peak near $A = 130$ and a sharp one at $A = 138$. This is repeated at $A = 194$ and $A = 208$; see figure 2.5. The explanation of these three sharp peaks is simple: tightly bound nuclei (having *magic numbers* of neutrons) have relatively small cross sections for neutron capture. When exposed to a *slow* flux of neutrons (so that beta-decays can occur before the next neutron capture), the matter flows into these bottlenecks, creating higher abundances for such nuclei. This gives the sharp *s-process* peaks at higher A; note that the *s* is for *slow*. The weak maximum at Sr ($Z, A, N = 38, 88, 50$), Y ($Z, A, N = 39, 89, 50$), and Zr ($Z, A, N = 40, 90, 50$) corresponds to a neutron magic number of 50. The next *s*-process maxima are at Ba ($Z, A, N = 56, 138, 82$) and Pb ($Z, A, N = 82, 208, 126$). Both have magic neutron numbers, and the Pb peak also has a magic proton number, as behooves its impressive alpine aspect.

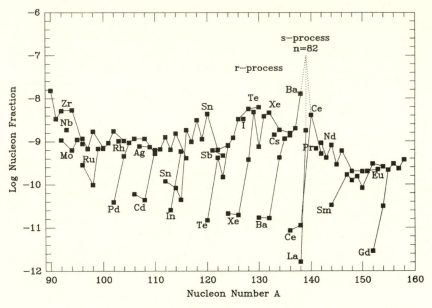

Fig. 2.4. Abundance ($A = 90, 160$)

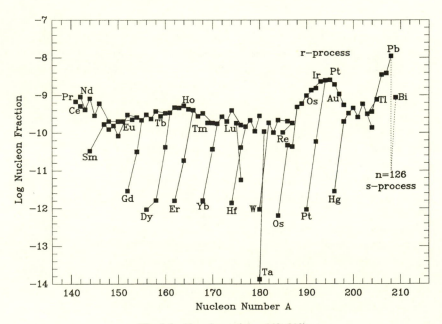

Fig. 2.5. Abundance ($A = 140, 210$)

With this clue we may ask what would happen if neutrons were added *rapidly*, so that the nuclei formed do not have enough time to β-decay all the way back to stable products. As nuclei are formed which are increasingly unstable, their corresponding decay times decrease, and a balance with capture times tends to develop. Upon removal from the neutron flux, the unstable nuclei would β-decay back to stable nuclei, keeping the same A but increasing Z. Further, the peaks would be broadened if several unstable progenitors were possible. For our three broad peaks, we have maxima at (roughly) Se ($Z, A, N = 34, 80, 46$), Te ($Z, A, N = 52, 130, 78$), and Pt ($Z, A, N = 78, 194, 116$). The first two would correspond to magic neutron numbers if they had experienced four decays ($\Delta Z = 4$), while for that last we need ten decays ($\Delta Z = 10$). These *r-process* peaks (*r* for *rapid*) clearly require more knowledge of both unstable nuclei and the astrophysical environment (which affects the balance between capture and decay) to be accurately interpreted. Still, the idea follows easily from the abundance pattern, and as a bonus suggests a natural way that the unstable elements Th (thorium) and U (uranium) originated.

See Clayton [173] for a detailed discussion of the original formulation. Interestingly enough, the yields of *r-process* and *s-process* nuclei seem comparable, which is curious in view of their different nature and presumably different origin (see also [596]).

There is a further implication of this picture: Cameron [139] suggested that further neutron capture results in fission. This process gives a distribution of fragments of high excitation energy [674], which is asymmetrically peaked at $A_0 \sim \frac{4}{2}(1 \pm 0.1)$. This could add complications to the range $90 \leq A \leq 170$, depending upon the nuclei that fission and the amount of cycling.

2.3 STELLAR ATMOSPHERES

Most data concerning the abundances of nuclei at various different locations in the universe come from interpretation of the properties of light emitted from the surface layers of stars. Because this is a complex subject in its own right, the abundances so derived have suffered from less than adequate accuracy, and have had less impact on the quantitative aspects of the subject than one might otherwise have expected. Some insight into the physics of this difficult problem can be obtained with a simple approach. There are two key points. First, the abundances depend sensitively upon the degree of ionization and excitation, which are related to observations through a temperature, and more weakly to a density, if we *assume* statistical equilibrium. Second, the detailed shape of the lines depends upon the structure of the outer layers of the star (this structure is always replaced by a simpler approximation, such as planar or spherical symmetry), and the strength of the lines depends upon atomic and molecular physics quantities which may be poorly known experimentally.

Most stars show a continuous spectrum with superposed dark lines. An early picture of the way these Fraunhoffer lines form in stellar atmospheres was due to A. Schuster and to K. Schwarzschild (see [6, 163]). It was suggested that the continuous spectrum of a star was emitted by a layer of unique temperature, overlaid by an absorbing layer responsible for the formation of absorption lines. This simplified picture is useful as a conceptual aid; actually, the absorbing and emitting regions are not separated so nicely. At the surface of stars the major source of energy transport is radiative diffusion. Let us now assume that, locally, in each thin layer above the photosphere, the matter is in equilibrium with blackbody radiation at the local temperature (this important approximation is called *local thermodynamic equilibrium*, or LTE). This implies that the radiation and matter are well coupled, and that the matter will also take on an equilibrium velocity distribution (which is maxwellian if the gas is nonrelativistic and nondegenerate). Emission and absorption coefficients are related locally by Kirchoff's law (i.e., detailed balance). Because the temperature varies with position above the photosphere, and the radiation processes are energy dependent, the emergent flux does not correspond to that of a blackbody.

Let us define the atmosphere to be that region of the star from which visual electromagnetic radiation can reach the observer directly. Roughly speaking, this can be divided into:

- **photosphere:** the source of continuum radiation,
- **reversing layer:** the source of absorption lines, and
- **chromosphere:** the source of emission lines.

For most stars line absorption is of secondary importance for heat transfer in the atmosphere, and thus has only small influence on the structure (*blanketing effects*). In this fortunate situation the solution for the structure can be separated from the solution for the spectrum. Suppose the photosphere radiates as a blackbody and that the reversing layer is approximated by a homogeneous slab of thickness Δx. Due to line absorption the photon flux $F(\nu)$ will be attenuated by an amount

$$\frac{\Delta F(\nu)}{F(\nu)} = -\Delta x N_\ell \sigma_\ell(\nu) \qquad (2.19)$$

where N_ℓ is the number density of line absorption particles and $\sigma_\ell(\nu)$ is the cross section for line absorption at frequency ν. The reversing layer can only be about one mean-free-path thick (on average over frequency), so

$$\Delta x \approx \frac{1}{N_c \sigma_c} = \lambda_c \qquad (2.20)$$

where c refers to an appropriate average over frequency, of the continuum (i.e., that region of frequency space lying between the lines). Thus the depth of the line,

$$\frac{-\Delta F(\nu)}{F(\nu)} \approx \frac{N_\ell \sigma_\ell(\nu)}{N_c \sigma_c} \qquad (2.21)$$

provides information about composition. Extracting this information is not simple. There are two dominant features of the problem. First is the question of ionization equilibria, or, what fraction of the particular atomic species is in the state from which the line absorption occurs? Second, what is the effective line absorption cross section, evaluated over the atmospheric structure? This involves both atomic properties and the nature of the formation of the line profile.

Ionization and excitation

It is instructive to consider the processes of ionization and excitation from a general viewpoint, which may also be applied to analogous nuclear problems, such as nuclear statistical equilibrium (NSE), quasiequilibrium (QE), Urca cooling, and the composition of a stellar core during gravitational collapse to the neutron star or black hole state. Let us recall two results of statistical physics (see [381], and Appendix B):

(1) For a Boltzmann gas of elementary particles, the chemical potential (including rest mass energy) is

$$\mu = mc^2 + kT \ln \left[\frac{n}{g} \left(\frac{h^2}{2\pi m kT} \right)^{3/2} \right], \qquad (2.22)$$

where $g = 2J + 1$ is the statistical factor for particles of spin J, and n is the number density and m the rest mass of these particles, and

(2) The chemical potential is the Lagrange multiplier that insures conservation of particles. For photons, particle conservation does not hold, so that the chemical potential for blackbody photons is zero.

Consider a system reacting by the schematic reaction $A + B = C$, which is in equilibrium and for which Equation 2.22 is valid for all particles.

Then

$$\mu_A + \mu_B = \mu_C, \tag{2.23}$$

which is a consequence of the conservation laws implied by the reaction equation, and

$$Q = (m_A + m_B - m_C)c^2. \tag{2.24}$$

Therefore,

$$\frac{n_A n_B}{n_C} = \left(\frac{g_A g_B}{g_C}\right)\left(\frac{m_A m_B}{m_C}\right)^{3/2}\left(\frac{2\pi kT}{h^2}\right)^{3/2} e^{-Q/kT}, \tag{2.25}$$

which is the Saha equation. For atomic ionization, let $B = e^-$, A and C be the neutral and ionized form, respectively, of atomic type Z. Because the binding energy Q is small compared to the rest mass, $m_A m_B / m_C$ is roughly m_B, so

$$\frac{n_+ n_e}{n_0} = \frac{2g_+}{g_0}\left(\frac{2\pi m_e kT}{h^2}\right)^{3/2} e^{-Q/kT} \tag{2.26}$$

where Q is just the ionization potential.

   For excitation, let B = photon so that $\mu_B = 0$, and let C be the excited state; then,

$$\frac{n_C}{n_A} = \frac{g_C}{g_A} e^{-Q/kT}, \tag{2.27}$$

where Q is now the excitation energy. The number density of free electrons n_e is equal to the charge-weighted sum of ion densities (by charge conservation). Note that instead of n_e, the equivalent quantity (electron pressure) $P_e = n_e kT$ is sometimes used in the astrophysical literature. This assumes that Dalton's law of partial pressures holds, which is usually true because the electron gas is almost *noninteracting*; the average kinetic energy of free electrons is large compared to their average coulomb interactions with each other and with ions. Then we have the form of the Saha equation:

$$\frac{P_e n_+}{n_0} = \frac{2g_+}{g_0}\left(\frac{2\pi m_e}{h^2}\right)^{3/2} (kT)^{5/2} e^{-Q/kT}, \tag{2.28}$$

which is common in the literature of stellar atmospheres. To get the total number of nuclei of element i, we sum over the number density n_{ij} of all excited states j,

$$n_i = \sum_j n_{ij} \tag{2.29}$$

$$= n_{i0} \frac{g_0 + g_1 e^{(-Q_1/kT+\cdots)}}{g_0} \tag{2.30}$$

$$= n_{i0} z(T)/g_0, \tag{2.31}$$

where $z(T)$ is the *partition function*.

Because of the enormously high abundance of hydrogen in most observed stars, the simple case of Balmer lines is of considerable interest. The Balmer lines are due to absorption by the first excited state of the hydrogen atom. The excitation energy is $E_1 = 10.15$ eV, and the statistical factors are $g_0 = 2$ and $g_1 = 8$. From Eq. 2.26,

$$\log(n_1/n_0) = \log(4) - 0.504/T_4 \tag{2.32}$$

where T_4 denotes $T/10^4$ degrees kelvin. For $T_4 = 0.3$, 1, and 3 respectively we find the ratio of excited-state to ground-state number densities to be $\log(n_1/n_0) = -16.4$, -4.5, and -1.1 (here log denotes base 10 logarithm, and ln will denote base e logarithm). Now in this range $n_1 \ll n_0$, so the total number density $n = n_0 + \cdots$ is approximately just n_0. Most atoms of hydrogen are in their ground state.

We must also consider ionization. From Eq. 2.25 we have

$$\log(n^+/n) = 9.52 + \log(2g^+/g) - 0.504I/T_4$$
$$+ 2.5 \log(T_4) - \log(P_e). \tag{2.33}$$

Now $2g^+/g = 1$ and the ionization potential is $I = 13.60$ eV; for simplicity we take a value of electron pressure roughly typical of stellar atmospheres, $P_e = 0.1$ dyne/cm^2. Then,

$$\log(n^+/n) = 10.5 - 6.82/T_4 + 2.5 \log(T_4). \tag{2.34}$$

This implies a ratio of ionized to neutral (all bound states) hydrogen of $\log(n^+/n) = -13.6$, 3.70, and 11.7 for $T_4 = 0.3$, 1, and 3 respectively.

The fractional ionization n^+/n changes faster with temperature than the fractional excitation n_1/n_0. This has an important effect on the fraction of the hydrogen in the first excited state of the atom. Denote the total number density of hydrogen nuclei (ionized and neutral) by N. At low temperature there are few ions, so $n_1/n_0 = n_1/N$. If we assume that the strength of absorption lines is proportional to $n(1)/N$ (roughly true), then at low T the line strength rises rapidly with temperature.

Consider the same situation, but at higher temperature. The number density of atoms n is just N minus the number density of ions n^+. Therefore,

$$\frac{n_1}{N} = \left(\frac{n_1}{n}\right)\left(\frac{n}{N}\right) \tag{2.35}$$

and since $n + n^+ = N$,

$$\frac{n_1}{N} = \frac{(n_1/n)}{(1 + n^+/n)}. \tag{2.36}$$

As T increases, the ionization fraction n^+/n increases faster than the excitation fraction n_1/n, so that n_1/N decreases after passing through a maximum. Clearly these arguments suggest that Balmer line absorption will be most pronounced at a certain temperature (actually about 9,000K), and become weaker for both higher and lower temperatures. This qualitative behavior is a common characteristic of absorption lines in stellar atmospheres.

A second example will illustrate the more complex situation in which the atom considered has several states of ionization. A simple and astronomically interesting case involves the resonant lines of Ca and Ca^+. For the calcium atom, $I = 6.09\,eV$ and $2g^+/g = 2.75$; for singly ionized calcium, $I = 11.87\,eV$ and $2g^{++}/g^+ = 0.562$. Taking $P(e) = 0.1$, we obtain for calcium,

$$\log\left(\frac{n^+}{n}\right) = 10.96 - \frac{3.06}{T_4} + 2.5\log T_4 \tag{2.37}$$

and,

$$\log\left(\frac{n^{++}}{n^+}\right) = 10.27 - \frac{5.98}{T_4} + 2.5\log T_4. \tag{2.38}$$

Considering only Ca, Ca^+, and Ca^{++}, mass conservation implies

$$N = \sum_i n(i) \approx n + n(+) + n(++). \tag{2.39}$$

The results are summarized in figure 2.6. Notice the sharp transitions between the different ionization stages. At low T, Ca is mostly neutral and goes over first to Ca^+ and then Ca^{++} as the temperature rises.

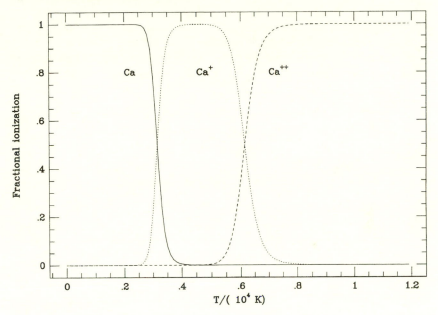

Fig. 2.6. Ionization of Ca

Therefore we see that for stars of the same composition and electron pressure, there will be a large variation in the strength of particular absorption lines due to different temperatures. Stated another way, a small error in the determination of the photospheric temperature can result in a large error in the estimated abundances of those species giving rise to absorption lines. Difficulties should be expected when dealing with states that have significantly different excitation energies, and therefore temperature dependences.

What effects result from variations in the electron pressure? First consider two stars with the same surface temperature and composition, but different electron pressures (and hence free electron densities) at their surfaces. This corresponds to different values of the *gravity* in stellar atmospheres terminology; lower gravity corresponds to lower density. The degrees of ionization of all species are coupled through the n_e factor in the Saha equation. For constant temperature, the degree of ionization is

$$\frac{n^+}{n} \propto \frac{1}{n_e}, \tag{2.40}$$

so that decreasing n_e makes ionization more complete. Such an effect occurs in the comparison of giant and main sequence stars of the same

TABLE 2.2

Stellar Spectra

Class	$T_e/10^4 K$	Comments
M	0.22–0.35	TiO, other molecular bands
K	0.35–0.5	neutral metals, molecular bands
G	0.5–0.6	neutral metals, ionized Ca; CH and CN
F	0.6–0.75	ionized metals, H and K lines of Ca^+, Fe
A	0.75–1.1	Balmer lines and ionized metals
B	1.1–2.5	Balmer lines, ionized C and O, neutral He
O	above 2.5	ionized He

temperature. Differences in their absorption lines, which puzzled early spectroscopists, contain information about the volume of a star and thus its evolutionary state.

Finally, it should be mentioned that at low temperatures the abundance of molecules increases to the extent that they can dominate the flow of radiation and consequently the atmospheric structure. The abundances of molecules can be computed by the same techniques used above, but even for the simple case of equilibrium at a single temperature the calculation is more complicated because a greater number of states must be included.

Some general properties of the spectra of stars are summarized in table 2.2 (see [7, 369]). This is basically an ionization (dissociation) sequence with ascending temperature: molecules, neutral metals, ionized metals, Balmer lines (H), neutral He, and ionized He.

Absorption lines

The line absorption coefficient is the number density of atoms in the given state times the absorption cross section σ_a. This is often written as $\sigma_a = f\sigma_{\text{classical}}$, lumping the interesting and more difficult quantum properties into the *oscillator strength* f. In principle, f is determined by quantum mechanical calculations, or for more complex systems by experiment. For a classical oscillator the absorption cross section is (see chapter 17 in [335])

$$\sigma_a = \frac{4\pi(e^2/mc)\omega^2\Gamma f}{(\omega_0^2 - \omega^2)^2 + \omega^2\Gamma^2},$$
(2.41)

where $\Gamma = 2e^2\omega_0^2/3mc^3$ is the natural (radiative) width of the state and $\omega_0 = 2\pi\nu_0$ is the angular frequency of the resonance. Considering absorptions near the line center, we have $\omega \approx \omega_0$ so that

$$\sigma_a \approx \frac{\pi(e^2/mc)\Gamma f}{(\omega_0 - \omega)^2 + (\Gamma/2)^2}. \tag{2.42}$$

Collisions with neighboring particles can perturb the system and shorten the lifetime of the state (sometimes called *pressure broadening*); this can be included by using $\Gamma = \Gamma(\text{natural}) + \Gamma(\text{collision})$. Doppler broadening of the line can also be important; for thermal motions the cross section for a sharp line is

$$\sigma_d = \frac{\pi e^2 f}{mc\nu_0}\left(\frac{Mc^2}{2\pi kT}\right)^{1/2} \exp[-(Mc^2/2kT)(\nu/\nu_0 - 1)^2], \tag{2.43}$$

where M is the mass of the atom in which the transition occurs (see [199], p. 217). Turbulent motion can also give doppler shifts, but proper inclusion of such effects is difficult.

As this discussion of cross sections begins to indicate, the detailed problem of absorption line profiles is complex. Because of theoretical uncertainties, and difficulties in obtaining accurate observations of dim sources, a common procedure is to study the total line strength rather than the line shape. The strength is measured in units of *equivalent width*, which is the width the line would have if its shape were rectangular but the same total energy were removed from the beam. The equivalent width is

$$W_\nu = \int_{line} (1 - I_\nu/I_c)\, d\nu, \tag{2.44}$$

where I_c is the intensity of continuum radiation. Equivalent widths are relatively easy to measure (compared to line shapes at least), and are supposed to be less sensitive to uncertainties in the theory, but of course potentially useful information is lost.

The abundance of an element in a stellar atmosphere can be estimated by first calculating a number of equivalent widths for various assumed abundances. Comparison with observation determines which abundance gives the best fit. If possible, a large number of lines of a given element are used in order to improve statistics and to check consistency. The general relation between equivalent width and the number of absorbing atoms producing it is called the *curve of growth*.

Consider the model atmosphere by Schuster and Schwarzschild which we used before. A flux of light of intensity I_c is emitted from the photosphere and falls on an absorbing slab of thickness L. For simplicity assume that there is no emission from the slab and that induced emission may be ignored. Then the intensity emerging from the slab is

$$I_\nu = I_c \exp(-\tau_\nu), \tag{2.45}$$

where $\tau_\nu = NL\sigma_\nu$ is the optical depth at frequency ν and N is the number density of absorbing atoms. Therefore

$$W_\nu = \int_{line} (1 - \exp(-\tau_\nu))\, d\nu. \tag{2.46}$$

First consider a weak line, so $\tau_\nu \ll 1$, and

$$W_\nu = \int_{line} \tau_\nu\, d\nu \tag{2.47}$$

$$= \int_{line} NL\sigma_\ell\, d\nu. \tag{2.48}$$

Now if we make the sometimes valid assumption that radiative scattering may be neglected,

$$\int_{line} \sigma_\ell\, d\nu = \pi \left(\frac{e^2}{mc}\right) f, \tag{2.49}$$

which is the dipole sum rule (see [335], p. 606). So we have

$$W_\nu = NL\pi \left(\frac{e^2}{mc}\right) f, \tag{2.50}$$

and the equivalent width is proportional to the number density of absorbing atoms. This is the *linear* region.

Now consider a strong line, $\tau_\nu \gg 1$, near the line center so $\nu \approx \nu_0$. Away from the line center absorption will again be small. There are two cases to investigate: dominance by (1) doppler broadening, or (2) damping (radiative and collisional broadening). In both cases the expression $1 - e^{-\tau_\nu}$ acts like a step function, being unity for $\tau_\nu \gg 1$ and small for $\tau_\nu \ll 1$. The switch occurs at $\nu = \nu_0$. Now this means that

$$W_\nu \approx \int_{line} d\nu = 2\Delta. \tag{2.51}$$

We can determine Δ by noting that

$$\tau_\nu \approx 1. \tag{2.52}$$

For the case of Doppler dominance,

$$\sigma_d = \left(\frac{\pi e^2 f}{mc\nu_0}\right)\left(\frac{Mc^2}{2\pi kT}\right)^{1/2}\exp\left[-\left(Mc^2/2\pi kT\right)\left(\Delta/\nu_0\right)^2\right]. \tag{2.53}$$

For conciseness, denote

$$\alpha = \nu_0\left(\frac{2\pi kT}{Mc^2}\right)^{1/2},$$

and

$$\beta = \left(\frac{\pi e^2 f}{mc}\right).$$

Then,

$$W_\nu \approx 2\alpha\sqrt{\ln(NL\beta/\alpha)}, \tag{2.54}$$

so that the equivalent width goes as $\sqrt{(\ln N)}$, a very weak dependence. This is the *saturation* regime.

If damping is dominant (a strong resonance),

$$\sigma_a \approx \frac{\beta\Gamma}{(2\pi\Delta)^2}, \tag{2.55}$$

and

$$W_\nu \approx \sqrt{(NL\beta\Gamma/\pi^2)}. \tag{2.56}$$

The equivalent width goes as the square root of N. This is the *damping* regime.

From these limiting cases we can understand the behavior of curves of growth such as shown in figure 2.7, in which the linear, saturation, and damping regions are labeled. The two curves represent different values for $\log a$, where a is the ratio of collisional and radiative broadening to Doppler broadening (see [6], §3.2). How might these be used to determine abundances?

If the equivalent width is measured (i.e., the vertical axis value is known), the reference curve of growth must be established before the

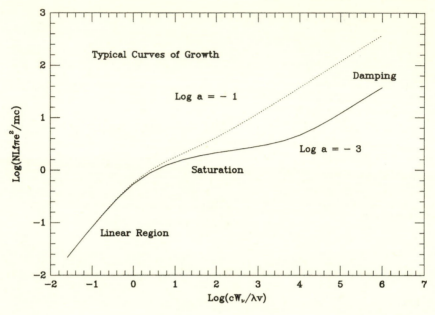

Fig. 2.7. Curves of Growth

abundance (horizontal axis value) can be inferred. Unless the line width is narrow enough to be in the linear region, the determination of the curve of growth requires a determination of the Doppler broadening (value of $\log a$ in figure 2.7), which is an additional complication. It also requires either (1) a choice among theoretical models to determine the shape of the curve, or alternatively (2) a choice of an empirical reference curve of growth for estimating relative abundances between stars of similar atmospheres.

It is the product Nf which is determined observationally. To find N requires that the oscillator strength be known. A revision of f-values for Fe led to an increase (by a factor of five to ten) in the determined values of the iron abundance in the solar photosphere [262, 273, 413], bringing them into agreement with meteoritic values. Prior to the revision the values were thought to be well known, and an unresolved discrepancy existed between the solar and the meteoritic determinations. Although we hope that such large revisions will no longer need to occur, it illustrates yet another difficulty in the thorny problem of stellar abundance determination.

A difficulty arises when this procedure is applied [701]; the measured equivalent widths are larger than they should be. Struve and Elvey [594] attributed the additional width to nonthermal motion in the stellar atmosphere and called it *microturbulence* because the velocities must vary only on a small spatial scale. Worrall and Wilson [701] criticized the use of microturbulence as an adjustable parameter not based upon a correct physical model. In particular, they argue that it incorrectly compensates for non-LTE effects. Pagel [478] countered that while there are difficulties on the flat (saturated) part of the curve of growth, these do not necessarily extend to the linear and square-root (damping) branches. Further, LTE gives an excellent approximation to the visible continuum in all but supergiants and the very hottest stars. Nevertheless, difficulties of this sort have spurred efforts to build more realistic procedures (see [671]).

If *differential coarse analysis* (as devised by Greenstein and collaborators) is used—that is, equivalent widths of weak lines are compared with those in a standard star of otherwise similar properties—many uncertainties should divide out. Such relative determinations are less prone to error than absolute ones. Of course there may be difficulty in obtaining a sufficient range of standard stars. Several important results were obtained by this method. Greenstein [272] summarized them as follows: (a) in no star is $X_{Fe} = 0$, even if [Fe] ≈ -2.5 to -3; (b) even if the ratio of α-particle nuclei to iron to heavy s-process elements is variable,[3] the factors seldom exceed ten, i.e., $[\alpha-\text{nuclei}] - [\text{Fe}] \leq +1$ and $[s-\text{process}] - [\text{Fe}] \geq -0.5$; (c) in unevolved stars, the range of [H/He] is less than 0.3, with values perhaps near 0.1. Recent reviews suggest that these conclusions are still intact; see [376, 680].

However, in 1969, Unsöld [644]—one of the pioneers of stellar spectroscopy and founder of the Kiel school—made an interesting critique of the results[4] obtained at that time. In a slightly paraphrased form, his points were: (1) the proposed abundance differences were mostly the same order of magnitude as the average error of analysis; (2) since a star highly deficient in metals (HD 140283, analyzed by Baschek [80]) shows no such effects, why should these effects be large in stars that are less deficient in heavy elements (more like solar)? And, (3) it is remarkable that the observed deficiency is the same for carbon, members of the iron group, and heavy elements like strontium and barium, and shows no relation to the various types of nuclear effects described.[5] Unsöld did not doubt that nuclear reactions are the source of stellar energy. Merrill's discovery [428] in 1952 of technetium in S stars (the most stable isotope

---

[3]See chapter 14.
[4]See Wallerstein [662] for a summary of the astronomical evidence for nucleosynthesis in stars, as known then.
[5]See chapter 14.

has a half-life of 2×10^5 years) was elegant evidence for the operation of something like the *s*-process. The question was whether or not nucleosynthesis in *stars* actually made the bulk of the abundance pattern seen in old (and young) stars.

What went wrong? Because the analysis errors were roughly a factor of two either way, only stars showing big effects could be convincing. Such stars are rare, and consequently one must look deep into space to find them. Because they are distant, they are dim. Only relatively bright stars had been used for the highest quality analysis. Thus there was a bias against those objects containing the most significant information. Greenstein [272] did make this response, which covers items (1) and (2) above. A star like HD 140283 could be deficient in heavy elements because (a) it was formed before most of the heavy elements were synthesized (as was assumed), or (b) it was formed from matter diluted by addition of gas not enriched in heavy elements. This untangling of variation in space and in time is still a problem in understanding galactic evolution. It seems, for example, that HD 140283 might be an example of the latter, and +39°4926, (see [361]) the former. Item (3) above is one of the themes of this book. It is interesting that the work on meteoritic anomalies in xenon [519] and neon [101] had already provided evidence—of a quite different sort—for stellar nucleosynthesis (see §2.4), but did not figure in the discussion.

As an actual tool for determining abundances, the curve-of-growth method has been superseded by direct *spectral synthesis*, in which the shape of the entire spectrum is calculated from a model of the stellar atmosphere. Comparison with observation allows a simultaneous determination of the best values for effective temperature, gravity, and abundances (at least in principle). The underlying physical problems remain those illustrated by the simpler procedures sketched above.

An example of the steady improvement is the observation of the Mg isotopes [111, 88, 625, 69]. General arguments of stellar nucleosynthesis [23] predicted a tendency for the neutron-rich isotopes of magnesium, ^{25,26}Mg, as well as Na and Al, to increase relative to ^{24}Mg as galactic evolution proceeds.[6] There is now observational evidence that such an effect exists in our galaxy, both from stellar spectra [625, 69] and from cosmic rays (see §2.5).

Abundance determination from stellar spectra is in the process of rapid improvement. The greater sensitivity (and spectral coverage) of charge-coupled device (CCD) detectors has greatly increased the distance to which high-quality spectra may be obtained. New, cost-effective

---

[6]See chapter 14.

technology for building telescopes promises significant improvements in speed and number of telescopes, and thereby a major increase in high-quality data. Improvements in interpretation are equally dramatic. The atomic and molecular data are steadily improving. Computer development has made it feasible to quickly simulate the spectra involving vast numbers of lines, which now agree with the observed spectra to an unprecedented degree. These developments should provide interesting new changes in a venerable subject.

2.4   METEORITES

Determination of abundances in stars by the general approach outlined in the preceding section allows us to explore a moderately large volume of space (although still woefully small), but at relatively low accuracy. If we can obtain specimens to analyze in the laboratory, more powerful techniques can be brought to bear. However, the question arises as to how representative are these specimens? Early work was based primarily upon the composition of the earth's crust. As time passed it became evident that meteorites were better objects for the study of the average abundance of the chemical elements in nature than were terrestrial (or lunar) rocks.

In using meteoritic abundances there is always the problem of a proper average of elements as they occur in the meteorites in the silicate phase, the troilite (sulfide) phase, and the metal phase. There are many types of meteorites: stones, irons, and stony-irons. While irons have a sufficiently dramatic appearance to the untrained eye to be popular as museum exhibits (and are generally bigger than stones), irons are relatively rare. Table 2.3 gives the relative proportions of the various types (accuracy less than the number of significant figures shown).

The term *chondrite* is based on the ancient Greek word *chondros*, meaning "grain of seed," a reference to the tiny rounded bodies occurring in these stones. The bodies themselves are called *chondrules*. The *achondrites* are quite different in composition and structure, resembling terrestrial igneous rocks for the most part.

Chondrites have come to be used as a proper average of the three phases since these objects are so obviously a heterogeneous mixture of materials from many sources, and hence may be a representative mixture themselves. In particular, the *carbonaceous* chondrites, which have relatively large abundances of gases and friable compounds, are often chosen (for example, see [148] or [12]) as giving the most representative abundances.

TABLE 2.3

Meteorite Types

Type	Subtype	Percentage
Stones		
	Chondrites	86.0
	Achondrites	7.0
Stony-irons		1.5
Irons		5.5

It has been evident for many years that there exist marked regularities in elemental abundances in meteorites. The rare earths are particularly important in influencing thought on this question. Because of their very similar chemical properties, the rare earths are resistant to chemical separation. Therefore they are likely to be relatively unfractionated, and hence should be (among themselves at least) a relatively unbiased sample. They exhibit a regular alternation of abundances between successive even-Z and odd-Z elements, and the successive odd (or even) elements change in abundance in a gradual and regular manner. These regularities reflect nuclear systematics (see chapter 3).

Suess and Urey [597] were the first to generalize these regularities to the isotopic abundances also, and generate a much improved abundance table based upon (1) these regularities, (2) meteoritic data, and in the last resort (3) astronomical data. This review of Suess and Urey was of fundamental importance in the development of nucleosynthesis theory.

With the use of mass spectrographs, isotopic abundance ratios can be determined with high accuracy. This greatly increases the amount of abundance data available. A number of regularities appear, which Suess [595] codified into *rules* for the abundances of stable isotopes,[7] and are discussed in detail in [597].

1. *Odd-A nuclei*: Abundances of odd-A species with $A > 50$ change steadily with mass number A. When isobars occur, the sum of their abundances must be used instead of individual abundances.

[7]Some terminology: *Isotopes* are nuclei with the same proton number Z, *isobars* have the same nucleon number A, and *isomers* have the same Z and A, but different excitation. A nucleus is defined by a given Z and A, and may exist in different isomeric states (different energies).

2. *Even-A nuclei*: (a) For $A > 90$, the sums of the abundances of the isobars with even-A change steadily with A. (b) For $A < 90$, abundances of nuclei with equal numbers of excess neutrons change steadily with A.

3. For $A < 70$, the isobar with the highest excess of neutrons is the less abundant for any given A. For $A > 70$, the neutron-poor isobar is the least abundant one.

4. Exceptions occur at *magic numbers* of neutrons (2, 8, 20, 28, 50, 82, 126, ...).

A more detailed discussion of these and other properties of nuclear abundance regularities will be presented after several other topics necessary for discussion have been introduced.

In addition to their interest as a sample of solar system composition, the nuclear abundances in meteorites give us a powerful tool for exploring the evolution of the solar system, stars, galaxies, and even cosmologies. The tool is nuclear radioactivity, and the study is called *cosmochronology*.

The question of radioactivity appears in three ways: (1) as long-lived radioactive nuclei which were part of the matter from which the solar system was formed, (2) as cosmic-ray induced radioactivities caused by irradiation of the meteorite prior to its entry into the earth's atmosphere, and (3) as decay products of *extinct radioactivities*, that is, of nuclei which have decayed to such a great extent that they are no longer directly observed.

By use of the abundance regularities discussed above—or better, by a theory which can explain those regularities—it is possible to estimate how much of a radioactive nucleus was initially present when the meteorite solidified. This, with the directly measured amount of decay, gives the time since the meteorite became a closed chemical system, i.e., its age. Further, different radioactive nuclei give different independent estimates (in principle at least) of this age. Finally, the abundance of a short-lived radioactivity (relative to its stable brethren) at the time of formation of the meteorite gives a measure of the rate of nucleosynthesis at that time. Short-lived radioactivities must be continually replenished in order to maintain any appreciable abundance. This gives important information for theories of galactic evolution.

Some important radioactive nuclei and their properties are listed in table 2.4.

Not all nuclei in this list have been used as *cosmic clocks* (see [239, 198]). For example, the initial abundance of ^{40}K cannot be predicted yet with sufficient accuracy to be useful at present. Podosek and Swindle [497] summarize the evidence for the presence of several extinct

TABLE 2.4

Some Cosmic Clocks

Radioactivity	Half-life (10^6 yrs)
^{41}Ca	0.13
^{26}Al$(\beta^+\nu)^{26}$Mg	0.75
^{53}Mn	3.7
^{107}Pd$(\beta^-\bar{\nu})^{107}$Ag	6.5
^{129}I$(\beta^+\nu)^{129}$Xe	15.7
^{247}Cu	16
^{244}Pu$(\alpha)^{240}$U	80.5
^{146}Sm	103
^{235}U$(\alpha, \beta^+\nu, \cdots)^{207}$Pb	703.8
^{40}K$(\beta^-\bar{\nu})^{40}$Ca	1,277 (0.893)
^{40}K(EC)^{40}Ar	1,277 (0.107)
^{238}U$(\alpha, \beta^+\nu, \cdots)^{206}$Pb	4,468.3
^{232}Th$(\alpha, \beta^+\nu, \cdots)^{208}$Pb	14,100
^{187}Re$(\beta^-\bar{\nu})^{187}$Os	45,000(?)
^{87}Rb$(\beta^-\bar{\nu})^{87}$Sr	47,200

radioactivities in the early solar system: ^{26}Al, ^{53}Mn, ^{107}Pd, ^{129}I, ^{244}Pu, and ^{146}Sm. This is an area of rapid progress, so that recent literature should be consulted for new results [150].

In order to better understand what cosmochronological information can be supplied by radioactivities, it is useful to examine some simple mathematical models. Consider an unstable nucleus with number density N and mean lifetime τ. Suppose it is produced at a rate R. Then its abundance is given by the solution of

$$\frac{dN}{dt} = R - \frac{N}{\tau} \tag{2.57}$$

where dN/dt denotes the time derivative of N.

By making the substitution $N = \phi \exp(-t/\tau)$, we obtain

$$\frac{d\phi}{dt} = R \exp(t/\tau). \tag{2.58}$$

In general there is another term on the right-hand side of Eq. 2.57 which is essentially a divergence of the flux due to flow of matter in and out of the volume of space being considered; these effects have been taken to be zero here and are discussed later in chapter 4 and in chapter 14.

Taking the origin in time, $t = 0$, to be just prior to element formation implies $N(0) = 0$ and therefore $\phi(0) = 0$. Then Eq. 2.58 may be integrated over a time interval Δ to give

$$\phi(\Delta) = \int_0^\Delta e^{t/\tau} R \, dt \qquad (2.59)$$

which implies (by the original substitution) that

$$N(\Delta) = e^{-\Delta/\tau} \int_0^\Delta e^{t/\tau} R \, dt. \qquad (2.60)$$

For a *stable* nucleus of abundance Z, $1/\tau = 0$ so

$$\frac{dZ}{dt} = R_Z \qquad (2.61)$$

where R_Z is the production rate for this nucleus, and

$$Z(\Delta) = \int_0^\Delta R_Z \, dt. \qquad (2.62)$$

For simplicity, two approximations[8] for the behavior of the production functions R will be examined: (1) the *uniform approximation* in which $R = $ constant, and (2) the *sudden approximation* in which $R = \delta(t) \times$ (constant), where $\delta(t)$ is the Dirac delta function. It might be hoped that a realistic behavior should lie between these extreme cases, but there seems to be no general argument that proves this to be so.

In the uniform approximation we have

$$N(\Delta) = R_N \tau (1 - e^{-\Delta/\tau}) \qquad (2.63)$$

and

$$Z(\Delta) = R_Z \Delta. \qquad (2.64)$$

[8] As we will see in chapter 14, the distribution of stellar ages in the solar neighborhood favors a more or less constant rate of star formation. This would imply that the uniform approximation is more realistic than the sudden.

The radioactive nucleus approaches a steady-state abundance $N \to R_N \tau$ for $\Delta \gg \tau$, while the stable nucleus increases linearly in Δ. Their ratio is

$$\frac{N(\Delta)}{Z(\Delta)} = \frac{R_N \tau}{R_Z \Delta}(1 - e^{-\Delta/\tau}) \tag{2.65}$$

$$\approx \frac{R_N \tau}{R_Z \Delta} \quad \text{for } \Delta \gg \tau.$$

Note that only the last τ years are effective in producing the radioactive nucleus.

In the sudden approximation we have[9]

$$N(\Delta) = N(0)\exp(-\Delta/\tau) \tag{2.66}$$

and

$$Z(\Delta) = Z(0). \tag{2.67}$$

The ratio of abundances is

$$\frac{N(\Delta)}{Z(\Delta)} = \frac{N(0)}{Z(0)}\exp(-\Delta/\tau) = \frac{R_N}{R_Z}\exp(-\Delta/\tau). \tag{2.68}$$

This exhibits an exponential decay with increasing Δ/τ while the previous case fell off like a hyperbola. For very short-lived radioactivities ($\tau \ll \delta$), the sudden approximation predicts more depletion of the radioactive species than the uniform case, assuming the same production ratios. In order to account for the evidence for short-lived radioactivities in meteorites, some production function less extreme than the sudden approximation is usually assumed (often relatively recent episodes called *spikes*).

To obtain an *age*, equation 2.66 or 2.68 is solved for Δ. While $N(\Delta)$, $Z(\Delta)$, and τ are measurable, the production ratio R_N/R_Z must be known. *The resulting age can be no better than this estimate.* Further, a choice of the *sudden, uniform,* or some other approximation must be made. To reduce the ambiguity it is assumed that the production ratios are independent of time, and several nuclei (for which this is thought valid) are used to obtain a self-consistent solution.

An important example is the pair ^{235}U and ^{238}U. Because they are isotopes of the same element, their abundance ratio at present is well

---

[9]Set $R_N(t) = N(0)\lambda \exp(-\lambda t)$ and let $\lambda \to \infty$; similarly for R_Z.

known: $^{235}U/^{238}U = 0.0072$. Because they are both likely to be produced in the r-process (rapid neutron capture), their production ratio
is thought to be constant in time. The estimate of its value is fraught
with complications (see [239, 198]). The abundance of each of these isotopes is contributed to by all nuclei that can decay to them by α-decay
chains.[10] Estimates range from simply counting those nuclei thought to
be progenitors to sophisticated reaction network calculations of conditions thought to be appropriate to the r-process. The values range from
about 1 to 2 (see [198]).

Since the formation of the solar system 4.55 billion years ago (t_{ss}), the
uranium isotopes have freely decayed, with no new production (note the
similarity to the *sudden* solution). At the formation of the solar system,
the ratio was

$$\left(^{235}U/^{238}U\right)_{ss} = \left(^{235}U/^{238}U\right)_{now} \exp\left(t_{ss}\left[\frac{1}{\tau_{235}} - \frac{1}{\tau_{238}}\right]\right) \quad (2.69)$$

$$= 0.0072 \times \exp\left(4.6 \times 0.8301\right)$$

$$= 0.328.$$

Observations of metal-poor stars, whose surface abundances are thought
to represent nucleosynthesis early in galactic history, suggest that the r-
process has continued from then to the present time. To illustrate the
general nature of such solutions we will use the *uniform* approximation
for the epoch before solar system formation. Then,

$$\left(^{235}U/^{238}U\right)_{ss} = \frac{\tau_{235}R_{235}(1 - \exp{-t_{ss}/\tau_{235}})}{\tau_{238}R_{238}(1 - \exp{-t_{ss}/\tau_{238}})}. \quad (2.70)$$

Figure 2.8 shows the behavior of the ratio $^{235}U/^{238}U$ as a function of
lookback time, from the present ($t = 0$), to the times of formation of
the solar system ($t = 4.55$ Gyrs) and of the onset of r-process nucleosynthesis (*galaxy formation*, shown for production ratios of 1.5 and 1.0).
Both the sudden and the uniform approximations are shown. The ratio at present is shown as a filled square on the left axis. From this
back to the point labeled "sun formed," there is simple decay, so that
the sudden (s) and the uniform (u) lines are identical. At earlier times,
when the U isotope ratio corresponds to that expected in the interstellar
medium, the sudden case continues this extrapolation; it intersects the
dashed lines representing production ratios at the corresponding *ages* (in

---

[10]For example, ^{238}U might be made as ^{238}U itself, ^{242}Pu, ^{246}Cm, and ^{250}Fm. At this point
there is severe competition from spontaneous fission; see chapter 3.

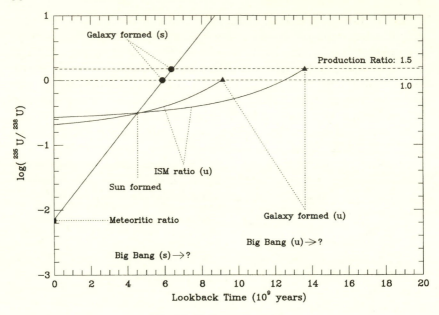

Fig. 2.8. U Cosmochronology

the *sudden* approximation). These intersections are shown as filled circles; their value on the horizontal axis is the inferred age of the uranium (here about 6 Gy).

For the possibly more realistic *uniform* approximation, ages of 9 to 14 Gyrs result. Allowing an additional 1 Gyr from the Big Bang to the onset of *r*-process nucleosynthesis [239] gives estimates in the range 10 to 15 Gyrs for the age of the Universe. In the uniform approximation, some of the ^{235}U is made relatively recently, keeping its abundance up; this requires that some of the uranium be much older to get the observed ratio. Thus the age of the oldest uranium is larger than in the sudden case.

The subject of cosmochronology is much more complex than these extremely simple models suggest [239, 198, 57, 552]. Additional radioactivities should be included, as well as other astronomical constraints. The solutions should take account of the significant errors in the input parameters from nuclear physics, and the uncertainty in both the site of the *r*-process and the quantitative nature of galactic evolution. Cosmochronology is a cup half-full or half-empty, depending upon one's degree of optimism. Nevertheless, such estimates do press some ages inferred from cosmology (chapter 5), as do those from stellar evolution (chapter 7), and it would be imprudent to ignore any of the set.

One of the most interesting facets of meteoritic research has been the recent isolation of intact interstellar grains that originated in stars. These grains permit the direct laboratory study of their chemical and physical nature, and in particular their isotopic patterns. The first such grains were carbon rich [393, 91, 11], but very recently [307, 308, 310] oxygen-rich grains have also been found. These are providing a fascinating and direct connection between observational astronomy, theoretical astrophysics, and laboratory geosciences, which was anticipated by D. D. Clayton [174]. The impact of these discoveries is only beginning to be understood, but promises to have important ramifications for the topic of this book.

2.5 COSMIC RAYS

In 1912, after observing the discharging behavior of electroscopes during a balloon ascent in Austria, the physicist V. F. Hess concluded that "a radiation of very great penetrating power enters our atmosphere from above." In 1936 Hess shared a Nobel Prize in physics for his discovery of cosmic rays. The cosmic radiation which impinges upon the atmosphere of the earth consists mostly of fast protons and alpha particles. Because of their net charge, the cosmic rays gyrate in magnetic fields. In a reference frame with no electric field,

$$\frac{d\mathbf{p}}{dt} = \frac{Ze}{c}(\mathbf{v} \times \mathbf{B}), \tag{2.71}$$

where $\mathbf{p}$ is the relativistic momentum, $m\mathbf{v}/\sqrt{1-(v/c)^2}$, $\mathbf{v}$ is the velocity, Ze the charge, and m the rest mass of the particle. For circular motion, $\mathbf{v}$ only changes direction, so

$$\frac{dp}{dt} = \left(\frac{p}{v}\right)\left(\frac{dv}{dt}\right), \tag{2.72}$$

$$= pv/r . \tag{2.73}$$

Therefore, in the extreme relativistic limit,

$$\frac{pc}{Ze} = rB, \tag{2.74}$$

or, pc is the energy ϵ, and

$$\frac{\epsilon}{ZrB} = 300 \text{ eV/gauss cm.} \tag{2.75}$$

The combination rB is called the *rigidity*. Thus, although the cosmic rays travel at essentially the speed of light, their net drift motion through cosmic magnetic fields is much slower; because the fields are tangled and dynamic, this motion has the character of a random walk.

For example, if the galaxy has an average field $B \approx 3 \times 10^{-6}$ gauss, the gyroradius is $r = 10^{14}$ cm for a 100 GeV proton. One circuit is completed in six hours. If the mean lifetime of the cosmic ray is 10^7 years, then more than 10^{10} circuits are completed. With fluid velocities of 10 km/s, significant changes could occur over a gyroradius in 3×10^4 circuits or three years. Thus the motion of the cosmic ray might be described by a random walk of its guiding center (see [335], ch. 12). For the radius of gyration to exceed that of the galaxy (10 kpc $\approx 3 \times 10^{22}$ cm), the proton energy must be greater than 3×10^{19} ev. Interestingly enough, the cosmic-ray energy spectrum has a break in this range [573].

For comparable rigidities above 2 GeV, the ratio of hydrogen to helium nuclei is about 7:1; isotopically the hydrogen is mainly protons, and the helium ^{4}He. The flux of cosmic-ray helium is about ten times that of all heavier nuclei combined. The latter constitute less than 0.02 of the particles and less than 0.15 of the rest mass in cosmic radiation.

The origin of cosmic rays has long been linked to ideas about nucleosynthesis. Baade and Zwicky [63] suggested that cosmic rays were produced (along with supernovae) by the formation of a neutron star. This connection with supernovae was also suggested on grounds of power requirements (see [265]). In this case, the composition of the cosmic rays might be expected to exhibit anomalies due to the nucleosynthesis expected to occur in supernovae (although matter accelerated to high velocity would not necessarily be the same as the bulk of the matter synthesized). On the other hand, Fermi [231] suggested that the cosmic rays were accelerated by the statistical effect of the dynamic magnetic fields associated with the interstellar medium. In this case the composition of the cosmic rays would reflect some sample of charged matter from an ensemble of nucleosynthesis sources contributing to the interstellar medium over some time-scale of the order of the cosmic-ray lifetime. Because that lifetime is short (relative to other galactic time-scales), the cosmic-ray abundances would reflect the recent interstellar medium, unlike solar system abundances which presumably reflect the interstellar medium at the time the sun was formed (4.55 billion years ago). A combination of these two ideas—that supernova shockwaves in the interstellar medium accelerate cosmic rays—is now in favor [62, 159, 103].

Cosmic rays are in a sense intermediate between stellar atmospheres and meteorites in accessibility. Because of their high energy, cosmic-ray nuclei interact violently with matter in the earth's atmosphere; by spallation and evaporation reactions their composition is altered. While they may be detected and analyzed by techniques of experimental nuclear physics, the astrophysically interesting information is their flux and composition when they are still above the atmosphere. This information (and its variation with particle energy) has been investigated by high-altitude balloon flights, satellites, and space probes. With large plastic balloons it is possible to get above most of the atmosphere, to where residual column densities are less than 5 g cm^{-2}. Even at these altitudes the composition or primary cosmic rays has been altered by interactions with the atmospheric matter. An extrapolation of observed abundances to the top of the atmosphere is required; this involves a knowledge of interaction cross sections and introduces some uncertainty. The use of satellites and space probes avoids this difficulty; with space probes, the terrestrial magnetosphere and to a lesser extent the heliosphere (which screen out very low-energy cosmic rays) can be left behind. There is internal evidence in the cosmic rays (see below) that they have traversed about 4 g cm^{-2} of matter, in interstellar space and perhaps in their source; the "top of the atmosphere" measurements must be extrapolated to get source abundances.

Cosmic-ray physicists use the expressive term *fragmentation* for a variety of nuclear spallation and evaporation processes. Direct measurements of those cross-sections that are important for cosmic-ray studies are incomplete. The cross sections for fragmentation into radioactive products have been easier to obtain than those for producing stable nuclei. Models of the fragmentation processes are used to concisely systematize the existing experimental data, and to establish those data not directly measured. The model used will change with incident energy being considered. At low energy ($E < 40\,\mathrm{MeV}$), the compound nucleus model is still valid. The energy of the incident particle is randomly distributed among various degrees of freedom, and eventually a fluctuation about the most probable distribution of this energy results in the *evaporation* of nucleons, or clusters of nucleons, from the excited nucleus. Most of the products are found in a narrow band of mass numbers near that of the target. At higher energies a different model is more appropriate. Since the energy of the incident particle is much larger than the interaction energy of the nucleons with each other, and since the wavelength of the incident particle is much less than the average internucleon spacing in the nucleus, the incident nucleon can be thought of as interacting with the nucleons one at a time. The struck nucleons may interact in the same manner with still other nucleons in the nucleus. The interaction becomes a multiple scattering problem which entails the idea of an intranuclear cascade of nucleons. Some of these cascade nucleons may

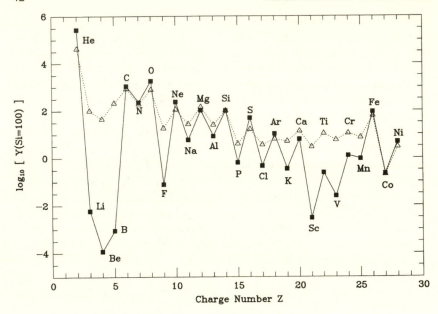

Fig. 2.9. Cosmic Ray Abundances

escape from the nucleus; in any case, finally there remains a nucleus with holes and excited particles. When the cascade is over, the de-excitation of the residual nucleus will probably involve evaporation of additional particles.

In order to use the fragmentation cross sections to estimate source abundances, it is still necessary to know how many cosmic rays traverse the various possible amounts of matter before entering the detector. That is, the *distribution of pathlength* must be specified. This requires a model of the transmutation and propagation of the cosmic rays from their source, through interstellar space, to the detector. From Eq. 2.75 we see that the radius of gyration of the cosmic rays, and probably their ease of transportation, increases with energy. A simple way to describe the production of *secondary* nuclei from the breakup of *primary* progenitors is to assume that (1) the cosmic-ray flux is roughly constant over time intervals $t < 10^9$ years, and (2) the propagation of cosmic rays can be approximated by a diffusion (random walk) from one magnetic irregularity to another. Energy change, for example by ionization, magnetic bremsstrahlung, statistical acceleration, nuclear collisions, etc., might be included. The assumed pathlength distribution and source composition are adjusted until the observed cosmic-ray abundances are obtained.

The elements (the charge number Z) versus their abundance in the galactic cosmic rays (as measured at the orbit of Earth) are plotted as open triangles in figure 2.9; see [573]. Also shown (as filled squares) are the solar system abundances; the cosmic-ray data are normalized to the solar system abundance of silicon (Si $=$ 100 scale). This data is from [573]; see also [402]. Both hydrogen and helium have low abundances in the cosmic rays, which is still regarded as a major puzzle [402]. While the low hydrogen abundance may be related to the different charge/mass ratio of hydrogen relative to the other elements, this does not apply to helium.

The deep trench at Li Be B is filled in, presumably by fragmentation; the value of $4\,\mathrm{g/cm^2}$ quoted above for the average pathlength in interstellar space is simply what is needed to do the filling. This is simple to understand. We note that the number ratio of Li Be B to CNO in the cosmic rays is about 0.1. Consider the fragmentation of CNO to Li Be B by hitting interstellar hydrogen. The number of events is $\Delta N_{\mathrm{Li Be B}} = N_{\mathrm{CNO}}N_{\mathrm{H}}\sigma c\Delta t$ where for energies of 50 GeV/nucleon or greater, the measured cross section (see [569]) is $\sigma \approx 70$ millibarns, and the pathlength is $\Delta x = c\Delta t$. Using solar abundances for hydrogen in the interstellar medium, $N_{\mathrm{H}} \approx 0.7N_A\rho$, and we have

$$\rho\Delta x \approx \frac{(\Delta N_{\mathrm{Li Be B}}/N_{\mathrm{CNO}})}{0.7N_A\sigma} \approx 4\mathrm{g/cm^2}. \qquad (2.76)$$

The valley in abundance just below iron (Sc Ti V Cr Mn) is also filled, and the odd-Z elements from F to K are considerably enhanced in the cosmic rays. These are most likely consequences of fragmentation and imply a natural limit on the cosmic-ray information we can obtain about these "buried" nuclei.

Measurement of the isotopic ratios of cosmic-ray beryllium provides an age for the cosmic rays. Since ^{7}Be can decay only by electron capture (and even if cosmic-ray nuclei could acquire electrons, they would be rapidly stripped), ^{7}Be survives as a secondary. In the rest frame, ^{10}Be has a half-life of 1.6×10^6 years. Present measurements indicate that the corresponding age for the cosmic rays is about 10^7 years [573], so that the cosmic rays are indeed a relatively modern sample of extra-solar-system matter.

Isotopic information on the cosmic-ray source composition implies an excess of neutron-rich isotopes (^{13}C, ^{22}Ne, ^{25,26}Mg, and perhaps ^{29,30}Si relative to ^{12}C, ^{20}Ne, ^{24}Mg, and ^{28}Si). There are a tantalizing number of other possibilities [429]. A general feature of galactic evolution is likely to be such an increase in the relative abundance of these neutron-rich isotopes with time ([23], §9.5). Woosley and Weaver [697] have given one specific model of this effect which captures some of the quantitative results. Prantzos et al. [504] have proposed another model in which the high velocity winds from Wolf-Rayet stars are favored for acceleration. See [429] for further discussion.

2.6   OTHER ASPECTS

As should now be clear, the subject of abundances in astronomical objects is complex, broad, and rapidly changing. This chapter is intended to provide conceptual connections between areas of science which have different histories and traditions. For example, the scientists cited here call themselves astronomers, physicists, astrophysicists, planetary scientists, chemists . . . This is far from a complete review; see the References section and recent literature for more detail. There are several additional topics of particular importance for later discussion.

Stellar ejecta

The matter ejected from stars directly contributes to the cycle of galactic evolution. Measurement of the abundances in these ejecta, and the amount of matter ejected, provides an empirical basis for our understanding.

Supernova remnants provide a direct measure of stellar nucleosynthesis, at least in principle. In practice, the reliable and quantitative determination of abundances from supernova ejecta is difficult. An early and influential result was Woltjer's [689] demonstration of an excess of helium relative to the solar value, in the Crab nebula (the remnant of the historical supernova of 1054). The Crab nebula is a prototype for a supernova remnant with an embedded pulsar. For supernova remnants containing such high-density cores, such as neutron stars or black holes, heating may occur due to the dense object, as is true for the Crab. The visible nebula has a mass of a few solar masses.

A supernova acts as an explosion in the interstellar medium. The energy of the explosion propagates as a shockwave, running ahead of the ejected matter. As the explosion encounters more material, it slows due to sharing of its energy with the new matter. The matter far behind the shock was accelerated to larger velocities, and is now coasting on ballistic trajectories. This inner matter collides with the slowed material behind the shock, setting up a *reflected shock*, which moves inward in mass coordinate. The deceleration of the inner matter is likely to cause a Rayleigh-Taylor instability to develop.

If the expansion encloses a sphere of radius r, then the mass swept up is $M = \frac{4\pi}{3}\rho r^3$, where ρ is the average density of the ambient matter. Setting this equal to the ejected mass from the supernova gives an estimate that the reverse shock will become important at a radius $r = 5.85 \times 10^{18} \ (m/n)^{\frac{1}{3}}$ cm, where $m = M/M_\odot$ and n is then number density in nucleons per cubic centimeter. Typical values might be $m/n \simeq 1$. For an average expansion velocity of 5,000 km s^{-1}, this would take about 400 years.

The development of a *reflected shock* is crucial for observations because it is the shock heating that provides the luminosity for the lines observed; we see that matter which is hot. If we wish to see matter ejected from the star, instead of shocked interstellar medium, we must look at a remnant in which the *reflected shock* is propagating through this stellar ejecta. Such an object is a *young* supernova remnant (SNR). If we wish to determine abundances in the interstellar medium, we examine *old* supernova remnants (with ages $>> 10^3$ years).

Cassiopeia A was first known as a radio source, and found to be associated with two types of nebulosity. The first was visible in emission lines of S, O, and Ar, with radial velocities of 3,000 to 9,000 km/s. The second was visible in Hα and N lines, with radial velocities of 100 km/s or less. The proper motion of the fast knots is linearly related to distance from the center of expansion, and converges at A.D. 1657±3 years [342]. The faster-moving knots are interpreted as supernova ejecta, and the slower-moving matter as the result of pre-explosion mass loss [164]. A high abundance of O, S, Ar, and Ca has been inferred [167], which is similar to the extensive region in a massive He-star which has done incomplete oxygen burning [22, 32].

Red giant stars are observed to be undergoing extensive mass loss [576], and provide a wealth of data relating to the early burning stages (H and He), primarily for stars of mass $M < 8M_\odot$. Also, early-type stars, Wolf-Rayet stars, planetary nebulae, and novae make contributions to the composition of the interstellar medium [8].

HII regions

Emission lines from HII regions, observed in the visible, radio, and now ultraviolet (UV) and infrared (IR) wavelengths, give information about the abundances in the *gas* in galaxies. An HII region is heated to about 10^4 degrees kelvin by UV radiation. The "II" is astronomical convention for singly ionized. Condensation onto grains affects gas-phase abundances; this is less severe than for HI gas (HI is atomic hydrogen), but Mg, Si, and Fe are depleted by large factors. Because the UV sources (usually massive stars) are luminous, HII regions can be observed far away in our galactic disk, and in external irregular and spiral galaxies not seen edge on. The depletion onto grains is not large for nonmetallic elements, so that HII regions are a good means to investigate abundance gradients in galaxies. Old supernova remnants give similar information. Meyer [430] gives an extensive review. The HII data range in O/H from 1/50 to 3 times solar, and provide information on the variation of abundance with metallicity which is independent of that from stars. The HII regions provide regions with different degrees of nucleosynthesis at a given *time* but varying in *space*, while a distribution of stars of different metallicity provides, at least to some extent, different degrees of nucleosynthesis as a function of *time* for a (perhaps) restricted region of *space*.

Molecular clouds

As the name implies, molecular clouds radiate mostly through molecular lines. From the isotope shifts of the molecular spectra, isotopic abundances may be inferred. At the low temperatures characteristic of molecular clouds, isotopic fractionation occurs due to *chemical evolution* of the matter.[11] Thus, until the chemistry of these clouds is well understood, their use for inferring nucleosynthetic information is probably premature.

Gamma-ray lines

As spectral lines identify atomic and molecular species, so do gamma-ray lines identify nuclei. The atmosphere of Earth severely attenuates gamma rays,[12] making their detection a project for space vehicles and balloons. Background caused by cosmic-ray interactions with the experimental apparatus and its carrier is a severe problem.

---

[11]Astronomical convention uses the term *chemical* evolution for the evolution of the elemental abundance in a galaxy, by *thermonuclear* processes; this seems an inappropriate choice. There is another, perhaps better, meaning for *chemical evolution* of galaxies, which involves the development of complex chemistry, and of life.

[12]The distribution of air near the surface of Earth is roughly isothermal, so $dP/dr = \Re Y T d\rho/dr = \rho g$. Let the density scale height be $\ell = (d\ln\rho/dr)^{-1}$. Then $\ell = Y\Re T/g$. For $g = 980$ cm s^{-2}, $T = 300$ K, and $Y \approx 1/28$ for an atmosphere dominated by N$_2$, we have $\ell \approx 9$ km. A gamma ray is most likely to interact by Klein-Nishina scattering [231] on atomic electrons; the cross section at 1 MeV is about 0.1 of the Thomson value ($\sigma_T \approx 2/3 \times 10^{-24}$ cm^2). We may speak of the atmosphere as having a thickness $\rho\ell \approx 900$ g cm^{-2}. The attenuation depth τ is $\ell\rho Y_e \sigma N_A \approx 9 \times 10^5$ cm $(1.29 \times 10^{-3}$ g cm$^{-3})(\frac{1}{2})(0.004) \approx 23$, so the attenuation is $e^{-\tau} \approx 8.2 \times 10^{-11}$.

How are gamma rays made? Charged particles of very high energy, if accelerated (i.e., if their trajectories are forced to bend), radiate high-energy photons. Collisions of particles can leave a nucleus in an excited state, and radiative de-excitation to the ground state will produce gamma-ray lines. Higher-energy particles (cosmic rays) may cause spallation, which leaves the spalled nuclei in excited states, which decay radiatively.

Naturally synthesized radioactive nuclei can decay to excited states of the daughter nucleus, again with radiative de-excitation. Detection of such lines both identifies ongoing nucleosynthesis processes and provides a quantitative measure of the yield of those processes. Examples are ^{26}Al (half-life of 0.75 Myr), ^{56}Co, and ^{57}Co which have been detected by satellite and balloons. The observation of the 1.809 MeV line from radioactive ^{26}Al was a milestone for gamma-ray astronomy; it was the first detection of a cosmic radioactive isotope. The discovery was made by the high-resolution germanium spectrometer on HEAO-C [407, 408]. The line has also been seen by many different instruments: besides HEAO-C, the SMM gamma-ray spectrometer [568], the balloon-borne germanium spectrometer of the Bell-Sandia group [405], the balloon-borne Compton telescope of the Max-Planck-Institute [654], by the two balloon-borne germanium spectrometers of the U.S./French collaboration [414] and of the GSFC/Bell–New Mexico collaboration [612], and more recently by COMPTEL and OSSE on the Comptom Gamma Ray Observatory.[13]

Supernova 1987A in the Large Magellanic Cloud provided the source for the lines of radioactive ^{56}Co and ^{57}Co. The ^{56}Co lines were first detected by the SMM satellite [416] and confirmed by several balloon flights [539, 409, 518, 613]. These detections were striking confirmation of quantitative theories of nucleosynthesis and of supernova explosions, as will become clear in subsequent chapters.

Positrons are formed either by high-energy processes giving rise to electron-positron pairs, or by the decay of unstable nuclei by positron emission. When they are slowed to thermal velocities, positrons combine with electrons to form *positronium*, which decays by gamma-ray emission. For antiparallel spins, two gammas of 0.511 MeV each are emitted. For parallel spins, three gammas are emitted, with a continuum of energies that sum to $2m_e c^2$.

---

[13]Recent observations with COMPTEL [211] suggest that ^{26}Al originated in massive stars.

3

Some Aspects of Nuclear Physics

Not from the stars do I judgment pluck,
And yet methinks I have astronomy ...
—William Shakespeare (1564–1616),
Sonnet 14

There are 286 nuclei which are considered to be "stable," at least to the extent that they appear in tables of solar system abundances. In addition there are roughly another thousand which have a sufficient number of measured properties to make their way into tabulations of nuclei. This crowd of nuclei becomes enormously more comprehensible with the aid of a few simple ideas. Cosmic abundances are a result of thermonuclear processes. To understand these abundances we must begin with an understanding of the systematic behavior of the properties of these nuclei. Because we seek to interpret clues from the effects of many events, we will emphasize the broad features of the subject of nuclear physics rather than its modern sophistication and complexity. We expect these sturdiest aspects to imprint their mark most indelibly on the astronomical phenomena.

Nuclear masses show amazingly regular behavior, to the extent that they may be reasonably well represented by an algebraic expression (a *mass law*) with only five coefficients. The nuclear masses are a direct reflection of the energies of the nuclei. In equilibrium a system tends toward the lowest energy state, so that (other things being equal) nuclear binding will be correlated with cosmic abundance. Also, the transition to lower energy states releases energy, providing a source to power—and to explode—stars.

Because all nuclei do not have the same binding energy, both spontaneous and induced transitions may occur to form those nuclei representing lower energy states. The stability of nuclei against the various modes of radioactive decay may be easily understood in terms of the mass law. Such decay provides both a source for more stable nuclei—and perhaps of most importance—a clock. The set of radioactivities which may be used as clocks provide uniquely powerful timing information for astronomical processes and cosmology. The most spectacular aspect of nuclear instability is the decay of ^{56}Ni to ^{56}Co to ^{56}Fe, the process that heats the ejecta of supernovae to power their prodigious luminosity.

The induced transitions among nuclei are simply nuclear reactions. Despite their complex detailed behavior, nuclear reaction rates have several systematic features. Although the greatest binding per nucleon occurs in the nickel isotopes, the cosmic abundance of nickel and iron is far less than that of oxygen or carbon. Why? One excellent reason is the coulomb barrier. Like charges repel. This coulomb repulsion between nuclei tends to prevent them from approaching closely enough for nuclear reactions to occur. The nuclear forces have a short range; they have essentially no effect beyond a very short distance. The first step toward a nuclear reaction must be penetration of this coulomb barrier. The coulomb penetrability is extremely sensitive to the velocities of the interacting particles (i.e., their temperature) and the product of their nuclear charges. Thus synthesis of nuclei of large charge like nickel or iron requires much more extreme (and rare) conditions than does the fusion of helium to carbon and oxygen.

Once the coulomb barrier is overcome, the nuclear forces come into play. Will a nuclear reaction occur, or can the coulomb barrier be penetrated again, as particles escape? The probability of a truly nuclear reaction is strongly affected by nuclear *resonance* phenomena. The nucleus is a wave-mechanical system which has resonances. These are analogous to the resonances of sound waves in an organ pipe, for example. For light nuclei ($A < 24$ or so), the spacing in energy of excited nuclear levels (and therefore of resonances) is sparse. There are not so many particles, so that there are not so many states to resonate with. As is usual with small numbers, statistical procedures fail, and each resonance must be individually measured. Much of our understanding of astronomy is based upon beautiful and difficult experiments by nuclear physicists. For heavier nuclei, the number of possible states is larger, so that the levels are more closely spaced, and many may contribute to the net rate. For such nuclei, semiempirical methods of estimating reaction rates, based upon statistical regularities, have been used with considerable success.

Cosmic abundances are written in the language of nuclear physics. Some study of that language greatly simplifies the decoding of the message.

3.1 NUCLEAR MASSES

Atomic nuclei are characterized by several quantum numbers. The charge number Z is equal to the number of protons in the nucleus; the mass number A is equal to the number of nucleons (neutrons plus protons). The net spin J, the parity π, and the excitation energy E_{ex} specify properties of particular nuclear states. The energy released, upon binding the A nucleons together into a neutral atom, is the *nuclear binding energy*:

$$B = (Z(m_p + m_e) + (A - Z)\, m_n - M(A, Z))c^2, \qquad (3.1)$$

where m_p, m_n, m_e, and $M(A, Z)$ are the rest masses of the proton, neutron, electron, and nucleus, respectively. Nuclear (and atomic) masses can be measured in terms of the atomic mass unit (amu), which is one-twelfth the mass of the neutral ^{12}C atom. If $M_u = 1/N_A = 1.660420 \times 10^{-24}$ grams denotes the atomic mass unit, the *atomic mass excess* is

$$\Delta M(A, Z) = (M(A, Z) - A)\, M_u c^2,$$
$$= 931.478\, \text{MeV}\, (M(A, Z) - A), \qquad (3.2)$$

where $M(A, Z)$ is expressed in amu. Here Avogadro's number is $N_A = 6.023 \times 10^{23}\,\text{g}^{-1}$.

Consider the reaction $^{12}\text{C} + {}^4\text{He} \rightarrow \gamma + {}^{16}\text{O}$, that is, $^{12}\text{C}(\alpha, \gamma)^{16}\text{O}$. The *entrance channel* is said to contain ^{12}C and ^{4}He; the *exit channel*, ^{16}O and the gamma ray. The *Q-value* for the reaction is defined as the energy released in the exit channel, in the limit that the kinetic energy of relative motion in the entrance channel becomes negligible:

$$Q = (M(^{12}\text{C}) + M(^4\text{He}) - M(^{16}\text{O}))c^2$$
$$= \Delta M(^{12}\text{C}) + \Delta M(^4\text{He}) - \Delta M(^{16}\text{O}). \qquad (3.3)$$

The energetics of nuclear reactions can be evaluated using tables of atomic mass excesses (see [173], pp. 289–91, for example). By convention, tables of masses use atomic masses. To obtain nuclear masses, subtract from the tabular values the rest mass of Z electrons (and to be exact, also the atomic binding energy of the electron cloud to the nucleus). Except for weak interactions these corrections to the Q-values cancel. For our example, $\Delta M(^{12}\text{C}) = 0$, $\Delta M(^4\text{He}) = 2.42475$, and $\Delta M(^{16}\text{O}) = 4.73655$, so that Q-value $= 7.16130\,\text{MeV}$.

To understand the behavior of nuclear reactions it is necessary to understand the systematic behavior of nuclear masses with A and Z. This behavior is concisely expressed by a mass formula [656, 93]. As a first step toward developing this semiempirical formula, consider the experimental behavior of nuclear radii. The definition of the radius of the nucleus depends to some extent upon the phenomenon used to measure it. For our purposes it is adequate to note that scattering of alpha particles, neutrons, protons, and electrons by nuclei give a radius

$$R \approx r_0 A^{1/3},$$ (3.4)

where

$$r_0 \approx (1.8 \text{ to } 1.1) \times 10^{-13} \text{cm}.$$ (3.5)

This means the number of nucleons A in the nucleus is proportional to the nuclear volume; so that the nuclear mass density is roughly constant,

$$\rho_{\text{nuclear}} \approx \frac{3M_u}{4\pi r_0^3},$$ (3.6)

$$\approx 2 \times 10^{14} \text{g cm}^{-3},$$

and the number density of nucleons in the nucleus is

$$N_{\text{nuclear}} = \frac{3}{4\pi r_0^3},$$ (3.7)

$$\approx 1.4 \times 10^{38} \text{cm}^{-3}.$$

Evidently the increase in attractive force due to added nucleons is not sufficient to cause contraction to higher densities; the nuclear force *saturates*. A similar situation occurs with water droplets; mass laws developed from N. Bohr's *liquid-drop* model of the nucleus.

The binding energy per nucleon B/A is roughly constant with A for $A \gg 1$ as may be seen in figure 3.1, which shows B/A for $A = 1$ to 70. The dominant feature of the curve is its tendency to be flat at $B/A \approx 8\text{MeV}$. Notice the more subtle maximum around $A \approx 56$. The release of nuclear energy in stars depends upon the evolution of the composition toward this maximum. Let us express the nuclear mass $M(A, Z)$ as a function of A and Z. The largest term will be

$$M(A, Z) = Z m_p + (A - Z) m_n$$ (3.8)

To this we add a volume term,

$$m_1 = -a_1 A.$$ (3.9)

The droplet analogy suggests the possibility of surface effects. A nucleon near the surface is not surrounded by neighbors and therefore feels an

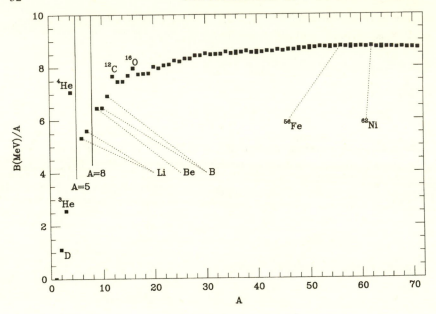

Fig. 3.1. Binding versus A

unbalanced force, attracting it toward those neighbors it does have (i.e., the center of the nucleus). This *surface tension* should be proportional to the number of nucleons so affected, that is, to the surface area,

$$4\pi R^2 = 4\pi \, r_0^2 \, A^{2/3}, \tag{3.10}$$

so we have

$$m_2 = a_2 \, A^{2/3} \tag{3.11}$$

where the sign is different from Eq. 3.9, which overestimates the binding energy; Eq. 3.11 corrects this error.

Stable nuclei tend to have roughly equal numbers of neutrons and protons, or $Z = A/2$. Thus those states symmetric in number of neutrons and protons are favored. To see how this *symmetry energy* could arise, consider the nucleons in the nucleus to be independent particles in a common potential well (a *shell* model). In the lowest energy state they represent two degenerate Fermi-Dirac gases. The Fermi energy for protons is

$$\epsilon_f = \frac{p_f^2}{2m} = C_0 \, (Z/A)^{2/3}, \tag{3.12}$$

where C_0 is a constant and the number density of protons is

$$N(Z) = \frac{Z}{\text{(nuclear volume)}}$$

$$= \frac{3(Z/A)}{4\pi r_0^3}. \tag{3.13}$$

The total energy for protons is

$$E(Z) = \frac{3}{5}N(Z)\,\epsilon_f$$

$$= C_1\, Z^{5/3}\, A^{-2/3}, \tag{3.14}$$

where C_1 is a constant. The combined energy of the neutron and proton "gases" is

$$E(Z, A) = C_1\, \frac{Z^{5/3} + (A - Z)^{5/3}}{A^{2/3}}. \tag{3.15}$$

Setting $\partial E(Z, A)/\partial Z\big|_A = 0$, we find a minimum at $Z = A/2$. The correction to the mass formula is

$$E(Z, A) - E(Z, A)_{\min} = C_1\, A^{-2/3}\left[\left(\frac{A}{2} + \Delta\right)^{5/3}\right.$$

$$\left. + \left(\frac{A}{2} - \Delta\right)^{5/3} - 2\left(\frac{A}{2}\right)^{5/3}\right] \tag{3.16}$$

where $\Delta = (A/2) - Z$. If we assume Δ to be small, and keep terms up to Δ^2 (the first nonzero terms),

$$E(Z, A) - E(Z, A)_{\min} \approx C_1\, A^{-2/3}\left[10\left(\frac{2}{A}\right)^{1/3}\frac{\Delta^2}{9}\right] \tag{3.17}$$

$$\propto \frac{\Delta^2}{A}.$$

In general, symmetry effects in potential energies would be included also. This is the historical form for the symmetry energy:

$$m_3 = a_3\, \frac{(A/2 - Z)^2}{A}. \tag{3.18}$$

Coulomb repulsion of protons gives an energy increase of the form

$$E_{\text{coul}} = (\text{constant}) \frac{Z^2}{R}. \tag{3.19}$$

This gives a term

$$m_4 = a_4 \frac{Z^2}{A^{1/3}}. \tag{3.20}$$

Nuclei with paired protons or paired neutrons tend to be more tightly bound; the *pairing energy* term is

$$m_5 = \delta(A), \tag{3.21}$$

where the usual form is

$$\begin{aligned}
\delta(A) &= \frac{a_5}{A^{3/4}} && \text{for odd-odd nuclei,} \\
&= 0 && \text{for odd-even nuclei,} \\
&= -\frac{a_5}{A^{3/4}} && \text{for even-even nuclei.}
\end{aligned}$$

Here *odd-odd* is short for *odd* Z and *odd* $(Z - A)$, and so on.

Thus for atomic masses,

$$M(Z, A) = (A - Z) m_n + Z (m_p + m_e) - a_1 A + a_2 A^{2/3}$$
$$+ \frac{a_3 \Delta^2}{A} + \frac{a_4 Z^2}{A^{1/3}} + \delta(A), \tag{3.22}$$

where the electrostatic binding energy of the electron cloud is ignored.

Let us consider how to estimate the coefficients a_i in the mass formula. For simplicity consider the nucleus to be a uniformly charged sphere so that (constant) = 3/5 in Eq. 3.19. Then $a_4 = 0.576$ MeV for $r_0 = 1.2$ fermi. The most stable nucleus will have the lowest mass. For a given A, we set

$$\left. \frac{\partial M(Z, A)}{\partial Z} \right|_A = 0. \tag{3.23}$$

Consider odd-even nuclei so that the pairing energy can be neglected. This gives

$$Z = A/(c + dA^{2/3}) \tag{3.24}$$

where

$$c = \frac{2}{1 + (q/a_3)},$$ \hfill (3.25)

$$d = c\, a_4/a_3,$$ \hfill (3.26)

and

$$q = \left(m_n - m_p - m_e\right) c^2 = 0.7825\,\text{MeV}.$$ \hfill (3.27)

Fermi [231] found that $c = 1.98$ and $d = 0.015$ for the stable nuclei. Figure 3.2 shows these nuclei in the Z vs. $(A - Z)$ plane; equation 3.24 is shown as the solid curve for Fermi's values of c and d; this is an approximation to the *valley of stability*. It works well, being buried by the symbols for stable nuclei in the figure. Notice that the long-dashed curve, which denotes $Z = N$, passes through abundant nuclei up to ^{40}Ca, after which coulomb repulsion of protons in the nucleus causes the most stable nuclei to shift to more neutron-rich isotopes. The short-dashed curve is for a fission parameter $Z^2/A = 37$. Above this line the lifetime for spontaneous fission is less than the age of the universe (10^{10} years); see below. The vertical lines at $Z = $ 2, 8, 20, 28, 50, 82, and 126 denote

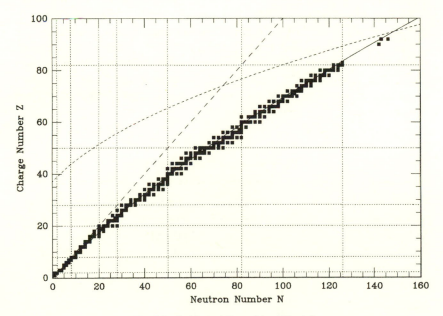

Fig. 3.2. Stable and Long-Lived Nuclei

neutron magic numbers; the horizontal lines denote the corresponding proton magic numbers. Notice the tendency for elements with magic proton number to have more isotopes which are stable. Finally, note the "island" of thorium and uranium, far above the rest of the nuclei, which stop at Pb and Bi.

Now using $c = 1.98$ we have

$$a_3 = \frac{q}{(2/c) - 1} = 77.4 \, \text{MeV}, \tag{3.28}$$

while $d = 0.015$ taken with $a_4 = 0.576 \, \text{MeV}$ gives $a_3 = 76.0 \, \text{MeV}$, which is fairly consistent for such a rough treatment. However, the slightly different choice of $c = 1.99$ and $d = 0.01466$ gives virtually the same curve in figure 3.2, but then we have $a_3 = 155.6$ and $78.2 \, \text{MeV}$ for these values of c and d respectively. The inconsistency could be due to the assumed a_4 which depends on r_0 and the charge distribution. Two points should be made here: (1) these a_i can be sensitive to fitting errors, and (2) all the a_i should be determined self-consistently. Table 3.1 gives Fermi's estimate for the a_i and a later set due to Green [271] for the same form for the mass law; note that the latter values give $c = 1.9836$ and $d = 0.014867$. The Fermi values were converted from amu using 1 amu = 931.478 MeV.

Semiempirical mass formulae are a method of last resort for estimating masses of nuclei that have not been determined experimentally. The versions of Cameron and Elkin [151] and Seeger [564] have been important in nuclear astrophysics. For a review of the subject, see Garvey [262]. Recent work [667] suggests promising new directions. The older mass laws are still valuable for the succinct representation of a large amount of data; they help us understand.

TABLE 3.1

Mass Formula Coefficients

Energy Term	Notation	Fermi	Green
Volume	a_1	14.04 MeV	15.53 MeV
Surface	a_2	13.04	17.804
Symmetry	a_3	77.3	94.77
Coulomb	a_4	0.584	0.7103
Pairing	a_5	33.5	33.6

A phenomenon of importance for nuclear masses, but not included in the mass formula, is the existence of *magic numbers*. These are the nuclear analogues of *noble gas* electronic configurations. In addition to the nuclear pairing effect, added binding occurs at closed shells. In nuclei the magic numbers are 2, 8, 20, 28, 50, 82, and 126, so the 3rd, 9th, 21st, 29th, ... nucleons are especially loosely bound. Doubly magic nuclei, such as ^{4}He with $Z = N = 2$, ^{16}O with $Z = N = 8$, and ^{56}Ni with $Z = N = 28$ figure prominently in nuclear astrophysics. Some of the magic numbers can be inferred from Figure 3.2 by noting the unusually large number of stable nuclei at these values of Z and N.

3.2   NUCLEAR STABILITY

Nuclei exhibit instability by decaying to nuclear states of lower energy. For example, very heavy nuclei are unstable. Energy is released by breaking the nucleus in two. However, in order to split, the nucleus must pass through intermediate states which have energy much higher than that of either the initial or the final state. It is only by quantum mechanical leakage through such potential barriers that some of the heavy nuclei do manage to decay; for many of them the decay time is too long to be detected. The last *stable* nucleus ^{209}Bi has an alpha-decay half-life of more than 2×10^{18} years, which is a hundred million times the estimated age of the universe.

In which nuclei are the nucleons most tightly bound? One answer is that under laboratory conditions these are the nuclei which have the largest binding energy per nucleon. Figure 3.1 gives B/A versus A; there is clearly a maximum in B/A around $A = 56$, for $B/A \approx 8.8$ MeV per nucleon. In the laboratory the nucleus with the largest value of B/A is ^{62}Ni, with ^{56}Fe a close second. The tightly bound nuclei near $A = 56$ are sometimes referred to as the *iron group*; as we saw in chapter 2, their abundances are relatively high.

Another way to answer this question is to consider the energy required to remove some part of the nucleus, the *separation energy S*. This depends upon what is pulled out as well as the nucleus from which it is pulled. For example, the alpha-particle separation energy for ^{16}O, which implies the reaction ^{16}O $\rightarrow$ ^{12}C $+$ ^{4}He, is

$$S_\alpha(^{16}\text{O}) = M(^{12}\text{C}) + \text{M}(^4\text{He}) - \text{M}(^{16}\text{O}) \tag{3.29}$$

$$= 7.1613 \text{ MeV}.$$

The proton separation energy (^{16}O $\rightarrow$ ^{15}N $+ p$) is $S_p = 12.126$ MeV and that for the neutron (^{16}O $\rightarrow$ ^{15}O $+ n$) is $S_n = 15.668$ MeV. These are relatively large (compared to $B/A \approx 8\,\text{MeV}$ per nucleon) as befits the doubly magic nucleus ^{16}O.

The separation energies for protons, neutrons, and alphas are defined in terms of mass excess ΔM and binding energy B by

$$S_p = \Delta M(Z - 1, A - 1) + \Delta M(1, 1) - \Delta M(Z, A), \qquad (3.30)$$
$$= B(Z, A) - B(Z - 1, A - 1),$$
$$S_n = \Delta M(Z, A - 1) + \Delta M(0, 1) - \Delta M(Z, A), \qquad (3.31)$$
$$= B(Z, A) - B(Z, A - 1),$$

and

$$S_\alpha = \Delta M(Z - 2, A - 4) + \Delta M(2, 4) - \Delta M(Z, A), \qquad (3.32)$$
$$= B(Z, A) - B(Z - 2, A - 4) - B(2, 4).$$

If the separation energy is negative, spontaneous emission of the corresponding particle is possible. For neutrons it might be expected that such decay occurs on a time-scale similar to the time for the neutron to cross the nucleus. The number density of neutrons in the nucleus is about $N(\text{nucleons})/2 = 5 \times 10^{37}\,\text{cm}^{-3}$ (see above), and the Fermi momentum is

$$p_f c = \left(\frac{3N}{\pi}\right)^{1/3} \frac{hc}{2} \qquad (3.33)$$
$$= 228\,\text{MeV},$$

so $\epsilon_f = 27.5\,\text{MeV}$ and $v = 0.245c = 7.34 \times 10^9\,\text{cm/s}$. The transit time across the nuclear radius is

$$t = 2\frac{R}{v} = 3.3 \times 10^{-22} A^{1/3}\,\text{s}. \qquad (3.34)$$

Such a decay is essentially instantaneous for our purposes. Some nuclei undergo spontaneous neutron emission after beta-decay (such as $^{12}\text{Be}, ^{16}\text{C}, ^{138}\text{I}$, etc.); the neutron emission occurs from an excited state after the weak interaction occurs.

For protons, the coulomb barrier will slow the decay, but not much. Because of the symmetry energy, only nuclei lying in a band around the stable nuclei will be stable against emission of protons or neutrons.

Alpha-decay can occur when the alpha-particle separation energy is negative. In figure 3.1 the binding energy per nucleon decreases for large

A. This is due to increasing coulomb repulsion between protons. Eventually the decrease is fast enough so that $B(Z, A) < B(Z - 2, A - 4) + B(2, 4)$, or $S_\alpha < 0$. Quantum mechanically there is a certain finite expectation of finding four nucleons existing as an alpha particle in the nucleus. For $S_\alpha < 0$, the rate of escape of such an alpha depends upon the probability that it can penetrate the coulomb barrier. Figure 3.3 shows this schematically. Inside the nucleus ($r < R$) the nuclear (attractive) forces dominate but outside ($r > R$) the coulomb (repulsive) forces do. This is due to the short range of the strong interaction. Gamow [258] showed that the probability of escape by quantum mechanical tunneling is

$$P \approx \frac{\exp(-2\pi\eta)}{\sqrt{E}}, \tag{3.35}$$

where

$$\eta = 2\pi\, Z_1\, Z_2\, e^2/hv. \tag{3.36}$$

Here $E = \mu v^2/2$ is the center of mass energy and μ the reduced mass. For larger energy (more negative S_α) the decay is exponentially faster. At small radii, the repulsive coulomb potential is overwhelmed by the attractive nuclear forces. A quantum mechanical particle can tunnel through

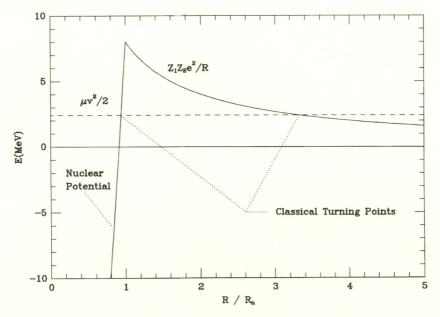

Fig. 3.3. Nuclear Potential

the barrier, between the classical turning points, causing a reaction that is not allowed classically.

Fission, like alpha-decay, can occur because the coulomb repulsion increases for large A. The *charged liquid drop* that is the nucleus can find a lower energy state by splitting in two. Let the difference in masses for symmetric fission be

$$Q = \Delta M(Z, A) - 2\Delta M \left(\frac{Z}{2}, \frac{A}{2} \right). \tag{3.37}$$

At the point of separation the two fragments have a mutual coulomb repulsion E_{coul}. For undeformed spheres of uniform charge in contact (a distance $2R$ between centers),

$$\begin{aligned} E_{coul} &= \frac{(Ze/2)^2}{2r_0} \left(\frac{A}{2} \right)^{1/3} \\ &= \frac{0.262 \, a_4 \, Z^2}{A^{1/3}}. \end{aligned} \tag{3.38}$$

For

$$Q > E_{coul}, \tag{3.39}$$

a nucleus will be unstable to fission. Eventually, this reduces to

$$\frac{Z^2}{A} > \frac{2.4 \, a_2}{a_4} \approx 54; \tag{3.40}$$

the quantity Z^2/A is sometimes called the *fission parameter*. Consideration of ellipsoidal deformation gives a better estimate of the critical value for fission,

$$\frac{Z^2}{A} > \frac{2 \, a_2}{a_4} = 45. \tag{3.41}$$

Other complications, such as quantum mechanical penetration of the fission barrier, asymmetry in fragment masses, and neutron emission should be considered (Fermi [231]). Nevertheless, Eq. 3.41 is qualitatively correct, although it may overestimate stability; this condition is shown in figure 3.2. For a fission parameter of 37, the half-life for spontaneous fission is (very roughly) of the order of 10^{10} years; for a value of 40 it is only one year (see Segre [565], p. 489).

Both fission and alpha-decay involve barrier penetration. This is more probable for small values of $\eta = 2\pi Z_1 Z_2 e^2/hv$; see equation 3.35. For a given energy $E = \mu v^2/2$ we can compare (roughly) the value of η for fission and for alpha-decay. Now

$$v = \left(\frac{2(m_1 + m_2) E}{m_1 m_2} \right)^{1/2}. \tag{3.42}$$

This is a minimum for $m_1 = m_2$ which might occur in fission but not in alpha-decay. Further, for fission the product $Z_1 Z_2$ approaches a maximum value at $Z_1 = Z_2$. Thus the penetrability is much less for fission than for alpha-decay with the same available energy. However, the fission energy depends upon (roughly) the Q-value, $Q_{fiss} = 2B(Z/2, A/2) - B(Z, A)$, which involves nuclei far apart in A, while the alpha-decay energy depends upon nuclei less widely separated. The difference in these energies depends upon the shape of the mass curve; for very large A the decrease in $B(Z, A)$ has become large enough so that $Q_{fiss} \gg S_\alpha$, and fission can compete with alpha-decay.

Gamma "decay," or more properly gamma *emission*, occurs when an excited nuclear state de-excites by electromagnetic radiation. This can occur after another reaction leaves the resulting nucleus in an excited state, or after collisionally induced excitation. Gamma-ray line radiation is also produced by particle-antiparticle annihilation; the electron-positron line at 0.511 MeV is an important example.

Beta-decay is of particular importance for astrophysics. These weak interaction processes can change protons into neutrons and vice versa. The modes of nuclear decay are:

- **positron emission:** $(Z, A) \to (Z - 1, A) + e^+ + \nu$,
- **electron capture:** $e^- + (Z, A) \to (Z - 1, A) + \nu$,
- **electron emission:** $(Z, A) \to (Z + 1, A) + e^- - \bar{\nu}$,
- **positron capture:** $e^+ + (Z, A) \to (Z + 1, A) + \bar{\nu}$,

Because the nucleon number A is constant in these reactions, let us examine the mass formula with this constraint imposed. Then

$$M(Z, A) - A m_n + a_1 A - a_2 A^{2/3} - \delta(A) = \begin{aligned} & Z (m_p + m_e - m_n) \\ & + a_3 (A/2 - Z)^2/A \\ & + a_4 Z^2/A^{1/3}, \end{aligned} \tag{3.43}$$

which is quadratic in Z, and

$$\begin{aligned} \partial M(Z, A)/\partial Z \big|_A = & (m_p + m_e - m_n) \\ & -2 a_3 (A/2 - Z)/A \\ & +2 a_4 Z/A^{1/3}, \\ = & 0, \end{aligned} \tag{3.44}$$

defines the most stable isobar. This condition gives the *valley of beta stability* in which must lie the most abundant nuclei found in nature.

3.3   COULOMB BARRIER

The nucleus is a quantum mechanical system. For classical behavior (the limit $h \to 0$; see [380]), the characteristic distance scale r should greatly exceed the de Broglie wavelength of the constituent particles. If the radius of the nucleus is $R \approx r_0 A^{1/3}$ (see above), then the interparticle spacing for nucleons is $r_0 = 1.2 \times 10^{-13}$ cm. If we take $r = r_0$, the condition for classical behavior is

$$r_0 \gg \lambda = (h/mc)(c/v). \tag{3.45}$$

The Compton wavelength for a nucleon is

$$\lambda_{\text{compton}} = h/mc = 1.13 \times 10^{-13}\,\text{cm}, \tag{3.46}$$

and in the nucleus,

$$v/c \approx 1/4, \tag{3.47}$$

so,

$$\lambda \approx 4.5 \times 10^{-13}\,\text{cm}, \tag{3.48}$$

which is actually larger than r_0. Nevertheless, some classical arguments will be used upon occasion if they are simpler and give roughly the correct answer.

Consider an idealized situation: *projectile* particles of uniform density N_p move uniformly to the right with equal velocity v. The flux passing through a unit area perpendicular to v is

$$(\text{Flux}) = N_p v. \tag{3.49}$$

Suppose the projectiles encounter a *target* of thickness x and area a, containing target particles of uniform number density N_t. The number of encounters (reactions) per unit time is proportional to the number of incident particles and to the number of targets. Thus the rate of reactions per unit volume is

$$\left(\frac{dN}{dt}\right)_{\text{reactions}} = \sigma \left(N_p\, va\Delta t\right)(N_t\, x)/(ax\,\Delta t)$$
$$= N_p\, N_t\, \sigma v, \tag{3.50}$$

where the proportionality constant σ has units of area; σ is the cross section. In general, the distribution of relative velocities of projectile and target is not a delta function as assumed above, but the expression

$$\frac{dN}{dt} = N_p \, N_t \langle \sigma v \rangle \tag{3.51}$$

is still valid if the cross section is properly averaged over relative velocities (see below). Using mole fractions as composition variables ($Y_i = N_i/\rho N_A$; see chapter 2), this becomes

$$\frac{dY}{dt} = Y_p \, Y_t \rho \, N_A \langle \sigma v \rangle \tag{3.52}$$

where dY/dt refers to the rate at which reactions occur; and the product $N_A \langle \sigma v \rangle$ is often called the *reaction rate*.

It is useful to separate the translational motion of a set of reacting particles from their relative motion (upon which reaction rate depends). As an example, consider a binary interaction of nonrelativistic particles. For relativistic particles, see the discussion in Landau and Lifshitz [379]. In nuclear astrophysics to date, triple collisions have been represented as sequential binary collisions. Let

$$\mathbf{x} = \mathbf{x}_1 - \mathbf{x}_2 \tag{3.53}$$

be the relative position vector of particles 1 and 2, and the position vector for the center of mass be

$$M\mathbf{X} = m_1 \, \mathbf{x}_1 + m_2 \, \mathbf{x}_2, \tag{3.54}$$

where $M = m_1 + m_2$. Using $\mathbf{v}_1 = d\,\mathbf{x}_1/dt$, etc., we find

$$\mathbf{v}_1 = \mathbf{V} + \mathbf{v}\,\frac{m_2}{M} \tag{3.55}$$

and

$$\mathbf{v}_2 = \mathbf{V} - \mathbf{v}\,\frac{m_1}{M} \tag{3.56}$$

where $\mathbf{V}$ is the translational velocity of the center of mass, and $\mathbf{v}$ is the relative velocity of the particles. The kinetic energy is

$$E = (m_1 v_1^2 + m_2 v_2^2)/2 \tag{3.57}$$

which becomes

$$E = (MV^2 + \mu v^2)/2 \tag{3.58}$$

where

$$\mu = m_1 m_2/(m_1 + m_2) \tag{3.59}$$

is the reduced mass.

Consider two interacting particles in the center of mass coordinate system. The impact parameter b is the distance of closest approach of the reduced mass to the origin if there were no interaction. For noninteracting classical particles, if

$$b < (r_1 + r_2), \tag{3.60}$$

a collision will occur; here r_1 and r_2 are the radii of the particles. The *area of collision* is

$$\sigma_{\text{classical}} = \pi (r_1 + r_2)^2, \tag{3.61}$$

the classical cross section, or

$$\sigma_{\text{classical}} = \pi r_0^2 \left(A_1^{1/3} + A_2^{1/3} \right)^2 \tag{3.62}$$

where $\pi r_0^2 = 4.5 \times 10^{-26}$ cm^2. For example, if $A_1 \ll A_2 = 27$, the classical cross section is 0.4 barns, where 1 barn is 10^{-24} cm^2. In the center of mass frame,

$$v = (2 E/\mu)^{1/2}$$
$$= \left(2E (A_1 + A_2)/A_1 A_2 M_u\right)^{1/2} \tag{3.63}$$

so

$$N_A \sigma_{\text{classical}} v =$$
$$3.8 \times 10^7 \left(A_1^{1/3} + A_2^{1/3}\right)^2$$
$$\times \left(E(\text{MeV})(A_1 + A_2)/A_1 A_2\right)^{1/2} \text{ sec}^{-1} \text{ cm}^3. \tag{3.64}$$

The numerical constant is (very) roughly the right order of magnitude. This expression requires that we assume that the classical particles do not interact until they approach nuclear separation; that is, only strong interactions occur. This might be roughly correct for neutrons interacting with nuclei, but not for charged particles. A mutual electrostatic repulsion will cause the particle trajectories to diverge, tending to avoid encounter. The cross section for an actual collision is reduced. The mutual potential energy of a pair of point charges Z_1 and Z_2 is

$$V_{\text{coul}} = Z_1 Z_2 \frac{e^2}{r}, \tag{3.65}$$

where r is their separation. For

$$r = r_{\text{nucleus}} = r_0 \left(A_1^{1/3} + A_2^{1/3} \right), \tag{3.66}$$

$$V_{\text{max}} = \frac{1.2 \, Z_1 \, Z_2}{A_1^{1/3} + A_2^{1/3}} \tag{3.67}$$

in MeV.

The classical turning point occurs at a radius for which the coulomb potential equals the kinetic energy of relative motion,

$$r_{tp} = \frac{Z_1 \, Z_2 \, e^2}{E}. \tag{3.68}$$

For

$$r_{tp} > r_{\text{nucleus}}, \tag{3.69}$$

the classical particles cannot approach close enough for their nuclear surfaces to touch; there is no nuclear *reaction*, and the classical cross section goes to zero. Quantum mechanical barrier penetration gives a qualitatively different result.

Because of the finite range of the strong interaction the nuclear forces dominate inside the nuclear surface, but give way to coulomb repulsion for larger radii. The effective potential was sketched in figure 3.3. The probability that a nuclear reaction occurs can be written as a product of two factors:

$$P_{\text{reaction}} = P_{\text{barrier penetration}} \, P_{\text{absorption}}. \tag{3.70}$$

For the moment, take the *sticking factor* $P_{\text{absorption}}$ to be roughly independent of energy E in the center of mass coordinate system. We will be most interested in the case in which the barrier is *thick*; that is,

$$E \ll V_{\text{max}}. \tag{3.71}$$

Then the probability of barrier penetration (see Landau and Lifshitz [380]) is

$$P_{\text{barrier}} = e^{-(4\pi/h)\int p\,dr} \qquad (3.72)$$

where

$$p = \left(2m(\frac{\alpha}{r} - E)\right)^{1/2} \qquad (3.73)$$

and

$$\frac{\alpha}{r} = Z_1 Z_2 e^2/r$$
$$= 1.44 Z_1 Z_2/r \qquad (3.74)$$

where the last expression is in MeV if r is in fermis. For $r_{\text{nuclear}} \ll r_{tp}$, $P_{\text{barrier}} = e^{-2\pi\eta}$, where $\eta = 2\pi Z_1 Z_2 e^2/hv$. This analysis assumes the angular momentum of target and projectile are zero; for the more general case, see Blatt and Weisskopf [104].

Because of the wave-particle duality, the *size* of the particle is the order of the de Broglie wavelength. The *cross section of the wave* is then $\pi\lambda^2 \approx (h/mv)^2 \approx 1/E$. Putting this factor together with P_{barrier} and the *sticking factor*, we expect

$$\sigma(E) = \frac{S(E)}{E} e^{-2\pi\eta} \qquad (3.75)$$

where $S(E)$ may be a quite complicated function of energy E (but we may hope that there are cases in which it is not). The factor $S(E)$ is sometimes slowly varying; it contains the intrinsically nuclear part of the cross section. As we have no complete theory of the nucleus at present, $S(E)$ is determined by experiment or estimated semiempirically.

Now we can examine exactly what is meant by the expression $N_A \langle \sigma v \rangle$ introduced above. In the domain of interest thus far explored, nuclei and nucleons participating in thermonuclear reactions can be treated as nonrelativistic and nondegenerate. Curiously enough, this seems to be violated only when nuclear statistical equilibrium holds (see below), in which case the precise reaction rates do not seem to be important. The distribution of particle velocities is therefore Maxwell-Boltzmann:

$$d^3N_1 = N_1 \left(\frac{m_1}{2\pi kT}\right)^{3/2} e^{-m_1 v_1^2/2kT} d^3 v_1 \qquad (3.76)$$

where d^3v_1 is the differential volume element in velocity space, and the expression for d^3N_2 is similar. Now,

$$N_1 N_2 \langle \sigma v \rangle = \int \int \sigma(E)v(E)d^3N_1 d^3N_2,$$

which becomes

$$N_1 N_2 \langle \sigma v \rangle = N_1 N_2 \int \int \sigma v \left[\frac{\sqrt{m_1 m_2}}{2\pi kT} \right]^3 e^{-\phi} d^3v_1 d^3v_2, \qquad (3.77)$$

where

$$\phi = (m_1 v_1^2 + m_2 v_2^2)/2kT \qquad (3.78)$$

or, transforming to the center of mass frame,

$$\phi = \frac{MV^2 + \mu v^2}{2kT}. \qquad (3.79)$$

We need to convert the differential volume element; we use the Jacobian determinant (see Clayton [173], p. 295) to find

$$\langle \sigma v \rangle = \int v\sigma(v) \left(\frac{\mu}{2\pi kT} \right)^{3/2} e^{-\mu v^2/2kT} d^3v, \qquad (3.80)$$

where the integral in d^3v is unity by normalization. Using Eq. 3.75 and converting to integration over energy $E = \mu v^2/2$,

$$\langle \sigma v \rangle = \sqrt{8/\pi\mu}(kT)^{3/2} \int S(E) e^{-E/kT - b/\sqrt{E}} dE, \qquad (3.81)$$

where

$$b = 31.28 \, Z_1 \, Z_2 \, A^{1/2} \sqrt{\text{KeV}} \qquad (3.82)$$

because

$$2\pi\eta = \frac{b}{\sqrt{E}}. \qquad (3.83)$$

The integral is dominated by the exponential; the integrand is a maximum when

$$\phi = \frac{E}{kT} + \frac{b}{\sqrt{E}} \qquad (3.84)$$

is a minimum, that is, at

$$E_0 = (bkT/2)^{1/2}$$
$$= 1.220 \left[Z_1^2 Z_2^2 A T_6^2 \right]^{1/3} \text{ KeV}, \tag{3.85}$$

where $T_6 = T/(10^6$ degrees kelvin). Approximating

$$e^{-\phi} = C \exp\left[-\left(\frac{E - E_0}{0.5\Delta} \right)^2 \right] \tag{3.86}$$

we require

$$C = \exp\left[-\frac{E_0}{kT} - \frac{b}{\sqrt{E_0}} \right]. \tag{3.87}$$

If we match the second derivatives at $E = E_0$, we find (see Clayton [173], p. 303)

$$\Delta = 4 \left(\frac{E_0\, kT}{3} \right)^{1/2},$$
$$= 0.75 \left[Z_1^2 Z_2^2 A T_6^5 \right]^{1/6} \text{ KeV}. \tag{3.88}$$

This maximum of the integrand is called the *Gamow peak*, and E_0 is the *optimum bombarding energy.*

For example, consider $^{12}\text{C}(p, \gamma)^{13}\text{N}$. Here $Z_1 = 6$, $Z_2 = 1$, and $A = A_1 A_2/(A_1 + A_2) = 12/13$, so

$$E_0 = 3.93\, T_6^{3/2} \text{ KeV}, \tag{3.89}$$

and,

$$\Delta = 1.35\, T_6^{5/6} \text{KeV}. \tag{3.90}$$

An approximate integration of equation 3.80 is possible using equation 3.86; the result is

$$N_A \langle \sigma v \rangle = \frac{4.34 \times 10^5 S_0 (\text{KeV barns}) \tau^2 e^{-\tau}}{A\, Z_1\, Z_2} \tag{3.91}$$

where

$$\tau = 42.48 \left(\frac{Z_1^2 Z_2^2 A}{T_6} \right)^{1/3}. \tag{3.92}$$

If we set

$$N_A \langle \sigma v \rangle = f(T) = f(T_0) \left(\frac{T}{T_0} \right)^n, \qquad (3.93)$$

then

$$n = \frac{\tau - 2}{3} \qquad (3.94)$$

as may be seen from a two-term Taylor series expansion of $\ln f(T)$ about $\ln T_0$. This indicates how sensitive such reaction rates are to variations in temperature.

3.4 RESONANCES

In the previous section the probability of absorption by the nucleus was assumed to vary slowly with energy, and lumped into a slowly varying cross-section factor $S(E)$. What happens if this is not true?

Suppose that the particle has penetrated the coulomb barrier separating it from the nucleus (see figure 3.3). If the channel energy E is positive, the particle is in an excited state which decays by barrier penetration with a mean lifetime τ. Suppose τ is much larger than the time for crossing the nucleus. For an exponential decay,

$$\int_{\substack{\text{nuclear} \\ \text{volume}}} \psi^* \psi \, dV = e^{-t/\tau} \qquad (3.95)$$

so the wave function of the quasi-stationary state has the form

$$\psi(t) = \psi(0) e^{-t/2\tau - iE_0 t/\hbar}. \qquad (3.96)$$

Energy and time are canonical conjugates, so the Heisenberg uncertainty principle can be written as

$$\Delta t \Delta E > \hbar. \qquad (3.97)$$

For $\Delta t = \tau$, the "fuzziness" in energy is $\Delta E > \hbar/\tau$. We can transform the wavefunction from a time representation $\psi(t)$ to an energy representation $\phi(E)$ by using Fourier transforms:

$$\psi(t) = \int_{-\infty}^{\infty} \phi(E) \exp\left(-\frac{i}{\hbar} E t \right) dE \qquad (3.98)$$

and

$$\phi(E) = \int_0^\infty \psi(t) \exp\left(\frac{i}{\hbar} Et\right) dt. \tag{3.99}$$

The latter expression becomes

$$\phi(E) \approx \int_0^\infty \exp\left[\frac{i}{\hbar}(E - E_0)t - \frac{t}{2\tau}\right] dt \tag{3.100}$$

which gives

$$\phi(E) \approx \frac{1}{\frac{i}{\hbar}(E - E_0) - \frac{1}{2\tau}}. \tag{3.101}$$

The probability $P(E)$ of finding the system at energy E in energy space is

$$P(E) = \phi^* \phi \approx 1 / \{[E - E_0]^2 + (\hbar/2\tau)^2\}. \tag{3.102}$$

The normalization condition,

$$\int_0^\infty P(E) \, dE = 1, \tag{3.103}$$

gives

$$P(E) \, dE = \frac{\hbar}{2\pi\tau} \frac{dE}{[E - E_0]^2 + (\hbar/2\tau)^2}. \tag{3.104}$$

The state has a *natural width* in energy,

$$\Gamma = \hbar/\tau \tag{3.105}$$

so we have the Breit-Wigner formula:

$$P(E) = \frac{\Gamma/2\pi}{[E - E_0]^2 + (\Gamma/2)^2} \tag{3.106}$$

where the resonance energy is E_0.

To use this to derive a cross section we must consider the concept of the *compound nucleus* introduced by N. Bohr. A reaction

$$A + a \rightarrow C \rightarrow B + b \tag{3.107}$$

proceeds through an intermediate state C which "forgets" the way it formed; it is specified only by its quantum numbers, not its mode of formation. The nucleus is a strongly interacting system; the virtual state is expected to share its energy among many different modes. The probability of the transformation is the product of the probabilities for each step:

$$P(A \rightarrow B) = P(A \rightarrow C)\,P(C \rightarrow B). \tag{3.108}$$

The rate r_i of a process of mean lifetime τ_i is

$$r_i = \frac{1}{\tau_i} = \frac{\Gamma_i}{\hbar}. \tag{3.109}$$

The net rate of destruction of state C by i independent processes is

$$r = \sum_i r_i = \frac{1}{\hbar} \sum_i \Gamma_i = \frac{\Gamma}{\hbar}. \tag{3.110}$$

If the reaction is initiated by a particle a, the cross section is proportional to

$$\sigma \propto \Gamma_a. \tag{3.111}$$

The probability of decay into an exit channel b is proportional to Γ_b/Γ, so

$$\sigma \propto \frac{\Gamma_a \Gamma_b P(E)}{\Gamma} \tag{3.112}$$

where the shape factor is included from equation 3.106.

At the resonance the cross section has a maximum value. This can be obtained by a geometrical argument although a correct treatment involves wave mechanics (see Blatt and Weisskopf [104]). Consider an incident plane wave of de Broglie wavelength λ for relative motion. Divide the beam into cylindrical zones. The inner zone contains particles with impact parameter less than $\lambda = \lambda/2\pi$; for the ℓth zone the impact parameters lie in the range $\ell\lambda$ to $(\ell+1)\lambda$. Roughly speaking, these particles have angular momenta of $\mu v \ell \lambda$. The cross-sectional area of the ℓth zone is $(2\ell + 1)\pi\lambda^2$. No more particles can leave the beam than are in it initially, so for the ℓth partial wave,

$$\sigma(\text{res}, \ell) < (2\ell + 1)\pi\lambda^2. \tag{3.113}$$

If we normalize Eq. 3.112 to unity at resonance and multiply by Eq. 3.113, we have

$$\sigma(\text{res}, \ell) = \frac{(2\ell + 1)\pi\lambda^2 \Gamma_a \Gamma_b \omega}{[E - E_0]^2 + (\Gamma/2)^2} \tag{3.114}$$

where $\omega = 2J_c + 1/(2J_a + 1)(2J_b + 1)$ corrects for our neglect of degeneracy of states due to nonzero angular momentum.

If the resonance is narrow,

$$\Gamma \ll E_0. \tag{3.115}$$

Then Eq. 3.114 resembles a delta function in the reaction-rate integral, so for one resonance,

$$N_A \langle \sigma v \rangle = \frac{2N_A}{\sqrt{\pi}} \frac{E_r}{(kT)^{3/2}} e^{-E_r/kT} \left(\frac{2}{\mu}\right)^{1/2} \int_0^\infty \sigma_{\text{res}} \, dE \tag{3.116}$$

where

$$\int_0^\infty \sigma_{\text{res}} \, dE = 2\pi^2 \lambda_r^{\,2}(2\ell + 1)\left(\frac{\Gamma_a \Gamma_b}{\Gamma}\right) \tag{3.117}$$

and the subscript r refers to evaluation at the resonance ($E = E_0$). For E_r in KeV,

$$N_A \langle \sigma v \rangle_{\text{res}} = 4.88 \times 10^{12} \frac{2\ell + 1}{(AT_6)^{3/2}} \frac{\Gamma_a \Gamma_b}{\Gamma} e^{-11.61E_r/T_6} \text{ cm}^{-3}\,\text{s}^{-1}. \tag{3.118}$$

If this is approximated by a power law in temperature , the effective exponent is

$$n = \frac{11.61E_r}{T_6} - \frac{3}{2}. \tag{3.119}$$

Again we see a sensitive dependence upon variation in temperature.

The approach used above is appropriate if the resonances are widely spaced. If no resonance lies in the Gamow peak, the contributions from distant resonances (the "wings" of equation 3.106; $(E - E_0)^2 \gg \Gamma^2$) may give a roughly constant cross section. In this case the approach of the previous section can be applied. At higher energy the density of states and the natural width of these states both increase; the states begin to overlap and merge into a continuum. In a given energy range, nuclei of higher A tend to have a higher density of states. In this case a statistical approach becomes more appropriate.

Consider the average $\langle \Gamma \rangle$ of levels of width Γ spaced at an average energy interval D. The average cross section is

$$\bar{\sigma} = 2\pi \lambda^2 (2\ell + 1)\frac{\Gamma_a \Gamma_b}{D\Gamma}. \tag{3.120}$$

For a heavy nucleus ($Z > 10$) we often find that the exit channel width is much larger than the width of the entrance channel,

$$\Gamma = \Gamma_b \gg \Gamma_a, \tag{3.121}$$

so

$$\bar{\sigma} = 2\pi\lambda^2(2\ell + 1)\left(\frac{\Gamma_a}{D}\right) \tag{3.122}$$

where the factor Γ_a/D is the *strength function*. For particle emission (for example (α, p), (n, p), (α, n) reactions), Γ/D can be as large as 0.1 or so. For radiative processes the width is much smaller than the particle width (if a particle channel is open); this is due to the weakness of the electromagnetic interactions relative to the strong interaction. It is important to note that the strength functions are statistical in nature, and that semiempirical procedures have been used to estimate them with some success (see [238, 631, 696, 615]).

3.5 REVERSE RATES

The rates of forward and reverse reactions are fundamentally related to each other by requirements of statistical physics and thermodynamics. Consider a reaction of the form

$$A + B \rightarrow C \tag{3.123}$$

which requires an energy Q. The inverse rate (which liberates energy Q) is

$$C \rightarrow A + B. \tag{3.124}$$

To be specific, suppose A, B, and C to be Maxwell-Boltzmann particles; the chemical potential of A is then

$$\mu_A = kT \ln\left[\frac{N(A)}{g_a}\left(\frac{2\pi\hbar^2}{m_AkT}\right)^{3/2}\right] + m_Ac^2 \tag{3.125}$$

and the chemical potentials of B and C are given by similar expressions. In *chemical equilibrium*,

$$\mu_A + \mu_B = \mu_C \tag{3.126}$$

or

$$\left(\frac{N_AN_B}{N_C}\right)_{eq} = \phi(T) \tag{3.127}$$

$$\phi(T) = \frac{g_Ag_B}{g_C}\left(\frac{m_Am_B}{m_C}\right)^{3/2}\left(\frac{kT}{2\pi\hbar^2}\right)^{3/2}e^{Q/kT}. \tag{3.128}$$

However, the rate of destruction of $A + B$ pairs by this process is

$$\frac{dN(AB)}{dt} = -N(A) N(B) \langle \sigma v \rangle_{AB} \tag{3.129}$$

while the rate of decay of C is

$$\frac{dN(C)}{dt} = -\lambda_C N(C). \tag{3.130}$$

For conservation of nucleons we have a constraint

$$\frac{dN(AB)}{dt} - \frac{dN(C)}{dt} = 0, \tag{3.131}$$

or

$$\frac{N(A) N(B)}{N(C)} = \frac{\lambda_C}{\langle \sigma v \rangle_{AB}}. \tag{3.132}$$

The right-hand side depends only on the microscopic properties of the reactants and thermodynamic variables, not on the abundances. In chemical equilibrium, we must also have

$$\left(\frac{N_A N_B}{N_C} \right)_{eq} = \frac{\lambda_C}{\langle \sigma v \rangle_{AB}} \tag{3.133}$$

so that in general,

$$\frac{\lambda_C}{\langle \sigma v \rangle_{AB}} = \phi(T) \tag{3.134}$$

in order for the equilibrium limit to be obtained when the system is in fact in equilibrium! Thus,

$$\lambda_C = \langle \sigma v \rangle_{AB} \frac{g_A g_B}{g_C} \left(\frac{m_A m_B}{m_C} \right)^{3/2} \left(\frac{kT}{2\pi\hbar^2} \right)^{3/2} e^{-Q/kT} \tag{3.135}$$

is the desired relation between forward and inverse rates for this reaction. Alternatively,

$$\lambda_C / \rho N_A \langle \sigma v \rangle_{AB} = 9.8678 \times 10^9 \frac{g_A g_B}{g_C} \left(\frac{A_A A_B}{A_C} \right)^{3/2}$$

$$\times \frac{T_9^{\frac{3}{2}}}{\rho} \exp\left(-11.605 \, Q(\text{MeV})/T_9 \right). \tag{3.136}$$

For two particles in the exit channel ($A + B \rightarrow C + D$),

$$\langle \sigma v \rangle_{CD} / \langle \sigma v \rangle_{AB} = \frac{g_A g_B}{g_C g_D} \left(\frac{A_A A_B}{A_C A_D} \right)^{3/2} \exp\left(-11.605\, Q(\text{MeV}) / T_9 \right).$$

(3.137)

The procedure is one of general validity.

3.6   HEAVY-ION REACTIONS

Reactions like ^{12}C + ^{12}C, ^{12}C + ^{16}O, and ^{16}O + ^{16}O are often called *heavy-ion* reactions by nuclear physicists; the terminology derives from the nature of the accelerated particles in the laboratory situation. The compound nucleus formed by such reactions is in a highly excited state. For example, the difference in binding energy between two ^{12}C nuclei and ^{24}Mg is about 14 MeV. To this should be added the effective bombarding energy attained in the given astrophysical environment, which is of the order of several MeV (see below). The decay of an excited state of ^{24}Mg at E $\approx$ 15 to 20 MeV is dominated by those particle-emission processes that are allowed by the high excitation energy. Radiative decay, which involves the weaker electromagnetic interaction, will be a rarer mode under these conditions. The emission of particles can be discussed in terms of a nuclear evaporation model (e.g., Fermi [231], p. 162, and Blatt and Weisskopf [104], p. 365).

The suggestion that the fusion of ^{12}C + ^{12}C would be of astronomical importance after helium burning seems first to have been made by Salpeter [535]. Hoyle [303] first discussed the nucleosynthesis following from such a process, and detailed numerical analyses of this problem were published almost simultaneously by Cameron [141], Reeves and Salpeter [512], and Hayashi, Nishida, Ohyama and Tsuda [286].

The dominant reaction in carbon burning involves the fusion of two ^{12}C nuclei:

$$^{12}\text{C} +^{12}\text{C} \rightarrow\, ^{23}\text{Na} + p + 2.2398\text{MeV} \tag{3.138}$$

$$\rightarrow\, ^{23}\text{Mg} + n - 2.5993\text{MeV}$$

$$\rightarrow\, ^{20}\text{Ne} + \alpha + 4.6168\text{MeV}$$

$$\rightarrow\, ^{24}\text{Mg} + \gamma + 13.9313\text{MeV}$$

$$\rightarrow\, ^{16}\text{O} + 2^4\text{He} - 0.1132\text{MeV} \tag{3.139}$$

$$\rightarrow\, ^{16}\text{O} +^8 \text{Be} - 0.208\text{MeV}. \tag{3.140}$$

Experimental data at higher energies suggests that the latter three channels are not important. Further, Koslovsky [361] estimated the enhancement of the $^{16}O + 2\alpha$ channel by direct transfer of an alpha particle; he found in the optimum case that this branch occurs only about 3×10^{-3} times as often as the first three channels listed. Neglect of the $^{16}O + 2\alpha$, $^{16}O + {}^8Be$, and the $^{24}Mg + \gamma$ channels should result in no significant error for astrophysical considerations. This is especially true since $^{12}C(^{12}C, \gamma)^{24}Mg$ is equivalent to $^{12}C(^{12}C, \alpha)^{20}Ne(\alpha, \gamma)^{24}Mg$ and $^{12}C(^{12}C, 2\alpha)^{16}O(\alpha, \gamma)^{20}Ne$ to $^{12}C(^{12}C, \alpha)^{20}Ne$.

Patterson, Winkler, and Zaidins [488] first measured these cross sections down to an astrophysically interesting energy of 3.25 MeV in the center of mass for the p and α channels. The optimum bombarding energy is

$$E_0 = 2.41 T_9^{2/3} \text{MeV} \tag{3.141}$$

and the energy spread is

$$\Delta = 1.055 T_9^{5/6} \text{MeV}. \tag{3.142}$$

Because of its lower reaction rate they were only able to measure the neutron channel down to 4.23 MeV. This was extended to 3.54 MeV by Dayras, Switkowski, and Woosley [205]. The branching ratio was found to be $b_n \approx 0.02$. Mazarakis and Stephens [421, 420] extended measurements lower to 2.45 MeV for the p and α channels. At the lowest energies the branching ratios are $b_p \approx 0.6$ and $b_\alpha \approx 0.4$. Therefore, for temperatures $T_9 \geq 1$ this gave direct measurement of the proton and alpha channels of the $^{12}C + {}^{12}C$ reaction.

Unfortunately these data exhibited a marked variation (an increase) in the cross-section factor at the lowest energies measured. Mazarakis and Stephens [421] noted that this might be another resonance phenomenon of the $^{12}C + {}^{12}C$ system. Michaud [434] suggested that it was due to "capture under the barrier" caused by the extensive tails of the wavefunctions implied by the optical model potential. A detailed analysis of the problem and possible uncertainties was given by Michaud and Vogt [436]. Using gamma-ray spectroscopy rather than particle counting, the cross section was measured down to 2.4 MeV in the center of mass [347, 293, 348]. The steep increase in the cross section factor at the lowest energies was not confirmed, but the resonant structure seemed to continue.

These experimental data are summarized in figure 3.4. The bottom panel shows the cross section σ in millibarns plotted versus the center-of-mass energy. The data of Mazarakis and Stephens [421, 420] for the α channel are shown as crosses, and the p channel as open triangles.

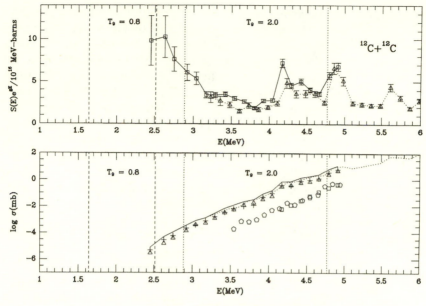

Fig. 3.4. $^{12}C + ^{12}C$

The neutron channel data of Patterson et al. [488] is shown as open squares; this data extends to energies higher than present astrophysical interest. The lower limit for the neutron channel data is well above threshold because of the small value of the cross section. The range $E_0 \pm \Delta$ around the optimum bombarding energy E_0 is shown for temperatures appropriate to hydrostatic carbon burning ($T_9 = 0.8$) and explosive burning ($T_9 = 2.0$). While the explosive case has measured cross sections throughout the relevant range of energy, only a few points are available for the hydrostatic range.

Notice that the cross section spans a range of 10^6. For a given statistical accuracy, a certain number of counts is needed. For a given experimental setup, the number of counts is proportional to the flux of incident particles and the time the target is exposed to this beam. Suppose the larger cross sections require only a tiny exposure, say three seconds. Then the smallest will require about two months! Getting the lowest energy points is a severely difficult task (see [525]).

The top panel of figure 3.4 shows the effective cross-section factor $\tilde{S} = \sigma E \exp(2\pi\eta + gE)$ plotted versus center-of-mass energy E; the units of $\tilde{S}$ are MeV barns. Here σ is the total cross section and $2\pi\eta = 87.254/\sqrt{E}$ if E is in MeV. For two nuclei of proton numbers Z_1 and Z_2, the coulomb barrier is $E_{coulomb} \approx Z_1 Z_2 e^2 / R$. If we take the interaction radius R to be about $1.44(A_1^{1/3} + A_2^{1/3})$ fermis, then for $^{12}\text{C} + {}^{12}\text{C}$, $E_{coulomb} \approx 8\text{MeV}$. The factor $\exp(-2\pi\eta)$ corresponds to the energy dependence of penetration of a coulomb barrier in the limit of low bombarding energy. Since E_0 is not much less than 8 MeV, this limit is *not* attained. For the case of a completely absorbing nucleus, a first order correction can be derived (see Evans [224], p. 874). Because one wishes S to be slowly varying with energy, a new quantity $\tilde{S} \equiv S \exp gE$ was introduced by Patterson, Winkler, and Zaidins [488]. The additional energy dependence was intended to account for the correction term, with the hope that $\tilde{S}$ would be nearly constant. The relevant physics is more complicated than this as the resonance-like structure in figure 3.4 shows. The data of Patterson, Winkler, and Zaidins [488] is shown as open triangles; that of Mazarakis and Stephens [421, 420] by open squares. The agreement is good in the common range of energy (see also [582]). At 3 MeV and below, the Mazarakis and Stephens data shows the increase mentioned above.

Reeves [511] fitted the early (higher energy) experimental results from Chalk River [653] with an optical model calculation. The experimental data involved energies from 12.5 down to 5 MeV. Earlier estimates by Fowler and Hoyle [238] and Reeves [510] predicted cross sections larger by a factor of 2.5. The Patterson et al. data agree with the earlier work at Chalk River for $E \geq 5\text{MeV}$, but falls below the Reeves extrapolation at the lowest energies measured. In the range of energy $6\text{MeV} \geq E \geq 3.23\text{MeV}$, Patterson et al. find $g \simeq 0.46\text{MeV}^{-1}$ and $S \simeq 2.3 \times 10^{16}$ MeV barns, giving

$$N_A\langle\sigma v\rangle = \exp(61.053 - \tau)/T_9^{2/3}, \tag{3.143}$$

and,

$$\tau = 84.173(1 + 0.0398T_9)^{1/3}/T_9^{1/3}. \tag{3.144}$$

These values of g and $\tilde{S}$ fall below the data of Mazarakis and Stephens for energies less than 3 MeV, but are consistent with latter experimental results [347, 293, 348].

It is clear that a better understanding of the physics of the reaction is necessary for extrapolating to energies below $E \approx 2.4$ MeV. What causes the "wiggles" in the data? At these high excitation energies (16–19 MeV) the ^{24}Mg compound nucleus has many states, with average level spacings of the order of KeV; however, the "wiggles" have a spacing of the order of, say, 0.5 MeV. Ordinary compound nuclear states are thus ruled out, but a clumping of such states might give rise to the observed phenomenon. Also, in passing from two ^{12}C nuclei to the ^{24}Mg compound nucleus, the system would pass through intermediate configurations that might give rise to the resonances observed. Finally, a giant resonance (that is, a single particle resonance in the optical model) might be the explanation.

The astrophysical importance of fusion of ^{16}O nuclei was discussed qualitatively by Hoyle [303]; early quantitative investigations were made by Cameron [143] and by Tsuda [637]. Because of the high excitation energy of the ^{32}S compound nucleus, the $^{16}O + ^{16}O$ reaction has many possible products. Consider a reaction, $2A \rightarrow B + C$. Not only must enough energy be available to form B and C, but B and C must leave their common potential well, and penetrate their common coulomb barrier. The coulomb energy is

$$E_{coulomb} \approx 1.4 Z_1 Z_2 / R \text{ MeV,} \tag{3.145}$$

if R is the distance between particles 1 and 2 measured in fermis. For $^{16}O + ^{16}O$, roughly,

$$R \approx 1.4 \ (A_1^{1/3} + A_2^{1/3}) \text{ fm,}$$
$$\approx 7.1 \text{ fm.} \tag{3.146}$$

From Rutherford scattering at a center-of-mass energy of 10.5 MeV, $R \approx 8.8$ fm; from a radius-dependent fit [580] to the low energy-absorption cross section, $R \approx 7.3$ fm. Fortunately this uncertainty is not a problem here. We have $E_{coulomb} \approx 0.2 Z_1 Z_2$ MeV, so $E_{coulomb} \approx 3$ MeV for a proton and 5.6 MeV for an alpha particle in the exit channel. At energies of this order or higher the protons and alphas escape the compound nucleus with relative ease.

In addition to the Q-value energy, the optimum bombarding energy E_0 is also available. For $^{16}O + ^{16}O$,

$$E_0 = 3.91 T_0^{2/3} \text{MeV,} \tag{3.147}$$

and,

$$\Delta = 1.34 T_9^{5/6} \text{MeV.} \tag{3.148}$$

For plausible hydrostatic and hydrodynamic situations in stars, the usual range is $4 \leq E_0 \leq 13$ MeV. To be specific we will take $E_0 = 6$ MeV.

To examine the energetics, define an effective Q-value by

$$Q_{eff} \equiv Q + E_0 - E_{coulomb}. \tag{3.149}$$

In table 3.2 (after Spinka [580]) are listed the dominant exit channels for $^{16}O + {}^{16}O$, with Q and Q_{eff}. The latter entries should not be taken too exactly, but they do suggest which channels will be most important. Note that direct electromagnetic de-excitation of ^{32}S will be slow compared to particle emission. Therefore the α, p, n, and d (deuteron) channels will be most likely to dominate. Also note that p and α channels have enough energy left to evaporate another nucleon (for example, $^{16}O(^{16}O, p)^{31}P^*$ followed by $^{31}P^*(p)^{30}Si$).

The first detailed experimental investigation of $^{16}O + {}^{16}O$ at astrophysically important energies was done by Spinka and Winkler [581, 582]. This was followed by experiments by Hulke, Rolfs, and Trautvetter [306], Wu and Barnes [702], and Thomas, et al. [619]. They measured the total cross section between $E = 6.8$ and 15 MeV in the center of mass frame, examining many different exit channels. The total cross section and the

TABLE 3.2

Energetics of $^{16}O + {}^{16}O$

Product	Q(MeV)	Q_{eff}(MeV)
$^{32}S + \gamma$	16.5410	$\cdots$
$^{31}P + p$	7.6770	≈ 11
$^{28}Si + \alpha$	9.5928	10
$^{31}S + n$	1.4531	1.5
$^{30}P + d$	-2.4058	0.6
$^{30}Si + 2p$	0.3795	0.4
$^{30}P + p + n$	-4.6304	-1.6
$^{29}Si + {}^3He$	-2.5118	-2.1
$^{24}Mg + {}^8Be$	-0.4841	-4.1
$^{29}P + {}^3H$	-7.4737	-4.4
$^{27}Al + p + \alpha$	-1.9925	-4.6
$^{24}Mg + 2\alpha$	-0.3920	-5.6
$^{27}Si + n + \alpha$	-7.5846	-7.2
$^{29}Si + p + d$	-8.0056	-8.0
$^{20}Ne + {}^{12}C$	-2.4317	-8.4
$^{16}O + {}^{12}C + \alpha$	-7.1616	-11
$^{20}Ne + 3\alpha$	-9.7065	-21

cross-section factor $S(E)$ are shown in figure 3.5; the cross-section factor is

$$S(E) \equiv \sigma E \exp{(179.1/\sqrt{E})}, \qquad (3.150)$$

where E is the center-of-mass energy in MeV. The open triangles are the Spinka-Winkler measurements; the crosses, Hulke et al., and the open squares, Wu and Barnes. The data of Thomas et al. were not presented in tabular form; they lie between Wu and Barnes, and Hulke et al. The total reaction cross section was a smooth function of $^{16}O + {}^{16}O$ energy, in contrast to the behavior for $^{12}C + {}^{16}O$, and $^{12}C + {}^{12}C$.

The branching ratios are not completely determined. Although the cross section for proton production has been measured, for example, the precise relative contributions of $^{16}O(^{16}O, p)^{31}P$ and $^{16}O(^{16}O, 2p)^{30}Si$ at low energy are not known. At lower energies, about 0.2 of the reactions involve three bodies in the exit channel. It appears that the most common exit channels are $p + {}^{31}P$ or $2p + {}^{30}Si$, which together comprise about 0.75 of the total. The alpha channel, $\alpha + {}^{28}Si$, happens about 0.2 of the time; while $d + {}^{30}P$ and $n + {}^{31}S$ each happen about 0.06 of the time.

Before the appearance of the Spinka-Winkler results, the two most commonly used estimates of $^{16}O + {}^{16}O$ were due to Fowler and

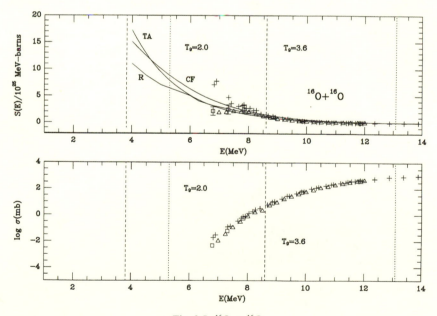

Fig. 3.5. $^{16}O + {}^{16}O$

Hoyle [238] and Reeves [511]. While a good fit to the early data (which involved only a few points below the coulomb barrier), these estimates are much too large at low energies.

Truran and Arnett [634] and Michaud [434] used a preliminary version of the Spinka-Winkler results to determine the reaction rate. Both these estimates are in error; final calibration of the experiment required that these rates be roughly half their value. A corrected version of the rate implies

$$S = 2.0 \times 10^{27} \exp(-0.613E) \text{ MeV-barns;} \qquad (3.151)$$

this is the solid line labeled TA. The corresponding reaction rate is

$$N_A \langle \sigma v \rangle = \exp(86.338 - \tau)/T_9^{2/3}, \qquad (3.152)$$

where

$$\tau = 135.958 \, (1 + 0.053T_9)^{1/3}/T_9^{1/3}. \qquad (3.153)$$

Reinhard et al., [513] have estimated the cross-section factor $S(E)$ with an adiabatic time-dependent Hartree-Fock calculation; this is shown as the solid line labeled R. Caughlan and Fowler [156] have fitted the experimental data referenced above; this is the solid line labeled CF.

The two vertical strips in figure 3.5 correspond to those energies which are most effective in determining the thermonuclear reaction rate. They denote $E_0 \pm \Delta/2$. The different estimates agree to well within a factor of two over the entire region covered by the strips. The estimates based upon the pioneering experimental work of the 1970s agree fairly well with the recent ones. For *explosive* oxygen burning, $T_9 \approx 3.6$, while for *hydrostatic* burning, $T_9 \approx 2$ (see chapters 9 and 10). The explosive case uses what is essentially an experimentally determined reaction rate for $^{16}O + {}^{16}O$. For hydrostatic burning, a small extrapolation is required. The Caughlan-Fowler rate is the best representation of the present experimental data. A few lower energy measurements are desirable to determine more accurately the rate at hydrostatic burning conditions.

The $^{12}C + {}^{16}O$ reaction rate is of some interest for nucleosynthesis, both near ^{12}C exhaustion during hydrostatic carbon burning ($T_9 \approx 1$) and during explosive oxygen burning $T_9 \approx 3.6$. Table 3.3 lists Q and Q_{eff} for some exit channels. The latter is calculated in the same manner as was done for $^{16}O + {}^{16}O$, and at $E_{CM} \approx 7.2$ MeV ($T_9 \approx 3.6$). Only the α, p, and n channels are expected to be effective.

The optimum bombarding energy is

$$E_0 = 3.06 \, T_9^{2/3} \text{ MeV,} \qquad (3.154)$$

TABLE 3.3
Energetics of $^{12}C + {}^{16}O$

Product	Q(MeV)	Q_{eff}(MeV)
$^{28}Si + \gamma$	16.7544	...
$^{27}Al + p$	5.1691	≈ 10
$^{24}Mg + \alpha$	6.7697	9
$^{27}Mg + n$	−0.4230	7
$^{26}Al + n$	−5.6642	−1.2
$^{26}Mg + 2p$	−3.1017	−1.3
$^{20}Ne + 2\alpha$	−2.5449	−5
$^{23}Na + p + \alpha$	−4.9218	−5

and the width of the Gamow peak is

$$\Delta = 1.19 \, T_9^{5/6} \text{ MeV.} \tag{3.155}$$

The total reaction cross section has been measured by Patterson, Nagorka, Symon, and Zuk [487]. Woosley, Arnett, and Clayton [692] have given a fit to the data; it is similar to that of Michaud [434]. For

$$S = 1.5 \times 10^{22} \exp(-0.98E) \text{ MeV-barns,} \tag{3.156}$$

the resulting rate is

$$N_A \langle \sigma v \rangle = \exp(74.492 - \tau)/T_9^{2/3}, \tag{3.157}$$

where

$$\tau = 106.610 \, (1 + 0.0848T_9)^{1/3}/T_9^{1/3}. \tag{3.158}$$

About half of the reactions use the proton channel and roughly an equal number the alpha channel. Spinka and Winkler (quoted by Patterson et al. [487]) estimate the neutron channel to result about 10 to 15 percent of the time. The $^{12}C + {}^{16}O$ reaction rate is directly measured at the energies important for explosive burning. The experimental data is probably adequate to show that it is too slow to be of importance during carbon burning. This reaction is important in determining at which thermodynamic conditions explosive oxygen burning makes part of the solar system abundance pattern.

The ^{12}C + ^{20}Ne reaction rate has been measured [306]; it seems much smaller at carbon burning temperatures than ^{12}C + ^{12}C. This suggests that it cannot compete with ^{12}C + ^{12}C at conditions at which there should be significant ^{12}C to burn.

3.7   WEAK INTERACTIONS IN NUCLEI

Weak interactions that occur in nuclei are of fundamental importance for nucleosynthesis theory. They determine the neutron-to-proton ratio which is a crucial parameter for abundance patterns. They cause the formation of Urca shells which dominate the cooling prior to carbon ignition under conditions of high electron degeneracy; this is thought to be vital for understanding the mechanism of Type I supernovae. They determine the nature of the core collapse for Type II and related supernovae.

Perhaps the simplest and clearest approach to nuclear weak interactions is to follow the lead of Fermi [229, 565, 171, 65]. Fermi treated the weak interactions in analogy to electromagnetic radiation from an atom; with hindsight, this certainly is justified by the unification [266, 534, 675] of these two interactions into an "electro-weak" interaction! The quantum mechanical treatment will be nonrelativistic, although the kinematics of the leptons (electrons and neutrinos) will not be restricted to nonrelativistic motion. In perturbation theory, the "Golden Rule" gives the rate of transition λ to be

$$\lambda = \frac{2\pi}{\hbar} |\mathscr{H}_{if}|^2 \rho(E_n), \tag{3.159}$$

where the matrix element is

$$\mathscr{H}_{if} = g \int u_f^* u_i \psi_e^* \psi_\nu \, d\tau, \tag{3.160}$$

the density of states accessible to the particles emerging from the reaction (exit channel) is $\rho(E_n)$, the nucleon wavefunctions are denoted by u and the lepton wavefunctions by ψ, and the weak interaction constant is $g \approx 10^{-49}$ erg cm^3.

Why is this interaction called "weak"? Consider the nuclear volume, $\Omega = 4\pi r_0^3/3 \approx 10^{-38}$ cm^3. A typical energy for the weak interaction will be $g/\Omega \approx 10^{-11}$ erg ≈ 6 eV. The nuclear binding energy of 8 MeV per nucleon is about six orders of magnitude larger. For a Fermi momentum typical of a nucleon in a nucleus, $pc \approx 200\,\text{MeV}$, the kinetic energy is about 20 MeV, again considerably larger.

Nuclear weak interactions are sometimes called *beta-decay* as a generic term, because historically it was the emission of electrons—beta rays—from radioactive nuclei by which the weak interactions were discovered. In beta-decay, the electron energies are often $\epsilon_e \approx pc \approx$ few MeV. Therefore their de Broglie wavelength is $h/p \approx$ few 10^{-11} cm, which is much larger than the nuclear size, $r_0 A^{\frac{1}{3}}$, with $r_0 = 1.2 \times 10^{-13}$ cm. This allows us to simplify the wavefunction. Consider a plane wave, for which $\psi = \exp(-i\mathbf{k} \cdot \mathbf{x})/\sqrt{\Omega}$, where Ω is the normalizing volume and $\mathbf{k} = \mathbf{p}/h$, the wavenumber. This uses the relativistic momentum as a first step toward a relativistic theory. Because the wavelengths are large, and the wavenumbers small, this wavefunction may be expanded using

$$\exp(-i\mathbf{k} \cdot \mathbf{x}) = 1 + i\mathbf{k} \cdot \mathbf{x} - \cdots, \tag{3.161}$$

where the first term generates the *allowed* transitions, and the following terms the *forbidden* transitions. When integrated over the nuclear volume, the higher-order terms have factors of $(r_0 k)$ which are small, hence the terminology. The forbidden terms become important when the matrix elements for the lower-order terms are exactly zero.

With this simplification, we may write

$$\mathcal{H}_{if} = gM_{if}/\Omega, \tag{3.162}$$

where $M_{if} = \int u_f^* u_i \, d\tau$ and the $\Omega = \int d\tau$ comes from integrating $\psi_e^* \psi_\nu$ over the nuclear volume.

This simple approach needs several improvements:

1. Dirac spinors in the wavefunctions, which will generate different interactions (vector, axial vector, tensor, scalar, and pseudo-scalar),
2. positrons and antineutrinos,
3. couplings for μ and τ leptons and their associated neutrinos,
4. parity nonconservation (V − A, or vector minus axial vector interaction), and
5. neutral current interactions.

Historically, these extensions have been added piece by piece; a beauty of the neutral current theory of the weak interactions is that these appear naturally.

Consider the reaction:

$$n \to p + e^- + \bar{\nu}. \tag{3.163}$$

If we ignore the small recoil of the nucleons (small because of their large rest mass), the energy in the exit channel is $W = E_\nu + E_e = p_{\bar{\nu}} c + (m^2 c^4 + c^2 p_e^2)^{\frac{1}{2}}$. Now the differential volume in momentum space of the electron and of the neutrino is

$$dN_e \, dN_{\bar{\nu}} = 4\pi\Omega h^{-3} p_{\bar{\nu}}^2 \, dp_{\bar{\nu}} \, 4\pi\Omega h^{-3} p_e^2 \, dp_e. \qquad (3.164)$$

This must be weighted according to the number of states implied by the spin orientations for the proposed interaction. For a given electron momentum p_e, $cp_{\bar{\nu}} = W - E_e$, so that $dW = c \, dp_{\bar{\nu}}$, and

$$dN_e \, dN_{\bar{\nu}}/dW = 16\pi^2\Omega^2 h^{-6} c^{-3} (W - E_e)^2 p_e^2 \, dp_e. \qquad (3.165)$$

The transition rate is

$$\lambda = \int_0^{p_{max}} w(p_e) \, dp_e, \qquad (3.166)$$

where $p_{max} = (\sqrt{W^2 - m^2 c^4})/c$ and

$$w(p_e) \, dp_e = \frac{g^2 |M_{if}|^2}{2\pi^3 \hbar^7 c^3} \int_0^{p_{max}} (W - E_e)^2 p_e^2 dp_e. \qquad (3.167)$$

How can this be related to experimental data? First, we introduce a complication of detail but not of principle by replacing our plane-wave wavefunction for electrons by a coulomb wavefunction. This introduces a correction factor $F(Z, \epsilon) \approx 2\pi\eta(1 - e^{-2\pi\eta})$, where $\eta = Ze^2/\hbar v_e$, with v_e being the velocity of the electron far from the nucleus, and $\epsilon = E_e mc^2$. If the electron energies are high, $v_e \to c$, and $\eta \to Ze^2/\hbar c \approx Z/137$. In general,

$$\lambda = \frac{g^2}{2\pi^3} \frac{m^5 c^4}{\hbar^7} |M_{if}|^2 f(Z, \epsilon_{max}), \qquad (3.168)$$

where $f(Z, \epsilon_{max})$ is a complicated function that includes the coulomb correction. For the simple case in which the coulomb correction is ignored,

$$f(\eta_0) = \int_0^{\eta_0} \left[(1 + \eta_0^2)^{\frac{1}{2}} - (1 + \eta^2)^{\frac{1}{2}} \right]^2 \eta^2 \, d\eta \qquad (3.169)$$

$$= -\frac{1}{4}\eta_0 - \frac{1}{12}\eta_0^3 + \frac{1}{30}\eta_0^5 + \frac{1}{4}(1 + \eta_0^2)\log(\eta_0 + (1 + \eta_0^2)^{\frac{1}{2}},$$

where $\eta_0 = p_{max}/m_e c^2$. The transition rate is simply the inverse of the mean lifetime, $\lambda = 1/\tau = \ln 2/t_{\frac{1}{2}}$, so

$$ft_{\frac{1}{2}} = \ln 2 / \left[\frac{g^2}{2\pi^3} \frac{m^5 c^4}{\hbar^7} |M_{if}|^2 \right], \tag{3.170}$$

which is a constant divided by the nuclear matrix element. Thus, with a measured half-life and a calculated coulomb correction $f(Z, \epsilon_{max})$, the nuclear matrix element is determined.

There are two possible spin orientations of the electron and the neutrino: the spins may be parallel or antiparallel. If antiparallel, the net angular momentum removed is $\Delta I = 0$, corresponding to *Fermi selection rules*. Otherwise, $\Delta I = \pm 1$ or 0, (but no $0 \rightarrow 0$), which are the *Gamow-Teller selection rules*. Free neutron decay involves a statistical combination of both. The fastest transitions (such as free neutron decay) are called *super-allowed*, with typical values of $\log_{10}(ft_{\frac{1}{2}}) \approx 3.0$.

Suppose the nucleons are not in free space, but surrounded by their peers in a nucleus:

$$(Z, A) \rightarrow (Z + 1, A) + e^- + \bar{\nu}. \tag{3.171}$$

The new restrictions upon the states available modify the process. First, the energetics are changed, so $M(Z, A) = M(Z + 1, A) + E_e + E_\nu$. For nuclei on the proton-rich side of the valley of beta stability, the processes of *electron capture*,

$$(Z + 1, A) + e^- \rightarrow (Z, A) + \nu, \tag{3.172}$$

and *positron emission*,

$$(Z + 1, A) \rightarrow (Z, A) + e^+ + \nu, \tag{3.173}$$

become possible. Further, because constraints upon angular momentum change between initial and final states, the matrix elements may be different. It may be that the transition must occur as a *forbidden* decay, and therefore be slower. This is illustrated in figure 3.6, for the astronomically interesting case of ^{26}Al decay.

The ground state of ^{26}Al has 5+ spin and parity, while the daughter nucleus ^{26}Mg has ground state 0+ spin and parity. The direct transition is "highly forbidden," and occurs at an inconsequential rate. It is overwhelmed by β-decay to the excited state at 2.9384 MeV (2.7% of the time), and by both β-decay and electron capture (EC) to the excited

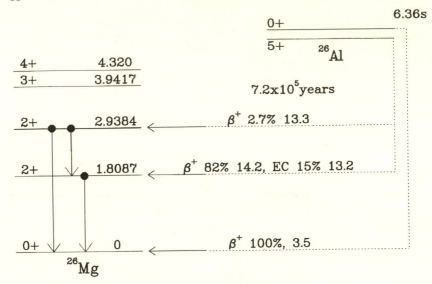

Fig. 3.6. ^{26}Al Decay

state at 1.8087 MeV (82% and 15% of the time, respectively). The transition rates are still slow (the half-life is 7.2×10^5 years), with values of $\log_{10} ft_{\frac{1}{2}} \approx 13$. The first excited state in ^{26}Al has 0+ spin and parity, so that it can decay to the ground state of ^{26}Mg by an allowed transition. The half-life is only 6.36 *seconds*, with $\log_{10} ft_{\frac{1}{2}} \approx 3.5$. If this state can be thermally excited, the decay rate changes by a factor of roughly 10^{12}.

By manipulating the reaction equation, other processes can be constructed. For example, running electron capture backwards gives *neutrino absorption*,

$$(Z, A) + \nu \rightarrow (Z + 1, A) + e^-. \tag{3.174}$$

Switching leptons across the arrow sign, and changing them to their antiparticles to maintain conservation of lepton number, gives other possibilities (we could also change baryons to antibaryons, but that is not known to be astronomically interesting). Antineutrino capture is

$$(Z + 1, A) + \bar{\nu} \rightarrow (Z, A) + e^+, \tag{3.175}$$

and positron capture is

$$(Z, A) + e^+ \rightarrow (Z + 1, A) + \bar{\nu}, \tag{3.176}$$

and so on.

A further way to change the environment for β-decay is to surround the nucleus in a sea of degenerate particles. Even in the laboratory, electron capture can occur from the K and L shells of the electron cloud surrounding the nucleus. In stars, these electrons are ionized at higher temperatures, but cooling processes during the evolution (especially emission of neutrino-antineutrino pairs) gives sufficiently high densities to make the electrons degenerate. During the collapse of a stellar core, the densities rise even higher, so that nucleons become degenerate too, and for a few seconds the neutrinos as well. Quantum mechanical degeneracy is a consequence of the Pauli exclusion principle for fermions. It may modify the weak interaction in both the entrance and the exit channels.

In a degenerate electron gas, with fermi energy ϵ_f which exceeds the energy of β-decay, the decay cannot occur because there are no electron states for the newly formed electron to occupy. In fact, it is energetically favorable to have electron capture, changing the nucleus which is stable in the laboratory into its parent nucleus which is unstable in the lab. Therefore the stability question is reversed! When the electron fermi energy is included, the lower energy state becomes the higher.

Consider electron capture:

$$(Z+1, A) + e^- \rightarrow (Z, A) + \nu. \tag{3.177}$$

Ignoring the small recoil of the nucleus, energy conservation now gives $E_e = E_\nu + W$, where W is the total energy from the beta-decay of nucleus (Z, A) to $(Z+1, A)$. The *end-point energy*, which is the maximum kinetic energy of the emitted electron, is $E_0 = W - m_e c^2$. The procedure for obtaining the reaction rate is similar to that used above. Because there is only one light particle in the exit channel (the neutrino), the density of states is

$$\rho(E_n) = dN_\nu/dE_\nu = 4\pi\Omega E_\nu^2/h^3. \tag{3.178}$$

The transition rate per electron is

$$\sigma v = \frac{2\pi}{\hbar} g^2 |M_{if}|^2 \frac{4\pi}{(hc)^3} E_\nu p_\nu c. \tag{3.179}$$

For the traditional case of electron capture of a K-shell electron, the electron wavefunction may be represented in the hydrogenic approximation; see [104]. This corresponds to $\psi = e^{-Zr/a_0}$, where $a_0 = \hbar^2/m_e e^2 = 0.529 \times 10^{-8}$ cm is the Bohr radius, and to an equivalent electron density at the nucleus of $Y_e \rho \approx Z^3 \times 11.2$ g cm^{-3}. If the electron is free, the transition probability per electron is [171]

$$\langle \sigma v \rangle = \sigma_0 c (E_{nu}/m_e c^2)^2, \tag{3.180}$$

where

$$\sigma_0 = \frac{2\pi^2 \ln 2}{(ft)_{\frac{1}{2}}} \left(\frac{\hbar}{m_e c}\right)^2 \left(\frac{\hbar}{m_e c^2}\right), \tag{3.181}$$

$$= 2.63 \times 10^{-41} \, \text{cm}^2 \, \text{s}/(ft_{\frac{1}{2}}). \tag{3.182}$$

This is appropriate for high-energy electrons. For lower energies, the electron wavefunction is not well approximated by a plane wave, and coulomb corrections must be made.

The absorption cross section for neutrinos (antineutrinos) on neutrons (protons) is formally similar, with differences in the values of the energetics and matrix elements:

$$\sigma c = \frac{2\pi}{\hbar} g^2 |M_{if}|^2 \frac{4\pi}{(hc)^3} E_e p_e c. \tag{3.183}$$

The neutrino absorption rates are reduced by a factor of 1/2 because the neutrino has only one helicity state. See also Bahcall [65], especially §8.1. Bahcall defines a different but related quantity as σ_0; he explicitly includes an atomic physics correction for the electron wavefunction (his $G(Z, w_e)$, see p. 198 in [65]), so σ_0(Bahcall) $\approx 2\pi\alpha\sigma_0$.

3.8 SOURCES OF RATES

The quantitative content of nucleosynthesis theory is built on a foundation of nuclear experiment and nuclear theory. Key developments have followed the publication of an improved set of reaction rates. There have been two approaches: (1) experimental efforts to directly measure the cross sections in the relevant range of energy (or determine the spectroscopic parameters), and (2) theoretical synthesis of experimental data to estimate the rates in regimes not directly accessible to experiment.

One of the most important resources has been the series of compilations by Fowler, Caughlan, and collaborators [236, 237, 282, 155, 156], which have emphasized the empirical basis of reaction rates, allowing only a little theory to slip in when all else fails.

The first extensive theoretical synthesis to generate large numbers of rates was conceived by A. G. W. Cameron and implemented in collaboration with Truran, Hansen, Gilbert, and Elkin [632, 631, 264, 151, 629].

This effort involved synthesizing an enormous amount of experimental data into a new mass formula (with shell and pairing corrections), a new level density formula (consistent with those corrections), thermonuclear rates for medium and heavy nuclei, and nuclear beta-decay and electron capture rates.

Wagoner, Fowler, and Hoyle [659] provided an extensive table of reaction rates useful for cosmological nucleosynthesis; see also [657, 553].

Michaud and Fowler [435] used the optical model to generate rates for silicon burning. Holmes, Woosley, Fowler, and Zimmerman used the Hauser-Feshbach statistical model to calculate neutron-induced reactions of heavy nuclei [300], and thermonuclear reaction rates for intermediate mass nuclei [696].

A discussion of reaction rates for processes occurring in explosive hydrogen burning (a hot, proton-rich environment) may be found in Wallace and Woosley [661].

The most extensive investigation of weak nuclear reaction rates is by Fuller, Fowler, and Newman [251, 253, 252, 254].

At present the most accurate theoretical estimates seem to be those of Thielemann, Arnould, and Truran [615], which are used in subsequent chapters on stellar nucleosynthesis.

4

Nuclear Reaction Networks

"I think I should understand that better," Alice
said very politely, "if I had it written down;
but I can't quite follow it as you say it."
"That's nothing to what I could say if I
chose," the Duchess replied, in a pleased tone.
—Lewis Carroll (1832–1898),
Alice's Adventures in Wonderland

To be scientific we must be quantitative; to be quantitative we must be mathematical. Nature seems to love mathematical behavior, and we would be wise not to ignore this. The set of differential equations which ties the physics of individual nuclei to an abundance pattern is a *reaction network*. The mathematical nature of these equations is challenging, so that routine solution of them has become possible only with the impressive development of computer technology. The rate at which reactions occur in a gas is proportional to products of the densities of reactants, so that the equations are *nonlinear*. Because of coulomb forces and nuclear resonances, cross sections may vary enormously from reaction to reaction. This results in a wide range of variation in the size of coefficients of different terms in the network equations; such equations are called *stiff*, and exhibit the mathematical problems common to such systems.

Despite these difficulties, there are some types of systematic behavior exhibited by solutions of the network equations. Two of the most important classes of solution are *steady state* and *equilibrium*. The steady-state solutions are characterized by an overall balance between creation and destruction of the nuclear species in question, so there is no net gain or loss. The equilibrium solutions are more fundamental and restrictive, and involve a "local" balance between forward and reverse rates for an important reaction. Such solutions are intimately connected with the concept of detailed balance in statistical physics and of invariance under time reversal for dynamical systems; they are the nuclear equivalent of ionization equilibrium of atoms. These two types of solutions provide powerful tools for understanding the often complex behavior of nuclear reaction networks.

The evolution of stars is intimately connected with nuclear reaction networks. It is these networks that determine the rate at which nuclear binding energy is made available to power the star. Once the nuclear reactions have destroyed the homogeneity of the stellar composition, new effects may occur as abundances are modified by mixing processes. As yet such phenomena are not really well understood. First, the flow is a hydrodynamic process and may be turbulent (chaotic), with all the difficulties which that implies. Second, inclusion of mixing in the network equations changes them from ordinary to partial differential equations, adding a new aspect to an already complex set of solutions. This mathematical difficulty reflects the more complex nature of the underlying physical processes.

None of the abundance patterns which we observe are themselves still in a region of active nuclear reactions. Thus all such patterns have gone through a stage in which the burning was terminated. Such termination can occur because the fuel was exhausted, or because of *freeze-out*. Freezeout occurs when the conditions are sufficiently changed for the reactions to be stopped. This can happen, for thermonuclear reactions, with only a slight lowering of the temperature. Similarly, density-dependent reactions may be slowed by changes in the density. In stars, the freezeout might be expected to occur because of the expansion following a stellar explosion, or a buoyant motion of a convective cell.

4.1   NETWORK EQUATIONS

Consider an element of matter at sufficiently high temperature and density so that thermonuclear reactions become important. In astrophysical situations a number of different types of reactions are sometimes important. For example consider reactions of the sort $X(p, \gamma)Y$, or equivalently $X + p \rightarrow Y + \gamma$, where X is the target nucleus, Y the product nucleus, p the bombarding particle (a proton) and γ an exit channel particle (a gamma ray). The reactions

$$(p, \gamma) \quad (p, n)$$
$$(p, \alpha) \quad (n, \gamma)$$
$$(n, \alpha) \quad (\alpha, \gamma)$$

and their reverse reactions

$$(\gamma, p) \quad (n, p)$$
$$(\alpha, p) \quad (\gamma, n)$$
$$(\alpha, n) \quad (\gamma, \alpha)$$

are often important. Several other types of reactions are important as well. During hydrogen burning, reactions involving D and
^{3}He must be considered (usually these abundances are so low that
they need not be followed explicitly). The "heavy-ion" reactions
^{12}C + ^{12}C, ^{12}C + ^{16}O, and ^{16}O + ^{16}O dominate carbon and oxygen
burning stages; the more important exit channels involve emission of
(1) a proton, (2) an alpha, (3) a neutron, and (4) a deuteron (for
oxygen burning). The most important three-body reaction is probably
$3(^4\text{He}) \rightarrow {}^{12}\text{C} + \gamma$; a number of more exotic three-body reactions have
to be considered in cosmological nucleosynthesis. The slower weak
interactions are of fundamental importance because they alone change
the ratio of neutrons to protons in the system.

Consider the rate of change in the number density N_j of species j. The
differential equation that describes the behavior of N_j may be written as

$$\frac{dN_j}{dt} = N_k N_\ell \langle \sigma v \rangle_{k\ell,j}$$
$$-N_j N_\ell \langle \sigma v \rangle_{j\ell,n}$$
$$+N_i \lambda_{i,j} - N_j \lambda_{j,m}$$
$$+ \dots \tag{4.1}$$

The expression $\langle \sigma v \rangle$ denotes the product of cross section and relative
velocity in the center-of-mass system, averaged over the appropriate distribution functions (see chapter 3). Two-body interactions and spontaneous decay are illustrated explicitly here. Note the convention that for
a reaction $k + \ell \rightarrow j + i$, we write $\langle \sigma v \rangle_{k\ell,j}$ in cases where the specification of i is redundant. Thus for ^{12}C(α, γ) the relevant result must be
^{16}O. For identical particles (e.g., ^{12}C + ^{12}C) a term of the form,

$$\frac{N_i^2}{2!} \langle \sigma v \rangle_{ii,j}$$

or for a three-body interaction, a term of the form

$$\frac{N_i^3}{3!} \langle \sigma v \rangle_{iii,j}$$

is needed. Using Eq. 2.9, we have

$$\frac{dY_j}{dt} = Y_k Y_\ell \rho N_A \langle \sigma v \rangle_{k\ell,j} - Y_j Y_\ell \rho N_A \langle \sigma v \rangle_{j\ell,n}$$
$$+Y_i \lambda_{i,j} - Y_j \lambda_{j,m}$$
$$+ \dots, \tag{4.2}$$

while the three-body term would become

$$\left(\frac{Y_i^3}{3!}\right)(\rho N_A)^2 \langle \sigma v \rangle_{iii,j}.$$

Unlike the number density N_j, the specific abundance (mole fraction) Y_j does not change when only an expansion or contraction occurs, but requires the action of nuclear processes or mixing. To shorten the notation it is useful to define the quantity

$$[k(\ell, j)] = \rho N_A \langle \sigma v \rangle_{k\ell, j}. \tag{4.3}$$

To make these ideas less abstract, consider a specific example: the abundance of ^{20}Ne during carbon burning. If Y_{20} represents the mole fraction of ^{20}Ne, then we have

$$
\begin{aligned}
\frac{dY_{20}}{dt} = \quad & Y_{16}\, Y_4[^{16}\mathrm{O}(\alpha, \gamma)] && (\text{for } {}^{16}\mathrm{O}(\alpha, \gamma)^{20}\mathrm{Ne}) \\
& -Y_{20}[^{20}\mathrm{Ne}(\gamma, \alpha)] && (\text{for } {}^{20}\mathrm{Ne}(\gamma, \alpha)^{16}\mathrm{O}) \\
& -Y_{20}\, Y_4[^{20}\mathrm{Ne}(\alpha, \gamma)] && (\text{for } {}^{20}\mathrm{Ne}(\alpha, \gamma)^{24}\mathrm{Mg}) \\
& +Y_{24}[^{24}\mathrm{Mg}(\gamma, \alpha)] && (\text{for } {}^{24}\mathrm{Mg}(\gamma, \alpha)^{20}\mathrm{Ne}) \\
& +(\text{similar terms such as } (\alpha, p),\ (p, n),\ (n, \gamma)) \\
& +Y_{12}\, Y_{12}[^{12}\mathrm{C}(^{12}\mathrm{C}, \alpha)]/2\ (\text{for } {}^{12}\mathrm{C}(^{12}\mathrm{C}, \alpha)^{20}\mathrm{Ne}) \\
& -Y_{20}\, Y_4[^{20}\mathrm{Ne}(\alpha, {}^{12}\mathrm{C})]\ (\text{for } {}^{20}\mathrm{Ne}(\alpha, {}^{12}\mathrm{C})^{12}\mathrm{C}) \\
& +\ldots
\end{aligned}
\tag{4.4}
$$

The list of terms could be extended, but many possible reactions may be ruled out because of their low rate under the conditions to be investigated. An equation like this must be written for all reacting species. Because of their many possible reactions, some species such as neutrons, alphas, and protons have an especially important role. As an example of the latter sort of equation, consider:

$$
\begin{aligned}
\frac{dY_p}{dt} = & -\sum_i Y_i\, Y_p[i(p, j)] + \sum_k Y_k[k(\gamma, p)] \\
& - (\text{other } p \text{ absorbing terms, such as } (p, \alpha),\ (p, n),\ \text{and so on}) \\
& + (\text{other } p \text{ releasing terms, such as } (\alpha, p),\ (n, p),\ \text{and so on}) \\
& + Y_{12}\, Y_{12}[^{12}\mathrm{C}(^{12}\mathrm{C}, p)]/2 \\
& - (\text{inverse of } {}^{12}\mathrm{C}(^{12}\mathrm{C}, p)^{23}\mathrm{Na}).
\end{aligned}
\tag{4.5}
$$

Thus we have a set of coupled, nonlinear ordinary differential equations. Such a system of equations is said to define a *nuclear reaction network*. In some cases it is desirable to add additional constraint equations, such as the first law of thermodynamics and an expansion law, as well as additional variables, such as temperature and density, to the system of coupled equations (see Müller [444]).

4.2 SOLUTIONS: STEADY STATE

Before proceeding to describe a general technique for solution of reaction network equations it is instructive to consider some solutions for special cases. In principle one could include all conceivable reactions and obtain the most general results for a given physical situation. The complexity of such a result would certainly inhibit easy understanding. As an aid to building such understanding, we consider some simple cases that exhibit behavior that is characteristic of more complex situations.

A steady-state solution exists when, for some section of a reaction network, all the time derivatives of abundances dY_j/dt are zero. In hydrogen burning by the CNO cycle such a condition is often referred to in the literature as a CNO *equilibrium* although *steady state* is a more accurate terminology; we will discuss various true equilibria below.

The network equations can be written in the form

$$\frac{dY_i}{dt} = -(\text{sum of destruction terms}) + (\text{sum of creation terms})$$

$$= -Y_i D_i + P_i \tag{4.6}$$

where

$$D_i = \sum_j Y_j[i(j, ?)] + \cdots \tag{4.7}$$

and

$$P_i = \sum_{k,\ell} Y_k Y_\ell[k(\ell, i)] + \cdots \tag{4.8}$$

where only the quadratic terms are explicitly shown, and the question marks refer to any allowed species. If the production and destruction terms are equal then $dY_i/dt = 0$. This is the "steady state" condition; it causes Eq. 4.6 to reduce to an algebraic equation. Even if dY_i/dt is not strictly zero but is much smaller in magnitude than the dominant production and destruction terms, the reduction of Eq. 4.6 to an algebraic equation is approximately valid. In this case the actual abundance

Y_i approaches the "steady-state abundance" defined by

$$Y_i^{ss} = \frac{P_i}{D_i}. \tag{4.9}$$

Because both P_i and D_i depend upon other abundances (Y_j, Y_k, Y_ℓ, etc.), we expect Y_i^{ss} to undergo secular variation as these change, even if the steady-state condition is well satisfied.

How fast will the steady-state abundance be approached? Using equations 4.6 and 4.9 we have

$$\frac{1}{Y_i^{ss}} \frac{dY_i}{dt} = D_i \left(1 - \frac{Y_i}{Y_i^{ss}} \right), \tag{4.10}$$

so that for $Y_i \ll Y_i^{ss}$ the steady-state value is approached in a time of the order of $1/D_i$, that is, the destruction time for Y_i. For $Y_i \gg Y_i^{ss}$, the steady-state value is approached in a time Y_i^{ss}/Y_i times shorter than the destruction time. Short destruction times favor the steady-state approximation. Because of Eq. 4.9 we expect low abundance to be a clue (but not a proof) as to the validity of the steady-state approximation.

For simplicity consider just the CN part of the CNO cycle; see Clayton [173] for a detailed description. The reaction sequence is

$$^{12}C(p, \gamma)^{13}N(\beta^+)^{13}C(p, \gamma)^{14}N(p, \gamma)^{15}O(\beta^+)^{15}N \tag{4.11}$$

at which point a branching occurs between $^{15}N(p, \alpha)^{12}C$ and $^{15}N(p, \gamma)^{16}O$. The (p, γ) branch occurs only about 0.4 percent of the time; thus this branch is a slow leak. The net effect of the sequence is

$$4p \rightarrow {}^4He + 2\nu + 2e^+. \tag{4.12}$$

At hydrogen-burning temperature, the coulomb barrier is too high for alpha reactions, and if no source of free neutrons is present then neutron capture may be neglected. If we denote $Y(^{12}C)$ by the symbol ^{12}C, then the network equations may be written:

$$\frac{d(^{12}C)}{dt} = -{}^{12}Cp[^{12}C(p, \gamma)] + {}^{15}Np[^{15}O(p, \alpha)] \tag{4.13}$$

$$\frac{d(^{13}N)}{dt} = -{}^{13}N[^{13}N(\beta^+)] + {}^{12}Cp[^{12}C(p, \gamma)] \tag{4.14}$$

$$\frac{d(^{13}C)}{dt} = -{}^{13}Cp[^{13}C(p, \gamma)] + {}^{13}N[^{13}N(\beta^+)] \tag{4.15}$$

$$\frac{d(^{14}\text{N})}{dt} = -^{14}\text{N}p[^{14}\text{N}(p, \gamma)] + ^{13}\text{C}p[^{13}\text{C}(p, \gamma)] \qquad (4.16)$$

$$\frac{d(^{15}\text{O})}{dt} = -^{15}\text{O}[^{15}\text{O}(\beta^+)] + ^{14}\text{N}p[^{14}\text{N}(p, \gamma)] \qquad (4.17)$$

$$\frac{d(^{15}\text{N})}{dt} = -^{15}\text{N}p\left([^{15}\text{N}(p, \alpha)] + [^{15}\text{N}(p, \gamma)]\right) + ^{15}\text{O}[^{15}\text{O}(\beta^+)](4.18)$$

$$\frac{d(^{16}\text{O})}{dt} = +^{15}\text{N}p[^{15}\text{N}(p, \gamma)] \qquad (4.19)$$

$$\frac{d(^4\text{He})}{dt} = +^{15}\text{N}p[^{15}\text{N}(p, \alpha)] \qquad (4.20)$$

$$\frac{d(^1\text{H})}{dt} = -^{12}\text{C}p[^{12}\text{C}(p, \gamma)] - ^{13}\text{C}p[^{13}\text{C}(p, \gamma)]$$
$$-^{14}\text{N}p[^{14}\text{N}(p, \gamma)] - ^{15}\text{N}p\left([^{15}\text{N}(p, \alpha)] + [^{15}\text{N}(p, \alpha)]\right). \quad (4.21)$$

Let us simplify this mess. Hydrogen burning occurs on time-scales $t > 10^6$ years even in massive, fast-evolving stars. Therefore the positron-decay reactions

$$^{13}\text{N}(\beta^+)^{13}\text{C} \qquad \tau = 9.96\,\text{minutes}$$
$$^{15}\text{O}(\beta^+)^{15}\text{N} \qquad \tau = 124\,\text{seconds}$$

will be regarded as "instantaneous." Then the "flow into" ^{13}N and ^{15}O should be followed immediately by an equal "flow out," that is, $d(^{13}\text{N})/dt$ and $d(^{15}\text{O})/dt$ are small so that the steady-state approximation is appropriate. Now equation 4.14 implies

$$^{12}\text{C}p[^{12}\text{C}(p, \gamma)] = ^{13}\text{N}[^{13}\text{N}(\beta^+)], \qquad (4.22)$$

so equation 4.15 becomes

$$\frac{d(^{12}\text{C})}{dt} = ^{12}\text{C}p[^{12}\text{C}(p, \gamma)] - ^{13}\text{C}p[^{13}\text{C}(p, \gamma)]. \qquad (4.23)$$

Note that the steady-state abundance of ^{13}N can be calculated from the algebraic relation. Similarly, $d(^{15}\text{O})/dt = 0$ may be used in Eq. 4.17. Then equation 4.18 becomes

$$\frac{d(^{15}\text{N})}{dt} = ^{14}\text{N}p[^{14}\text{N}(P, \gamma)] - ^{15}\text{N}p\left([^{15}\text{N}(p, \alpha)] + [^{15}\text{N}(p, \gamma)]\right). \quad (4.24)$$

Clayton [173] gives a detailed discussion; be warned that his symbols denote the number density N_j rather than Y_j. By considering a specific stellar environment the (p, γ) rates are specified also. Given sufficiently long time intervals, a steady state is approached in which the rate of any reaction equals the rate of all the other reactions in the CN cycle:

$$^{12}C(p, \gamma)^{13}C = {}^{13}N(\beta^+)^{13}C =$$

$$^{13}C(p, \gamma)^{14}N = {}^{14}N(p, \gamma)^{15}O =$$

$$^{15}O(\beta^+)^{15}N = {}^{15}N(p, \alpha)^{12}C.$$

This means that the time derivatives for ^{12}C, ^{13}N, ^{13}C, ^{14}N, ^{15}O, and ^{15}N are small, at least to the extent that $^{15}N(p, \gamma)^{16}O$ can be neglected. From equation 4.23,

$$\frac{^{12}C}{^{13}C} = \frac{[^{13}C(p, \gamma)]}{[^{12}C(p, \gamma)]} \tag{4.25}$$

and so on; the abundance ratios of these nuclei depend only on the ratio of reaction rates (i.e., on nuclear cross sections, density, and temperature). All abundances in the network are not constant in time however. Besides the slow leak of CN nuclei through $^{15}N(p, \gamma)^{16}O$, we also have $d(^1H)/dt < 0$ and $d(^4He)/dt > 0$. Every time around the cycle, $4H \to {}^4He$.

4.3 SOLUTIONS: EQUILIBRIA

The steady-state condition discussed above is the result of a global balance between all production and destruction processes for a given species. Consider a reaction that is fast (it has mean reaction times that are short compared to the interval of interest), and is just balanced by the reverse or inverse reaction. For example, $^{12}C(p, \gamma)^{13}N$ is balanced by $^{13}N(\gamma, p)^{12}C$ for stellar temperatures $T > 0.6 \times 10^9$ K at the onset of carbon burning. Because the reactions are fast, the distribution of abundances in the entrance and exit channels simply reflects the phase space available. Thus the abundance ratios may be calculated using chemical potentials in the usual way (see Landau and Lifshitz [381]; Appendix B). This reaction link is in equilibrium in the thermodynamic sense. The abundances so joined may be said to be in a steady state, but in this case the balance is dominated by individual forward-reverse reactions, link by link.

What is the distinction between the steady-state approximation and the various equilibrium approximations used in discussions of nucleosynthesis? This may be shown most clearly and succinctly in terms of chemical potentials and the law of mass action. Consider a bit of stellar matter so hot and dense that reactions involving strong, weak, and electromagnetic interactions beyond electrons, neutrons, protons, and all of their antiparticles must be considered (see Tsuruta and Cameron [638] for a discussion of equilibrium of "hyperonic mixtures"). For a given temperature and volume this gives us eight unknowns to seek: the abundances of eight species e^-, e^+, ν, $\bar{\nu}$, n, $\bar{n}$, p, $\bar{p}$). We minimize the free energy for ensembles containing all eight species, subject to constraints implied by conservation laws; this procedure generates the derivatives of the free energies of each species, which are the chemical potentials. Because particles and antiparticles annihilate, their relativistic chemical potentials are opposite in sign (Landau and Lifshitz [381]); this relationship may be used to halve the number of unknowns to four. Three constraints we require are conservation of nucleon number, lepton number, and charge; each implies a constant that must be specified. The strength and range of the electromagnetic force causes macroscopic systems to tend toward charge neutrality, so we consider a net charge of zero. Without loss of generality we may go from a fixed coordinate system to one comoving with the nucleons, so that our specific volume is $V = 1/\rho$; here the number of nucleons in a unit volume is $n = \rho N_A$ where N_A is Avogadro's number. The remaining constant may be expressed as Y_ℓ, the ratio of leptons to nucleons. Thus the triad (T, ρ, Y_ℓ) is sufficient to determine the equilibrium abundances of the eight species. We will call this "thermal equilibrium" (TE); it may be generalized to include more particles. If the new particles require new conservation laws, then for each new conservation law a value of the corresponding conserved quantity must be specified. If no new laws are required, then there is an equation for the chemical potential of the new particle in terms of the chemical potentials of the others.

Let us now examine thermodynamic equilibria which are successively further away from TE. Suppose that neutrinos can escape freely from the comoving volume but are not replaced; the net negative electric charge due to leptons now equals the net lepton number, $Y_\ell = Y_e = Y(e^-) - Y(e^+)$. In this case we suppose the weak interactions to be too slow to establish the thermal equilibrium (TE), and consider *nuclear statistical equilibrium* (NSE), which is determined by the triad (T, ρ, Y_e). Note that some authors find it convenient to use other quantities which are equivalent to Y_e. The nucleus (Z, A), having Z protons and $A - Z$ neutrons, is linked to the nucleus $(Z - 1, A - 1)$ by (γ, p) and (p, γ)

reactions, and their chemical potentials are related by

$$\mu(Z, A) = \mu(Z - 1, A - 1) + \mu_p, \tag{4.26}$$

where μ_p is the chemical potential of free protons. For brevity in notation, these chemical potentials are intended to *include* rest mass. Similarly, for (γ, n) and (n, γ) reactions we have

$$\mu(Z, A) = \mu(Z, A - 1) + \mu_n, \tag{4.27}$$

where μ_n is the chemical potential of free neutrons. Because all (p, γ)
and (n, γ) links are assumed to be in equilibrium, these expressions may
be used as recursion relations to derive

$$\mu(Z, A) = Z\mu_p + (A - Z)\mu_n, \tag{4.28}$$

and

$$\mu_\alpha = 2(\mu_p + \mu_n) \tag{4.29}$$

for alpha particles (see "B²FH" [135]).

There are still other sorts of equilibria of importance in nucleosynthesis. Suppose there to be no equilibrium link between nucleus (Z_1, A_1)
and nucleus (Z_2, A_2). In particular we can take (Z_2, A_2) to represent
free neutrons and protons. Then equations 4.26, 4.27, and 4.29 may still
be valid but not 4.28. If we introduce (Z_1, A_1) as a base nucleus for
reference purposes, then

$$\mu(Z, A) = \mu(Z_1, A_1) + \mu_p(Z - Z_1) + \mu_n(A - A_1 - Z + Z_1) \quad (4.30)$$

so that we may solve for the abundances if we specify four quantities
(T, ρ, Y_e, Y_1), or equivalently (T, ρ, Y_e, μ_1). Note that the choice of a
base nucleus is arbitrary to the extent that we could use any nucleus in
the set which is coupled by equilibrium links. Such a *quasiequilibrium*
(QE) occurs in silicon burning (Bodansky, Clayton, and Fowler [109]),
oxygen burning (Woosley, Arnett, and Clayton [692]), and alpha-rich
freezeout (Arnett, Truran, and Woosley [49]).

We can go still further. Suppose there to be another set of nuclei, represented by another base nucleus (Z_3, A_3), which are connected among
themselves by equilibrium links, so that for (z, a) in this set,

$$\mu(z, a) = \mu(Z_3, A_3) + \mu_p(z - Z_3) + \mu_n(a - A_3 - z + Z_3). \tag{4.31}$$

Then we need five quantities (T, ρ, Y_e, Y_1, Y_3) to solve for all the abundances. Here we have two quasiequilibrium clusters (see Woosley, Arnett, and Clayton [692]). As the number of such clusters increases, complexity reduces the usefulness of the quasiequilibrium approach. *Reaction-link* equilibrium (RLE) may still be useful for pairs of nuclei, giving equations such as 4.26, 4.27 or

$$\mu(Z, A) = \mu(Z - 2, A - 4) + \mu_\alpha, \tag{4.32}$$

in special cases, even though quasiequilibrium has broken down in general.

Fortunately the steady-state approximation becomes valid at about this point. At lower temperatures the inverse rate no longer balances the forward rate of a given reaction link. The most important inverse rates are the photodissociation rates (γ, p), (γ, n), and (γ, α) which are especially sensitive to temperature. Thus there exists a regime of temperature in which many forward rates are active but the corresponding inverse rates are not. Because of the enormous variation in cross section these forward rates are mostly either much faster or much slower than that rate which controls the destruction of fuel. Such a situation is ideal for systematic use of the steady-state approximation previously discussed.

4.4 SOLUTIONS: GENERAL METHOD

The reaction network equations often behave as *stiff* differential equations. In general they have rate terms which contain second and higher powers of the abundances being sought. Thus they are nonlinear. Only numerical techniques can provide solutions to a general set of reaction network equations. Consider the equation

$$\frac{dY_j}{dt} = f(\rho, T, Y_j, Y_k, Y_\ell, \ldots) \tag{4.33}$$

where

$$f(\rho, T, Y_j, \ldots) = -Y_j Y_k [j(k, \ell)] + Y_\ell [k(\gamma, j)] + \ldots \tag{4.34}$$

This is one of the equations for the general network discussed in §4.1. For simplicity in notation we will suppress subscripts until they are needed again. Suppose we approximate the time derivative by a finite difference

$$\frac{dY}{dt} = \frac{Y(t + \delta t) - Y(t)}{\delta t}. \tag{4.35}$$

This implicitly assumes the change in variables to be small during the interval δt. Two simple choices for evaluating f are apparent:

$$\frac{Y(t + \delta t) - Y(t)}{\delta t} = f(t) \tag{4.36}$$

$$= f(t + \delta t), \tag{4.37}$$

that is, we can evaluate the reaction rate terms at time t, or at time $t + \delta t$. Both are said to be accurate to first order in δt; 4.36 results from a Taylor expansion from time t, while 4.37 results from a similar expansion about time $t + \delta t$. We might also use some average of the two, or estimate $f(t + \delta t/2)$, and arrive at an expression accurate to second order in δt. We must distinguish between *accuracy* and *stability*; both are needed for a correct solution. Because we approximate a derivative using a truncated Taylor expansion, the resulting difference equation may have extra solutions in addition to those of the differential equation (Potter [502]). One could say that the resulting numerical garbage is due to *numerical instability* and consider it to be a mathematical problem, or regard the difference equations to be incorrectly formulated to represent the physics involved. The latter point of view is rarer, but often of more practical value.

With the choice denoted by 4.36, called *forward* or *explicit* differencing, we have n uncoupled equations for n unknowns. All the Y's and density and temperature at $t + \delta t$ may be found individually in terms of the Y's and density and temperature at time t. The latter quantities are just those we start with. Although the solution of these equations is truly elementary, difficulties are encountered near steady state and equilibrium. We write

$$f = -D + P, \tag{4.38}$$

where D denotes here the sum of *all* terms representing destruction (of species j) and P is a similar sum for all processes producing that species. At a steady state, $D = P$, so that near such a state

$$D \gg |f|; \tag{4.39}$$

$$P \gg |f|. \tag{4.40}$$

In such a situation small errors in the values of the Y's, or density and temperature, cause large errors in f. To avoid this having a large effect, time steps must be small:

$$\delta t \ll \frac{Y}{D} \quad \text{and} \quad \delta t \ll \frac{Y}{P}. \tag{4.41}$$

For calculations approaching any of the special solutions discussed in the previous sections, such constraints on the time step can be a disaster for numerical work.

To illustrate this point, consider the simpler differential equation,

$$\frac{dx}{dt} = -xa + b, \tag{4.42}$$

which has the solution[1]

$$x(t) = \frac{b}{a} + \left(x(0) - \frac{b}{a}\right) e^{-at}. \tag{4.43}$$

Using forward differencing, 4.42 becomes

$$\frac{x(\delta t) - x(0)}{\delta t} \approx -x(0)a + b. \tag{4.44}$$

Let us compare the solutions for a specific case. Take $x(0) = 2\left(\frac{b}{a}\right)$, and assume that $\delta t \gg \frac{1}{a}$ and $\frac{x}{b}$ (contrary to 4.41). Then, from 4.44,

$$x(\delta t) \approx 2b/a - b\delta t, \tag{4.45}$$

while from 4.43,

$$x(\delta t) \approx b/a. \tag{4.46}$$

Clearly these are inconsistent in general, and as the latter is correct, the procedure leading to the former must be wrong. However, if $\delta t \ll 1/a$ and x/b, then 4.44 gives

$$x(\delta t) \approx (2 - a\delta t)b/a, \tag{4.47}$$

while 4.43 gives

$$x(\delta t) = (1 + \exp(-a\delta t))b/a$$
$$\approx (2 - a\delta t)b/a \tag{4.48}$$

in the limit. Here the inconsistency disappears if the time step is small enough.

---

[1]To see this, substitute $y = x - b/a$.

Consider the approximation implied by 4.37. Then 4.42 becomes

$$\frac{x(\delta t) - x(0)}{\delta t} \approx -x(\delta t)a + b. \tag{4.49}$$

This is called the *backwards* or *implicit* difference form. Solving for $x(\delta t)$ gives

$$x(\delta t) \approx \frac{x(0) + b\delta t}{1 + a\delta t}. \tag{4.50}$$

For the previous example,

$$x(\delta t) \approx \frac{b}{a} \tag{4.51}$$

if $b + \delta t$ and $a + \delta t \ll 1$. This form of differencing works in both limits (although it is still accurate only to first order).

The particular algebraic manipulation needed to derive 4.50 is rather specialized. Solution of nonlinear equations and of coupled equations is more complicated. Consider a simple generalization of 4.42 which illustrates the effects of coupling such equations.

$$\frac{dx}{dt} = -xa + yb + c, \tag{4.52}$$

$$\frac{dy}{dt} = -yd + xe + f. \tag{4.53}$$

Using an *implicit* differencing scheme, we have

$$\begin{aligned} x(t + \delta t)(1 + a\delta t) + y(t + \delta t)(-b\delta t) &= x(t) + c\delta t, \\ x(t + \delta t)(-e\delta t) + y(t + \delta t)(1 + d\delta t) &= y(t) + f\delta t. \end{aligned} \tag{4.54}$$

This may be written in matrix form as

$$\mathbf{YA} = \mathbf{B} \tag{4.55}$$

where

$$\mathbf{Y} = \begin{bmatrix} x(t + \delta t) \\ y(t + \delta t) \end{bmatrix}, \tag{4.56}$$

$$\mathbf{A} = \begin{bmatrix} 1 + a\delta t & -b\delta t \\ -e\delta t & 1 + d\delta t \end{bmatrix}, \tag{4.57}$$

and

$$\mathbf{B} = \begin{bmatrix} x(t) + c\delta t \\ y(t) + f\delta t \end{bmatrix}. \tag{4.58}$$

Formally, the solution is

$$\mathbf{Y} = \mathbf{B}\mathbf{A}^{-1}. \tag{4.59}$$

The necessity of matrix manipulation is a result of using an implicit difference scheme on the system of coupled equations.

Let us return to our more complex set of coupled equations that make up the reaction network. If we use implicit differencing, then in $f(t+\delta t)$ there are terms that are at least quadratic in the unknowns. Products of the sort $Y_j(t + \delta t) Y_k(t + \delta t)$ give a nonlinearity which means that no simple general solution can be obtained. However in introducing finite differences to replace time derivatives, we had to assume that

$$\left| \frac{dY_j}{dt} \delta t \right| \ll Y_j(t) \quad \text{or} \quad Y_j(t + \delta t). \tag{4.60}$$

Denote

$$\Delta_j = Y_j(t + \delta t) - Y_j(t). \tag{4.61}$$

Then,

$$Y_j(t + \delta t) Y_k(t + \delta t) = \left(Y_j(t) + \Delta_j \right) \left(Y_k(t) + \Delta k \right)$$
$$\approx Y_j(t)\Delta_k + Y_k(t)\Delta_j + Y_j(t)Y_k(t) \tag{4.62}$$
$$\approx Y_j(t + \delta t) Y_k(t) + Y_j(t) Y_k(t + \delta t)$$
$$- Y_j(t) Y_k(t) \tag{4.63}$$

if we neglect terms above first order in Δ_j and Δ_k. Because this approximation is only that which would be required for reasonable accuracy of numerical integration, the process of linearizing the reaction network equations does not really introduce any significant new restrictions. Essentially we move forward in time by analytic continuation; at each new time interval we update the nonlinear coefficients of the A and B matrices, so that we obtain the nonlinear solutions. For three-body reactions this linearization can be more restrictive, but predictor-corrector schemes or multistep iterations can solve these problems [516].

Now upon linearization, we obtain

$$\Delta_j(1 + Y_k[jk, \ell m]\delta t + \ldots)$$
$$+ \Delta_k(Y_j[jk, \ell m]\delta t)$$
$$+ \Delta_\ell(-Y_m[\ell m, jk]\delta t)$$
$$+ \Delta_m(-Y_\ell[\ell m, jk]\delta t) + \ldots$$
$$= -Y_j Y_k[jk, \ell m] + Y_\ell Y_m[\ell m, jk] + \ldots \qquad (4.64)$$

Such an equation, linear in the unknowns (here the Δ's), may be written for all species. The equation for a species j contains a term in Δ_k for each species k that is connected to j by a reaction. Clearly the equations for *mediating* particles such as p, n, and alphas have many terms. We may write all these equations succinctly in matrix form:

$$\Delta \mathbf{A} = \mathbf{B}, \qquad (4.65)$$

where Δ is a column matrix in the Δ's, $\mathbf{B}$ a column matrix in constants such as the right-hand side of 4.64, and $\mathbf{A}$ a square matrix. See equations 4.53 through 4.59 for comparison.

To better understand the structure of $\mathbf{A}$, consider a system in which there is only an *alpha particle chain*; that is

$$C(\alpha, \gamma)\, D(\alpha, \gamma)\, E(\alpha, \gamma)\, F(\alpha, \gamma)\, G(\alpha, \gamma)H. \qquad (4.66)$$

If the order is alphas first, followed by six nuclei, the structure is illustrated visually as follows.

$$\mathbf{A} = \begin{pmatrix} 1 & 1 & 1 & 1 & 1 & 1 & 1 \\ 1 & 1 & 1 & 0 & 0 & 0 & 0 \\ 1 & 1 & 1 & 1 & 0 & 0 & 0 \\ 1 & 0 & 1 & 1 & 1 & 0 & 0 \\ 1 & 0 & 0 & 1 & 1 & 1 & 0 \\ 1 & 0 & 0 & 0 & 1 & 1 & 1 \\ 1 & 0 & 0 & 0 & 0 & 1 & 1 \end{pmatrix} \qquad (4.67)$$

Here the 1's denote nonzero elements, not quantitative values; notice the structure of the zero elements. Because of the coupling of the alphas to all the nuclei, the matrix is not *tridiagonal* (i.e., all elements zero except those on the diagonal and their nearest neighbors). The fast solution schemes for tridiagonal matrices are not directly applicable (although the structure of the matrix can be used to assist solution). Note that as the dimensions increase for a matrix with this or many similar structures, the fraction of zero elements increases also, resulting in a sparse matrix. Because the time needed to solve a general matrix equation such as 4.65 increases rapidly with the dimension, it is best to use as few reactions in a network as physically plausible. This method was first used in astrophysics for silicon burning (Truran, Arnett, and Cameron [630]); and presented in detail by Arnett and Truran [54]. Wagoner [657] used a second order version for Big Bang nucleosynthesis. Using standard methods [516] for solving linear equations, we calculate

$$\Delta = \mathbf{B}\mathbf{A}^{-1}. \tag{4.68}$$

As the new abundances are $Y_j(t+\delta t) = Y_j(t)+\Delta_j$ for each species j, we have advanced one time step δt. Examining the fractional change Δ_j/Y_j and other criteria such as change in density, temperature, or size of last time step, a new time step is chosen. New temperatures and densities are calculated (if they were not included in the matrix solution), and new reaction rates. The new matrices $\mathbf{A}$ and $\mathbf{B}$ are constructed, and solved, and so on around the cycle until the evolution of the abundances has gone on as far as desired.

4.5 ENERGY GENERATION

We have considered one effect of nuclear reactions: the changing of abundances. A second effect of considerable importance is the generation of energy. Consider the first law of thermodynamics,

$$dQ = dE + PdV, \tag{4.69}$$

where dQ is the gain of energy due to flow of energy in and out of the system, dE is the change in internal energy, and PdV is the work done by the system by its pressure on its surroundings. This form is valid with or without reactions which change abundances.

In stars, the average interaction energies of particles[2] are usually small compared to their kinetic energies, so the total energy is linear in the energies of the components:

$$E = E(\text{ion}) + E(\text{electron}) + E(\text{photon}) + \ldots \tag{4.70}$$

---

[2]In this sense, "particle" might refer to a nucleus, molecule, or atom, depending upon the environment, as well as to an electron or a nucleon.

Now, the relativistic energy of a particle is

$$w^2 = (pc)^2 + (mc^2)^2. \tag{4.71}$$

In the limit, $p << mc$, this becomes

$$w = \frac{p^2}{2m} + mc^2. \tag{4.72}$$

This internal energy has the form

$$\text{(internal energy)} = \text{(kinetic energy)} + \text{(rest mass energy)}. \tag{4.73}$$

The rest mass term can be subtracted out (rezeroing the energy scale), if the composition is constant in time. We are interested in the case that the composition does change, so that we must carefully consider the effect of the rest mass terms. The energy density per unit gram is E, so if V is the specific volume, then VE is the energy density per unit volume. Now,

$$E = \frac{1}{V} \int_0^\infty w \frac{dN}{dp} dp \tag{4.74}$$

and

$$N = \int_0^\infty \frac{dN}{dp} dp. \tag{4.75}$$

Thus using 4.74 we have

$$E = E_0 + VNmc^2 \tag{4.76}$$

where

$$E_0 = \frac{1}{V} \int_0^\infty \frac{p^2}{2m} \frac{dN}{dp} dp \tag{4.77}$$

is the usual expression for nonrelativistic internal energy. Note that the separation of rest mass energy does not require the nonrelativistic approximation. Suppose we have a mixture of species. If we can represent the contributions of each species as a single term in a summation like

4.70, then

$$E = \sum_i E_i$$
$$= \sum_i (E_0)_i + \sum_i N_A m_i c^2 Y_i. \tag{4.78}$$

Usually we have $E_0 \ll N_A mc^2$ so that it is desirable to subtract out (most of) the rest mass. We choose to reset the zero of the energy scale by subtracting out the rest mass of one atomic mass unit (amu) per nucleon. Then

$$E \rightarrow E_0 - N_A M_u c^2, \tag{4.79}$$

where M_u is the atomic mass unit, or

$$E - E_0 = c^2 N_A \sum_i Y_i \Delta M_i \tag{4.80}$$

where ΔM_i is the atomic mass excess (in mass units). In terms of binding energies B_i (see §3.1), we have

$$E - E_0 = Y_e N_A (m_p - m_n)c^2 + N_A (m_n - M_u)c^2 - N_A \sum_i Y_i B_i. \tag{4.81}$$

The proton and neutron masses m_p and m_n appear here.

Now we will assume for simplicity that the nuclei and nucleons are nonrelativistic, nondegenerate, ideal gases (usually a reasonable approximation). Then $E = \frac{3}{2} N_A kTY$, where $Y = \sum_i Y_i$. Now,

$$\frac{dE}{dt} = \frac{3}{2} N_A kY \frac{dT}{dt} + \sum_i N_A (\frac{3}{2} kT - B_i) \frac{dY_i}{dt}. \tag{4.82}$$

In the astrophysical literature, one uses

$$\frac{dE}{dt} = \frac{3}{2} N_A kY \frac{dT}{dt} \tag{4.83}$$

so that the dY_i/dt terms must appear elsewhere. They are added to the energy flow term dQ/dt to the detriment of clarity. In stars,

$$\frac{dQ}{dt} = -s_\nu - V\nabla \cdot \mathbf{F}, \tag{4.84}$$

where $\mathbf{F}$ contains the radiative, convective, and conductive fluxes, and s_ν is the neutrino energy emissivity. If we define the "extra" dY_i/dt term to be

$$\epsilon = \sum_i \frac{dY_i}{dt} N_A \left(B_i - \frac{3}{2}kT\right), \tag{4.85}$$

then we have

$$\frac{dE}{dt} + P\frac{dV}{dt} = \epsilon - s_\nu - V\nabla \cdot \mathbf{F}, \tag{4.86}$$

which is the form usually quoted. The second order terms (e.g., from the variation of pressure) are neglected.

Note that for most nuclei, $B = (8\,\text{MeV})A = 1.3 \times 10^{-5}\,A$ erg. The right-hand side of 4.85 can be approximately evaluated: the two terms are equal if

$$T = \frac{2B_i}{3k} = (6 \times 10^{10}\,\text{K})A. \tag{4.87}$$

Actually, even for an extreme case (oxygen burning),

$$T/A = 2 \times 10^9\,\text{K}/16 \ll 6 \times 10^{10}\,\text{K},$$

so that the $\frac{3}{2}N_A kT$ term may usually be neglected, giving

$$\epsilon = \sum_i \frac{dY_i}{dt} N_A B_i. \tag{4.88}$$

Now we consider a reaction converting $c + C \rightarrow d + D$. Then by nucleon conservation,

$$A = A_c + A_C = A_d + A_D. \tag{4.89}$$

Also,

$$\frac{dY_c}{dt} = \frac{dY_C}{dt} = -\frac{dY_d}{dt} = -\frac{dY_D}{dt} \tag{4.90}$$

and

$$\frac{dY_c}{dt}A_c + \frac{dY_C}{dt}A_C = \frac{dY_c}{dt}A. \tag{4.91}$$

Therefore,

$$\epsilon = \frac{dY}{dt} N_A Q, \tag{4.92}$$

where

$$Q = B_c + B_C - B_d - B_D, \qquad (4.93)$$

and $dY/dt = -dY_c/dt$ is the rate of formation of the compound nucleus. The rate at which nucleons are processed by this reaction is proportional to

$$\frac{dX}{dt} = A\frac{dY}{dt}. \qquad (4.94)$$

It is convenient to define the quantity

$$q = \frac{QN_A}{A}, \qquad (4.95)$$

where A is defined in 4.89. The quantity q is essentially the energy released upon consumption of a unit mass of fuel by the process in question. Some examples are given in table 4.1.

Using equations 4.95 and 4.92 gives

$$\epsilon = q\frac{dX}{dt}, \qquad (4.96)$$

which is a common form for ϵ. It is *not* generally exact; equation 4.85 is.

TABLE 4.1

Energy Release for Burning Stages

Process	$q(10^{18}$ erg/g)	q(MeV/nucleon)
H $\rightarrow$ ^{4}He	5 to 7	5 to 7
$3\alpha \rightarrow$ ^{12}C	0.585	0.606
$4\alpha \rightarrow$ ^{16}O	0.870	0.902
2 ^{12}C $\rightarrow$ ^{24}Mg	0.5	0.52
2 ^{20}Ne $\rightarrow$ ^{16}O + ^{24}Mg	0.11	0.11
2 ^{16}O $\rightarrow$ ^{32}S	0.5	0.52
^{28}Si $\rightarrow$ ^{56}Ni	0 to 0.3	0 to 0.31

Note: 1 MeV/Nucleon = 0.964844 × 10^{18} erg/g

4.6   MIXING AND HYDRODYNAMICS

In an Eulerian coordinate system (i.e., one fixed in space), the equation of continuity is

$$\frac{\partial \rho}{\partial t} + \nabla \cdot (\rho \mathbf{v}) = 0, \tag{4.97}$$

which from the discussion in §2.1 is equivalent to the requirement that nucleon number is conserved. For the number density N_j of nuclear species j, we have

$$\frac{\partial N_j}{\partial t} + \nabla \cdot (N_j \mathbf{v}) = \text{RHS}, \tag{4.98}$$

where RHS denotes the right-hand side of 4.1, that is, the production and destruction terms due to nuclear reactions. Using $N_j = Y_j \rho N_A$, the differential operations on N_j in 4.98 may be split into operations on ρ and upon Y_j. Making use of 4.97, this becomes

$$\frac{\partial Y_j}{\partial t} + \mathbf{v} \cdot (\nabla Y_j) = \frac{\text{RHS}}{N_A \rho}, \tag{4.99}$$

where this right-hand side is just the reaction terms expressed in Y_j notation. For conciseness we denote this as $\mathfrak{R}$ temporarily. The left side is the comoving time derivative of Y_j. If we use the comoving coordinate system, the resulting differential equations,

$$\frac{dY_j}{dt} = \mathfrak{R}, \tag{4.100}$$

are ordinary rather than partial, and easier to solve.

This result is elegantly simple; if our coordinate system is not strictly comoving, or if diffusive processes occur, the problem becomes more complex. First we consider the sort of "comoving" coordinate systems usually used in stellar evolutionary calculations. These "Lagrangian" coordinates do not move with the matter, strictly speaking, but only to the extent that the *amount* of mass enclosed by the coordinate cells is constant. To illustrate the difference, consider adding to velocity $\mathbf{v}$ an additional velocity $\mathbf{u}$ chosen so that

$$\nabla \cdot (\rho \mathbf{u}) = 0. \tag{4.101}$$

We recover 4.97 again, but using the same procedure as before, 4.99 now becomes

$$\frac{\partial Y_j}{\partial t} + \mathbf{v} \cdot (\nabla Y_j) = \Re - \mathbf{u} \cdot (\nabla Y_j). \tag{4.102}$$

The left-hand side is the apparent comoving time derivative in this coordinate system, but we now have a new term on the right-hand side which couples the change in abundance in the apparent comoving system to the "mixing currents" implied by the velocity $\mathbf{u}$. The reduction from partial to ordinary differential equations no longer holds, and the procedures for solution are more complex in general.

There are other sorts of motion which conserve "contained mass (nucleon number)" but give rise to additional terms in the abundance equations. Consider a number flux of nuclei of species j given by

$$F_j = -D\nabla Y_j, \tag{4.103}$$

for example, where the factor D ("diffusion coefficient") is independent of which nucleus is considered. The total nucleon flux is then

$$F = \sum_j F_j A_j = 0, \tag{4.104}$$

because if we take A_j, the constant number of nucleons per nucleus j, and the summation through the gradient operator we then find that it operates on a constant. Any such addition to the mass flux preserves 4.97 but modifies 4.100 by adding a term involving the divergence of a flux of the form given in 4.103.

The time-scale for actual microscopic diffusion of nuclei is

$$\tau_d \approx \frac{(\Delta R)^2}{\lambda} v_d, \tag{4.105}$$

where λ is the mean free path for a nucleus to scatter, v_d is the mean velocity of the nucleus relative to the fluid reference frame, and ΔR is the characteristic dimension over which the diffusion takes place. If we compare this to the time-scale for radiative diffusion over the same dimension, we find

$$\frac{\tau_d}{\tau_r} \approx \left(\frac{\sigma_d}{\sigma_r} \right) \left(\frac{c}{v_d} \right). \tag{4.106}$$

If we take the ratio of cross sections to be roughly that given by coulomb collision of nuclei to values typical of stellar opacity,

$$\frac{\sigma_d}{\sigma_r} \approx 10^8, \tag{4.107}$$

and consider thermal velocities at a stellar temperature T, we find

$$\frac{\tau_d}{\tau_r} \approx 10^{11}\sqrt{10^7 K/T}. \tag{4.108}$$

The microscopic diffusion time-scale is long compared to radiative diffusion; for the sun this corresponds to 10^{16} years. *Unless some other process is at work, stars will not be well mixed.* Fluid motions, of the sort described by velocity $\mathbf{u}$ above, will change this situation if they are turbulent (see Tennekes and Lumley [614], esp. ch. 7). Turbulent motion gives macroscopic mixing of matter on successively smaller length scales. With the decrease in the characteristic dimension ΔR separating different compositions Y_j, the diffusion time drops rapidly, as 4.105 indicates.

Because of the sensitivity of nuclear reaction rates to temperature and density, nuclear burning occurs selectively in parts of a star. This drives the star toward an inhomogeneous composition. Mixing resists this tendency. The net effect is fundamental to our understanding of stellar evolution. The usual astrophysical argument is as follows. Consider that species j has a characteristic time-scale τ_j for production (or destruction) by nuclear reactions. If we consider a mixing region of characteristic dimension ΔR, then the turnover time is $\tau_t = \Delta R/u$. If the flow is turbulent, then one claims that the mixing time is $\tau_{\text{mix}} \approx \tau_t$. We may rewrite 4.99 in the schematic form,

$$\frac{d(\ln Y_j)}{dt} \approx \frac{1}{\tau_j} - \frac{|\Delta Y_j|}{Y_j \tau_{\text{mix}}}, \tag{4.109}$$

where ΔY_j is the variation of Y_j over the mixing region. Suppose $\tau_{\text{mix}} \gg \tau_j$. Then the mixing has little effect on the nucleosynthesis and the second term may be neglected. Now consider $\tau_{\text{mix}} \ll \tau_j$. The sign of the second term is always such as to decrease ΔY_j. Eventually ΔY_j gets small enough so that the mixing and nuclear terms are comparable. Because $|\Delta Y_j| = Y_j \tau_{\text{mix}}/\tau_j \ll Y_j$ under these conditions, a complete mixing approximation (uniform abundance) is suggested. However, nuclear lifetimes may vary widely between species, so that in some cases one may have $\tau_{\text{mix}} \ll \tau_j$ for one nuclear species but $\tau_{\text{mix}} \gg \tau_i$ for a different species i. This situation has seldom been considered because of its difficulty. It does occur in the advanced stages of stellar nucleosynthesis.

4.7 FREEZEOUT

The patterns of abundances now observed were once in actively burning
regions, but no longer are. In what manner does this transition occur?
Does it affect the abundance pattern? Burning can end due to exhaustion of fuel, or by quenching of the flame (*freezeout*). If the burning
releases energy that is needed to maintain a thermal balance against energy losses, exhaustion of fuel can be a complex process. Most nuclear
reaction rates can be considerably reduced by changes in temperature
and density. Expansion, whether in a buoyant convective blob, an exploding star, or an expanding universe, can stop nuclear burning.

In general, freezeout is a complicated phenomenon. Some insight into
its nature may be gleaned by considering a simple model equation for
the destruction of a chosen nucleus:

$$\frac{dY}{dt} = -D(t)Y. \tag{4.110}$$

Here $D(t)$ is a sum for destruction terms (see §4.2) with Y factored out;
the production terms are assumed to be negligible in this special case.
The actual time dependence of the terms reflects both the changes in reaction rates (due to changing temperature and density) and the changes
in composition of the other reacting species. The product $D(t)Y$ makes
the equation nonlinear in general. We will examine the (deceptively?)
simple differential equation for the case that the decrease in $D(t)$ is
given by

$$D(t) = D(0)(T(t)/T(0))^{\alpha}, \tag{4.111}$$

where,

$$T(t) = T(0)\exp(-\lambda t). \tag{4.112}$$

Therefore,

$$d\ln Y/dt = -D(0)e^{-\alpha\lambda t}dt, \tag{4.113}$$

which has the solution

$$\ln(Y(t)/Y(0)) = -(D(0)/\alpha\lambda)(1 - e^{-\alpha\lambda t}). \tag{4.114}$$

For $t \to \infty$,

$$Y(\infty)/Y(0) = e^{-(D(0)/\alpha\lambda)}. \tag{4.115}$$

If we define a time for freezeout, $t_f = 1/\alpha\lambda$, then direct substitution shows that,

$$Y(\infty) = Y(t_f)(Y(0)/Y(t_f))^{\frac{1}{e}}. \qquad (4.116)$$

If $Y(0) \approx Y(t_f)$, then $Y(\infty) = Y(t_f)$, which is no help as yet. However, now consider the case in which production terms are not negligible:

$$\frac{dY}{dt} = P(t) - D(t)Y, \qquad (4.117)$$

which has the steady-state solution,

$$Y(t)_{ss} = P(t)/D(t). \qquad (4.118)$$

These two results suggest a crude procedure for examining freezeout. This "freezeout approximation" proceeds as follows:

1. define a freezeout time t_f, and
2. estimate the abundance as the steady-state value at t_f.

A price must be paid for this relative simplicity. While possibly useful as a guide, such a model cannot be expected to contain the complexity of the full equations. The difficulties arising in even this simple case reflect the complex nature of the problem; any such procedure should be used with considerable care. In practice, similar approaches are widely used, but the choice of the freezeout time t_f is somewhat arbitrary. While such approximations may be extremely useful in practice, freezeout is such a complex phenomenon that accurate simple solutions are rare. *Caveat emptor!*

5

Cosmological Nucleosynthesis

And I shall reply that this might perhaps be fabled
to have occurred in primordial chaos, where vague
substances wandered confusedly in disorder.
—Galileo Galilei (1564–1642),
Dialogue Concerning the Two Chief World Systems

The relative abundances of nuclei are remarkably constant in those cosmic objects in which they have been determined. Prior to the discovery of *metal-poor stars* (also called Population II by astronomers), it seemed that the abundances of nuclei were a truly universal property of cosmic matter (except insofar as atomic processes cause fractionation on smaller scales). This universality of abundances, coupled with (1) a finite age for the universe (interpreted from the data of nuclear geochemistry on radioactive nuclei, and that of astronomy on the Hubble expansion), and (2) the recognition that all nuclei were just combinations of neutrons and protons, seemed to support the idea that the abundances we observe are the ashes of the fiery birth of the universe (see Alpher and Herman [10] for an interesting review of this period of science).

The observation in stellar spectra by Merrill [428] of technetium—which has no stable isotopes, and the discovery of evidence for unstable nuclei at the formation of the solar system (see chapter 2), require that the process of nucleosynthesis continue to the present. Recent detection of gamma-ray lines from the radioactive decay of ^{26}Al in interstellar gas, and ^{56}Co and ^{57}Co in supernovae, reinforce this conclusion.

The lack of stable nuclei with nucleon number $A = 5$ and $A = 8$ was thought be a barrier to the synthesis of any but the lightest elements. In stars this gap is bridged by three-body reactions (the triple-alpha reaction in particular). As we have seen, the rates at which thermonuclear reactions occur are dependent upon both temperature and density. As will be discussed, in both cosmologies and stars there is an important relation between these variables; they tend to vary together, so that higher densities go with higher temperatures. A convenient way of expressing this is to categorize the conditions in terms of a particular combination of temperature and density. The density tends to scale with the cube of the temperature. Another way of saying the same thing is to say that the radiation entropy is nearly constant as a star evolves, or a cosmology expands. There is still the question of scale: for a given temperature, which is denser, the star or the cosmology? Space is empty and cold. In this sense it is emptier than it is cold—that is, its entropy is high! In the Big Bang cosmologies the entropy is much higher than in a star—so that for a given temperature the mean density of the cosmology is much lower— making three-body reactions more unlikely. With a scarcity of particles, a collision of three at once is even rarer than of two.

Nucleosynthesis in stars and galaxies is related to cosmological theory primarily in three ways. First, the initial conditions for stellar nucleosynthesis, in particular the abundance of H and He, are determined by cosmology. Second, the observed abundances of hydrogen, deuterium, ^{3}He, ^{4}He, and ^{7}Li may be taken as relics of the Big Bang—and hence providing cosmological information—to the extent that they are not produced in other ways. Thus stellar effects may obscure the cosmological information. Third, the age of the universe must be as large as that inferred for those contents of the universe—such as star clusters and radioactive nuclei—for which ages can be determined.

In a curious sense, cosmological nucleosynthesis is a relatively easy problem. Historically it has dealt primarily with isotropic and homogeneous universes, so that space variations were ignored. The nuclear reaction networks were then reduced from partial to ordinary differential equations, as were the equations for conservation of mass, momentum, and energy. Although at a given temperature the cosmological densities are less than those in stars, they are high enough so that two important simplifications are valid: the radiation is that of a blackbody and the particle velocity distribution is maxwellian. Both radiation and gas can be characterized simply by a temperature. As the evolution proceeds at nearly constant entropy, the time behavior of temperature and density are related in a relatively simple way. Because few nuclei are made; reaction networks are small. Finally, it seems that the abundances of ^{4}He and D are difficult to explain in other ways, so that they provide what seems to be a relatively clear observational test for the theory. The staggering fact is that this simple theory works so well, on such a grand scale.

5.1   KINEMATICS

For isotropic and homogeneous systems, change can occur only in time. What is the time dependence of the Big Bang?

Consider the motion of matter in a coordinate system chosen so that the matter is at rest at the origin. Suppose that the velocity is directed radially outward from the origin, and proportional to the distance $\mathbf{r}$ from the origin. Then,

$$\mathbf{u} = H\mathbf{r}, \tag{5.1}$$

where H is Hubble's constant. The term *constant* means here that H is independent of $\mathbf{r}$, but may depend upon time. Obviously such a distribution of velocities is isotropic. At an arbitrary point A, the radius vector of which is $\mathbf{r}_A$, matter has a velocity $\mathbf{u}_A$. Let us translate the origin to A and perform a Galilean transformation on the system. Quantities in the new system will be capitalized. Now,

$$\mathbf{R} = \mathbf{r} - \mathbf{r}_A, \tag{5.2}$$

so,

$$\mathbf{U} = \mathbf{u} - \mathbf{u}_A = H\mathbf{R}.$$

The law for the distribution of velocities is invariant for this procedure. This velocity distribution is remarkable because it does not single out any particular point, hence its obvious attraction as a dynamic cosmology. An observer at any point who is moving with the matter will see a recession of all surrounding matter. Here we have considered a small region of space so that the velocities were small compared to that of light. If we use the general theory of relativity we can relax this constraint, but as pointed out by McCrea and Milne [425, 676] we obtain equivalent results.

How does the density behave? Consider a sphere of radius $R(t)$ containing mass M. Then

$$\rho = \frac{3M}{4\pi R^3}, \tag{5.3}$$

so differentiating this with respect to time and using

$$\frac{d\mathbf{R}}{dt} = \mathbf{U},$$

we have

$$\frac{d\rho}{dt} = -3H\rho. \tag{5.4}$$

In the Newtonian approximation, the acceleration of a particle on the surface of the sphere is

$$\frac{dU}{dt} = -\frac{GM}{R^2}. \tag{5.5}$$

With pressure neglected, this is a "dust-filled" universe, and is appropriate for describing our universe back to about the epoch of decoupling of radiation and matter (see below). The matter outside the sphere does not affect this acceleration (the universe is assumed to be globally homogeneous and isotropic). Using 5.1, 5.3, and 5.5 gives

$$\frac{d(HR)}{dt} = R\frac{dH}{dt} + RH^2$$

$$= -\frac{4\pi G\rho R}{3},$$

so,

$$\frac{dH}{dt} = -H^2 - \frac{4\pi G\rho}{3}. \tag{5.6}$$

Equations 5.4 and 5.6 form a system that fully determines the evolution of local properties in a universe. Note that the radius R of the arbitrary mass M does not appear. These quantities will be useful in obtaining a graphic idea of the nature of the solutions. Multiply 5.5 by dR/dt and integrate:

$$\frac{(dR/dt)^2}{2} - \frac{GM}{R} = K, \tag{5.7}$$

where K is a constant. This equation corresponds to an energy conservation law for a Newtonian system, a fact that can aid intuition. Consider H_0 and ρ_0 to be the values of Hubble's constant and of the density at time t_0. Note that t_0 is *now*, the time of observation. If we choose a radius R_0 then the mass enclosed is $M_0 = 4\pi\rho_0 R_0^3/3$. Now,

$$\left(\frac{dR}{dt}\right)_0 = H_0 R_0,$$

so 5.7 may be written as,

$$\left(\frac{dR}{dt}\right)^2 = a\left[\frac{R_0}{R} - b\right] \tag{5.8}$$

where

$$a = 8\pi\rho_0 \, GR_0^2/3,$$

and

$$b = 1 - 3H_0^2/8\pi G\rho_0.$$

What is the general behavior of the solution? At present we observe the expansion of the universe, so dR/dt is positive. Thus R was smaller in the past and R_0/R was greater. Consequently, from 5.8, dR/dt was even greater in the past. Thus at some previous time t_∞, $R = 0$, $dR/dt = +\infty$, and $\rho = \infty$. This is the *cosmological singularity*, or the *Big Bang*.

The nature of future evolution depends upon the sign of b; this is the analog of Newtonian solutions of positive and negative total energy. Let us denote

$$\rho_c = \frac{3H_0^2}{8\pi G}. \tag{5.9}$$

This is the *critical density* separating solutions of qualitatively different nature. If $\rho_0 > \rho_c$, b is positive. As R increases, the right-hand side of 5.8 will decrease, eventually going to zero. Then $dR/dt = 0$ and expansion will cease, and contraction begins (dU/dt is negative). This finally results in $R \to 0$ and density becoming infinite at some later time. Universes with such behavior are called *closed*.

If $\rho_0 < \rho_c$, expansion continues indefinitely. As t goes to infinity, so does R, and

$$\frac{dR}{dt} \to a[\rho_c - \rho_0] \tag{5.10}$$

approaches a constant value. This is an *open* universe .

The value of the Hubble constant is still a matter of debate; it probably lies in the range $H_0 = 75 \pm 25 \, \text{km s}^{-1} \, \text{Mpc}^{-1}$, which includes most competing estimates. This corresponds to $H_0 = 2.5 \times 10^{-18} \, \text{s}^{-1}$ or $1/H_0 = (1.3 \pm 0.4) \times 10^{10}$ years, if the cancelling units of length, km and Mpc, are eliminated. J. Oort [467] estimated the cosmic density $\rho_0 = 3 \times 10^{-31} \, \text{g cm}^{-3}$ based on $H_0 = 75 \, \text{km s}^{-1} \, \text{Mpc}^{-1}$ (see Peebles [490]). This gives $\rho_0/\rho_c = 0.03$; the actual value seems still uncertain today. Significant amounts of unseen mass would raise this value, and there is increasing evidence [628] that such *dark matter* does exist. Let us estimate t_∞, the past time when the density was infinite. If dR/dt were constant at its present value—taken at time $t = t_0$,

$$R_0 = \left(\frac{dR}{dt}\right)_0 (t_0 - t_\infty),$$
$$= H_0 R_0 T, \tag{5.11}$$

where

$$T = t_0 - t_\infty = \frac{1}{H_0} \tag{5.12}$$

is the age of the universe. This linear extrapolation back to the singularity is roughly correct for cosmologies with small $\Omega = \rho_0/\rho_c$, but overestimates the age otherwise, as may be seen in figure 5.1. It is amazing that this value is only slightly more than twice the accurately determined age of the solar system, 4.55×10^9 years. The evolution of the cosmological length scale is shown, with the present time indicated for selected cosmological models. Time is measured in units of inverse Hubble constant (e.g., for $H_0 = 50$, we have $1/50$ km s^{-1} Mpc^{-1} $= 2 \times 10^{10}$ years). Three values of the density are shown; in units of the critical density for closure of the universe, they are $\Omega = \rho/\rho_c = 0.03$, 1.0, and 1.5. The solid line shows the extrapolation back from the present, using the measured Hubble constant; this puts the Big Bang 50 percent further back than it really was for the critical universe, for example.

For $\rho_0 = \rho_c$ we have the critical solution ($\Omega = 1$) which is particularly simple, and especially attractive for inflationary models of cosmology (see [359, 360]),

$$\left(\frac{dR}{dt}\right)^2 = a\left(\frac{R_0}{R}\right), \qquad (5.13)$$

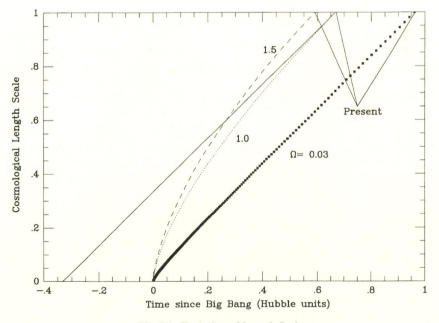

Fig. 5.1. Evolution of Length Scale

so the solutions are

$$R = R_0 \left[\frac{t - t_\infty}{t_0 - t_\infty} \right]^{2/3} , \tag{5.14}$$

and

$$T = \frac{2}{3H_0} = 13 \text{ Gyrs} \left(50 \text{ km s}^{-1} \text{ Mpc}^{-1}/H_0 \right), \tag{5.15}$$

and

$$\rho = \frac{1}{6\pi G \left(t_0 - t_\infty \right)^2},$$
$$= \frac{8 \times 10^5 \text{ s}^2 \text{ g cm}^{-3}}{\left(t_0 - t_\infty \right)^2}. \tag{5.16}$$

The age of the universe is only 7 Gyrs for $H_0 = 100$. If estimates of the ages of globular clusters, and of nucleo-cosmochronology are correct,[1] then either $\Omega \ll 1$ or $H_0 \ll 100$.

The analysis so far has ignored the presence of relativistic particles because pressure has been neglected relative to rest mass energy. Such an approximation is certainly invalid for particles with zero (or small) rest mass such as photons and neutrinos. For extremely relativistic particles, the energy per unit volume is

$$\epsilon = \rho c^2, \tag{5.17}$$

and the pressure is

$$P = \frac{\epsilon}{3} = \frac{1}{3}\rho c^2, \tag{5.18}$$

so that $P \ll \rho c^2$ does not hold. For nonrelativistic matter, the mass density is essentially the number density of particles times the rest mass per particle, so

$$\rho_{nr} \propto \frac{1}{R^3}. \tag{5.19}$$

For extremely relativistic particles the situation is different. If the particles undergo no reactions to change their number, then their number

[1] With Helmholtz and Kelvin, and with Hubble, astronomy has argued with geology for smaller cosmic ages and has been wrong both times. Do the higher values of the Hubble constant indicate a chance for the geologists to make it three out of three?

density does fall off as $1/R^3$ for expansion, but the energy per particle is also red-shifted by a factor $1/R$, resulting in

$$\rho_{er} \propto \frac{1}{R^4}. \tag{5.20}$$

If the reactions are so strong that we may consider the adiabatic expansion of an extremely relativistic gas,

$$d(\epsilon V) = -PdV. \tag{5.21}$$

The specific volume scales like R^3, so that from this and 5.18,

$$d\ln(\epsilon) = -\frac{4}{3}d\ln(V), \tag{5.22}$$

so we again find

$$\rho_{er} \propto \frac{1}{R^4}. \tag{5.23}$$

Therefore as $R \to 0$, ρ_{er} increases faster than ρ_{nr}. As a result, even if most of the energy density in the universe is currently in the form of nonrelativistic matter, so that we can assume $P \ll \rho c^2$ at present, in the past extremely relativistic particles were still the dominant form of mass-energy. The microwave background is a blackbody at $\approx 3\,\text{K}$, the corresponding mass-energy is $\approx 6.9 \times 10^{-34}\,\text{g cm}^{-3}$, which is much less than the Oort density of $3 \times 10^{-31}\,\text{g cm}^{-3}$. Thus at present our dust universe solutions seem appropriate. However from 5.19 and 5.23 we find that these densities were comparable at $R_0/R = 600$ (a redshift z of 600). At that epoch, the blackbody temperature extrapolates back to 1800 K. For such and earlier times, the relativistic particles dominate.

The effect of pressure is subtle; other things being equal, it decelerates rather than speeds the expansion. In such a homogeneous system, no pressure gradient exists, so that there is no corresponding acceleration term. Given a finite pressure, a change in the volume results in a change in the energy within the volume because of the work done. The gravitational acceleration results not just from the rest mass of the particles, but from the total mass-energy. In this case, the total mass-energy is not constant because of the work done by the pressure. *The ultimate cause of the expansion is not related to repulsive force of pressure but to the assumed velocity distribution in the initial state.*

Tolman [624] has shown that the general relativistic acceleration for the gravitation of our sphere is

$$g = -\frac{4\pi G(\epsilon + 3P)R}{3c^2}.$$ (5.24)

Thus, 5.5 becomes

$$\frac{dU}{dt} = -\frac{4\pi G(\epsilon + 3P)R}{3c^2}.$$ (5.25)

If we assume that baryon conservation is valid, the total number of baryons n inside radius R is constant. The number density of baryons is

$$N = \frac{3n}{4\pi R^3}.$$ (5.26)

The relativistic energy density E satisfies

$$dE = d\left(4\pi R^3 \frac{\epsilon}{3}\right) = -PdV.$$ (5.27)

Multipling 5.25 by dR/dt, and using $U = dR/dt$ gives

$$\frac{1}{2}\frac{d(U^2)}{dt} = -\left(\frac{4\pi G}{3c^2}\right)(R\epsilon U + 3PRU).$$ (5.28)

We may write 5.27 as

$$R\frac{d\epsilon}{dt} = -(\epsilon + P)3U,$$ (5.29)

so that the expression in parentheses in 5.28 becomes

$$R\frac{dR}{dt}(\epsilon + 3P) = -\frac{d\left(R^2\epsilon\right)}{dt}.$$ (5.30)

Finally we obtain

$$\frac{U^2}{2} - \left(\frac{G}{R}\right)\left(\frac{4\pi\epsilon R^3}{3c^2}\right) = K,$$ (5.31)

where K is a constant; this corresponds to 5.7. As before, the constant K can be evaluated in terms of quantities evaluated at the present time t_0.

Consider extremely relativistic particles (such as neutrinos and photons) undergoing adiabatic expansion. We further assume that the chemical potential for the particles is "close" to zero, a result we will discuss below. For all such particles we have $\epsilon = aT^4$, where for photons a is the usual radiation density coefficient and for neutrinos the constant is smaller by a factor of $\frac{7}{8}$ per family. The entropy is

$$S = \frac{4aT^3}{3\rho_b},\qquad (5.32)$$

and the baryon density scales as $\rho_b(t) = \rho_b(t_0)\,(R_0/R)^3$, so that

$$\epsilon = \frac{B}{R^4},\qquad (5.33)$$

where

$$B = \left[\frac{3S_\gamma\rho_b(t_0)}{4}\right]^{4/3}\left[\frac{R_0^4}{a^{1/3}}\right].$$

While this is explicitly written for photons only, the generalization is not difficult. If most of the energy density is due to these extremely relativistic particles, as is expected in the earlier stages of expansion, then 5.25 becomes

$$\frac{dU}{dt} = -\frac{8\pi GB}{3c^2R^3},\qquad (5.34)$$

and 5.31 becomes

$$\frac{U^2}{2} - \frac{4\pi GB}{3c^2R^2} = K.\qquad (5.35)$$

As R decreases, the $1/R^2$ term increases in magnitude; this must be balanced by a corresponding change in the $U^2/2$ term. Thus at early times, the constant K is negligible.

The Hubble constant evaluated at present is

$$H_0 = \frac{U_0}{R_0}\qquad (5.36)$$

and the deceleration parameter is

$$q_0 = -\frac{R_0}{U_0^2}\frac{dU_0}{dt},\qquad (5.37)$$

(see Weinberg [676], §14.6). Using these, we find

$$K = (1 - q_0) U_0^2/2, \tag{5.38}$$

which from observations cannot be drastically different in magnitude from the $1/R^2$ term in 5.31. This supports our neglect of K for earlier times. Now,

$$R \frac{dR}{dt} = \sqrt{8\pi GB/3c^2}, \tag{5.39}$$

which has the solution,

$$R = \left(32\pi GB/3c^2\right)^{1/4} \sqrt{t - t_\infty}, \tag{5.40}$$

so that

$$\frac{\epsilon}{c^2} = \frac{(3/32\pi G)}{[t - t_\infty]^2}, \tag{5.41}$$

where $3/32\pi G = 4.48 \times 10^5 \, \text{g} \, \text{s}^2 \, \text{cm}^{-3}$.

This, then, provides some insight into the kinematic properties of an expanding universe—in particular the expansion time-scale—which we will need for a discussion of thermodynamics and nucleosynthesis.

5.2 RADIATION AND PARTICLES

Consider an expanding universe containing a few galaxies. The mean distance between galaxies scales like R as defined in the previous section. Photons traveling between galaxies suffer a cosmological redshift, where the wavelength of a photon is defined by local observers at rest relative to the universe around them. In a time interval dt a photon moves between two such observers separated by a distance $c \, dt$. Because lengths scale as R, the observers move apart a distance $(dR/dt) \, dt\lambda/R$ during dt. The photons are redshifted, therefore,

$$\frac{d\lambda}{\lambda} = \frac{dR/dt}{R} dt, \tag{5.42}$$

and the redshift increases with distance of the source $(c \, dt)$. This also implies for a given photon that

$$\lambda \propto R. \tag{5.43}$$

Consider the expansion of a universe having such a low matter density that there is little interaction between photons and matter; photon number is then approximately conserved. For a blackbody distribution the number of photons in a wavelength range $d\lambda$ is

$$dN = n_\lambda = 8\pi \frac{d\lambda}{\exp(hc/kT\lambda) - 1}. \tag{5.44}$$

As the universe expands by a factor R the new wavelength of these photons is

$$\lambda' = R\lambda. \tag{5.45}$$

If the number of photons is conserved, then their number density goes as

$$N \propto \frac{1}{R^3}. \tag{5.46}$$

so,

$$dN' = \frac{dN}{R^3}. \tag{5.47}$$

The new distribution law is

$$n'(\lambda)d\lambda = \frac{n\left(\frac{\lambda}{R}\right) d\left(\frac{\lambda}{R}\right)}{R^3}$$
$$= n(\lambda')d\lambda'. \tag{5.48}$$

The new distribution is also that of a blackbody, with a new temperature

$$T' = \frac{T}{R}. \tag{5.49}$$

This is the adiabatic cooling law for a uniformly expanding photon gas. Cosmological expansion does not affect the shape of the photon distribution, and the intensity changes in just such a way as to be that of the blackbody at the new temperature [490]. The argument follows equally well for a Fermi-Dirac distribution; the result may be applied to neutrino gases as well.

What about the effects of interaction with matter? To explore this question we must consider expansion from the singularity. In universes dominated by extremely relativistic particles, their entropy is constant except in those epochs when important reactions come out of equilibrium (freezeout); the most obvious example concerns electron-positron pairs. The entropy per nucleon of blackbody photons is

$$\frac{S_\gamma}{N_A k} = \frac{1.213 \times 10^{-22} T^3}{\rho_b}, \tag{5.50}$$

with T in degrees kelvin and ρ_b in g cm^{-3}. For a temperature $T = 3K$,

$$\frac{S_\gamma}{N_A k} = 3 \times 10^8 \left(\frac{\rho_c}{\rho_0}\right), \tag{5.51}$$

which is large; in contrast, the corresponding stellar entropies we shall consider later will be smaller (ranging from 100 to 10^{-5} in extremes). The ratio of photon to nucleon number is

$$\frac{n_\gamma}{n_b} = 0.2776 \left(\frac{S_\gamma}{N_A k}\right). \tag{5.52}$$

Kolb and Turner [360] use a parameter $\eta \equiv n_b/n_\gamma$. A still different notation for this parameter is due to Wagoner [658], who uses

$$h_{wagoner} = \rho_b \left(10^9 K/T\right)^3, \tag{5.53}$$

so that,

$$h_{wagoner} = 1.213 \times 10^5 \, N_A k/S_\gamma. \tag{5.54}$$

The dominant process tending to keep the matter temperature equal to the radiation temperature in the epoch of interest is Thomson scattering by free electrons (Peebles [490], ch. 7). To estimate the conditions at the epoch at which the radiation and the matter decouple we will compare expansion rates to mean reaction rates for Thomson scattering. The mean rate of scatter for a typical photon is

$$\frac{1}{N_\gamma}\frac{dN_\gamma}{dt} = N_e \sigma_{Th} v, \tag{5.55}$$

$$\approx \left(1.2 \times 10^{10}\right) \rho_b Y_e,$$

in cgs units, where $\sigma_{Th} = 6.65 \times 10^{-25}$cm^{-2} is the Thomson cross section, v is essentially equal to c, the velocity of light, and Y_e is the number of *free* electrons per baryon.

As we saw in the previous section, the expressions for expansion differ little for the dust- or the radiation-dominated universes. This similarity is fortunate because decoupling seems to occur near the epoch at which the switch occurs between radiation domination and rest mass domination. We will simply use 5.41 here. Thus,

$$\frac{1}{\rho}\frac{d\rho}{dt} = -\frac{2}{t}$$

$$= -(\rho/1.8 \times 10^6)^{\frac{1}{2}}, \tag{5.56}$$

in cgs units. If we arbitrarily assume that the matter is totally ionized, then $N_e \approx N_b$ so $Y_e \approx 1$. With equations 5.55 and 5.56, and using the Oort density, we find expansion by a factor of 60 since that time. The corresponding temperature was therefore 180 K, which is much too small for the assumption of total ionization to be valid, so we must try again!

We may with little error simply consider the ionization of pure hydrogen; this should provide most of the free electrons. For equilibrium, the reaction $H^0 = H^+ + e^-$ implies a relation between chemical potentials (see chapter 2 or Appendix B),

$$\frac{x^2}{1-x} = \left[\frac{(2\pi m_e kT)^{3/2}}{N_A \rho_b h^3}\right] e^{-Q/kT} \tag{5.57}$$

where x denotes the fraction of hydrogen which is ionized. We may now identify the fraction of free electrons per baryon with x, and solve 5.55, 5.56, and 5.57 by iteration. We find $x \approx 0.01$, $T \approx 3,000$ K, and redshift $z \approx 1,200$. This is the epoch of *decoupling* or *recombination*.

Can recombination actually occur fast enough for these equilibrium arguments to be valid? The recombination cross section is of the order of $\sigma_{rec} \approx \pi a_0^2 \approx 10^{-16}$ cm^2 where a_0 is the Bohr radius for hydrogen. The mean time for recombination per nucleon is

$$\tau_{rec} = \frac{1}{\rho_b N_A \sigma_{rec} v}, \tag{5.58}$$

where v is roughly the thermal velocity for electrons, or

$$v \approx \sqrt{3kT/m_e} \approx 4 \times 10^7 \text{cm/s}. \tag{5.59}$$

Note that while the recombination cross section exceeds the Thomson cross section by a factor of 10^8, the electron thermal velocity is less than the velocity of light by only 10^{-3}. The ratio of recombination time to scattering time is then proportional to the inverse of this product, i.e., 10^{-5}. If we correct for only 0.01 of the nucleons being ionized, the factor is 10^{-3}. Recombination time is therefore still shorter than the scattering time (or the expansion time at decoupling).

As we have seen before, at this epoch the mass energy density is roughly the same as the radiation energy density. Therefore the ratio of the *thermal* energy density of the matter to the *thermal* energy of the radiation is essentially $kT/m_b c^2 \approx 10^{-7}$. Because of this enormous difference in the thermal capacity of the photons and of the matter, it is difficult for the photon distribution to be altered by the matter as they decouple.

As we consider earlier epochs we consider increasingly higher temperatures. For particles of finite rest mass m, there are two limiting cases, the nonrelativistic ($kT \ll mc^2$) and the relativistic ($kT \gg mc^2$). We will consider the behavior of electrons to illustrate the relativistic limit. The transition occurs at a temperature $T = mc^2/k = m$ 11.6×10^9 K (MeV)$^{-1}$, which corresponds to $T \approx 6 \times 10^9$ K for electrons and $T \approx 10^{13}$ K for nucleons. If we take the electron energy to be well represented by the relativistic expression $\epsilon \approx pc + mc^2$ where p is the electron momentum, then we find the number density (relativistic Fermi-Dirac gas) to be

$$N(\eta) = 8\pi(kT/hc)^3 F_2(\eta) \qquad (5.60)$$

where the (usual) Fermi-Dirac functions $F_2(\eta)$ are defined in Appendix B. Here,

$$\eta = \frac{\mu - mc^2}{kT} \qquad (5.61)$$

for the electrons, and since $\mu(e^-) = -\mu(e^+)$,

$$\eta(e^+) = \eta - 2mc^2 \qquad (5.62)$$

is the corresponding value for positrons. At extremely high temperatures, electron-positron pairs are copiously produced; at low temperatures these recombine, leaving a net charge difference which balances that of the nucleons. Thus it is useful to define a *net* electron number density by

$$N_e = N_{e^-} - N_{e^+},$$
$$= 16\pi(kT/hc)^3 e^{-2mc^2} (e^\eta - e^{-\eta}). \qquad (5.63)$$

For small η,

$$e^{\eta} - e^{-\eta} \to 2\eta. \tag{5.64}$$

Note that the right-hand side of 5.63 has many factors in common with the radiation entropy, so that we may rewrite it with 5.64 as

$$\frac{N_e}{N_b} = \left(\frac{45}{\pi^4}\right) \eta \left(\frac{S_\gamma}{N_A k}\right) e^{-2mc^2/kT}. \tag{5.65}$$

For a universe with no net electric charge, $N_e \approx N_b$, so if $kT \approx mc^2$ or greater,

$$\eta \approx \frac{N_A k}{S_\gamma} \approx 10^{-8}, \tag{5.66}$$

or less. This supports our use of the nondegenerate approximation, which is not too bad for $\eta \approx 0$. The small value of the chemical potential (relative to kT) implies that the number of positrons is large and almost equal to that of electrons. In fact, for $\eta = 0$, the Fermi-Dirac and Bose-Einstein functions of integral order are simply related. Thus the energy density of photons is related to that for positrons and for electrons by

$$E_\gamma = \frac{7}{8}E_{e^+} = \frac{7}{8}E_{e^-}. \tag{5.67}$$

Extremely relativistic particles act as a radiation gas. Upon cooling, the exponential factor in 5.65 causes the equilibrium to shift away from a large number of pairs; they annihilate to add to the photon gas.

Neutrinos may be discussed in a similar way, except for our use of electric charge to relate the net electron number to the baryon number. The corresponding neutrino number is not so well restricted by observation or experiment (see Kolb and Turner [360]). If the lepton number is small compared to the number of photons, there are three more pairs of terms, of the form $\frac{7}{8}E_\nu$ and $\frac{7}{8}E_{\bar{\nu}}$, one pair for each lepton family: electron, muon, and tau. These *light* particles will dominate the energy density, and the expansion rate (equation 5.41).

Finally, if we go back to more extreme temperatures, similar behavior may be expected from baryon-antibaryon pairs . This brings us to an important question which borders our discussion here: why does the Universe have a slight excess of baryons, not antibaryons? Why is there no evidence for antistars, antiplanets, and antipeople? The standard model for *baryogenesis* dates to Sakharov [533]; the basic idea is that at more extreme conditions processes become important which exhibit C and CP

as well as baryon nonconservation. As these freeze out, a skew toward baryons results. The challenge becomes the development of a theory of interactions (strong, weak, and electromagnetic) which explains baryogenesis as well as more conventional experimental data; see [360, 359].

5.3 WEAK INTERACTION FREEZEOUT

Basically there are three relics of the Big Bang which are of particular interest here:

1. the cosmic microwave background radiation, and
2. the large abundance of ^{4}He (which is the form of roughly a quarter of the observed matter in the universe), and
3. the rarer nuclei—deuterium, ^{3}He, and ^{7}Li—which seem also to be synthesized cosmologically.

In each case it must be determined that the relic was not in fact produced in some other way. At present the nucleosynthesis argument for ^{4}He and deuterium is most persuasive [553]; for the other nuclei it seems not beyond the power of the theorist to invent alternatives, although ^{3}He and ^{7}Li probably have a cosmic connection.

As we have seen in the previous section, the freezeout of the electromagnetic interactions at recombination determines the nature of the cosmic microwave background. The abundance of ^{4}He is strongly influenced by the freezeout of the weak interactions. As with the electromagnetic interactions, we proceed by comparing the rate of expansion to the rate of those weak interactions which maintain equilibrium. This sets the stage for the subsequent nucleosynthesis.

For temperatures $T \gg m_e c^2 / k \approx 6 \times 10^9$ K, the number of electron-positron pairs is comparable to that of photons, and greatly exceeds that of baryons. The reaction

$$e^- + e^+ = \nu + \bar{\nu} \tag{5.68}$$

equilibrates the electron and neutrino gases. The reactions

$$e^- + p = \nu + n \tag{5.69}$$

and

$$e^+ + n = \bar{\nu} + p \tag{5.70}$$

provide a coupling to the baryon gas. In the last section we saw that the chemical potentials for electrons and positrons were a tiny fraction of

kT. From 5.68 this implies

$$\mu(\nu) + \mu(\bar{\nu}) \approx 0 \tag{5.71}$$

which is satisfied if

$$\mu(\nu) = \mu(\bar{\nu}) \approx 0. \tag{5.72}$$

However, 5.71 does not *require* 5.72. The standard cosmological nucleosynthesis will result if we *assume* 5.72 in addition to the derived result 5.71. If the standard cosmological model is correct, then presumably 5.72 is the natural result of the particle physics occurring at still higher temperature and density [359, 360].

Now 5.69 and 5.70 imply (at weak interaction equilibrium)

$$\mu(e^-) + \mu(p) = \mu(\nu) + \mu(n) \tag{5.73}$$

and

$$\mu(e^+) + \mu(n) = \mu(\bar{\nu}) + \mu(p). \tag{5.74}$$

Therefore,

$$\mu(n) - \mu(p) = \mu(e^-) - \mu(\nu) \tag{5.75}$$
$$\approx -\mu(\nu).$$

Using 5.72, this gives the ratio of neutrons to protons,

$$\frac{N_n}{N_p} = e^{-Q/kT} \tag{5.76}$$

where $Q = 1.293\,\text{MeV}$. This ratio will be attained if the weak interactions are sufficiently fast to maintain equilibrium. At lower temperatures, the equilibrium neutron number tends toward zero; this may be attained by neutron decay, or by neutron capture.

The cross section for maintaining weak interaction equilibrium is

$$\sigma_{\text{weak}} \approx 2 \times 10^{-44}\text{cm}^2 \left(\frac{\epsilon}{m_e c^2} \right)^2 . \tag{5.77}$$

The mean time for a neutron to undergo an electron capture is

$$\frac{1}{N_n} \frac{dN_n}{dt} = -N_\nu \sigma_{\text{weak}} c, \tag{5.78}$$

which may be compared to equation 5.55. Now 5.72 implies that $N_\nu \approx$ N_{e^-} which is 3/4 of N_γ, or

$$\frac{N_\nu}{N_b} = 0.225 \left(\frac{S_\gamma}{N_A k} \right). \tag{5.79}$$

We use 5.77 and 5.78 to write

$$\frac{1}{N_n} \frac{dN_n}{dt} = -3.6 \times 10^{-10} \left(\frac{N_\nu}{N_b} \right) \rho_b \epsilon^2 \tag{5.80}$$

in cgs units, where the center-of-momentum energy ϵ is measured in electron rest mass units. For the expansion time-scale we use 5.56 with the *radiation* density ρ_γ; equating this to 5.80 gives

$$\rho_\gamma = 1.18 \times 10^6 \rho_b^2 \epsilon^4 \tag{5.81}$$

in cgs units. If we take $\epsilon = 3kT/m_e c^2 = T/(2 \times 10^9 \, \text{K})$, and scale from the ratio of baryon-to-photon energy density at present ($S/N_A k \approx 1.2 \times 10^{10}$), we find

$$T \approx 9.9 \times 10^9 \, \text{K} \tag{5.82}$$

for the weak interaction freezeout temperature. The redshift factor is $1 + z \approx 3 \times 10^9$. The corresponding mass density is $\rho_b = 2.7 \times 10^{28} \rho_{\text{oort}} \approx 0.008 \, \text{g cm}^{-3}$. From 5.76,

$$\frac{N_n}{N_p} = 0.22 \tag{5.83}$$

or, $Y_p = Y_e = 0.82$ and $Y_n = 0.18$.

Table 5.1 summarizes the variation in neutron abundance during the weak interaction freezeout. The equilibrium value, given by 5.76, and the estimated value at freezeout, given by 5.83, are compared to the numerical integrations of Peebles [489] and a set by the author (see below). The freezeout approximation is inexact for reasons discussed in Chapter 4, but the real discrepancy is that neutrons can decay during the time between weak interaction freezeout and subsequent capture to form nuclei.

The strong and electromagnetic interactions do not freeze out until a much lower temperature ($T < 10^9 \, \text{K}$) is reached. Consequently the freezeout of the weak interactions may be treated separately, greatly simplifying analysis. Following the pioneering work of Peebles [489], and especially that of Wagoner, Fowler, and Hoyle [659, 553], we directly integrate the reaction rate equations for the weak interaction rates listed

TABLE 5.1

Neutron Abundance in the Big Bang (Weak Freezeout)

$T/10^{10}$ K	Y_n(T.Eq.)	Y_n(freezeout)	Y_n(Peebles)	Y_n
3	0.377	⋯	0.380	0.380
2	0.320	⋯	0.330	0.329
1	0.181	⋯	0.238	0.235
0.1	2.87×10^{-7}	⋯	0.130	0.130
0	0	0.179	⋯	⋯

above. Consider a "standard" set [705] of parameters : three lepton families ($n_\nu = 3$), a conventional baryon density ($N_b/N_\gamma = 3 \times 10^{-10}$, or, $S_\gamma/N_A k = 1.2 \times 10^{10}$), and a neutron half-life of $\tau_{\frac{1}{2}} = 10.6$ minutes. More detail will be given in the next section.

The solid lines in figure 5.2 show the evolution of the abundances of neutrons and protons with cooling from $T = 4 \times 10^{10}$K to $T = 5 \times 10^9$K. Also shown, as dotted lines, are the equilibrium estimates

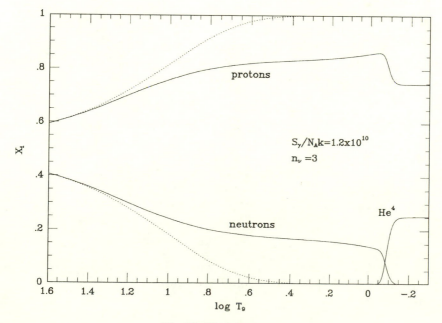

Fig. 5.2. Weak Freezeout

(equation 5.76) for these abundances. At the highest temperatures, the weak interactions maintain equilibrium, but over a wide range centered around $T = 10^{10}$K, equilibrium is lost. The free neutrons decay, increasing the proton number but lagging behind the equilibrium value. Only at much lower temperatures are subsequent reactions effective in producing more complex nuclei; ^{4}He is abundant below $T \simeq 8 \times 10^8$K.

If, as we shall see, all other nuclei are rare, it is easy to estimate the *primordial* ^{4}He abundance in terms of Y_e, the electron mole fraction: $X_\alpha = 2(1 - Y_e)$, where Y_e equals Y_p, the proton mole fraction at weak interaction freezeout. The numbers are roughly consistent with the simpler estimate above. The primordial ^{4}He abundance is essentially determined by the *freezeout of the weak interactions*.

5.4 COSMOLOGICAL NUCLEOSYNTHESIS

What about the abundances of nuclei other than ^{4}He? They involve more complex and subtle effects.

Between the epoch of weak freezeout ($T \approx 1 \times 10^{10}$K) and cosmological nucleosynthesis ($T \approx 0.8 \times 10^9$K), there is a transition in which electron-positron pairs recombine. This is of quantitative importance for nucleosynthesis because it changes the relation between baryon density ρ_b and temperature. The neutrino gases expand, conserving their entropy, because they have decoupled from the matter. As the electron-positron pairs recombine, their energy is equilibrated by electromagnetic interactions into the photon gas. Before this occurs, the entropy for the combined gas of photons and $e^\pm$ pairs is

$$S = \frac{4}{3} \frac{aT^3}{\rho} \left(1 + \frac{7}{4}\right), \tag{5.84}$$

where the temperature is the same as that of the neutrino gases T_ν, and the $\frac{7}{4}$ term is due to the $e^\pm$ pairs. Because the electromagnetic processes are fast, $e^\pm$ annihilation and photon equilibration generates little additional entropy; afterward the combined entropy is

$$S = \frac{4}{3} \frac{aT^3}{\rho}. \tag{5.85}$$

Equating the two relates the photon and neutrino temperatures after $e^\pm$ annihilation:

$$T = \left(\frac{11}{4}\right)^{\frac{1}{3}} T_\nu. \tag{5.86}$$

This affects the expansion time-scale through the total mass-energy ρ.

Following the approach of Wagoner [659, 658, 553], during nucleosynthesis we have

$$V^{-1}dV/dt = (24\pi G\rho)^{\frac{1}{2}}. \tag{5.87}$$

The first law of thermodynamics, in a comoving frame, is

$$\frac{d}{dt}(\rho V) + \frac{P}{c^2}\frac{dV}{dt} = 0, \tag{5.88}$$

where P is the pressure, and conservation of nucleons implies

$$\frac{d}{dt}(\rho_b V) = 0, \tag{5.89}$$

where ρ_b is the baryon mass-energy density. The total mass-energy ρ is dominated by the sum of contributions from photons, $e^{\pm}$ pairs, and neutrino gases:

$$\rho = \rho_\gamma + \rho_e + \rho_\nu. \tag{5.90}$$

The complicated expression for ρ_e, involving $e^{\pm}$ pairs, may be found in [659, 658]. If entropy is conserved,

$$T_\nu = T\left[(\rho_\gamma + P_\gamma)/(\rho_e + \rho_\gamma + P_e + P_\gamma)\right]^{\frac{1}{3}}, \tag{5.91}$$

and

$$\rho_\nu = \frac{a}{c^2}T_\nu^4\left(n_\nu\frac{7}{8}\right), \tag{5.92}$$

where $n_\nu = 3$ for electron, muon, and tau families of neutrino and antineutrinos. This provides an algorithm for constructing the total mass-energy ρ (nucleon contributions are tiny), which with equations 5.87, 5.88, and 5.89 can be used to generate the thermodynamic history of the Big Bang during the epochs of weak freezeout through nucleosynthesis.

From the discussion earlier, we see that the nucleons and nuclei will be nondegenerate, so that we may use the corresponding chemical potentials (Appendix B). The chemical potentials, and therefore the NSE abundances, are strongly affected by the mass differences of the different nuclear species. Because the weak interactions are frozen out, both proton and neutron number are conserved in reactions. This "isospin conservation" implies that the relevant combination of chemical potentials will generate differences of masses which may be written simply as nuclear binding energies (see §4.3). Table 5.2 gives both the mass excess, and the nuclear binding energy per nucleon, for those species which will

TABLE 5.2

Binding Energy per Nucleon and Mass Excess for Nuclei up to ^{12}C

Z	A	Symbol	B(MeV)/A	Mass Excess (MeV)
0	1	n	0.	8.07144
1	1	H	0.	7.28899
	2	D	1.112260	13.13591
	3	T	2.827307	14.94995
2	3	He	2.572693	14.93134
	4		7.074027	2.42475
3	6	Li	5.332148	14.08840
	7		5.606490	14.90730
4	9	Be	6.462518	11.35050
5	10	B	6.474995	12.05220
	11		6.927810	8.66768
6	12	C	7.680215	0.

be most relevant here (the particle-stable nuclei up to ^{12}C). The binding of ^{4}He is considerably larger than that of other light nuclei. The high entropies involved here imply that the translational phase-space factors in the chemical potential (the $T^{3/2}/\rho$ factors) will tend to favor free particles over bound, and hence nuclei of smaller A. These two effects, taken with the lack of an adequately fast reaction to synthesize carbon (see below), cause ^{4}He to be the major product of Big Bang nucleosynthesis. To be specific, consider the nuclei with $Z = N = A/2$. Then equation 4.28 can be written as

$$X_A = A^{\frac{5}{2}} \left(Y_p Y_n\right)^{\frac{A}{2}} \{\rho_b N_A \{2\pi \hbar^2/M_u kT\}^{\frac{3}{2}}\}^{A-1} \exp\left(B_A/kT\right). \quad (5.93)$$

The expression in braces { } is proportional to $T^{\frac{3}{2}(A-1)} S_\gamma^{-(A-1)}$, so that for a given temperature, it decreases strongly with increasing photon entropy S_γ, and more strongly with increasing nucleon number A. For a given small range of A, the exponential dependence on the binding energy B_A dominates.

The ^{4}He is not formed by a direct fusion of four nucleons but by a sequence of two-body reactions. The probability of reaction in a gas is proportional to the product of the probabilities of each reactant being in the volume where the reaction occurs. For nuclear reactions this is roughly proportional to factors of the ratio of the nuclear volume to the specific volume per reactant. These factors are small for densities less than nuclear ($\rho \sim 10^{14}$ g/cc); here each factor is of order 10^{-20}. Thus the rate of a four-body (three-body) reaction is reduced by a factor of order 10^{-40} (10^{-20}) relative to a binary interaction, *provided nuclear properties are equivalent.*

The obvious two-body reaction is deuterium formation:

$$p + n = D + \gamma. \tag{5.94}$$

In equilibrium,

$$\mu(p) + \mu(n) = \mu(D) \tag{5.95}$$

which gives

$$Y_D = Y_p Y_n f(T) \tag{5.96}$$

where

$$f(T) \approx (\rho_b / T_9^{3/2}) \exp(-23.01 + 25.81/T_9). \tag{5.97}$$

Using our previous value of $S_\gamma / N_A k$,

$$\rho_b = 8 \times 10^{-6} T_9^3, \tag{5.98}$$

so

$$f(T) \approx T_9^{3/2} \exp(-34.75 + 25.81/T_9). \tag{5.99}$$

The exponential is small for $T_9 > 0.75$, so that at such temperatures there is little deuterium; even at $T_9 = 1$, $f \approx 1.31 \times 10^{-4}$.

This behavior may be seen in figure 5.3. The top panel shows the history of the abundances of the major constituents—neutrons, protons, and ^{4}He. Note the sharp transition at $T_9 \approx 0.8$. In the lower panel are shown rarer nuclei. Deuterium is the first to rise as the temperature is lowered, and reaches a peak near $T_9 \approx 0.8$. At this temperature those light nuclei with larger binding energies successfully compete, and the deuterium abundance decreases slowly with temperature. The reaction rates are fast enough to make as much ^{4}He as is possible; its abundance is essentially constrained by the number of neutrons available. Heavier nuclei, such as ^{12}C, have low equilibrium abundances because of the entropy factors mentioned above; their abundance remains low even though the reaction rates are rapid enough to approach the equilibrium values. The key to the low production of heavier nuclei is not the *deuterium bottleneck* but the high entropy of the cosmology [360].

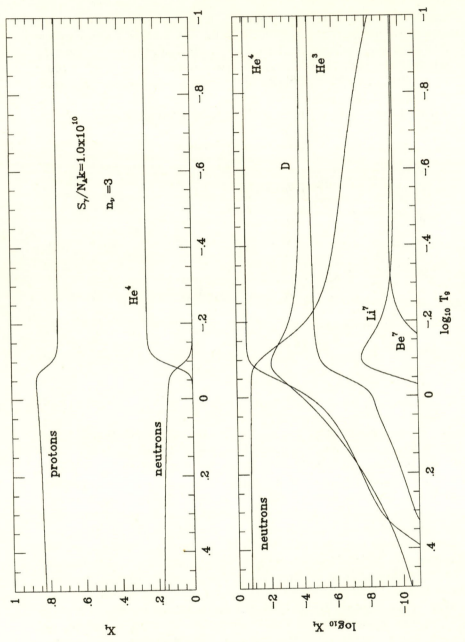

Fig. 5.3. Cosmological Burning

The final abundances of the rarer nuclei produced from cosmological nucleosynthesis are sensitive to the entropy of the universe [659]. Figure 5.4 shows these abundances as a function of the entropy of the universe after nucleosynthesis. The reaction networks were continued until the temperature dropped to $T_9 = 0.1$. The top panel shows the abundance of ^{4}He, along with two dashed lines which indicate the range of primordial He abundances inferred from observations [360, 553]. In the lower panel are the rarer species. The small dashed lines bound the region inferred for primordial ^{3}He + D. The ^{7}Li is made as itself in lower nucleon-density universes, and as the unstable ^{7}Be for higher-density ones. The large dashed line indicates the mean of estimates for the primordial Li abundance [583, 360].

It is suggestive that, despite the considerable uncertainties in inferring any primordial parameter, that there is a range of entropy in which ^{4}He, ^{3}He+D, and ^{7}Li predictions agree with observed values. This is a striking success for cosmological nucleosynthesis theory.

5.5 FURTHER IMPLICATIONS

In order to use cosmological nucleosynthesis in interpreting observations, it is necessary to know how these abundances may be further modified. What, if any, are the processes of subsequent production or destruction of these nuclei?

Deuterium is especially interesting in that it seems unlikely to be produced in any other way. The high abundance of D is due to the high entropy of the universe. Lower entropy objects, such as stars, tend to favor reactions that convert D to more tightly bound nuclei. Because of its low nuclear charge, $Z = 1$, deuterium is readily destroyed by reactions with protons, even at modest stellar temperatures. Cosmic matter, as it is formed into stars and then ejected, will tend to be deprived of its deuterium. This process, called *astration*, must be considered in order to compare stellar abundances to predicted cosmological ones. The cases of ^{3}He and ^{7}Li are more complex. Both of these nuclei are certainly produced in some stars, although the extent to which they are then ejected to pollute the interstellar medium is less clear. Because of their larger nuclear charges, they are more resistent than D to thermonuclear destruction as well. It is striking that their predicted abundances from the Big Bang are close to those observed, and suggestive.

If ^{1}H, D, ^{3}He, ^{4}He, and ^{7}Li are formed by cosmological nucleosynthesis, what about their near neighbors ^{6}Li, ^{9}Be, ^{10}B, and ^{11}B? In their influential review, Burbidge, Burbidge, Fowler, and Hoyle [135] attributed their formation to spallation reactions, which we have already encoun-

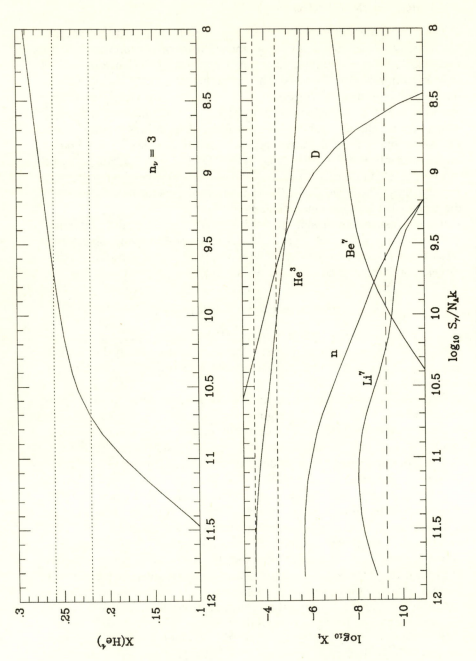

Fig. 5.4. Cosmological Ashes and Entropy

tered in our discussion of cosmic rays in chapter 2. As stressed by Audouze and Reeves [59], a simple argument makes this plausible. First, the spallation cross sections for the formation of mass $A = 6, 7, 8, 9, 10, 11$ nuclei are not very different, varying by less than an order of magnitude above 200 MeV. Second, with the suggestive exception of ^{7}Li, the ordering of the cross sections is similar to the ordering of the solar system abundances. Finally, consider this simple calculation. Take the ^{9}Be production cross section to have a mean value of 5 mb (5×10^{-27}cm^2). For a flux of galactic cosmic-ray protons of ≈ 5 cm^{-2} s^{-1}, a ratio of target nuclei to hydrogen of 10^{-3} (for CNO), and the age of the Galaxy to be $\approx 10^{10}$ years, then the expected production is B/H $\approx 10^{-11}$, which is essentially the solar system value.

If this picture is true, then it has important implications for (1) production of CNO nuclei over Galactic history, (2) production of cosmic rays over that same time interval, and (3) the abundances of these isotopes relative to ^{7}Li in the oldest stars.

6

Some Properties of Stars

With that this eagle 'gan to cry,
"Let be," quoth he, "thy fantasy!"
"Wilt thou learn of stars aught?"
"Nay, certainly," quoth I, "right naught."
"And why?" "For I am now too old."
—Geoffry Chaucer (1343–1400),
The House of Fame

Cosmological nucleosynthesis is an astounding success—it explains the abundances of 98% of the observed matter in the Universe. However, it explains the production of only five nuclei, H, D, ^{3}He, ^{4}He, and ^{7}Li. Of these, only hydrogen is a major component in our terrestrial environment, or in our selves. To explore the rest of nucleosynthesis, we must learn about stars—how they are born, how they evolve, and how they die. Even a sketchy version of this complex and still unfolding story will take the rest of this book.

Because of the complicated interplay of the various physical processes involved, stellar evolutionary theory involves the construction of complex mathematical models. Because of this complexity, the equations which result are almost never soluble in a simple way. To proceed it was necessary to resort to brute force, and build numerical models of stars in a computer. This was one of the areas in which the application of computers to scientific research was pioneered. Even the most modern computers are very limited in comparison to nature, so that the art involves creating a system much simpler than the real one, but which captures its essence. In this book, complex numerical results are often described in terms of *analytic approximations*: simpler equations are solved by conventional means, and then woven together to represent the actual solution. Such an approach could be justified simply as a mnemonic device: it is easier to remember a simple theory (an idea) than a table of numbers. There is a more fundamental reason for this approach. Stellar evolutionary calculations may profitably be viewed as numerical experiments. Representation of these reams of numbers by simple approximations is a step toward understanding the interplay of the various physical processes involved. The choice of approximations to describe a complex physical system is often not an easy one. This analytic approach is a valuable aid in suggesting new numerical experiments and in understanding old ones.

Stars last so long because of a delicate state of balance. The crush of gravity which pulls them in upon themselves is balanced by the increasingly high pressures in their interiors. Their prodigious radiation of energy, as light and neutrinos, is balanced by a prodigous generation of energy, by nuclear burning and gravitational contraction. Much of stellar evolution may be understood in terms of these balances, their breakdown, and their reestablishment. An elegant and venerable contribution by Eddington, the *Standard Model*, combined with a sturdy approximation invented by Fowler and Hoyle, gives a simple and powerful way to treat together both the balance of forces, and of energy flows.

Stellar fires, like terrestrial ones, burn out. The stellar case is curious in that the ashes of one burning stage may become the fuel for the next. There is an escape from this sequence: if the star is not massive, its gravity may be balanced by degeneracy pressure of its electrons, and settle into a long decline as a cooling cinder—a white dwarf. The possible burnings may be illustrated by the concept of an *ignition mass*; if the star considered is progressively more massive, then it will proceed progressively further through the sequence of nuclear burning stages. But finally, it must eject its excess matter, perhaps leaving a neutron star—supported by the degeneracy pressure of neutrons—or become a black hole.

6.1 STELLAR EVOLUTION EQUATIONS

The conventional form of the equations of stellar evolution involve some approximations which are best made explicit. Let us examine the equations of hydrodynamics, with newtonian gravity. The conservation of mass gives

$$\frac{\partial \rho}{\partial t} + \nabla \cdot \rho \mathbf{v} = 0, \tag{6.1}$$

and the conservation of momentum gives

$$\frac{\partial \mathbf{v}}{\partial t} + (\mathbf{v} \cdot \nabla)\mathbf{v} = -\frac{1}{\rho}\nabla P - \mathbf{g}, \tag{6.2}$$

where the gravitational acceleration is

$$\mathbf{g} = -\nabla \phi, \tag{6.3}$$

and

$$\nabla^2 \phi = 4\pi G \rho. \tag{6.4}$$

The conservation of energy gives

$$\frac{\partial}{\partial t}\left(\frac{1}{2}\rho v^2 + \rho E\right) = -\nabla \cdot \left[\rho \mathbf{v}(\frac{1}{2}v^2 + E + P/\rho) + \mathbf{F}\right], \tag{6.5}$$

where $\mathbf{F}$ is the flux of energy by radiative processes. For example, the radiative flux caused by diffusion of thermal photons is

$$\mathbf{F} = -(\lambda c/3)\nabla(aT^4), \tag{6.6}$$

where λ is the appropriate mean free path. For freely escaping neutrinos the $-\nabla \cdot \mathbf{F}$ term is augmented by a volume emissivity $\rho \epsilon_\nu$.

Following the discussion in §4.5, we include the energetics of nuclear burning in the time derivatives of E for brevity at present.

These equations are expressed in terms of some volume element fixed in space, and are called "Eulerian." Alternatively they may be expressed from a comoving coordinate system, i.e., in "Lagrangian" form. For conservation of mass,

$$\frac{d\rho}{dt} + \rho \nabla \cdot \mathbf{v} = 0, \tag{6.7}$$

and for conservation of momentum,

$$\frac{d\mathbf{v}}{dt} = -\frac{1}{\rho}\nabla P - \mathbf{g}. \tag{6.8}$$

Conservation of energy in the comoving frame is just the first law of thermodynamics,

$$\frac{dE}{dt} + P\frac{dV}{dt} = -\frac{1}{\rho}\nabla \cdot \mathbf{F}. \tag{6.9}$$

See Landau and Lifshitz [378], for detail.

The usual next step is to assume spherical symmetry, and work in a Lagrangian frame. Then the momentum equation is

$$\frac{dv}{dt} = -\frac{1}{\rho}\frac{dP}{dr} - g, \tag{6.10}$$

where v and g are now the *radial* components of fluid velocity and gravitational acceleration, respectively. Also,

$$g(r) = -Gm(r)/r^2, \tag{6.11}$$

where

$$m(r) = \int_0^r 4\pi x^2 \rho(x) dx. \tag{6.12}$$

Now conservation of mass is simply

$$\frac{dm}{dr} = 4\pi r^2 \rho. \tag{6.13}$$

The radiative flux becomes

$$F = -(\lambda c/3)\frac{d}{dt}(aT^4). \tag{6.14}$$

If the motion is very subsonic ($v \ll v_{sound}$), then the momentum equation becomes the hydrostatic equation,

$$\frac{dP}{dr} = -\frac{Gm\rho}{r^2}. \tag{6.15}$$

This last approximation was originally demanded because of the inability of numerical algorithms to remain stable for time steps large compared to sound travel times over a zone (this is the CFL condition of Courant, Fredrichs, and Lewy [196]).

These equations are incorrect if convective motions are present, i.e., for most stars. Suppose that the motion is decomposed into a radial velocity of spherical surfaces (Lagrange coordinates) which contain a constant *amount* of mass, and a nonradial motion which balances the net inflow and outflow over those spherical surfaces. See §4.6. If we denote this latter motion by **u**, then

$$\nabla \cdot \rho\mathbf{u} = 0 \tag{6.16}$$

in the frame of the spherical surfaces. Notice that these surfaces are no longer thought to move with the matter in detail, but only in an average sense, so that the amount of mass contained by them is constant in time.

If the up-welling parts of the spherical surface carry different quantities than the down-welling areas, a net transfer of those quantities results. In §4.6 this was examined for differences in composition. Similarly, this gives a "convective" flux of energy

$$F_{convective} = \langle \rho(uE + \frac{1}{2}u^2) \rangle, \tag{6.17}$$

where the average is taken over the spherical surface (compare to Eq. 7.3 in Kippenhahn and Weigert [355]).

Let us trace the derivation of the expression for the convective flux that is most widely used in stellar evolution computations. If we assume that motions are sufficiently slow that fluctuations of pressure are quickly evened out, then we have the *anelastic approximation*, which filters out sound waves. This is not valid at the surfaces of red giant stars, where the convective velocities are comparable to the sound speed. Further, if we also assume that the density fluctuations are small $\delta\rho/\rho \ll 1$, we have the *Boussinesq approximation* (see Spiegel [579]). Numerical experiments by Potter and Woodward (private communication) suggest that this is not strictly valid either. Following Spiegel, we may separate the mean and the fluctuating parts of the variables for the convective fluid. The viscosity of stellar matter is tiny and the characteristic length scale is large, so that we expect the Reynolds number to be large and the flow turbulent. Let us further assume that the spherical curvature is not important for the flow, and consider the plane-parallel case. Taking horizontal means of the Boussinesq equations gives a momentum equation

$$\frac{\partial}{\partial z}\left(\overline{P} + \overline{\rho w^2}\right) = -g\rho, \tag{6.18}$$

and an energy equation

$$\frac{\partial}{\partial t}\overline{T} + \frac{\partial}{\partial z}\overline{w\theta} = \kappa\frac{\partial^2\overline{T}}{\partial z^2}, \tag{6.19}$$

where κ is the thermal diffusivity, θ the fluctuating part of the temperature, w the vertical component of the velocity field $\mathbf{u}$, and z the vertical coordinate. The $\overline{\rho w^2}$ is the turbulent pressure and $\overline{w\theta}$ the convective flux. Notice that the anelastic approximation causes the kinetic energy part of this energy flux to be small compared to the internal energy part. The turbulent pressure is ignored because it tends to cause numerical difficulties. It is *not* correct to argue[1] that if $w < v_{sound}$, then the turbulent pressure is necessarily negligible, because its *gradient* may still be significant compared to the pressure gradient, especially at the edge of a convection zone.

[1]However, see Cox and Guili [201].

The equations for the fluctuating quantities are

$$\left(\frac{\partial \mathbf{u}}{\partial t} + \mathbf{u} \cdot \nabla \mathbf{u}\right) = -\frac{1}{\rho}\nabla \varpi + g\alpha\theta\mathbf{k} + \nu\nabla^2\mathbf{u}, \tag{6.20}$$

and

$$\frac{\partial \theta}{\partial t} + \mathbf{u} \cdot \nabla \theta = \kappa\nabla^2\theta + \beta\mathbf{u} \cdot \mathbf{k} \tag{6.21}$$

and

$$\nabla \cdot \mathbf{u} = 0, \tag{6.22}$$

where $\mathbf{k}$ is a unit vector in the vertical, $\varpi = (P - \overline{P})$, β is the excess temperature gradient relative to the adiabatic, and

$$R_a = \frac{g\alpha d^3}{\kappa\nu}\left(\Delta T - \frac{gd}{C_P}\right), \quad \text{and} \quad \sigma = \nu/\kappa, \tag{6.23}$$

where R_a and σ are the Rayleigh and Prandtl numbers, and α is the coefficient of thermal expansion. Here d is the vertical extent of the convective fluid, ΔT the imposed temperature difference across this layer, ν the viscosity and C_P is the specific heat at constant pressure. These relatively simple equations are difficult to solve, and in view of the extreme assumptions already made, approximate solutions may be all that is appropriate. We seek a steady state, and so set the time derivatives to zero. We replace the spatial gradient operators by ℓ^{-1}, where ℓ is a characteristic dimension called the "mixing length." We drop the pressure derivatives. Then 6.20 becomes

$$g\alpha\ell^2\theta = \nu(1 + R_e)w, \tag{6.24}$$

and 6.21 becomes

$$w\left(\frac{\partial \overline{T}}{\partial z} + \frac{g}{C_P}\right)\ell^2 = \kappa(1 + P_e)\theta. \tag{6.25}$$

The dimensionless ratios

$$P_e = w\ell/\kappa \quad \text{and} \quad R_e = w\ell/\nu \tag{6.26}$$

are known as the Peclet and Reynolds numbers, and measure the ratio of the turbulent diffusivity to the radiative and to the molecular diffusivities. Let

$$\theta \approx \ell\frac{\partial}{\partial z}(T - \overline{T}), \tag{6.27}$$

and take $R_e \gg 1$ everywhere. Then

$$w^2 = g\alpha \ell^2 \theta \tag{6.28}$$

and using 6.27 and noting that $\alpha = 1/T$ and $(T - \overline{T})/T = -(\rho - \overline{\rho})/\rho$, we have

$$w^2 = g\ell^2 \frac{\partial}{\partial z} \left(\frac{\rho - \overline{\rho}}{\rho} \right), \tag{6.29}$$

which is equivalent to equation (3-274) of Clayton [173], except for geometric factors of order unity, if we identify his $\Delta \nabla \rho$ with our $\frac{\partial}{\partial z} \left(\frac{\rho - \overline{\rho}}{\rho} \right)$. The heat carried is $\Delta Q = C_P \Delta T$, where

$$\Delta T = \ell \left(\frac{dT}{dz} + \frac{g}{C_P} \right) \equiv \Delta \nabla T. \tag{6.30}$$

Thus in this approximation (see also §7.3 and §7.4 in [355]) the convective heat flux is

$$F = \rho w C_P \Delta \nabla T. \tag{6.31}$$

Even if all the other approximations were true, we have paid a high price for replacing spatial gradient operators by ℓ^{-1} and averaging the adiabatic excess over a mixing length. This causes us to lose information about the edges of the convection zone, and hence its extent. Mixing is an irreversible process, so that such crude treatment is a cause for worry in an evolutionary calculation. Errors might build, or worse, different branches along the evolutionary paths might be taken.

Fortunately it seems that in practice that a limiting case—apparently independent of the details of convection—is often relevant. Consider a parcel of fluid displaced vertically and adiabatically; its energy changes by an amount $\delta m (C_P \, dT + g \, dz)$ where δm is the mass of the parcel. We expect instability if this change is negative, so

$$\frac{dT}{dz} < -\frac{g}{C_P} \tag{6.32}$$

is the *Schwarzschild criterion* for instability. If the mixing length ℓ is approximately equal to the pressure scale height, as usually supposed, then $P/\rho \approx Gm\ell/r^2$, and since the sound speed v_{sound} satisfies the relationship $v_{sound}^2 = \gamma P/\rho$, we have $(w/v_{sound})^2 = \Delta \nabla \rho/\gamma$. For times long compared to the sound travel time, the large specific heat insures that the motion can be subsonic (except at the surface of red giants!) and the gradient just slightly superadiabatic.

Therefore we may approximate the convective region as an adiabatic structure which can carry energy as fast as we wish. Thus we use 6.31 and do not explicitly specify w, $\Delta \nabla \rho$, or $\Delta \nabla T$. The mixing is as fast as necessary to maintain homogeneity in the convective region. The limits of accuracy of this limiting case are not presently known. It is not a valid approximation for the later stages of stellar evolution, when the star changes rapidly in time. It may give errors for the extent of the mixed region (*convective overshoot*) even for early stages of evolution.

6.2 STANDARD MODEL

Considerable understanding of stellar properties may be gleaned from a relatively simple model suggested by Eddington [218]. Consider a static, spherically symmetric star with no hydrodynamic motion. The equations used to describe such an object are:

$$\frac{dP}{dr} = -\frac{Gm\rho}{r^2} \tag{6.33}$$

$$\frac{dm}{dr} = 4\pi\rho r^2 \tag{6.34}$$

$$L = -4\pi r^2 \left(\frac{ac}{3\rho\kappa}\right)\frac{dT^4}{dr} \tag{6.35}$$

$$\epsilon = \frac{dL}{dm} \tag{6.36}$$

The Standard Model represents a solution of the first three; Eq. 6.36 is not explicitly considered.

For quantitative work it is desirable to use some results from the theory of polytropic, self-gravitating spheres; these are derived in Appendix C. However, to better understand the physical meaning of subsequent mathematical expressions, consider a crude finite-difference approximation to these equations. For example, for 6.33 and 6.34 we have $\Delta P/\Delta R = -Gm\rho/r^2$ and $\Delta m = \rho\Delta(4\pi r^3/3)$. Consider just two points in the star, the center and the surface. Then $\Delta P = P_s - P_c \approx -P_c$, and $\Delta m = M$, where M denotes the total mass. We may write

$$\frac{P_c}{\rho_c} \approx \frac{GM}{R}, \tag{6.37}$$

and

$$M = \frac{4\pi\rho R^3}{3}. \tag{6.38}$$

Note that there is some ambiguity as to the appropriate "average" value of Gm/r and of ρ to use in these expressions. Ignorance of these "form factors" is the price we pay for simplicity. From the more precise analysis in Appendix C we have

$$\frac{P_c}{\rho_c} = C_n \frac{GM}{R} = D_n GM^{2/3}\rho_c^{1/3} \tag{6.39}$$

where the subscript c refers to the central values of pressure and density, n is the polytropic index, and

$$M = \frac{4\pi\langle\rho\rangle R^3}{3}$$

$$= \left(\frac{\langle\rho\rangle}{\rho_c}\right)\frac{4\pi R^3 \rho_c}{3}, \tag{6.40}$$

which may be regarded as the definition of $\langle\rho\rangle$, the density of a uniform sphere of the same M and R. The values of C_n and D_n are given in table C.1 of Appendix C; they are of order unity.

For the earlier (and longer) stages of stellar evolution, it is often reasonable to approximate the equation of state as that of a Maxwell-Boltzmann gas plus a radiation gas. Let us introduce

$$\beta = \frac{\Re Y\rho T}{P}$$

$$= \frac{P_{\text{gas}}}{P},$$

$$(1 - \beta) = \frac{aT^4}{3P}$$

$$= \frac{P_{\text{rad}}}{P} \tag{6.41}$$

where $P = P_{\text{gas}} + P_{\text{rad}}$ is the total pressure. Therefore,

$$\frac{\beta}{1-\beta} = \frac{3\Re Y\rho}{aT^3}. \tag{6.42}$$

Note that the entropy of radiation is

$$S_\gamma = \frac{4aT^3}{3\rho}, \tag{6.43}$$

so that—for fixed composition—there is a one-to-one relationship between radiation entropy and β:

$$\frac{S_\gamma}{4\Re Y} = \frac{1 - \beta}{\beta}. \tag{6.44}$$

Taken with equation 6.39, 6.41 gives

$$1 - \beta = \left(\frac{a}{3\Re}\right)\left(\frac{D_n G}{\Re}\right)^3\left(\frac{\beta}{Y}\right)^4 M^2 \tag{6.45}$$

$$\approx 0.00298(m/Y^2)^2\beta^4,$$

where $m = M/M_\odot$ and the numerical value is for $n = 3$. This is the *quartic equation* of Eddington. In the preceding derivation it has been implicitly assumed that β is the same throughout the star. This assumption gives the Standard Model of Eddington; it is useful because often β really is varying slowly in space (a notable exception would be thermonuclear burning shells). Using Eq. 6.45, and choosing a mass M and a composition (actually just Y), we can determine β (or equivalently, S_γ). Some values of both β and photon entropy are given in table 6.1. For an ionized gas of pure hydrogen, $Y = 2$; for pure ionized ${}^4\text{He}$, $Y = 3/4$. The radiation pressure equals the gas pressure ($\beta = 0.5$) only for rather large mass, $m \approx 200$ and 29 (and $S_\gamma = 8$ and 3) for pure H and ${}^4\text{He}$ respectively.

Having solved for S_γ, the density and temperature in the Standard Model are related by

$$\rho = (4a/3\Re)(\Re/S_\gamma)T^3 \tag{6.46}$$

$$= 0.1213\,\text{g cm}^{-3}(\Re/S_\gamma)(T/10^7\,\text{K})^3.$$

This applies at any point in the Standard Model.

For most stars, β lies in the range near unity $(1 - 10^{-6})$ to 0.2, so that S_γ ranges from 10^{-5} to 30 in dimensionless units, depending upon the composition. Note that the total entropy is not so low; this is only the radiation component, which has a special role in Eddington's Standard

TABLE 6.1
Eddington's Standard Model

β	m/Y^2	$S_\gamma/\Re Y$	β	m/Y^2	$S_\gamma/\Re Y$
1.	0.	0.	0.7	20.5	1.71
0.999	0.580	4.00×10^{-3}	0.65	25.7	2.15
0.995	1.31	2.01×10^{-2}	0.6	32.2	2.67
0.99	1.87	4.04×10^{-2}	0.5	51.8	4.
0.95	4.54	2.11×10^{-1}	0.4	88.7	6.
0.9	7.15	4.44×10^{-1}	0.3	1.70×10^2	8.33
0.85	9.82	7.06×10^{-1}	0.2	4.10×10^2	1.60×10^1
0.80	12.8	1.	0.1	1.74×10^3	3.60×10^1
0.75	16.3	1.33	0.0	∞	∞

Model. In later stages of stellar evolution, the development of *red giant* envelopes gives the higher values of S_γ. For the Big Bang cosmology, we found $S_\gamma/\Re \approx 10^8$ to 10^{10} (see chapter 5). For the same temperature, cosmological nucleosynthesis involves densities which are about 10^8 to 10^{10} times lower than the stellar case. As we saw, this prevents cosmological nucleosynthesis from consuming all the nuclear fuel, thus allowing stars to burn it.[2]

Equation 6.35 is a quantitative statement that photons escape from stars by a diffusion-like process. The mean free path for photons is

$$\lambda = \frac{1}{\sum_i \sum_j N_i \sigma_{ij}}, \tag{6.47}$$

where i refers to the type of interacting object having number density N_i, and j refers to the type of interaction involved. It is convenient to define the opacity κ as

$$\kappa = N_A \sum_i \sum_j Y_i \sigma_{ij} = \frac{1}{\rho\lambda}, \tag{6.48}$$

[2]Requiring a standard model to have the cosmological value of S_γ implies a "stellar" mass of $10^{17} M_\odot$, which is a million times the mass of a large galaxy.

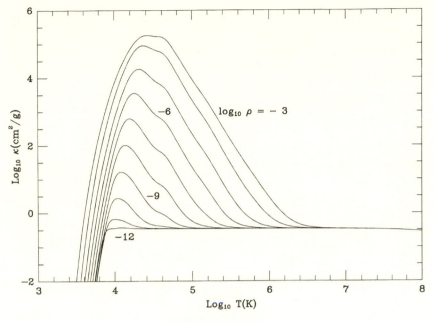

Fig. 6.1. Opacity

where N_A is Avogadro's number. One important process is Thomson scattering on electrons;[3] $\sigma_{th} = 0.67 \times 10^{-24}\, cm^2$, or

$$\kappa_{th} = 0.40\, Y_e\, cm^2\, g^{-1}. \tag{6.49}$$

Another important process is the photo-ejection of a bound electron into a continuum state; this gives rise to the *bound-free* opacity. Similarly, the interaction of a photon with an unbound electron in the field of a nearby ion causes transitions between continuum states and gives rise to the *free-free* opacity. Transitions between bound states (*bound-bound*) can also make an important contribution to the opacity. While the details of the opacity problem are notoriously complex, the qualitative features are fairly simple. Figure 6.1 gives the approximate opacity of a gas of Population I composition as a function of temperature; several representative densities are shown.

At higher temperatures and lower density the opacity approaches a constant value κ_{th}, where it is the number of free electrons which is relevant. As T and $1/\rho$ decrease, the opacity becomes proportional to $\rho/T^{3.5}$ (at least roughly); such an opacity is called a *Kramers* opacity. This

[3]See [531], p. 90.

effect is due to bound-free and free-free transitions. As the temperature decreases further, the bound-free transitions are reduced because fewer photons have enough energy to free an electron. The free-free transitions are reduced because of the decreasing number of free electrons due to recombination. Hence the precipitous drop in opacity near $T \approx 10^4$ K.

For temperatures well above 10^4 K, the opacity of a gas of Population I composition is given roughly by

$$\kappa \approx \kappa_{th} + \kappa_K, \tag{6.50}$$

where

$$\kappa_K \approx 1.2 \times 10^{24} \left(1 + X_H\right) \left(\frac{0.1}{\rho}\right)^k \frac{\rho}{T^{3.5}} \tag{6.51}$$

in $cm^2\,g^{-1}$, where X_H is the ^{1}H abundance by mass and the exponent k is

$$k = 0 \quad \text{for} \quad \rho < 0.1\,g/cm^3,$$

and

$$= 0.3 \quad \text{for} \quad \rho > 0.1\,g/cm^3. \tag{6.52}$$

While not adequate for accurate work, this crude approximation will be useful in the discussion below.

In order to evaluate the relative importance in stars of these two types of opacity we will need to know "typical" temperatures and densities. Let us take

$$\langle aT_d^4 \rangle = \frac{\int_0^R aT^4 4\pi r^2 dr}{\int_0^R 4\pi r^2 dr}, \tag{6.53}$$

which is a volume average of the radiation energy density (this is what diffuses); T_d will be our "typical" temperature insofar as radiative diffusion is concerned. In the finite difference approximation, $dT/dr \approx \Delta T/\Delta r \approx -T_c/R$, and $T_c \approx T_c(1 - r/R)$. The integration is elementary, giving $T_d \approx 0.41 T_c$. A numerical integration over an $n = 3$ polytrope gives

$$T_d \approx 0.322\, T_c. \tag{6.54}$$

The corresponding density is obtained from Eq. 6.47. Using 6.45,

$$\frac{\kappa_K}{\kappa_{th}} = \left(\frac{3.86}{m^2}\right) \left(\frac{Y}{\beta}\right)^3 \left(10^7 K/T\right)^{\frac{1}{2}} (0.1/\rho)^k. \tag{6.55}$$

As an example, consider the Sun. From table 6.1 it can be seen that for $m \approx 1$, we have $\beta \approx 1$. The observed radius of the Sun is $R = 6.965 \times 10^{10}$ cm, so that from 6.40 we have $\langle \rho \rangle = 1.404 \, \mathrm{g \, cm^{-3}}$. Using this radius, with $\beta = 1$ and $Y = 1.634$ for a composition of 27% helium and 2% heavier nuclei, gives $T_c = 1.20 \times 10^7$ K for the central temperature of the Sun. We have used Eq. 6.47 and the ratio of central-to-average density for an $n = 3$ polytrope.[4] At such a high temperature, ionization will be nearly complete, at least for hydrogen and helium. As a first guess, let us use these characteristic temperatures and densities in 6.55; the relatively weak dependence on T and ρ suggests this approximation. In particular, $T_d = 0.322 \times T_c = 3.85 \times 10^6$ K, and $\rho_d = 2.56 \, \mathrm{g \, cm^{-3}}$. Then,

$$\frac{\kappa_K}{\kappa_{th}} \approx \left(\frac{5.88}{m^2} \right) \left(\frac{Y}{\beta} \right)^3 , \tag{6.56}$$

$$= 25.7/m^2 ,$$

for Population I abundances and $\beta = 1$. This ratio is unity for $m \approx 5$, so that electron scattering is the dominant source of opacity for massive main sequence stars. In a realistic stellar model the ratio does not have the same value throughout the star; Schwarzschild [555] gives the run of opacity through main sequence stars of 10 and 2.5 solar masses; see also Clayton [173]. The values of κ_K/κ_{th} derived above should be thought of as "typical," i.e., as an average over the star.

The characteristic time for the energy of a photon to escape a star of radius R by radiative diffusion is roughly

$$\tau \approx \frac{3R^2}{c\lambda} . \tag{6.57}$$

For the Sun, we have $\tau \approx 6.6 \times 10^5$ years. The mean time interval between interactions is only $t \approx \lambda/c \approx 7.8 \times 10^{-13}$ s however! The typical number of interactions undergone before the energy of the photon eventually escapes the star is then roughly $\tau/t \approx 2.7 \times 10^{25}$. This reluctance to allow photons to escape is the property that causes stars to have such long lifetimes.

---

[4]Notice that an observed quantity, the radius, was required to close the set of equations. A different constraint, namely that nuclear burning releases as much energy as is radiated, will be introduced later. It will give a slightly higher central temperature and, consequently, underestimate the solar radius. This indicates that the Sun has a higher ratio of central-to-average density than the $n = 3$ polytrope.

Let us estimate the luminosity of stars. It should be roughly

$$L \approx \frac{\text{total energy in photons}}{\text{mean escape time}}$$

$$\approx \frac{4\pi R^3}{3} \frac{\langle a T_d^4 \rangle}{\tau} \tag{6.58}$$

where 6.53 has been used. By using equations 6.39, 6.41 and 6.57 we have

$$L \approx \left(\frac{4\pi c G(1 - \beta) M}{\kappa} \right) \left(C_n 3^{-5} \frac{\rho_c}{\langle \rho \rangle} \right), \tag{6.59}$$

where the second factor is 0.190 for a polytrope of $n = 3$. Having illustrated the conceptual basis for this expression we can derive a more accurate expression by an algebraic trick. We may write 6.35 as

$$L = -4\pi r^2 \left(\frac{c}{\kappa \rho} \right) \frac{dP_{\text{rad}}}{dr}. \tag{6.60}$$

If β does not vary through the star (the Standard Model), then as $P(1 - \beta) = P_{\text{rad}}$, we have

$$\frac{dP_{\text{rad}}}{dr} = -\frac{(1 - \beta)\rho G M(r)}{r^2}. \tag{6.61}$$

Eliminating the derivatives gives

$$L(r) = \frac{4\pi c G(1 - \beta) M(r)}{\kappa(r)} \tag{6.62}$$

at *any* point with radius r. If we take $M(r) = M(R)$, the stellar mass, and $\kappa(r)$ to be some appropriate average value, 6.62 allows us to estimate the total radiative luminosity of the star. Even though it was derived with the assumption that $d\beta/dr = 0$, Eq. 6.62 is often a useful approximation even if this is not strictly true.

For massive stars, Thomson scattering dominates the opacity, so 6.62 becomes

$$L = 1.25 \times 10^{38} \left(\frac{0.4}{\kappa} \right) m(1 - \beta) \, \text{erg/s}. \tag{6.63}$$

Because $1 - \beta \ll 1$ for all but the most massive stars, it is useful to eliminate $(1 - \beta)$ by use of equation 6.45, so

$$L = 3.72 \times 10^{35} \left(\frac{0.4}{\kappa} \right) m^3 \left(\frac{\beta}{Y} \right)^4 \, \text{erg/s}. \tag{6.64}$$

For lower masses ($m < 3$), $\beta \approx 1$ and Kramers opacity is dominant,

$$\frac{\kappa}{0.4} \approx Y_e \left(1 + \frac{5.88}{m^2}(Y/\beta)^3\right), \tag{6.65}$$

and this gives

$$L \approx 3.72 \times 10^{35} \mathrm{erg\,s^{-1}} m^5 (\beta/Y)^7 / \left(5.88 + m^2(\beta/Y)^3\right) Y_e. \tag{6.66}$$

We used T_d and ρ_d estimated above for the Sun to evaluate the opacity factors. Let us evaluate this for the Sun. We take $X(\mathrm{H}) = 0.71$ and $X(\mathrm{He}) = 0.27$. This gives $L \approx 2.29 \times 10^{33}$ erg/s. The present solar luminosity is 3.86×10^{33} erg/s. Schwarzschild [555] gives a zero-age solar model which has a luminosity of 2.2×10^{33} erg/s. Solar models by Iben [317] suggest an increase in brightness by 1.4 in 4.5 billion years, or an initial luminosity of about 2.75×10^{33} erg/s. Our estimate is accurate in an *ab initio* sense to about a factor of 1.5; the discrepancy is mostly due to our cavalier treatment of average opacity, rather than our choice of a particular polytropic structure. Notice the sensitivity to composition (Y^7 factor). See Bahcall [65] for a critical discussion of solar models.

We can now derive the mass-luminosity relation, a result of considerable significance for galactic evolution (see chapter 14). For very massive stars, $\beta \approx 0$ and Eq. 6.63 gives

$$\frac{d \ln L}{d \ln M} = 1. \tag{6.67}$$

For moderately massive stars, $\beta \approx 1$ but $\kappa \approx \kappa(th)$, so Eq. 6.64 gives

$$\frac{d \ln L}{d \ln M} = 3. \tag{6.68}$$

For low mass stars, equation 6.66 gives

$$\frac{d \ln L}{d \ln M} = 5, \tag{6.69}$$

where we have used our solar estimate to fix the weak temperature and density dependence.

6.3   NUCLEAR ENERGY

The evolution of stars is just the story of their nuclear transmutations and the consequent effects on the stellar structure. The thermonuclear evolution of stellar matter may be thought of as consisting of a sequence of stages, in which the ashes of one stage become the fuel of the next. To describe such a sequence the stellar evolutionist must know at least two things:

1. the rate of energy generation by the thermonuclear consumption of a given fuel, and
2. the composition of the ashes which will become the fuel for the next stage.

The evolutionary change in stellar composition is, mathematically speaking, an initial-value problem; in general, errors can be amplified with subsequent evolution. Consequently it is vital to represent accurately the earlier burning stages if we wish to explore the later ones.

Even now the vast bulk of stellar evolutionary work involves no stage beyond hydrogen and helium burning. There is only one primary product of hydrogen burning, ^{4}He. Clearly the next major burning stage after hydrogen burning just involves the consumption of ^{4}He. Since the energy generation rate for helium burning is fairly insensitive to the nature of the ashes formed, both hydrogen and helium burning demand only a rather crude treatment of nucleosynthesis, missing only some fairly subtle but conceptually important effects (for example, the production of ^{3}He and ^{14}N, the rearrangement of other CNO isotopes, and production of Na and Al). However, if for example, no ^{12}C is formed, then there is no carbon-burning stage at all! This is a qualitative as well as a quantitative difference. For the later stages of stellar evolution the question of the composition produced as well as that of the energy generated must be carefully considered.

Table 6.2 summarizes the primary thermonuclear burning stages in stars. Because a direct e–ν coupling in the weak interaction does exist, stages after helium burning are dominated by neutrino cooling rather than photon diffusion. Subsequent chapters will explore the detailed nature and sites of these processes.

There is another point that should be stressed. Massive stars ($M \geq 3M_\odot$) spend so little time in late burning stages (carbon burning and beyond) that the HR diagram is no longer so useful a test of evolutionary theory of these objects; this is due to poor statistics. However, these stars all have a pronounced characteristic: they burn nuclear fuel at a prodigious rate. Their thermonuclear ashes may reveal their history. In order to even attempt to read this history, and in a sense replace the HR diagram with an abundance table as our observational constraint, we must calculate abundances correctly.

The solution of the coupled nonlinear differential equations, which govern abundances of nuclei undergoing thermonuclear reactions, was

TABLE 6.2

Thermonuclear Burning Stages

Fuel	$T/10^9(K)$	Ashes	q(erg/g fuel)	Cooling
^{1}H	0.02	^{4}He, ^{14}N	$(5 \sim 8) \times 10^{18}$	photons
^{4}He	0.2	^{12}C, ^{16}O, ^{22}Ne	$7 \cdot 10^{17}$	photons
^{12}C	0.8	^{20}Ne, ^{24}Mg, ^{16}O	$5 \cdot 10^{17}$	neutrinos
		^{23}Na, ^{25,26}Mg		neutrinos
^{20}Ne	1.5	^{16}O, ^{24}Mg, ^{28}Si, ...	$1.1 \cdot 10^{17}$	neutrinos
^{16}O	2	^{28}Si, ^{32}S, $\cdots$	$5 \cdot 10^{17}$	neutrinos
^{28}Si	3.5	^{56}Ni, $A \approx 56$ nuclei	$(0 \sim 3) \cdot 10^{17}$	neutrinos
^{56}Ni	$6 \sim 10$	n, ^{4}He, ^{1}H	$-8 \cdot 10^{18}$	neutrinos
$A \approx 56$ nuclei	(depends on ρ)	photodisintegration and neutronization		

examined in chapter 4. As a star evolves to higher temperature and density, an increase in the number of possible reactions requires more complex reaction networks. In principle all nuclei should be included, but in practice the size of the network can be determined by an accuracy criterion (such as all nuclei having abundances greater than ε are to be calculated to an accuracy of δ, where ε and δ are some chosen numbers). Clearly the accuracy needed depends upon the use to be made of the results. Considerable computational economy can be obtained by judicious choice of the reaction network to be used. Consider a set containing all networks giving an error of size δ or less for any species having an abundance ε or greater. Any member of this set will be called an "equivalent" network to any other member of the set. For efficiency we wish to find the minimum equivalent network, that is the one with the fewest reactions and nuclear species, and thus the minimum expenditure of computational resources. It should be noted that the question of accuracy will imply in practice the calibration of a smaller network by a larger, more general one. Guided by these ideas we may construct the simplest acceptable network for subsequent burning stages. However, the "acceptable" is a time-dependent quantity; as we learn more we will require better treatments of the physics.

In the previous section, we used Eddington's Standard Model to obtain an approximation to the luminosity of a star; we considered the *loss* of energy due to radiative diffusion. For a star to be in thermal balance, this loss must be countered by a gain in energy. Consider a nuclear energy generation rate that may be approximated by

$$\epsilon \approx \epsilon_c \left(\frac{\rho}{\rho_c}\right)^{u-1} \left(\frac{T}{T_c}\right)^s, \qquad (6.70)$$

where the subscript c denotes central values. In Appendix C.3 it is shown how to approximate the average energy generation rate,

$$\langle \epsilon \rangle = \frac{1}{M} \int_0^M \epsilon \, dm, \qquad (6.71)$$

in such a case. If we equate the luminosity per unit mass to this average energy generation rate, we impose the condition of *global thermal balance*, so that

$$L = \langle \epsilon \rangle M. \qquad (6.72)$$

This result, taken with the standard model, gives a powerful approximate tool for examining stellar evolution, and one that will be used extensively in subsequent discussion.

For the Sun, $L_\odot \approx 3.86 \times 10^{33}$ ergs/s and $M_\odot \approx 2 \times 10^{33}$ g, so that $\langle \epsilon \rangle \approx$ 2 erg/g s . It is interesting to compare this to the corresponding value for a familiar power-generating entity, the human body. If we estimate a consumption of 10^3 Calories/day, this corresponds to 10^6 calories—or 4.2×10^6 joules—per day. Recall that 1 joule $= 10^7$ ergs. For a body of 50 kilograms (110 pounds), this gives a power per unit mass of 0.8×10^4 ergs/g s, which is actually larger than for the Sun. We radiate at a lower temperature because we have more surface area per unit mass. The densities are comparable, with $\langle \rho \rangle \approx 1.5$ g cm^{-3} for the Sun and slightly less than 1 for those of us who can float in fresh water. The radiated energy scales as T^4 while the ratio of volume (hence mass) to surface area scales as the linear size. So,

$$(T_{body}/T_\odot)^4 = (R_{body}/R_\odot)(\langle \epsilon \rangle_{body}/\langle \epsilon \rangle_\odot), \qquad (6.73)$$

so that if we use the effective temperature of the Sun to be $T_e \approx 6 \times 10^3$ K, we find $T_{body} \approx 300$ K, or about 80 degrees Fahrenheit, which seems embarrassingly reasonable, if not medically exact. This estimate could be improved because (1) the evaporative cooling of the body from the lungs and skin was ignored, (2) the question of conductive and convective cooling from the skin has not been considered, and (3) the human body has a more interesting shape, and consequently more surface area per unit volume, than a sphere. As we saw in §6.2, $L/L\odot \approx (M/M_\odot)^n$, where n ranges from 3 to 5, so $(L/M)/(L/M)_\odot \approx (M/M_\odot)^3$. The stars that have the same power production per unit mass as the human body are more massive than the Sun, $M \approx 20M_\odot$, which is similar to the initial mass of the brightest blue star (Rigel) in Orion, or the progenitor of SN 1987A.

6.4 NEUTRINO PROCESSES

A fact of fundamental importance for astrophysics is the existence of a direct e-ν coupling in the weak interaction. This implies that, for example, an electron-positron pair can also decay by the weak interaction

$$e^+ + e^- \rightarrow \nu + \bar{\nu}. \tag{6.74}$$

This process is rarer by a factor of about 10^{-20} than the electromagnetic channel

$$e^+ + e^- \rightarrow 2\gamma. \tag{6.75}$$

However, for an evolved star of, say, $M = 5M_\odot$ and $R = 10^{10}$ cm, a photon would take a time of the order of

$$\tau_\gamma \approx \frac{R^2}{\lambda c} \approx 5 \cdot 10^6 \, \text{y} \tag{6.76}$$

to diffuse out to the surface (if Thomson scattering of photons by free electrons is dominant), while only a single 1 MeV neutrino in several million would scatter even once (neutrino scattering by electrons). Thus the weakly interacting neutrinos, once formed, escape more easily than photons.

As a star evolves it contracts and heats. The photon diffusion time

$$\tau_\gamma \approx R^2 \frac{N\sigma}{c} \approx \rho^{\frac{1}{3}} M^{\frac{2}{3}} \tag{6.77}$$

increases with increasing density; the neutrino emission rate increases with temperature.

Unlike radiative diffusion of photons, or convection, neutrino radiation is a *local* process: once a neutrino is formed it has a high probability of escaping the star before interaction.[5] The transition from photon to

---

[5] The exception occurs during core collapse, during which the extreme temperature and density enhance the probability of interaction to the extent that neutrinos are *trapped*, and diffuse away on a time-scale of seconds rather than milliseconds. SN 1987A provided a dramatic confirmation of this idea. See chapter 12.

neutrino cooling occurs as the temperature increases past $T \approx 5 \cdot 10^8$ K (i.e., after helium burning and prior to carbon burning). It is now possible to produce intense beams of neutrinos (especially muon neutrinos). A common method is to let a high-energy, monoenergetic beam of pions (or kaons) decay to muons ($\pi^+ \to \mu^+ \nu_\mu$, for example), and then absorb the muons by a thick shield, letting only the neutrinos through. Given the success of the electroweak theory in explaining the experimental results, its predictions for stellar cooling may be considered quite reliable. Given the great difficulty in calculating radiative (electromagnetic) opacities, the neutrino cooling may be known to better accuracy than that for photons.

In quantum electrodynamics, an electromagnetic disturbance may be represented by virtual electron-positron pairs. Here, virtual means that they exist only for a time limited by the uncertainty relation $\Delta t \leq \hbar / \Delta E$, with $\Delta E < 2m_e c^2$. Virtual electron-positron pairs may also decay to produce neutrino-antineutrino pairs. As a consequence a host of electromagnetic processes can have a weak analog. Some important ones are plasmon decay (a plasma excitation decays into a neutrino-antineutrino pair), the photoneutrino process (a neutrino-antineutrino pair replaces the scattered photon in a photon-electron interaction), and neutrino-nuclear bremsstrahlung (the photons of the braking radiation are replaced by neutrino-antineutrino pairs). Pontecorvo [500] first suggested that a direct $e-\nu$ coupling would imply efficient cooling processes for late stages of stellar evolution. Such an interaction was implied by the conserved-vector-current (CVC) theory of weak interactions proposed by Feynman and Gell-Mann [233] and by Marshak and Sudarshan [411]. Neutrino cooling soon became a key part of the picture of the late stages of stellar evolution [238, 528, 170, 171].

In 1967 Beaudet, Petrosian, and Salpeter [86] used the CVC theory to numerically evaluate the composition-independent neutrino emission rates; they gave analytic fits for the rate of radiation of energy by these processes. Also using the CVC theory, Festa and Ruderman [232] examined neutrino-pair bremsstrahlung in a plasma consisting of degenerate electrons and nondegenerate ions; this process is important at high density and low temperature. More modern estimates of the rates differ little from these. Dicus [210] showed that the theory of leptons of Weinberg [675], Salam [534], and Glashow [266] gives cooling processes similar to those predicted by CVC, with a small uncertainty related to the precise value of the mass of the charged vector mesons ($M_W \approx 80$ GeV) which mediate the interaction.

Figure 6.2 displays the rate of energy loss due to neutrino emission per gram ϵ_ν. For compositions typical of conditions in which neutrino

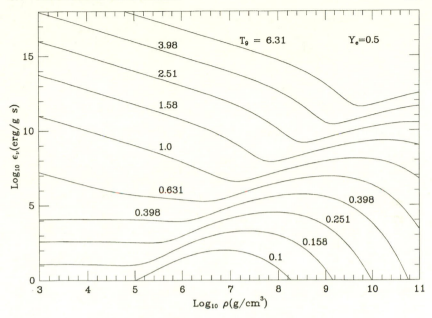

Fig. 6.2. Neutrino Cooling Rates

cooling is important, the number of electrons per nucleon is close to one-half, so $Y_e = 0.5$ was used in figure 6.2. The emissivity ϵ_ν is plotted versus $\log_{10} \rho$, where ρ is the usual nucleon mass density. The curves are parameterized by temperature T, and labeled by $\log_{10} T(K)$. The solid lines refer to the sum of the pair, plasmon and photoneutrino rates. At low temperature and density, the photoneutrino rates dominate, and the cooling per unit mass is almost independent of density. For higher temperatures, the electron-positron annihilation rate dominates, and it is suppressed with increasing density. At high density and lower temperature, electron degeneracy inhibits pair formation, so that the plasmon rate dominates. At still higher density it too is suppressed, causing the curves to turn down.

When does neutrino cooling become important in stars? Consider the rate of energy loss at low density and temperature. The rate of energy loss shown in figure 6.2 is approximately

$$\epsilon_\nu \approx 1.1 \times 10^7 T_8^8 \, \text{erg g}^{-1} \text{s}^{-1}, \qquad (6.78)$$

for $T_8 \leq 6$ and $\rho \leq 3 \times 10^5 \, \text{g cm}^{-3}$. Thus, $< \epsilon_\nu > /(\epsilon_\nu)_c \approx 0.1$. Under what conditions does the average cooling by neutrinos $< \epsilon_\nu >$ equal that from photons, L/M? For an opacity appropriate for matter processed past hydrogen burning, $\kappa \approx 0.2$, we find $L/M \approx 3.1 \times 10^4 S_\gamma / \Re$ erg g^{-1} s^{-1}. The cooling rates are equal at $T_8 = 0.62$ for $S_\gamma / \Re = 1$, and $T_8 = 0.46$ for $S_\gamma / \Re = 0.1$. As we shall see in later chapters, these values of radiation entropy approximate the evolutionary conditions for the central regions of stars of interesting masses. Roughly speaking, stars cool by photon emission so long as their core temperatures are less than $T = 5 \times 10^8$ K, above which neutrino cooling dominates.

At the high densities encountered in the late states of stellar evolution, electron capture on nuclei becomes important. In the *Urca process* of Gamow and Schönberg [259], a nucleus alternatively captures an electron and undergoes a beta-decay, meanwhile emitting a neutrino and an antineutrino. Thus a cyclic (but nonreversible) process occurs. In *neutronization*, increasing density induces electron capture and causes a diminution in the number of electrons per nucleon present in the plasma. In general this is a noncyclic process. In both cases the nature of the process depends upon previous evolution. For the Urca process, the abundance of Urca-active nuclei is of vital importance and depends upon previous thermonuclear processing which destroys or produces such nuclei. Neutronization might be less sensitive to previous evolution than the Urca process, but it too depends upon the composition, as well as the size and nature, of the stellar core.

As a star evolves, the density increases and so does the mean energy of electrons. If in the laboratory we observe a beta-decay process

$$(Z, A) \rightarrow (Z + 1, A) + e^- + \bar{\nu}, \qquad (6.79)$$

then in the star the energy of electrons can rise so as to force

$$(Z + 1, A) + e^- \rightarrow (Z, A) + \nu \qquad (6.80)$$

to occur. A particularly interesting case is that of electron degeneracy. All electron states are filled below the Fermi energy ϵ_F. If the end-point energy for beta-decay ϵ_0 is less than ϵ_F, then the decay cannot occur; in such an environment (Z, A) becomes the "stable" nucleus (see §3.7). The threshold occurs for $\epsilon_0 = \epsilon_F$: this specifies a density above which the electron capture occurs.

In table 6.3 the thresholds for electron capture are given for nuclei that are prominent initially or as the result of nuclear burning in stars. The threshold density is given as $2Y_e\rho$; the case of interest is usually $Y_e \approx 0.5$. Note that the primary products of hydrogen and helium burning have high thresholds; they are far more resistant to electron capture than the products of carbon, oxygen, and silicon burning.

TABLE 6.3

Threshold Density for Electron Capture

Nuc.	ϵ_0 (MeV)	$2Y_e\rho$ (g/cm^3)	Nuc.	ϵ_0 (MeV)	$2Y_e\rho$ (g/cm^3)
^{1}H	0.782	2.44×10^7	^{28}Si	4.643	1.97×10^9
^{3}He	0.0186	3.94×10^4	^{29}Si	3.681	1.05×10^9
^{4}He	20.6	1.37×10^{11}	^{30}Si	8.539	1.08×10^{10}
^{12}C	13.37	3.89×10^{10}	^{31}P	1.491	1.06×10^8
^{13}C	13.44	3.95×10^{10}	^{32}S	1.710	1.47×10^8
^{14}N	0.156	1.15×10^6	^{33}S	0.249	2.60×10^6
^{15}N	9.772	1.58×10^{10}	^{34}S	5.38	2.95×10^9
^{16}O	10.42	1.90×10^{10}	^{35}Cl	4.854	2.22×10^9
^{17}O	8.480	1.06×10^{10}	^{36}A	0.7096	1.99×10^7
^{18}O	14.06	4.51×10^{10}	^{37}Cl	0.1675	1.30×10^6
^{19}F	4.819	2.18×10^9	^{38}A	4.917	2.30×10^9
^{20}Ne	7.026	6.20×10^9	^{39}K	0.565	1.24×10^7
^{21}Ne	5.686	3.44×10^9	^{40}Ca	1.312	7.85×10^7
^{22}Ne	10.85	2.13×10^{10}	^{41}K	2.492	3.78×10^8
^{23}Na	4.374	1.67×10^9	^{42}Ca	3.521	9.34×10^8
^{24}Mg	5.513	3.16×10^9	^{44}Ca	5.659	3.39×10^9
^{25}Mg	3.833	1.17×10^9	^{48}Ti	3.990	1.30×10^9
^{26}Mg	9.325	1.38×10^{10}	^{52}Cr	3.976	1.29×10^9
^{27}Al	2.609	4.25×10^8	^{56}Fe	3.695	1.06×10^9

6.5 STELLAR ENERGY

The internal energy of a star is:

$$U = \int_0^M E \, dm, \tag{6.81}$$

where E is the internal energy per unit mass. In falling in upon themselves to form, stars generate energy that is shared between heating their own matter and losing radiation to infinity. The gravitational potential energy is just the total work done by the gravitational field as each element of mass is added to the star:

$$\Omega = \int_0^M \left(\frac{GM'}{r}\right) dM'. \tag{6.82}$$

The gravitational binding energy B (that is, the negative of the total energy) is

$$B = \Omega - U. \tag{6.83}$$

With this sign convention Ω is intrinsically positive, hence the explicit sign difference between U and Ω. As shown in Appendix C.2, if hydrostatic equilibrium prevails,

$$\Omega = \int_0^M \frac{3P}{\rho} \, dm, \tag{6.84}$$

where we assume the pressure goes to zero at the surface. For a given form for the run of density and pressure versus radius (a *polytrope* of index n), Ω assumes a simpler form,

$$\Omega = \left(\frac{3}{5-n}\right)\left(\frac{GM^2}{R}\right). \tag{6.85}$$

If $E = PV/(\gamma - 1)$ where γ is a constant, then

$$\Omega = 3(\gamma - 1)U, \tag{6.86}$$

and

$$B = 3\left(\gamma - \frac{4}{3}\right)\left(\frac{GM^2/R}{(5-n)(\gamma-1)}\right). \tag{6.87}$$

Ignore energies of ionization, excitation, and nuclear binding for the moment, so that the total energy of the star is just $-B$. The stability of the star depends upon the sign of the factor $\gamma - \frac{4}{3}$. Consider an adiabatic contraction (so γ is really constant) of magnitude δR. The change in gravitational binding energy is

$$\delta B = \frac{\partial B}{\partial R}\delta R = -B\delta R/R. \tag{6.88}$$

For $\gamma = \frac{4}{3}$, the gravitation binding energy B (and therefore the total energy) is unchanged, regardless of the value of δR. This is a condition of marginal stability.

For $\gamma > \frac{4}{3}$, contraction requires the hydrostatic star to have increased binding energy, or lower total energy. The star must await loss of this excess energy, by radiation for example, to accomplish this hypothetical contraction. However, for $\gamma < \frac{4}{3}$, contraction requires a smaller binding energy, or a larger total energy. The deficit can be made up by *kinetic* energy. Further contraction gives more kinetic energy, and a collapse ensues. This is a dynamic instability, occurring on a gravitational time-scale (note the gravitational acceleration factor GM/R^2 in B). See Zel'dovich and Novikov [708], and Shapiro and Teukolsky [567] for further discussion of the energy method of evaluating dynamical stability, and Cox [200] for a detailed examination of pulsational stability.

For a mixture of ideal gas and radiation (Chandrasekhar [162], ch. 7), Eq. 6.87 reduces to

$$B = \left(\frac{\beta}{2} \right) \Omega, \qquad (6.89)$$

a surprisingly simple result.

Consider some special cases. For large mass, $M \gg 100 M_\odot$, $\beta \to 0$, and the gravitational binding energy B goes to zero! Such objects are loosely bound, despite their large mass. Because their restoring forces are small, they are sensitive to distorting forces—such as rotation.

For an extremely relativistic, degenerate electron gas,

$$3PV - E = N_A Y_e m_e c^2, \qquad (6.90)$$

so that the gravitational binding energy of the star approaches the rest mass energy of all the electrons supporting it by their pressure (this is easily derived from the formulae in Appendix B). This corresponds to about 0.25 MeV per *nucleon*, and is less than many of the Q-values for the relevant thermonuclear reactions (see §6.3).

Finally, for an ideal gas with negligible radiation pressure, $\gamma = 5/3$ and $\beta = 1$, so that

$$B = \Omega/2, \qquad (6.91)$$

which is the classical result of the virial theorem. For such a star (the Sun is an example), this implies that the energy released by slow gravitational contraction is equally divided between heating the interior of the star and supplying its radiation to space.

Stars are mostly composed of nuclear fuels; the *nuclear* energy of a star may be defined as

$$E_{nuclear} = \int_0^\infty \left(\sum_i (M_i - M_u A_i) c^2 Y_i \right) N_A dm. \tag{6.92}$$

Here we measure the mass energy relative to M_u, the atomic mass unit (see §3.1). The energy available to a given nuclear process may be written as

$$E_{available} = \sum_i M_i q_i, \tag{6.93}$$

where M_i is the mass in the form of species i, and q_i is the energy per unit mass which is available upon consumption of this fuel. From table 6.2, we have $q_H \approx 7 \times 10^{18}$ erg g^{-1}, so that burning a fraction f of the hydrogen in a star of m solar masses, will release 1.4×10^{51} $m(f/0.1)$ ergs. For comparison, the total rest mass energy of a star of m solar masses is 1.8×10^{54} m ergs. Hydrogen burning is not a very explosive process because its rate is limited by beta-decays; helium, carbon, and oxygen burning are prime candidates for powering explosions, although the specific energy release q is less than a tenth that for hydrogen.

Most of the matter in stars is ionized. For a star of uniform Population I composition (i.e., solar), the ionization energy is dominated by H and He because of their large abundance. The ionization energy would be

$$E_{ion} \approx M N_A (X_H I_H + X_{He} I_{He}). \tag{6.94}$$

For H, He, and He$^+$ the ionization potentials are 13.598, 24.587, and 54.42 eV, respectively. This gives about 15 eV per nucleon, or 1.5×10^{13} erg/g for solar abundances, so that

$$E_{ion} \approx 3.0 \times 10^{46} m \text{ ergs.} \tag{6.95}$$

How much contraction must a star undergo in order to release enough gravitational binding energy to ionize itself? Taking $n = 3$ and $\gamma = 5/3$ as typical values, using $B = E_{ion}$, and Eq. 6.87 gives

$$R = GM^2/2E_{ion}, \tag{6.96}$$

$$= 0.4 \times 10^{13} \ m \text{ cm.} \tag{6.97}$$

This is similar to the radii of red giant stars; in hydrostatic equilibrium, they are on the verge of recombining. With recombination comes transparency (see §6.2), so that this is essentially a limit on the radius. The precise value will depend upon the exact structure of the star. Perhaps the fact that the red giants with the largest radii tend to irregular variability in their light output, and to active mass loss, is related to this.

Time-scales may be constructed from a stellar energy E and a power— or more precisely a luminosity L—so that

$$\tau = E/L. \tag{6.98}$$

Estimates for luminosity L were made in §6.2. For a massive star of $m = 20$ and solar abundances, $Y = 1.63$ and $\beta = 0.95$. From 6.64 we have $L = 3.9 \times 10^{34} m^3$ erg/s or 3.2×10^{38} erg/s . If gravitational binding is the energy source, we get the *Helmholtz-Kelvin* time-scale

$$\tau_{H-K} \approx 0.5 \times 10^5 (R_\odot/R) \text{ y}. \tag{6.99}$$

This is a measure of how long it takes such a star to settle down, ignite hydrogen, and start its life on the main sequence.

The main sequence lifetime itself may be estimated by using the nuclear release from hydrogen to helium conversion, which gives

$$\tau_{H-burning} \approx 1.0 \times 10^7 \text{ y}. \tag{6.100}$$

We have assumed here that such a star will consume the fuel in about 30 percent of its mass (and $X_H = 0.71$) during this stage (see chapter 7).

For a star of nearly solar mass, $m \approx 1$ and $\beta \approx 1$, so from Eq. 6.66 we have

$$\tau_{H-K} \approx 2 \times 10^7 m^{-3} (R_\odot/R) \text{ years} \tag{6.101}$$

as the Helmholtz-Kelvin time-scale, and

$$\tau_{H-burning} \approx 1.3 \times 10^{10} m^{-4} \text{ years} \tag{6.102}$$

as the main sequence lifetime, assuming that the ashes comprise the inner 10 percent of the stellar mass, and that such stars brighten during their main sequence life (we have taken the average luminosity to be twice the zero-age main sequence value; see also chapter 14).

In the limit of very large mass, the lifetime approaches a constant value (because the mass dependence in the luminosity, Eq. 6.63, cancels that in the fuel supply),

$$\tau_{H-burning} \rightarrow 1.8 \times 10^6 (1 + X_H) X_H \text{ years.} \tag{6.103}$$

In this limit, the star is fully convective and β is small (radiation pressure dominates).

6.6 IGNITION MASSES

Which stars can burn? A useful concept for understanding the advanced evolution of massive stars is the *ignition mass*. Consider a hydrostatic object of mass M and radius R.

$$\frac{1}{\rho} \frac{dP}{dr} = -\frac{GM}{r^2}. \tag{6.104}$$

As shown in §6.2 and Appendix C we have

$$\frac{P_c}{\rho_c} = D_n GM^{2/3} \rho_c^{1/3} \tag{6.105}$$

where the subscript c refers to center, and D_n is a form factor which is of order unity. For advanced evolution, the $n = 3$ polytrope is a useful approximation; then $D_3 = 0.364$.

The pressure P depends upon the thermodynamic state of the gas, and in general is a function of temperature, density, and composition:

$$P = P(\rho, T, Y_i). \tag{6.106}$$

Qualitatively the equation of state of a partially degenerate, relativistic electron gas and its associated ions and radiation field can be represented as

$$\frac{P}{\rho} \approx \Re Y_e T + K_\gamma Y_e^\gamma \rho^{\gamma-1}, \tag{6.107}$$

at least for this discussion. For serious calculations one must come to grips with the nasty details.[6] Thus we have a relation between temperature and density, for hydrostatic objects of an assumed structure, parameterized by the mass M:

$$\Re Y_e T = DGM^{2/3} \rho^{1/3} - K_\gamma Y_e^\gamma \rho^{\gamma-1}. \tag{6.108}$$

---

[6]For example, the use of Y_e instead of $Y_e + \sum_j Y_j$, where the summation is over all the nuclei, gives a more accurate result because it compensates for an overestimate of the pressure in the partially degenerate region. This subtlety is significant for hydrogen ignition.

For $\gamma > \frac{4}{3}$ (stability!) and "low" ρ we have

$$T \propto \rho^{1/3} M^{2/3}. \tag{6.109}$$

This behavior is shown in figure 6.3 by the diagonal, straight solid lines at the upper part of the graph. Increasing density results in increased temperature; all fuels can eventually ignite.

For high density the electron gas becomes relativistic, and $\gamma \to 4/3$. Then, for $T = 0$, the density dependence factors out and we have

$$M_{\text{Ch}} = \left[\frac{K_{4/3}}{DG}\right]^{3/2} Y_e^2 \approx 5.85 Y_e^2 M_{\odot}, \tag{6.110}$$

which is the Chandrasekhar mass. The figure was constructed from an accurate equation of state for electrons (and positrons), and an ideal gas of ^{16}O ions. Coulomb effects and general relativistic corrections to gravity were ignored; these small effects may be easily included as correction factors (see [567], §2.4 and §6.10). The curves are labeled with the ratio of the actual mass to the critical mass (M/M_{cr}). Here $M_{cr} \approx 1.477 M_{\odot}$, slightly larger than the Chandrasekhar mass because of the finite temperature. This critical value separates qualitatively different behavior. The dotted line represents the value

$$kT = m_e c^2 / \sqrt{2}\pi \approx 1.335 \times 10^9 \text{ K}$$

to which the temperature tends as the mass approaches M_{cr}.

For $M < M_{cr}$,

$$K_\gamma Y_e^\gamma \rho^{\gamma-1} > DG M^{2/3} \rho^{1/3} \tag{6.111}$$

for sufficiently high density, that is, as $T \to 0$. Near this density both these terms are large compared to $\Re Y_e T$. Thus

$$\rho \approx \left(\frac{DG M^{2/3}}{K_\gamma Y_e^\gamma}\right)^{\frac{1}{\gamma-4/3}}. \tag{6.112}$$

This behavior is shown by the vertical lines in figure 6.3. For $M > M_{cr}$, there is no maximum in the $T(\rho)$ curve for finite ρ; increasing ρ always gives increasing T.

Imagine the center of a star of mass M, evolving to a state of higher density and gravitational binding energy. It would trace out a curve like those in figure 6.3. Suppose it contains a nuclear fuel that must be heated to a temperature T_{ign} for ignition. For $M < M_{Ch}$ the object evolves up to a maximum temperature T_{max}. If $T_{max} < T_{ign}$ the object will not ignite the fuel, but cool down to a degenerate state. For $T_{max} \geq T_{ign}$ the fuel will ignite. Because T_{max} increases with increasing mass M, there will be a minimum mass for which the ignition of any given fuel will occur (in this simple framework). This will be called the *ignition mass*.

A crude estimate for the maximum temperature may be found simply by differentiating Eq. 6.108 to construct $dT/d\rho$ and set it to zero. This gives

$$DGM^{2/3}\rho^{1/3} = 3(\gamma - 1)K_{\gamma}Y_e^{\gamma}\rho^{\gamma-1}, \tag{6.113}$$

which, when inserted in Eq. 6.108, defines the maximum temperature. The variation of γ with density makes this awkward for practical use, for which numerical results are preferable. Table 6.4 gives T_{max} and the corresponding density for various masses (M/M_{Ch}). The values are in part from the elegant analysis of Takarada, Sato, and Hayashi [603].

The ignition temperature T_{ign} can be estimated by equating the rate of nuclear energy release with the relevant cooling rate. These values will usually be slightly below the characteristic burning given in table 6.2.

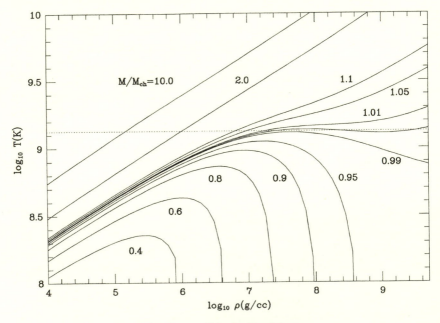

Fig. 6.3. $\rho - T$ for Ignition Masses

TABLE 6.4

Ignition Masses: Maximum Temperatures

M/M_{Ch}	$T_{max}/10^9 K$	$Y_e\rho$ gcm^{-3}	M/M_{Ch}	$T_{max}/10^9 K$	$Y_e\rho$ gcm^{-3}
0.986	3.43	1.13×10^9	0.794	0.929	1.07×10^7
0.983	3.06	8.08×10^8	0.631	0.586	2.77×10^6
0.966	2.17	2.84×10^8	0.501	0.399	1.12×10^6
0.931	1.53	9.66×10^7	0.316	0.195	2.88×10^5
0.891	1.25	3.65×10^7	0.1	0.0392	1.79×10^4

Thus, if $^{12}C + {}^{12}C$ ignites at, say, $T_{ign} \approx 7 \cdot 10^8$ K, then the ignition mass is $M_{ign} \approx 1.0 M_\odot$.

An important result is that to ignite $^{16}O + {}^{16}O$ ($T_{ign} \approx 1.7 \times 10^9$ K) requires a mass $M_{ign} \approx 1.39 M_\odot \approx 0.95 M_{Ch}$. An oxygen core can release more than 5×10^{17} ergs per gram of fuel consumed, which is larger than the gravitational binding energy. Even in the limiting case, the white dwarf gravitational binding per nucleon is $Y_e m_e c^2 \approx 0.26$MeV, or $\approx 2.6 \times 10^{17}$ ergs/g. If oxygen is not consumed hydrostatically, the core has more than enough nuclear energy to blow itself apart.

During hydrostatic oxygen burning, electron capture easily occurs on ^{33}S and ^{35}Cl because of their exceptionally low thresholds. As the capture proceeds, these nuclei become more abundant as Y_e drops through 0.485 because they have $Z/A = 16/33 = 0.4849$ and $Z/A = 17/35 = 0.4857$ respectively. Further electron capture drives the abundances in favor of more neutron-rich nuclei, which have higher electron capture thresholds. This shuts off the process at $Y_e \approx 0.48$, which is relevant for the ignition of silicon. For that process to occur, a temperature of $T_{ign} \approx 3.4 \times 10^9$ K is required. Thus we have

$$M_{ign} \approx 0.986 \times 1.46 M_\odot (2 \times 0.48)^2 = 1.33 M_\odot. \qquad (6.114)$$

Notice that this is *below* the ignition mass for oxygen burning, so that it would seem that once oxygen burning begins, silicon burning will be difficult to avoid.

For more precise values, other effects should be included. Differing thermal histories can modify the ignition conditions, for example, and the effects of coulomb interactions on the equation of state would lower the values of mass given above by a few percent, being larger for nuclei with larger Z; see §56 in Landau and Lifshitz [381].

6.7 FINAL STATES

A landmark achievement in astrophysics was the realization by Chandra-sekhar [161] that while sufficiently small mass stars could burn out to cold cinders, larger mass stars must contract to a more exotic and inter-esting end. With the discovery of the neutron arrived the possibility of a denser state for stars. Because of their larger rest mass, neutrons require greater density to suffer the softening effects of near-light speeds on their resistance to compression. While white dwarf stars are supported by the degeneracy pressure of electrons, neutron stars are supported by the degeneracy pressure of neutrons. Oppenheimer, with Volkoff [470] and with Snyder [469], gave the first quantitative models for both the neutron star, and the end result of still more massive stars—the black hole. Figure 6.4 shows the gravitational mass of white dwarfs and neu-tron stars, versus central density (rest mass only). The gravitational mass is what would be measured by binary motion of the neutron star and a companion, for example. The white dwarf sequence extends from low density up to $\rho \approx 10^{11} \, \mathrm{g \, cm^{-3}}$, at which point the electron fermi energy exceeds the threshold for electron capture on the most recalcitrant nu-clei (see §6.4). This point of instability is denoted by a filled square.

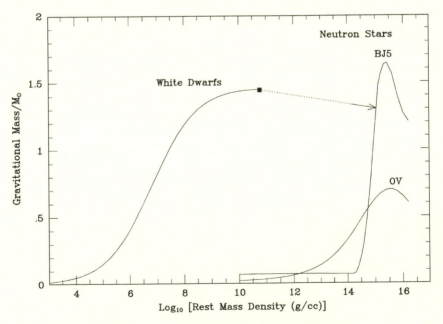

Fig. 6.4. White Dwarf and Neutron Star Masses

Addition of even a slight amount of mass would result in contraction to higher density, which would drive electron capture. This would cause the pressure to increase less rapidly upon contraction than gravity, destroying hydrostatic balance. A dynamic collapse would result; this is shown schematically by the dotted line and arrow. The exact trajectory would depend upon details of the dynamics.

Two curves are shown for neutron stars. The curve labeled OV represents the Oppenheimer and Volkoff models, in which the nucleons are represented as ideal, noninteracting fermions at zero temperature. At first sight, it is surprising that this curve has a maximum at $M_g \approx 0.7 M_\odot$, which is far smaller than the white dwarf limit at $M_g \approx 1.4 M_\odot$. White dwarfs have one degenerate fermion (the electron) per two nucleons, while a neutron star has a degenerate fermion for every nucleon. This would give *more* pressure per unit of gravitating mass. In fact, in Newtonian theory, the maximum mass for a star of ideal neutrons is a factor of four larger. However, as the nucleons become more relativistic with higher density, their energy becomes significantly more than just that due to their rest mass. Gravity is no longer approximately Newtonian, but is even stronger. In general relativity, this correction reduces the limiting mass from 5.8 to $0.7 M_\odot$.

Suppose the Oppenheimer-Volkoff limit were the relevant one (it is not). Adding a slight amount of matter to a white dwarf at the limiting mass—*accretion-induced collapse*—would result in the formation of a black hole, because, as we shall see, it is difficult to use efficiently the binding energy of the neutron star for ejection of mass.

However, nucleons are strongly interacting particles, so that the ideal, noninteracting fermion equation of state is a poor approximation. In particular, the nucleon-nucleon interaction is zero for large separations, becomes attractive at distances comparable to the interparticle spacing in a nucleus (see chapter 3), and repulsive for closer distances. Therefore, the real equation of state will be dominated by nuclei and electrons at the lowest densities, be softer below nuclear density, and stiffer above nuclear density. That is the behavior of the curve denoted BJ5, for the "Bethe-Johnson V" equation of state [95], which is representative of modern calculations that include these effects. More detail may be found in Shapiro and Teukolsky [567]. With such equations of state, there exist neutron star configurations that are accessible from accretion-induced collapse.

As we shall see later, a stellar core cooled by neutrino emission approaches a degenerate state, and is somewhat similar to a white dwarf, but with a finite temperature. The mass curve for white dwarfs depends upon the composition, and upon the actual temperature (see the discussion of ignition masses in §6.6). Thus for stellar cores, a variety of masses might begin collapse. The maximum mass of neutron stars for BJ5 is $M_g \approx 1.64 M_\odot$: this requires the collapse of $M_{amu} = 1.84 M_\odot$ of nucleons. The difference in mass is the gravitational binding energy (about 3.5×10^{53} ergs), which is radiated away. There seems to be no obvious reason why some stellar cores should not exceed this maximum mass, M_{amu}, and collapse directly to form a black hole.

By gathering some of the results of this chapter, some of the main features of stellar nucleosynthesis are revealed. These themes will be elaborated in the following chapters, and synthesized in chapter 14. From 6.102, we saw that stars of mass much less than that of the Sun, cannot finish their evolution in the age of the universe. Without some effective process to eject mass—which is not observed—this matter remains locked in their interiors, and no longer participates in stellar death and birth.

If these low-mass stars are least effective at nucleosynthesis, which are most? For a given star, the rate of nucleosynthesis per unit mass is proportional to the inverse of its lifetime, that is, $L/M \sim M^\alpha$, where $\alpha \approx 2$ to 4. Massive stars burn at the most prodigous rate. However, the probability of forming such stars (the *initial mass function*, see chapter 14) must be factored in, and stars with increasing mass become increasingly rare. The compromise between these competing effects is that stars of 20 to $30 M_\odot$ are probably most effective for stellar nucleosynthesis; such stars will receive much attention in what follows. The recent supernova in the Large Magellanic Cloud, SN1987A, was thought to have a progenitor of $20 M_\odot$, at the lower edge of this range. Its character provides pertinent checks upon the theory of such stars, which was well developed (in many respects) prior to its explosion.

As we saw in chapter 2, the empirical data for nucleosynthesis almost always requires that the freshly cooked material be dispersed from the site of its production. Although hydrogen—which is converted into helium, then into carbon and oxygen, and finally buried in the core of a white dwarf—may be said to have undergone nucleosynthesis, it is unable to participate in further adventures which allow it to be directly observed. Most stars capable of burning fast enough to do much nucleosynthesis do bury their products in a condensed remnant. It has long been known [5] that the average mass of white dwarf stars is less (0.5 to $0.6 M_\odot$) than the mass of their progenitors. It appears that stars as massive as 6 to $8 M_\odot$ may end their lives by producing white dwarfs [526, 394, 672]. Lower-mass stars are observed to produce planetary nebulae, which are ejected gas illuminated by UV radiation from a newly forming white dwarf. For this to happen, a significant amount of matter, $M_{ej} = M - M_{wd}$, must be ejected. Although primarily unprocessed hydrogen and helium, this matter may be enriched in rare nuclei (e.g., s-process, CNO isotopes). It acts to dilute more extensively processed matter from more massive stars. Even for higher masses, $8 < M/M_\odot < 15$, most of the synthesized matter is thought to end in the neutron star which is supposed to be made as the star dies. Currently measured masses [383] of neutron stars all lie in the range $M_g = 1.0$ to $1.8M_\odot$; the error bars allow the range to be at least as tight as $M_g = 1.1$ to $1.55M_\odot$.

7

Hydrogen-Burning Stars

In clines beyond the solar road ...
—Thomas Gray (1716–1771),
The Progress of Poesy

Despite their long lives, stars are not eternal. They shine only as long as their energy source can supply the radiation they emit. Almost all the stars we see are bright because they are consuming either hydrogen or helium. There are two reasons for this. First, the nuclear energy released by hydrogen and helium burning is converted almost entirely into light. These stages occur at temperatures sufficiently low that the amount of energy carried away by neutrino radiation is relatively small. In later, hotter burning stages, the thermonuclear energy released is radiated almost entirely in the invisible form of neutrino-antineutrino pairs. Second, the energy release—per unit fuel burned—is larger for hydrogen and helium burning, drastically lengthening their duration. Hydrogen burning releases ten times more energy per gram of fuel burned than does helium burning (about 8×10^{18} erg/g versus 8×10^{17} erg/g), so that about 90 percent of the observed stars are probably burning hydrogen. For comparison, carbon burning and oxygen burning—the two most energetic of the remaining burning stages—only release about 5×10^{17} erg/g. While much of any star's life is spent in this initial stage of hydrogen burning—in which the Sun now exists—it is the later stages that are more active in producing complex nuclei.

For the first and longest burning stages, the stars that are intrinsically brightest are burning their fuel most rapidly, and are of most interest for nucleosynthesis. Stellar luminosity increases with the mass of the star. It is the bright, rare, rapidly evolving stars upon which we focus. There still remain unresolved questions about their nature and their behavior. The arcane subjects of semiconvection, mass loss, and interaction with a binary companion star appear.

The complexity of stellar behavior can be tamed by judicious approximation. Combining the ideas that stars are static—they balance gravity by internal pressure—and that they derive the light they radiate from the energy of nuclear burning, gives a quantitative estimate of stellar brightness and surface temperature. It compares successfully to observations of the nearby stars.

Stars, ranging from 50 percent more massive than the Sun to the most massive known, develop convection in their central regions as they consume fuel there. These convective cores determine how the stars subsequently behave. Because of the mixing by convection, the core becomes ashes as the fuel is depleted. The mass of the core is the mass of ashes produced by that stage of burning, so that it will be relevant for estimating the stellar yield of newly synthesized nuclei. The ashes of one burning stage become the fuel for the next, so that the size of the convective core affects how that next stage will begin. A simple means of estimating the mass of convective cores is derived, which can be a powerful tool for developing a coherent understanding of how stellar evolution works.

After core burning ceases, the fuel which overlays the ashes may itself heat to ignition: then shell burning occurs. Because there is a sequence of burning stages, there can be a sequence of burning shells. The evolution of these shells, and their interactions, can be complex. The general trends are for the flame zones of the shells to "narrow" in mass, often burning more vigorously, and to add ashes to the core below, possibly causing it to grow to ignition of its next burning stage.

While schematic networks of nuclear reactions may suffice for exploring the evolution of stars, more detail is necessary for predictions of nucleosynthesis. An approximation to the conditions in burning zones— both cores and shells—allows a relatively complete treatment of the nuclear evolution to be computed. Here "relatively complete" means a network of 300 nuclei, up to and including krypton (proton number $Z = 36$). Applied to hydrogen burning, this gives interestingly high production of the radioactive nucleus ^{26}Al, which is famous both for being involved with the ^{26}Mg anomaly in meteorites, and having its gamma-ray line detected in space. Population II stars have low abundances of the catalysts that allow the CNO cycle to burn hydrogen; their burning is driven to higher temperatures to compensate. This gives additional thermonuclear processing of Na and Al, for example, and seems to be related to Na-O anticorrelation observed in globular cluster giant stars. What a nice bonus for a modest effort!

7.1 BIRTH OF STARS

Because much of our empirical knowledge of the abundances of nuclei is based upon solar system matter, the birth of the solar system is directly related to our ideas of nucleosynthesis. The way in which stars are made is of obvious interest for understanding the evolution of galaxies and the probability for planetary systems elsewhere. It is an unresolved problem at present. Our own solar system provides important clues and constraints on the process [149]. Observations of interstellar clouds, primarily in the infrared and millimeter wavelengths, are beginning to probe the process as it presently occurs in our galaxy [571, 540]. The theoretical interpretations resulting from these different sources of information are not now consistent (see discussion of ^{26}Al by Cameron [149]).

A large, transparent cloud tends to have a cooling time that is short compared to the hydrodynamic time. Because the cloud is transparent, cooling is a local property, depending upon density, temperature, and composition. The hydrodynamic time depends upon sound speed, a local property, but also on the size of the cloud; the larger the cloud, the longer the hydrodynamic time. For an opaque cloud this tendency is reversed. Cooling occurs not simply by radiation but the diffusion of that radiation. The diffusive process tends to slow with increasing density, so that the hydrodynamic time can become shorter than the diffusive time. The transparent cloud tends to be uncorrelated and messy. The opaque cloud develops strong symmetry in shape and structure. This qualitative transition is made as an interstellar cloud forms a star.

Any theory of star formation must address the angular momentum problem [145]. The galactic year at the position of the Sun is about 2×10^8 years. It is unlikely that an interstellar cloud will have an angular velocity less than this ($\Omega \approx 10^{-15}$ s^{-1}). Consider a cloud of one solar mass at a density typical of molecular clouds ($n \approx 10^4$ cm^{-3}). The radius would be $r \approx 3 \times 10^{17}$ cm and the specific angular momentum $J = r^2\Omega \approx 10^{20}$ cm^2 s^{-1}. If the cloud contracts conserving angular momentum, centrifugal force balances gravity at a radius $r \approx 0.8 \times 10^{13}$ cm. In chapter 6 we found that a star of one solar mass must contract to a radius of about this size in order to release enough gravitational binding energy to ionize itself. Because real clouds are expected to have larger angular momenta than this, they cannot immediately contract to a stellar state.

The dynamical behavior of magnetic fields is a subtle and complex subject [486], and our understanding of its application to star formation is probably incomplete. The interstellar gas is threaded by magnetic fields; even cold clouds have some small ionization due to cosmic rays or radioactivity, which tends to tie these fields to the matter. Because of its angular momentum, the cloud tends to develop differential rotation. Because of its tension along field lines, the magnetic field tends to resist shearing. The outcome of these competing effects depends upon dynamics and topology, and is difficult to calculate convincingly. It is thought that the fields do transport angular momentum outward, and then decouple from the matter, allowing clouds to collapse to stellar dimensions (see [571] and references therein).

Suppose the collapsing cloud were initially of uniform density, and angular momentum unimportant. The acceleration would be $dv/dt = -GM/r^2$. However, $M = 4\pi\rho r^3/3$, so $dv/dt = -4\pi G\rho r/3$. In this limit, the acceleration at any instant of time is proportional to the radius. This is the Hubble law of velocity (see §5.1), but with a reversed sign for the direction of velocity. The choice of the origin of the coordinate system is arbitrary. Suppose there is a perturbation in the uniformity of the density. Denser regions will have accelerated collapse, less dense regions will lag behind. The collapse is unstable toward growth-of-density perturbations. This will continue until pressure gradient or centrifugal forces become important. What about matter in the less dense regions? It still falls toward the nearest dense "core," causing it to grow by accretion. This is the "inside-out" collapse of Shu et al. [571]. What might stop the process? Exhaustion of the supply of gas, heating of the in-falling gas by the protostar, or interaction with neighboring protostars are possibilities.

In chapter 6 we estimated the time for a newly ionized protostar to contract to hydrogen-burning conditions (the Helmholtz-Kelvin time-scale), and the time for a star to burn hydrogen on the main sequence. The hydrogen-burning time for a star of 20 $M_\odot$ was found to be comparable to the Helmholtz-Kelvin time for a star of one solar mass. Therefore the process of star formation may involve interaction between massive stars, which ionize the gas around them, and protostars that are still contracting. The final result could depend upon the interactions of the newly forming stars upon each other, and with their gas supply.

7.2 BURNING PROCESSES

The thermonuclear aspects of hydrogen burning have been discussed in detail by Clayton [173], and more recently by Bahcall [65]. It is only necessary to summarize these treatments here.

At low temperatures, hydrogen burning proceeds by the *pp chains*. The PPI chain occurs at the lowest temperatures, and is shown in table 7.1. The first—and crucial—reaction may be thought of as the spontaneous weak decay of a proton in the field of a second proton; the nuclear interaction gives a binding that makes this energetically possible. Such a process is slow because it involves the weak interaction, and because the number of close proton pairs is small. Because of coulomb repulsion, the probability of finding such a close pair depends on the temperature as well as the density. Higher temperatures are required for close approach. This is the coulomb penetrability we discussed in chapter 3. Because the

TABLE 7.1

The PP Chains

Reaction	Q	ϵ_ν
PPI:		
$p + p \rightarrow d + e^+ + \nu$	1.442	≤ 0.420
$d + p \rightarrow\ ^3\text{He} + \gamma$	5.493	
$^3\text{He} +\ ^3\text{He} \rightarrow\ ^4\text{He} + 2p$	12.859	
PPII:		
$^3\text{He} +\ ^4\text{He} \rightarrow\ ^7\text{Be} + \gamma$	1.586	
$^7\text{Be} + e^- \rightarrow\ ^7\text{Li} + \nu$	0.861	$\leq 0.861(90\%)$
		$\leq 0.383(10\%)$
$^7\text{Li} + p \rightarrow\ ^4\text{He} +\ ^4\text{He}$	17.347	
PPIII:		
$^7\text{Be} + p \rightarrow\ ^8\text{B} + \gamma$	0.135	
$^8\text{B} \rightarrow\ 2\ ^4\text{He} + e^+ + \nu$	18.074	< 15

charge product $Z_1 Z_2$ is small (unity), the temperature dependence is relatively weak.

The second reaction, $d + p$, has the same coulomb penetrability but involves the strong rather than weak interaction. It is much faster, so that it is usually possible to assume that deuterium is near its steady-state abundance (see chapter 4 for a discussion of this approximation, or chapter 5 of Clayton [173]). Notice that the formation of ^3He releases about four times as much energy as the formation of deuterium.

To proceed to the last reaction in this chain, a larger coulomb barrier must be penetrated. For $^3\text{He} +\ ^3\text{He}$, the charge product $Z_1 Z_2$ is four, not unity, and it appears in the exponential function (see §3.3). Consider a range of temperature, as will be encountered from the center to the surface of a star. If there is no convective mixing, a ^3He "ring" may be formed. At the lowest temperature, no reactions will proceed fast enough to change the abundances. At the highest temperatures, the star will presumably be able to burn with the full PPI chain. There must be an intermediate range of temperatures for which the first two reactions can occur, but $^3\text{He} +\ ^3\text{He}$ is inhibited, giving production of ^3He without corresponding destruction. Whether or not this ^{3}He will survive and be ejected to become part of the cosmic abundance pattern is unclear. The possibility of forming it in low temperature (and therefore common) environments does make the identification of the ^{3}He abundance with cosmological nucleosynthesis (§5.4) less secure. Since the solar system ratio of ^{3}He/^{4}He is only about 2×10^{-4}, even a small production of ^{3}He may be significant.

The net effect of PPI is to convert two protons each to a neutron, a positron, and a neutrino, and to consume four protons by radiative capture, spitting back two of the protons and a ^{4}He nucleus. Schematically,

$$4p \rightarrow {}^4\text{He} + 2\,e^+ + 2\,\nu. \tag{7.1}$$

The average energy of the neutrinos is small, only 263 KeV.

At higher temperatures the ^{3}He may be destroyed by reaction with ^{4}He. This gives the PPII chain, which is shown in table 7.1. Again the average neutrino energy is small. As the ^{4}He builds up, this mode is favored.

At still higher temperature another chain occurs, PPIII. Here ^{7}Be is destroyed by proton capture rather than electron capture. A higher temperature is required because of the larger coulomb barrier. The average neutrino energy is large here, 7.2 MeV. This branch has been crucial for the solar neutrino problem, because of the larger detection efficiency for higher-energy neutrinos. For the Sun, this branch is calculated to be rare, and from the low counting rate observed for these neutrinos, it may be even rarer than that! Unfortunately, there seems to be no other solar property besides the flux of relatively high-energy neutrinos, which is sensitive to the strength of the PPIII branch in the Sun.

Stars of only one solar mass do not contribute much to nucleosynthesis. However, the solar neutrino problem does have implications for the accuracy with which stellar models can be computed, and perhaps for the nature of the weak interaction (which is crucially important for the later stages of stellar evolution). Bahcall [65] has reviewed the problem, and provides extensive references. The Davis experiment involves the capture of electron-type neutrinos in an enormous vat (10^5 gallons) of perchloroethelene (C_2Cl_4) by the reaction $\nu + {}^{37}\text{Cl} \rightarrow {}^{37}\text{Ar} + e^-$ to a mirror state in argon, which is then collected by repeated flushing of the vat with helium gas, and detected by its decay. The ^{37}Cl experiments of Davis and collaborators gave a capture rate of 2.05 ± 0.3 SNU (solar neutrino units), which is to be compared with a predicted value of 7.9 SNU. About 75 percent of this is contributed by ^{8}B neutrinos. The later experiments in the (1970–1988) epoch gave values larger by about a factor of 1.7, which are still inconsistently low. The Kamiokande II experiment [296], which directly counts scintillations caused by the weak interactions as they occur in water, has yielded a result of 0.45 of the Standard Model flux of ^{8}B neutrinos. This inconsistency between the two independent experiments and the standard theoretical model remains a nagging worry. At the time of writing, preliminary data is emerging from the Gallium experiments [14] which suggests that the p-p neutrinos have been detected; the result is consistent with the full p-p neutrino flux together with a reduced flux of ^{8}B and ^{7}Be neutrinos as observed in the Homestake and Kamiokande experiments.

Returning to our story: if we apply the steady-state approximation to d, Li, Be, and B, we reduce the network equations to three, involving only p, ^{3}He, and ^{4}He. In the p–p chains, ^{3}He is a laggard and for accurate results should be followed separately. Following Clayton [173], for ^{3}He in steady state,

$$\epsilon_{\text{PPI}} = \epsilon_{\text{PPI}}^0 \, \rho X_H{}^2 T_6{}^{-4/3} \exp(-33.81/T_6{}^{1/3})$$
$$\times \left(1 + 0.0123 T_6{}^{1/3} + 0.0109 T_6{}^{2/3} + 0.00095 T_6\right), \qquad (7.2)$$

where $\epsilon_{\text{PPI}}^0 = 2.32 \times 10^6 \text{erg g}^{-1}\text{s}^{-1}$. More approximately, we have

$$\epsilon = \rho X_H{}^2 \epsilon_0 \left(\frac{T}{T_0}\right)^n \qquad (7.3)$$

where for $T_0 \approx 1.0 \times 10^7 \, \text{K}$, $\epsilon_0 = 0.068 \, \text{erg g}^{-1}\text{s}^{-1}$ and $n = 4.6$.

Higher temperatures allow the coulomb barriers of nuclei of larger charge to be penetrated, so that the slow $p + p$ reaction may be replaced by other, faster weak interaction processes. The first to become important is the CN cycle, shown in table 7.2. It consists of (p, γ) reactions on ^{12}C, ^{13}C, and ^{14}N, interspersed with β-decays of ^{13}N and ^{15}O, and finished by a (p, α) reaction on ^{15}N. The beta-decays have mean lifetimes of 870 and 178 seconds, respectively. Except in explosive situations, these may be assumed to be fast. In novae explosions, these times are sufficiently long that the energy release is inhibited. This situation is termed the *hot CNO process*; see [661, 160]. The highest coulomb barriers in the CN cycle are for proton capture on the N isotopes; these rates are slower than those on the C isotopes. Further, the (p, γ) on ^{14}N involves an electromagnetic rather than a strong interaction, so that it is slower than the (p, α) reaction on ^{15}N. Because ^{14}N is the nucleus most resistant to destruction, it will become the most abundant as the cycle proceeds. In fact, all the C and N isotopes act as catalysts, and tend

TABLE 7.2

The CNOF Cycles.

Reaction	Q	ϵ_ν
CN:		
$^{12}C + p \rightarrow \ ^{13}N + \gamma$	1.944	
$^{13}N \rightarrow \ ^{13}C + e^+ + \nu$	2.221	≤ 1.199
$^{13}C + p \rightarrow \ ^{14}N + \gamma$	7.550	
$^{14}N + p \rightarrow \ ^{15}O + \gamma$	7.293	
$^{15}O \rightarrow \ ^{15}N + e^+ + \nu$	2.761	≤ 1.732
$^{15}N + p \rightarrow \ ^{12}C + \alpha$	4.965	
$^{15}N + p \rightarrow \ ^{16}O + \gamma$	12.126	
NO:		
$^{16}O + p \rightarrow \ ^{17}F + \gamma$	0.601	
$^{17}F \rightarrow \ ^{17}O + e^+ + \nu$	2.762	≤ 1.740
$^{17}O + p \rightarrow \ ^{14}N + \alpha$	1.193	
OF:		
$^{17}O + p \rightarrow \ ^{18}F + \gamma$	5.609	
$^{18}F + e^- \rightarrow \ ^{18}O + \nu$	1.6555	≤ 0.6335
$^{18}O + p \rightarrow \ ^{19}F + \gamma$	7.993	
$^{19}F + p \rightarrow \ ^{16}O + \alpha$	8.115	

toward a *steady-state* abundance pattern, which is called the *equilibrium*[1] CNO abundance. The abundance of ^{13}C can be increased, especially relative to ^{12}C, but only up to approximately 0.01 of the steady-state abundance of ^{14}N, because of the greater resistance to destruction of N.

If the CNO cycle is in a steady state, the rate of energy release is

$$\epsilon_{CNO} \approx \epsilon_{CNO}^0 \rho X_H X_{CNO} f_N T_6^{-2/3} \exp\left(\frac{-152.31}{T_6^{1/3}}\right), \qquad (7.4)$$

where X_{CNO} is the mass fraction of all isotopes of carbon, nitrogen, and oxygen at the time of interest, $\epsilon_{CNO}^0 = 8 \times 10^{27} \mathrm{erg\,g^{-1}\,s^{-1}}$ and f_N is the screening factor for $^{14}N(p, \gamma)^{15}O$. Near $T_6 = 25$,

$$\epsilon_{CNO} \approx \epsilon_0 \rho X_H X_{CNO} f_N (T_6/25)^{16.7}, \qquad (7.5)$$

[1]This unfortunate choice of terminology is widely used; see chapter 4.

where $\epsilon_0 \approx 2.2 \times 10^4 \, \mathrm{erg\,g^{-1}s^{-1}}$. Especially in lower-mass stars, ^{16}O lags in the approach to a steady state, so that for accurate work its abundance should be followed separately.

The second set of reactions in table 7.2 represents a slow leakage out of the CN cycle, which connects it with a new sequence of reactions called the NO cycle. This NO cycle becomes operational at higher temperatures, again because of higher coulomb barriers. Notice that it connects to the CN cycle by taking the leakage into ^{16}O and eventually transferring it back to ^{14}N. The mean lifetime for the beta-decay is 95 seconds.

At this point we diverge from the discussion of Clayton. Experimental work by Rolfs and Rodney [523, 524] indicates that the earlier estimates of the rate of the $^{17}O(p, \alpha)^{14}N$ reaction was overestimated by a factor of 60. This suggests that a third cycle should be considered. The OF cycle is illustrated in table 7.2. The mean lifetime for the beta-decay is 158.4 minutes. The subtleties of the NO and OF cycles are important for accuracy, but are not necessary in what follows.

Equating 7.3 and 7.5 gives

$$T \approx 1.7 \times 10^7 \left(\frac{X_{\mathrm{H}}}{50 X_{\mathrm{CN}}} \right)^{\frac{1}{12.1}} \mathrm{K}. \qquad (7.6)$$

Below this temperature the p–p chain is most important, above it the CNO cycle is. This transition occurs in stars slightly more massive than the Sun (see table 7.3); the different temperature dependence of the two burning processes makes the local $(\epsilon_{pp}/\epsilon_{CN})$ and the global $(<\epsilon>_{pp} / <\epsilon>_{CN})$ comparisons slightly different (see §6.3).

7.3   MAIN SEQUENCE

We are now in a position to calculate the characteristics of stars burning hydrogen in thermal and hydrostatic equilibrium. If the hydrogen burning occurs in the central regions, the stars occupy a particular band on a graph of luminosities and effective temperatures. Such an *HR diagram* (for Hertzsprung and Russell) shows most stars lying along this *main sequence* because this is the longest stage in the life of a typical star.

We will now construct models of stars of homogeneous composition, which are beginning to burn hydrogen; this is the *zero-age main sequence* (ZAMS). To begin we choose a star of m solar masses, or $M = 1.987 \times 10^{33} \mathrm{g} \, m$. For the composition we take the nucleon fractions of hydrogen, helium, and heavier elements to be $X_H = 0.71$, $X_{He} = 0.27$, and $X_Z = 0.02$. For solar system ratios, the fraction for C and N alone is $X_{CN} =$

$X_Z/4$, and for O included, $X_{CNO} = \frac{2}{3}X_Z$. The electron mole fraction is

$$Y_e \approx X_H + \frac{1}{2}(X_{He} + X_Z) = 0.855. \tag{7.7}$$

This will be needed to evaluate the opacity. The mole fraction of nuclei is

$$Y_{nuc} \approx X_H + \frac{1}{4}X_{He} + \frac{1}{16}X_Z = 0.779, \tag{7.8}$$

so that the total mole fraction of "ideal gas particles" is

$$Y = Y_e + Y_n = 1.634.$$

This will be needed to evaluate the equation of state for the pressure. We take the equation of state to be that of radiation plus ideal gas.

Next we assume the structure is that of a polytrope of $n = 3$, and use Eddington's quartic equation (6.45) to evaluate β, the ratio of ideal gas to total pressure. This presumes hydrostatic equilibrium, and requires choice of composition for Y and a choice of stellar mass m. It may be solved by iteration.

Having β, we can relate temperature and density by

$$T_c = \beta D_n\, GM^{2/3}\rho_c^{1/3}/\Re Y, \tag{7.9}$$

$$= 0.4614 \times 10^7 \text{K}\ \beta\rho_c^{1/3}/Y. \tag{7.10}$$

The last version assumes that density is expressed in cgs units. This will be useful for evaluation of energy generation and of opacity.

The luminosity that can be supplied by nuclear burning is

$$L = M\langle\epsilon\rangle$$
$$= M\frac{\langle\epsilon\rangle}{\epsilon_c}\epsilon_c \tag{7.11}$$

where $\langle\epsilon\rangle/\epsilon_c$ is defined in Appendix C:

$$\frac{\langle\epsilon\rangle}{\epsilon_c} = \frac{3.23}{(3u + s)^{3/2}}. \tag{7.12}$$

The luminosity due to photon diffusion is given by the Standard Model values (see §6.2),

$$L \approx 3.72 \times 10^{35} \text{erg s}^{-1} m^5(\beta/Y)^7/\left(5.88 + m^2(\beta/Y)^3\right)Y_e. \tag{7.13}$$

As we have specified X_H, X_{CNO}, and m, these equations give a unique solution. The composition enters in the energy generation rate ϵ and the opacity κ. For low-mass stars, $\epsilon_{pp} > \epsilon_{CNO}$, so the effect of the composition upon opacity is the most important. For high-mass stars, $\epsilon_{CNO} > \epsilon_{pp}$, and Thomson scattering dominates, so the energy generation rate is the quantity sensitive to composition.

Eliminating density from the energy generation rate ϵ_c allows the thermal balance equation to be expressed in terms of powers of central temperature T_c. The solution may be obtained by iteration. Except for the awkwardness of the two iteration steps, the procedure is simple. The results are summarized in table 7.3, which may be used to choose iterative guesses for modified calculations. The units of central temperature are 10^7 K, and of radius are 10^{11} cm. Notice that we use a global balance for heating and cooling rather than a local one. This allows us to avoid a spatial integration, and is a key to the simplification.

The Hertzsprung-Russell diagram for these stars is shown in figure 7.1, and compared to observed stars which are thought to be on the zero-age main sequence.

The open triangles represent a set of well-observed spectroscopic binaries (Harris [281]). The inverted open triangles represent an observational estimate of the zero-age main sequence (Blaauw [100]). The crosses represent the set of very-low-mass stars of Liebert and Probst [396]. The Sun is shown as an open circle. The solid line is from the approximate models just presented. The numbers refer to several selected values of stellar mass (in solar units) for the theoretical models. These models work well for all but the lowest-mass stars. A second, hotter line represents the corresponding curve for helium stars beginning helium burning (see chapter 8).

Consider a star of a given luminosity. For core burning by a given process, the central conditions are constrained. The position of that star in the HR diagram is still not entirely fixed; it depends upon the structure of the star in the sense that the central conditions map into a surface radius. For polytropes this parameter is equivalent to the ratio of central to average density, $\rho_c/\langle\rho\rangle$. To compare our models to observed quantities that are insensitive to this parameter, we use the mass-luminosity diagram, which is shown in figure 7.2. It is difficult to weigh stars. For binary stars, we can infer the combined mass if we can measure the period and the semi-major axis of the orbit. The latter appears as a cube in the expression for the mass, and is more difficult to determine accurately as well (see Batten [82], ch. 5).

Solid triangles represent the set of well-observed spectroscopic binaries (Harris [281]), the crosses represent the set of very-low-mass stars

TABLE 7.3
Zero-Age Main Sequence

$M/M_\odot$	β	$L/L_\odot$	$(T_7)_c$	$\rho_c(g/cc)$	R_{11}	T_{eff}
0.25	1.0000	6.003E-4	0.4428	61.64	0.4706	1.101E+3
0.50	0.9999	1.906E-2	0.7646	79.34	0.5451	2.428E+3
0.75	0.9998	1.429E-1	1.0510	91.65	0.5947	3.847E+3
1.00	0.9996	5.915E-1	1.3140	100.8	0.6342	5.313E+3
1.25	0.9993	1.765E+0	1.5480	105.5	0.6728	6.780E+3
1.50	0.9991	4.276E+0	1.7320	102.8	0.7211	8.170E+3
2.00	0.9983	1.688E+1	1.9720	85.47	0.8441	1.064E+4
3.00	0.9963	1.083E+2	2.2530	56.97	1.106	1.480E+4
5.00	0.9900	9.223E+2	2.5870	31.67	1.595	2.105E+4
7.00	0.9810	3.230E+3	2.8020	21.11	2.043	2.545E+4
10.0	0.9639	1.061E+4	3.0240	13.70	2.657	3.004E+4
16.0	0.9225	4.157E+4	3.2970	7.916	3.731	3.566E+4
20.0	0.8934	7.390E+4	3.4170	6.207	4.359	3.810E+4
25.0	0.8582	1.258E+5	3.5290	4.936	5.068	4.036E+4
30.0	0.8253	1.884E+5	3.6140	4.140	5.711	4.206E+4
40.0	0.7676	3.392E+5	3.7370	3.199	6.849	4.449E+4
50.0	0.7196	5.155E+5	3.8230	2.661	7.846	4.615E+4
70.0	0.6451	9.207E+5	3.9390	2.061	9.557	4.834E+4
100	0.5672	1.613E+6	4.0470	1.611	11.68	5.030E+4
150	0.4839	2.896E+6	4.1540	1.247	14.57	5.214E+4
300	0.3610	7.199E+6	4.3090	0.8388	20.95	5.460E+4
1000	0.2086	2.982E+7	4.5290	0.4543	38.39	5.754E+4
5000	0.0964	1.705E+8	4.7740	0.2155	84.18	6.009E+4

of Liebert and Probst [396], and the Sun is shown as an open circle. The solid squares are from Popper [501]. The solid line is from the approximate models described above. Now that the average-density problem is avoided, it is clear that something is wrong for stars less massive than the Sun. The error is that we used radiative luminosity to balance energy generation. For lower-mass stars, the opacity is larger due to the increasingly larger Kramers component, demanding a steeper tempera-

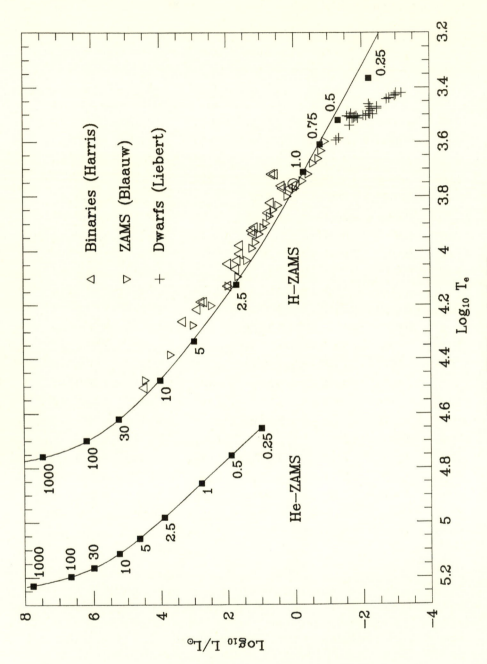

Fig. 7.1. HR Diagram for the MS.

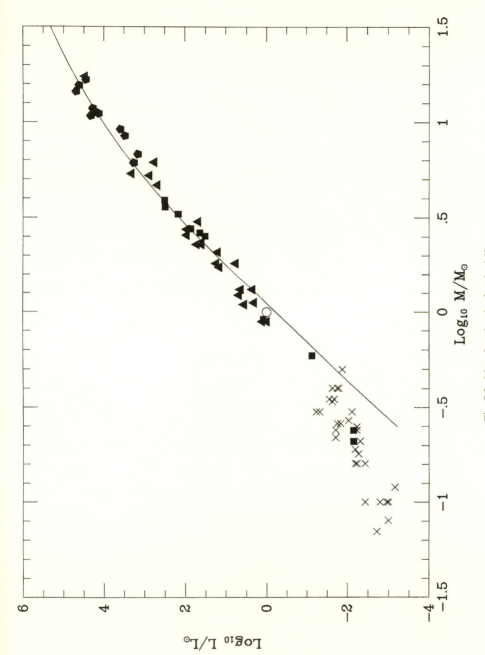

Fig. 7.2. Mass-Luminosity for the MS.

ture gradient to drive the luminosity. The region of high opacity becomes convectively unstable, *so that this opacity then has no effect on the energy flow*. Thus, for a given mass the luminosity is not so low as we would predict. If the opacity is limited to be less than about 30 times the Thomson value, the luminosities will fall in the observed region. Apparently the convection in these stars "short-circuits" the heat flow at about this value. See [275, 650, 204] for more detail.

What is the minimum mass which can ignite hydrogen burning? For sufficiently low mass, stars will cool down to degenerate configurations (or planets) without igniting hydrogen. Following the procedure outlined in §6.6, we estimate the ignition temperature T_{ign} by equating the luminosity per unit mass L/M to the average energy generation rate $\langle \epsilon \rangle$. The result depends upon stellar mass; for $M \approx 0.1 M_\odot$, $T_{ign} \approx 5 \times 10^6$ degrees kelvin. For this maximum temperature, the corresponding ignition mass is $M \approx 0.1 M_\odot$, which is the consistent solution. The estimate assumes nonrelativistic electrons (i.e., $M \ll 1.4 M_\odot$); a more accurate value [204, 590] is $0.08 M_\odot$. Below this mass, the temperatures are too low for $\langle \epsilon \rangle$ to balance L/M.

Stars less massive than this (the *brown dwarfs* [590]) will be very faint, and hence difficult to observe. We do not know how much of the matter in the Galaxy is tied up in such stars; this question is important for theories of galactic nucleosynthesis (see chapter 14).

7.4 CONVECTIVE CORES

On the zero-age main sequence, stars are still homogeneous in composition, but this cannot last. The rate of nuclear burning is unevenly distributed, and tends to make composition gradients. To what extent, if any, are these removed or modified by mixing?

In order to investigate in more detail how hydrogen burning occurs in stars, let us consider the question of convective cores. A simple analysis will lead to sweeping conclusions. For a hydrostatic thermal balance of the star as a whole,

$$L = \int \epsilon \, dm. \tag{7.14}$$

From the Standard Model (chapter 6), we have a global expression for the luminosity carried by photon diffusion, L_d. However, if we have

$$\frac{\partial L_d}{\partial m} < \epsilon \tag{7.15}$$

locally, then the radiative flux must be supplemented by convection. This is the local condition for convection instability[2] in a hydrostatic star (we neglect conductive luminosity for simplicity). The condition for a core to be convective is then

$$\left(\frac{\partial L_d}{\partial m}\right)_c < \epsilon_c \tag{7.16}$$

at the center of the star. How might the luminosity vary in the presence of nuclear burning? Suppose the star were marginally convective over some core of mass M_{core}. For Thomson opacity and β slowly varying, the radiative luminosity would grow linearly as the mass coordinate increases. Figure 7.3 shows such an idealized case, with a convective core of mass M_{core}; the luminosity carried by radiation increases in an approximately linear fashion with mass. At M_{core}, the radiative luminosity just equals the total luminosity, so that the convective luminosity can go to zero. Outside this region there is no source of energy, so the "envelope" luminosity is constant. The Standard Model gives a luminosity

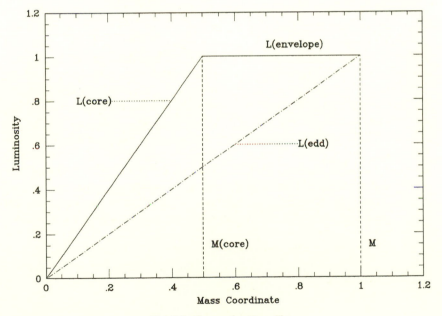

Fig. 7.3. Convective Core Size

[2]The excess heating will build a negative entropy gradient, so that this will imply the usual dynamic condition (see [378], p. 8).

(dot-dashed curve) which increases linearly with mass from center to surface, a rather different behavior.

To deal with this situation we proceed as follows:

First, we set the actual luminosity L equal to the Eddington value at the surface $(m = M)$. This implies $\langle \epsilon \rangle = 4\pi c G(1 - \beta)/\kappa$.

Second, we evaluate $(1 - \beta)$ more carefully. In §6.2, we derived the quartic equation for β assuming the structure of an $n = 3$ polytrope. If we now allow for other polytropic indices n, we may formally write a quartic equation for each choice of n: $(1 - \beta_c) = (D_n/D_3)^3 (M/M_\odot)^2 (\beta/Y)^4$. A useful approximation is $D_n/D_3 \approx 1 + 0.184(3 - n)$. Therefore,

$$(1 - \beta_c)/(1 - \beta_{edd}) = (D_n/D_3)^3. \qquad (7.17)$$

If we define $(n + 1) \equiv (\nabla \ln P/\nabla \ln T)$, then we may derive an *exact* expression

$$L(m)/m = \left(4\pi c G(1 - \beta)/\kappa\right)\left(\frac{4}{n+1}\right). \qquad (7.18)$$

Because this $(n + 1)$ is not constant in general, we have merely renamed variables. However, if we identify this n with the polytropic index n used above, this expression gains new meaning. Notice that for the Standard Model case, $n = 3$, this reduces to the Eddington expression. For smaller n, and hence steeper temperature gradients, the luminosity is larger than for the Standard Model. In such a case, the luminosity would resemble L_{core} in figure 7.3.

Notice that we play a dangerous game with the mathematics here, in order to obtain an important result. If n is not three, the β is not really constant through the star. The idea is to split the star into two aspects and use the method of successive approximation to improve our treatment of the aspects separately. An exact solution would be a composite polytropic structure, with the convective core having a polytropic index equal to n.

Finally we estimate the fractional size of the core region q_{cc} by equating the core luminosity at $m = M_{core}$ to the Eddington value at the surface $m = M$, so that

$$q_{cc} = \frac{(n_{core} + 1)}{4}(D_3/D_n)^3. \qquad (7.19)$$

We get the same constraint if we require that the outer edge of the convective core be radiative with the adiabatic gradient, and equate this to the nuclear luminosity contained inside. The exact solution involves specifying a structure outside the core, which gives constant luminosity.

A simpler expression is obtained if we expand $\log q_{cc}$ around the polytropic index value $n = 3$. This gives

$$q_{cc} \approx \exp\left(-0.802[3 - n]\right). \tag{7.20}$$

For $n = 1.5$ this gives $q_{cc} = 0.300$; the accurate value for such a polytrope with a Thomson opacity is 0.312. Any constant opacity gives the same result in this scale-free form. As n approaches 3, the fraction of the star which is in the convective core approaches unity. Although we have expanded about the central conditions, this expression works correctly in the opposite limit! If we use the Chandrasekhar [162] expression relating adiabatic index and β, and identify the adiabatic index in the core with the polytropic index, then

$$n \geq \frac{24 - 18\beta - 3\beta^2}{8 - 6\beta}, \tag{7.21}$$

which allows us to examine cases of intermediate β with some accuracy.

It will be important to have the solution for the case in which the opacity is not constant; this is more complicated. We can expand the $1/\kappa$ factor in the radiative luminosity expression, using

$$\kappa = \kappa_0 \rho^r / T^k = \kappa_0 \rho_c^r T_c^k (\tilde{T}/T_c)^{-k+nr}, \tag{7.22}$$

where we must evaluate $(\tilde{T}/T_c)$, $\tilde{T}$ being the temperature at the edge of the convective core. This dependence suggests an iterative solution.

Suppose we approximate[3] the temperature distribution $T(r)$ by

$$T(r) = T_c(1 - x) \tag{7.23}$$

where $x \equiv r/R$ and T_c is the central temperature. Then,

$$\rho = \rho_c(T/T_c)^n. \tag{7.24}$$

---

[3]The choice of a linear rather than a quadratic approximation has to do with which part of the stellar object is most important in the situation considered. For nuclear burning, the high temperature dependence makes the central regions most important; they are best approximated by a quadratic dependence of temperature upon radius. However, a linear dependence gives a better representation than the quadratic for global properties (for example, the mean density).

Since

$$m(r) = \int_0^r 4\pi\rho(y)y^2 dy, \tag{7.25}$$

we have

$$m(x) = 4\pi\rho_c R^3 \left[\frac{x^3}{3} - \frac{3x^4}{4} + \frac{3x^5}{5} - \frac{x^6}{6} \right], \tag{7.26}$$

with

$$M = 4\pi\rho_c R^3 / 60. \tag{7.27}$$

For $x = 0.1$, this gives $m/M = 0.01585$ and $\tilde{T}/T_c \approx 0.9$ while for $x = 0.3$, $m/M = 0.2557$ and $\tilde{T}/T_c \approx 0.7$. We note that 0.7 is a typical value for $\tilde{T}/T_c$ and use this good guess in what follows to avoid iteration. Thus,

$$q_{cc} = e^{-\alpha}, \tag{7.28}$$

where $\alpha = 0.802(3 - n) + 0.357(k - nr)$. Compare this to equation 7.20. By construction we see from figure 7.3 that in the core region,

$$\frac{\partial L}{\partial m} = L/M_{core} = M\langle\epsilon\rangle/M_{core}, \tag{7.29}$$

so that

$$\langle\epsilon\rangle/\epsilon_c = q_{cc}. \tag{7.30}$$

Now we have a powerful and simple result. Let us consider a few numerical examples. For stellar masses $M > 5M_\odot$ (see chapter 6), the opacity is dominated by Thomson scattering, so that $r = k = 0$, and 7.28 implies that for $s > 0$ a convective core will result. Indeed, this is the result of a vast number of more complex calculations of hydrogen burning in massive stars.

For lower-mass stars, Kramers opacity dominates, and $r = 1$ and $k = 3.5$, so if $n = 3/2$ we find $s \geq 4.7$ gives a convective core. This is near—but greater than—that value for s actually found for the p–p chain, so that these cores should be convectively stable. More accurate calculations show, for stars with sufficiently low mass ($M \leq 1M_\odot$), that Kramers opacity and the p–p chain are dominant, and radiative cores do result.

Can a convective core develop, if no nuclear fuel is burning but energy is supplied by gravitational contraction? We may define a local *gravitational energy generation rate*

$$\epsilon_g = \frac{\partial \dot{B}}{\partial m} = \frac{1}{2} \frac{\Re T Y}{\beta} \frac{\dot{\rho}}{\rho}, \tag{7.31}$$

where B is the gravitational binding energy. For a homologous contraction, $\dot{\rho}/\rho$ is independent of radius, so

$$\epsilon_g \propto T. \tag{7.32}$$

For Kramers opacity,

$$n > 2.45 \tag{7.33}$$

gives a convective core, but n is that large only for massive stars. This implies that low-mass stars ($M \leq 1M_\odot$) do not develop convective cores from pre-main-sequence contraction (their surfaces develop very low entropy, which results in convective *envelopes* that may extend deep into the interior; this convection has a different cause). For massive stars, their times for pre-main-sequence contraction are comparable to their formation times, and their entropy distribution—and therefore their convective state—will probably depend upon the formation process.

A key question for nucleosynthesis is: what is the extent of these convective cores which occur during hydrogen burning? Using Eq. 7.28, for relatively low-mass stars ($1.5 > m$), we take $n = 3/2$ and Kramers opacity, so that

$$q \approx 0.1. \tag{7.34}$$

For higher-mass stars, we take Thomson opacity, and find

$$q \approx e^{-0.802(3-n)} \tag{7.35}$$

where

$$n \geq \frac{24 - 18\beta - 3\beta^2}{8 - 6\beta}. \tag{7.36}$$

For $\beta \approx 1$,

$$q \approx 0.30 \tag{7.37}$$

while for the most massive stars, $\beta \approx 0$, and

$$q \approx 1. \tag{7.38}$$

More massive stars burn a larger fraction of their hydrogen on the main sequence. For example, on the zero-age main sequence a star of $M \approx 150M_\odot$ has $\beta = 0.5$, which implies $n = 2.85$ and $q = 0.89$. These results were used in our estimates of stellar lifetimes in chapter 6.

The existence of large convective cores during hydrogen burning in massive stars is thought to imply—because of the convective mixing—that this large amount of matter has been processed under relatively homogeneous conditions to a uniform composition. If the mass of the convective core is well above the Chandrasekhar mass, it cannot all contract to become a white dwarf or a neutron star. There remain two interesting possibilities: (1) synthesized matter is ejected, or (2) a black hole is formed. Consider a star with the mass of its convective core equal to the Chandrasekhar mass. Such stars are in the transition region from Kramers to Thomson opacity, so $q \approx 0.1$ to 0.3, giving $M_{Ch}/q \approx$ 5 to $15\,M_{\odot}$. Even allowing for some interesting processing during shell burning, *the bulk of stellar nucleosynthesis does seem to occur above this range of mass* $(M > 9\,M_{\odot})$.

7.5   SHELL BURNING

How does a star behave as its hydrogen fuel is depleted? The radiative (diffusive) luminosity (see Eq. 6.63) is

$$L_d \propto M(1 - \beta)/\kappa \tag{7.39}$$

and the nuclear luminosity (see equations 6.72 and 7.5) is

$$L_n \propto MT^n X_H X_i \tag{7.40}$$

where $X_i = X_H$ for the p–p chain and $X_i = X_{CNO}$ for the CNO cycle. For $L_d = L_n$,

$$X_H X_i \propto \frac{(1 - \beta)}{\kappa T^n}. \tag{7.41}$$

Hydrogen depletion implies that $X_H X_i$ decreases, so that $\kappa T^n/(1 - \beta)$ must increase. Even for the worst case, which is the p–p chain and Kramers opacity, $\kappa T^n \propto T^k$ where $k > 0$. Therefore the central temperature increases. Hydrogen depletion robs the gas of two electrons per ^{4}He formed, and four protons go into one ^{4}He nucleus. This decreases the number of pressure-supplying particles in the center where hydrogen is being depleted. It also reduces the Thomson opacity, which is dominant at these high temperatures. This allows radiation to escape more freely, and increases the luminosity. This gives a stage of "main sequence brightening."

Following hydrogen depletion in a convective core, contraction ensues. Since the whole star contracts, the surface temperature increases. If the

star's mass is sufficiently small, the core mass will be less than the ignition mass for helium burning ($M < 2.25 M_\odot$; see chapter 8). In this case,
hydrogen burning begins in the unburned fuel surrounding the depleted
core; this is the beginning of the hydrogen *shell-burning* phase. More
massive stars proceed directly to a stage of core helium and shell hydrogen burning. In less massive stars, which burn H by the p-p chain, the
core is not convective, and the transition to shell burning is smoother.

Let us examine the character of shell burning more closely. Consider a
thin shell of width ℓ, lying over a core of radius R_c and mass M_c. Define
a shell density by

$$\rho_s \ell \equiv \int_{R_c}^{R_c + \ell} \rho \, dr. \tag{7.42}$$

If the pressure overlying the shell is assumed to be small, the shell pressure is

$$P_s = \frac{GM_c}{R_c^2} \rho_s \ell, \tag{7.43}$$

and the mass of the shell is

$$M_s = 4\pi R_c^2 \rho_s \ell. \tag{7.44}$$

Eliminating ℓ, we have

$$P_s = \frac{GM_c M_s}{4\pi R_c^4}. \tag{7.45}$$

We imagine the shell to be evolving in a slow, steady manner. How does
it react to perturbations, if M_c and M_s vary little? Then

$$\frac{\delta P_s}{P_s} = -4 \frac{\delta R_c}{R_c}, \tag{7.46}$$

and

$$\frac{\delta \rho_s}{\rho_s} = -2 \frac{\delta R_c}{R_c} - \frac{\delta \ell}{\ell}. \tag{7.47}$$

From Eq. 6.35, the luminosity carried out of the shell by radiation is

$$L_s = \frac{16\pi^2 ac}{3\kappa} \frac{R_c^4 T_s^4}{M_s}, \tag{7.48}$$

where T_s is the temperature in the shell. Therefore,

$$\frac{\delta L_s}{L_s} = 4 \frac{\delta R_c}{R_c} + 4 \frac{\delta T_s}{T_s}. \tag{7.49}$$

Suppose first that there is no nuclear burning in the shell, and that it is in thermal equilibrium with its surroundings. Then $\delta L_s = 0$ to maintain thermal balance, and

$$\frac{\delta T_s}{T_s} = -\frac{\delta R_c}{R_c}. \tag{7.50}$$

If the equation of state is $P = \Re Y \rho T / \beta$, with β slowly varying,

$$\frac{\delta P_s}{P_s} = \frac{\delta T_s}{T_s} + \frac{\delta \rho_s}{\rho_s}. \tag{7.51}$$

Using the results above, this implies

$$\frac{\delta R_c}{R_c} = \frac{\delta \ell}{\ell}, \tag{7.52}$$

which means that the response is *homologous*, or shape-preserving.

   If the shell is burning, the response is different. The variation in the luminosity lost by the shell must compensate for the change in energy generated, so

$$\delta L_s = \delta \epsilon \, M_s, \tag{7.53}$$

and if $\epsilon \propto T^s \rho^u$, then

$$\frac{\delta L_s}{L_s} = s \frac{\delta T_s}{T_s} + u \frac{\delta \rho_s}{\rho_s}. \tag{7.54}$$

This must equal the previous expression for $\delta L_s / L_s$. Therefore,

$$\frac{\delta \ell}{\ell} = \frac{2s + u - 4}{s + u - 4} \frac{\delta R_c}{R_c}. \tag{7.55}$$

The temperature exponent s for energy generation is often large, so that the change in the width of the shell is twice the perturbation in the core radius. The response of the shell temperature is

$$\frac{\delta T_s}{T_s} = \frac{4 - u}{s + u - 4} \frac{\delta R_c}{R_c}, \tag{7.56}$$

which for large s approaches zero; the shell reacts so that it maintains its temperature. For shell luminosity,

$$\frac{\delta L_s}{L_s} = 4\frac{s + 2u - 8}{s + u - 4}\frac{\delta R_c}{R_c}. \tag{7.57}$$

To obtain larger shell luminosity the shell needs to move to larger radii. Finally, for shell density,

$$\frac{\delta \rho_s}{\rho_s} = -\frac{4s + 3u - 12}{s + u - 4}\frac{\delta R_c}{R_c}, \tag{7.58}$$

and $\delta P_s/P_s$ is the same for large s. Increasing shell luminosity means increasing R_c, and decreasing density for constant temperature. This means increasing radiation entropy. Active burning shells will be associated with jumps in radiation entropy. Because of their strong negative feedback, *active* shell sources act as *nodes*; see Stein [588].

Why do stars become red giants? The underlying cause [516] seems to be a thermal instability resulting from the shape of the opacity as a function of temperature and density (see however [324]). Consider the envelope of a star, which has at least some layers in which the luminosity is transported by radiative diffusion. At the inner edge, which might be a burning shell or the layer above a convective core, the luminosity is

$$L_i = 4\pi cGM_i(1 - \beta_i)/\kappa_i. \tag{7.59}$$

A similar expression holds for that luminosity which the envelope transmits,

$$L_e = 4\pi cGM_e(1 - \beta_e)/\kappa_e, \tag{7.60}$$

where the subscript e refers to values typical of the envelope. The inner temperatures are higher, so that κ_i tends to approach the Thomson value. Because the mass coordinate of the envelope includes the inner regions, $M_e > M_i$, and $\kappa_e > \kappa_i$ is possible, even if we have thermal balance, $L_i = L_e$. However, there is a threshold luminosity, above which the envelope cannot transmit the energy it is fed without changing its structure drastically. Suppose L_i is increased. Then, the energy fed into the envelope exceeds that which it is capable of passing on, so that its gravitational binding energy decreases, and it expands. The expansion gives lower densities and temperatures. However, at all but the lowest densities, this produces higher opacities. Less luminosity is transmitted, and more expansion occurs. The positive feedback makes a runaway possible, causing the envelope to expand as fast as energy is trapped by it. As figure 6.1 shows, the opacity reaches a maximum at intermediate temperatures, and then drops steeply as recombination occurs. The coolest regions can now radiate as well as the inner ones. The middle zone of higher opacity cannot, a steep temperature (entropy) gradient develops, and convection carries the energy. A new thermal balance is possible. The star is now a red giant.

Suppose the inner luminosity L_i now decreases. The envelope radiates more than it is given from below. Its gravitational binding energy increases and it contracts. Its opacity decreases, it radiates still more, and the instability brings the envelope back to smaller radii, and hotter temperatures. The *blue loops* sometimes found in stellar evolutionary sequences are due to the reaction of the envelope to changes in luminosity which pass the threshold value for that particular envelope structure.

Given the existence of the instability, what drives it? The basic answer is, anything that causes L_i to vary. In particular, interactions between hydrogen shell burning and helium burning in core or shell give variation in total luminosity supplied to the envelope.

In the first stages of shell burning, the properties of the shell sources are affected by the composition gradient left by the previous stage of core burning. This is called by Iben [316] the *thick shell* stage. This is followed by a narrowing of the burning shell in mass coordinate.

Schwarzschild [555] introduced a useful approximate expression for the luminosity required by a thin shell source if energy transport is by radiative diffusion. Calculations show that through such a region $(1 - \beta)$ is roughly constant[4] if Thomson scattering is the dominant source of opacity. By ignoring the change in β through the shell region, and applying the Eddington procedure at the burning shell (see chapter 6), we have the shell luminosity:

$$L_s = \frac{4\pi c}{\kappa} GM_c \left[\frac{4}{3} aT^3 \frac{dT}{dP} \right]$$

$$= \frac{4\pi cG}{\kappa} M_c (1 - \beta)$$

$$\approx 2.51 \times 10^{38} \frac{M_c}{M_\odot} \frac{1 - \beta}{1 + X_H}, \tag{7.61}$$

[4]This implies that S_γ / Y is also constant. As Y increases, so does S_γ, so radiation entropy increases through the shell.

where M_c is the mass interior to the burning shell. The manipulations are similar to those in chapter 6 for estimation of total luminosities.

This expression is useful in understanding how burning shell sources evolve and why they become narrow. If ϵ is the average energy generation rate in a shell source of mass Δm, then $L_s \approx \epsilon \Delta m$ again, so that $\epsilon \Delta m \propto M_c$, the core mass. By the hydrostatic equation, the pressure in the shell P_s, which is necessary for support, is $P_s/\Delta m \approx GM_s/4R_s^4$. The core can be degenerate or nondegenerate. If the core were initially nondegenerate, thermal diffusion tends to bring its temperature to that of the burning shell. Schönberg and Chandrasekhar [560] showed that an isothermal nondegenerate core can contain no more than about 10 percent of the stellar mass without contraction. Such contraction in a nondegenerate core causes an increase in temperature, so that the next burning stage could be ignited. The phase of burning in only a single shell would be brief for such stars.

Suppose the core to be at least partially degenerate. For a nonrelativistic degenerate Fermi gas, $P \propto \rho^{5/3}$. Using this and the hydrostatic equation, we find that the radius of the core is related to the core mass by $R_c \propto M_c^{-1/3}$, and therefore $P_s/\Delta m \propto M_c^{7/3}$. Both $P_s/\Delta m$ and $\epsilon \Delta m$ are increasing functions of the core mass. Suppose M_c increases. While $\epsilon \Delta m$ must increase only like M_c, $P_s/\Delta m$ must increase like $M_c^{7/3}$. Since ϵ is more sensitive to T than is P_s, and the density dependence is roughly the same for ϵ and P_s, the additional variation must be made up by a decrease in Δm. Shell *narrowing* is primarily concerned with this narrowing in the mass coordinate m rather than the spatial coordinate r.

This has an important effect on nucleosynthesis. As the mass of the core grows, L_s increases and Δm decreases. As $\epsilon = L_s/\Delta m$, the energy generation rate must increase even faster. Thus *shell burning will occur at significantly higher temperature than core burning.*

By the process of shell burning around a degenerate helium-rich core, low-mass stars can consume an appreciable amount of their hydrogen fuel. Because of the large numerical coefficient in Eq. 7.61, the luminosity of such objects is much larger than on the main sequence. The corresponding lifetimes are correspondingly shorter for the same amount of mass consumed.

As the shell narrows, the luminosity approaches the value given by Eq. 7.61, which is a slowly varying function of time. The slow growth of the core produces an increasing shell luminosity, and this may trigger the red giant instability discussed above, and give a large increase in the photospheric radius. This expansion requires an energy supply, and causes the slight decrease in luminosity. Iben [316] finds that for a $5M_\odot$ Population I star, expansion takes about half the energy release. Expansion stops when recombination limits the radius, and the extra energy now escapes, increasing the observed luminosity. The core mass grows, also causing the luminosity to rise. Eventually the core is too large to be well supported by degeneracy pressure, contraction sets in, the temperature rises, and helium burning begins with the *helium flash* [355].

For still more massive stars, the opacity is dominated by Thomson scattering of radiation by unbound electrons, so $\kappa \approx 0.2(1 + X_H)$. As hydrogen begins to burn, the fraction of the star which is in the convective core ranges from 0.3 to almost all for the most massive stars (see equations 7.28 and 7.36, and table 7.3). Hydrogen burning reduces the number of free particles per nucleon, so that β, which measures the importance of gas pressure to radiation pressure, must decrease. From eq. 7.28, this implies that the fractional size of the convective core q_{cc} must increase. At the edge of the convective core, the luminosity is just that which can be carried by radiation,

$$L_r = 4\pi c G M_r (1 - \beta)/0.2(1 + X_H). \qquad (7.62)$$

As pointed out by Schwarzschild and Härm [556], this presents a dilemma (see also [611]). The matter just inside the convective core will have a lower hydrogen abundance, and therefore opacity, than the matter just outside. The temperature gradient in the matter just outside must be steeper, so it will be convective too, contrary to what we assumed. Schwarzschild and Härm argued that a static solution existed, and described how it could be found. Consider a short time interval in which the convective core moves out a small amount. There is a thin layer which was outside the core, but is now unstable. As helium from the core is mixed into this region, its opacity is decreased, as is its temperature gradient, and its degree of convective instability. When the addition of helium proceeds to the point that convective neutrality is reached, convection will cease, and the composition is fixed. In this intermediate layer the convection is throttled down by its own effects, so that it contributes little to the energy transport, but does modify the composition to maintain convective neutrality. Such a layer is called *semiconvective* and was first studied by Ledoux [390]. See §6 in [355] for a clear discussion of this and related instabilities. In the astrophysical literature the term *semiconvection* has become a catchall phrase which refers to a variety of instabilities that are thought to move mass but provide little energy transport.

An interesting aspect of the Schwarzschild-Härm solution is that the intermediate layer increases in mass, so that while the outer edge of the semiconvective zone does move outward in mass, the outer edge of the completely convective core moves inward. The result is a distribution of hydrogen abundance with mass that looks like a *ramp*. The literature on the evolution of massive stars with semiconvection is extensive; some of the pioneering efforts are [532, 593, 169, 522]. Semiconvection during core hydrogen burning begins to become effective for stars of mass $M >$ $10M_\odot$; see [556, 522].

This solution, and its justification by a plausible physical argument, has dominated astronomical thinking on this topic. However, it is not unique. The hydrogen abundance could vary with mass as *stairsteps*. The flat parts would be set up by convective mixing, and separated by thin regions of rapid increase in hydrogen abundance and in entropy. Because of their steep gradients, these intermediate regions could carry luminosity by radiative diffusion. Spruit [586] has pointed out that analogous terrestrial examples use the *stairstep* solution.

Figure 7.4 illustrates the helium distribution in a star of $48M_\odot$ for several treatments of semiconvection. In each case, hydrogen is nearly exhausted in the core, but shell burning has not begun. The curve labeled I is the method of Lamb, Iben, and Howard [375], in which semiconvection is not explicitly calculated, but the convective condition is the Schwarzschild criterion that

$$\nabla_r = \nabla_a, \tag{7.63}$$

with abundance gradients ignored. The calculation shown used slightly over 300 zones, which seems adequate to illustrate the method. There is a general gradient which resembles the *ramp* solution; it arises because of a series of intermittent convective episodes. There are vestiges of several of these which may be seen as "steps" containing 3 to 10 zones (each dot shown represents a computational zone).

The curve labeled L represents almost exactly the same procedure, but with the convective condition being that of Ledoux,

$$\nabla_r = \nabla_a - \left(\frac{\partial \ln Y/dr}{\partial \ln P/dr}\right). \tag{7.64}$$

Unlike the I case, this is logically self-consistent because the convection criterion is correct in gradients of composition as well as homogeneous regions. The *stairstep* nature of the solution is obvious. The jumps in composition are accompanied by jumps in entropy, so that (formally at least) the convective motions will not penetrate them. Addition of more zones does not seem to change the result. The profiles resemble those of Spruit [586].

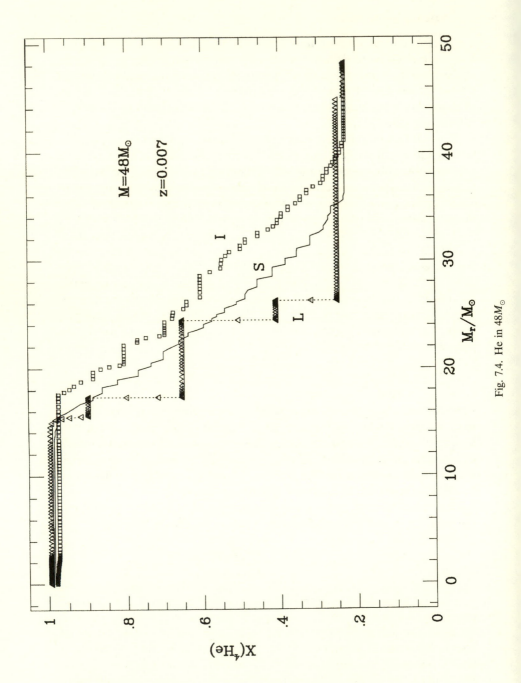

Fig. 7.4. He in $48M_\odot$

The third curve, labeled S, is based upon a Ledoux criterion for convection, but with a semiconvective velocity which gives a net transport of mass but not energy. In a region which was stable by Ledoux but unstable by Schwarzschild, the matter was mixed at a speed that went to zero as neutrality was approached. This approaches the Schwarzschild condition asymptotically. In the case shown, the semiconvective velocity was taken such that its kinetic energy was equal to half the energy difference between the actual state and neutrality, so $v_s^2 = C_p T (\nabla_r - \nabla_a)$. This was used merely as a convenient scaling. As the semiconvective speed was reduced, the abundance curves became more steplike, until they coincided with curve L. Even with the high semiconvective speed used in S, abundance steps appear due to convective regions.

How significant would such effects be on nucleosynthesis? The three curves show only about a 10 percent range about the middle value (case S). Uncertainties in the effects of mass loss [168], and/or interaction with a binary companion [152], are probably larger. According to Chiosi and Maeder [168], a star this massive is expected to lose all of its envelope and become a Wolf-Rayet star (see also [383]). Below about $25M_\odot$, mass loss probably does not remove all the hydrogen. The progenitor of SN1987A still had about $10M_\odot$ of hydrogen left when it exploded. Its pre-explosion luminosity suggests that it had a helium core of $6M_\odot$, so that it probably began life with $20M_\odot$. This corresponds to a mass loss of only a few solar masses.

The implications for stellar evolution are more serious. It has long been known that the hydrogen profile affects the position of the star in the HR diagram. This position would affect the rate of mass loss, which would affect the evolution, which would affect the position in the HR diagram! Unraveling this interrelated problem will be difficult. We leave these uncertainties and focus on the evolution of the core, which—once formed—seems to be relatively little affected by them.

7.6 NUCLEOSYNTHESIS

The evolution of stars involves the development of composition gradients, burning shells, and structure quite different from that of a polytrope. If we focus on the central region of the star (the *core*), it may be approximated by polytropes which are then attached to an *envelope*. The mass of the polytrope used will *decrease* as the star changes from one burning stage to the next; at first it represents the whole star, but later only some of the central part. However, in a given burning stage, shell burning may add mass to the core, so that the effective mass of the polytrope should *increase*. Neutrino emission, which dominates cooling in the later stages, has effects which make the structural representation even more complex.

We need a simple, conceptually clear procedure which captures the general features of this complicated behavior. Fortunately there is one. Consider some region of the star—the flame zone—in which the temperature T_f and density ρ_f are such that

$$\epsilon(T_f, \rho_f) = < \epsilon > = L/M. \qquad (7.65)$$

Having the temperature T_f and density ρ_f, along with the composition, is sufficient to calculate the abundance changes. To uniquely specify the solution, we need a constraint which relates T_f and ρ_f. Previously we used Eddington's Standard Model, which is adequate for stars of homogeneous composition. Once nuclear burning begins to change the composition, application of the Standard Model becomes complicated. For example, only the fuel in a fraction of the mass q_{cc} (the convective core) will be depleted, not throughout the whole star. At this compositional interface, shell burning may occur.

We will examine the nucleosynthesis in a sequence of models having constant radiation entropy:

$$S_\gamma/\Re = 4aT^3/3\rho$$

$$= 0.1213 \, T_9^3/\rho_6. \qquad (7.66)$$

Notice that this is independent of composition. The approximation is suggested by the results of many numerical calculations of stellar evolution, in which the average behavior is that $\rho \sim T^3$, at least for hydrogen and helium burning. This is shown in figure 7.5, which gives the temperature-density structure of a $25M_\odot$ star [43] at two epochs: core hydrogen burning and core helium burning. The structure at core hydrogen burning lies close to a line of constant radiation entropy $S_\gamma/\Re = 1$. Two bounding lines, for $S_\gamma/\Re = 10$ and 0.1, are also shown. The slight upturn at the center reflects the adiabatic gradient in the convective core. The structure at core helium burning is more complex. The core region lies close to the $S_\gamma/\Re = 1$ line. Again there is an upturn at the center, but smaller now because the adiabatic exponent γ is closer to 4/3. The major change is due to the hydrogen burning shell, which has a significantly higher radiation entropy, $S_\gamma/\Re \approx 3$ to 5.

Suppose we consider a polytropic model whose mass corresponds to a *part* of a complete star, usually its core. We need not specify this mass

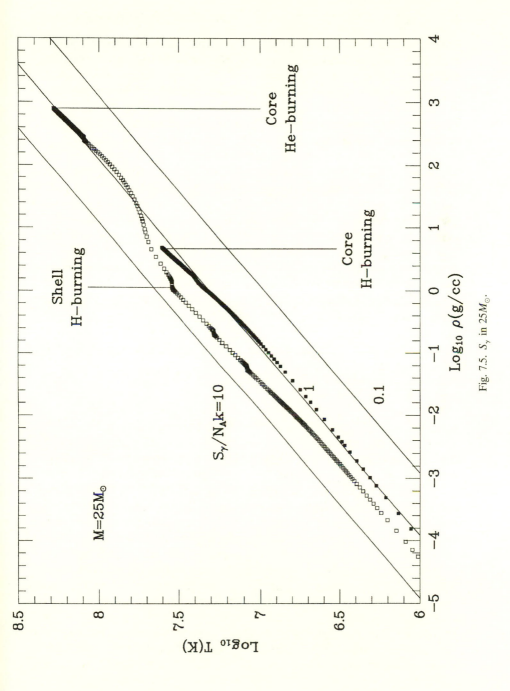

Fig. 7.5. S_γ in $25 M_\odot$.

at present. For a given composition, a choice of S_γ determines β, so that L/M may be estimated using $L/M \approx 4\pi cG(1 - \beta)/\kappa$. Equating this to the energy generation rate $< \epsilon >$ closes the system of equations, and evolutionary sequences may be constructed. To the extent that this is a useful approximation, these sequences will resemble the results of complete stellar evolutionary calculations. That correspondence defines which "mass" the sequence represents. Since this mass is a function of composition, which is changing, the mass also changes. It may be thought of as an estimate of the mass in the region in which fuel is consumed. Later we will add neutrino cooling in order to extend the procedure to the later stages.

Other than direct comparison with the results of stellar evolutionary codes, there are two simple ways to define a "mass" for such sequences. The simplest is to use the instantaneous value for β and composition to define a Standard Model.[5] This is a polytrope of index 3. Alternatively, the L/M value may be regarded as an estimate, and the structure given by a polytrope of index n, where $n = (24 - 18\beta - 3\beta^2)/(8 - 6\beta)$. The generalized version of Eddington's quartic equation is

$$1 - \beta = E_n m^2 (\beta/Y)^4, \tag{7.67}$$

where $E_n = 0.00298(D_n/D_3)^3$ and $D_n/D_3 \approx 1 + 0.183(n - 3)$; see §6.2 and §7.4.

As we saw in §7.4, a solar abundance pattern gives a mole fraction of ideal gas particles of $Y = 1.634$. Hydrogen burning converts the hydrogen to ^{4}He; for the burned hydrogen, this reduces the electrons by two, and the ions from four to one, per He nucleus. This gives $Y = 0.745$. Table 7.4 gives, for a sequence of $S_\gamma/\Re$ values, the values of β, polytropic mass $M_n/M_\odot$, and Standard Model mass $M_3/M_\odot$, for both the solar system composition (H) and for the hydrogen-burned composition (α). The difference in $M_n/M_\odot$ and $M_3/M_\odot$ is a measure of the ambiguity in our approximation; it is larger for the smaller masses, which are less like an $n = 3$ polytrope. Notice, for example, that for a solar composition star of $15M_\odot$, the corresponding helium star has a mass of about $5M_\odot$; accurate evolutionary calculations give a helium core of $4M_\odot$ for such stars. Our approximation does not account for this slight reduction in S_γ during evolution.

Because they are thought to be major contributors to stellar nucleosynthesis, we will concentrate upon stars of initial mass $M \approx 20$ to $25M_\odot$, or $S_\gamma \approx 1.0$. The resulting evolution of abundances, for $S_\gamma = 1.0$, is

---

[5] As β and composition change, the mass changes, so this is an unusual Standard Model.

TABLE 7.4

Mass and Entropy of Stars (Mass-Entropy Relation)

$S_\gamma/\mathfrak{R}$	β_H	$(M_n/M_\odot)_H$	$(M_3/M_\odot)_H$	β_α	$(M_n/M_\odot)_\alpha$	$(M_3/M_\odot)_\alpha$
0.02	0.997	1.895	2.718	0.993	0.590	0.841
0.04	0.994	2.705	3.861	0.987	0.851	1.202
0.10	0.985	4.396	6.189	0.968	1.425	1.957
0.20	0.970	6.494	8.951	0.937	2.198	2.904
0.30	0.956	8.287	11.208	0.909	2.911	3.725
0.50	0.929	11.547	15.109	0.856	4.310	5.256
0.70	0.903	14.648	18.645	0.810	5.742	6.762
1.00	0.867	19.244	23.686	0.749	8.008	9.091
2.00	0.766	35.240	40.380	0.598	16.849	17.994
3.00	0.685	53.005	58.396	0.498	27.848	28.999
4.00	0.620	72.819	78.310	0.427	41.081	42.226
5.00	0.567	94.781	100.309	0.373	56.573	57.711
7.00	0.483	145.321	150.854	0.299	94.378	95.503
10.00	0.395	237.964	243.454	0.230	168.201	169.311

shown in Figure 7.6. This corresponds to a star of $19.2M_\odot$ and $23.7M_\odot$, respectively, in our approximations, using 71% hydrogen and 27% helium as initial abundances. Odd-A nuclei are shown as dotted lines, even-A nuclei with solid. The initial luminosity is $L = 4.59 \times 10^{38}\,\mathrm{erg\,s^{-1}}$. It takes $2.62 \times 10^{14}\,\mathrm{s}$ $(8.30 \times 10^6\,\mathrm{y})$ to evolve through hydrogen burning (to $T_9 = 0.05$).

Below $T_9 = 0.02$ $(T < 2.0 \times 10^7$ K) little happens except for very light nuclei. Deuterium, ^{3}He, and the isotopes of lithium, beryllium, and boron are destroyed. As $T_9 \to 0.0225$ $(\log T_9 \to -1.65)$, the CN cycle begins, as may be seen from the changes in ^{13}C, ^{14}N, and ^{15}N. For $T_9 \to 0.025$ $(\log T_9 \to -1.6)$, ^{12}C, ^{17}O, ^{18}O, and ^{26}Al begin to participate. Hydrogen consumption begins in earnest as $T_9 \to 0.03$ $(\log T_9 \to -1.52)$, at which point there is considerable activity. The CNOF isotopes approach their steady-state values. This increases the number of CNOF catalysts for the cycle, and increases the burning rate. The stellar core has excess energy production, so that it expands, cools, and maintains

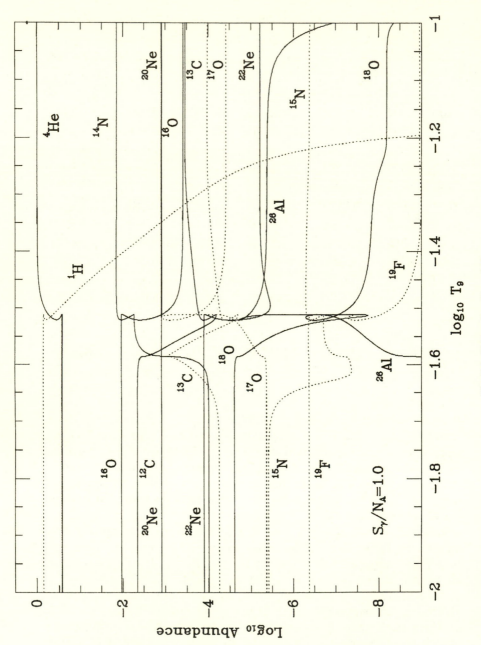

Fig. 7.6. H Burning ($S_\gamma = 1.0$)

thermal energy balance.[6] This causes the complex behavior just as hydrogen consumption begins. In addition to the rearrangement of the usual CNO catalysts, ^{19}F is depleted, ^{22}Ne is reduced in abundance, and a significant amount of the radioactive ^{26}Al is made. Hydrogen is depleted as $T_9 \to 0.05$ ($\log T_9 \to -1.2$). Finally, as $T_9 \to 0.1$ ($\log T_9 \to -1.0$), ^{18}O and ^{26}Al begin to be depleted. The ^{18}O is destroyed by the neutrons produced by ^{13}C$(\alpha, n)^{16}$O reaction; because of the low abundance of ^{13}C, this is a minor effect.

The energy-level structure of the low-lying states of ^{26}Al was shown in Figure 3.6. The ground state of ^{26}Al has a spin and parity of 5^+, so that its decay to the ground state of ^{26}Mg, which is 0^+, is inhibited, and has a half-life of 7.5×10^5 y. However, there is a low-lying excited state of ^{26}Al at 0.2282 MeV which is 0^+; this is not inhibited by a large angular momentum change, and decays with a half-life of 6.36 s. The ratio of half-lives is 3.72×10^{12}, while the fractional population of the excited state is $(\omega_0 + \omega_1 e^{-E_x/kT})/(\omega_0 + \omega_1) \approx 2.63 \times 10^{-13}$, where $\omega \equiv 2J + 1$, so that at $T_9 > 0.1$, the decay from this excited level dominates. Figure 5-15 of Clayton [173] gives a similar evolution of abundances (up to ^{17}O), but at constant temperature, and corresponding to a lower-mass star.

The dominant effect of hydrogen burning is the obvious one: ^{1}H is transformed to ^{4}He. The other major effect is the transformation of CNO isotopes to ^{14}N. This is shown in figure 7.7. The initial abundances for nuclei in the range $12 \le A \le 30$ are shown as filled squares; isotopes of a given element are connected by solid lines. Solar abundances were used for the initial composition. The abundances following hydrogen burning are shown as crosses, with isotopes of an element connected by dotted lines. The abundances of ^{20}Ne, ^{24}Mg, ^{27}Al, and all the Si isotopes are *not* changed. There is some production of ^{14}N, ^{17}O, and ^{23}Na; the enhancements are roughly a factor of ten. Some ^{13}C is made, but ^{12}C, ^{15}N, ^{16}O, ^{18}O, ^{19}F, ^{21}Ne, ^{22}Ne, and ^{25}Mg are depleted. A significant amount of ^{26}Al is produced, but is destroyed by decay as the temperature rises. For lower values of S_γ (lower-mass stars), the duration of hydrogen burning becomes so long that even decay from the ground state of ^{26}Al is sufficient to reduce its abundance.

The production of ^{26}Al is interesting in its own right, but especially so since the gamma-ray line from its decay has been observed (see §2.6). If ^{26}Al is ejected from the star before the end of hydrogen burning, it may be a significant part of the ^{26}Al yield (see also §2.4). More ^{26}Al is produced for larger values of S_γ. This is shown in figure 7.8, which gives

---

[6]These calculations include a thermal energy capacity, so that for fast changes there is a finite time needed to reestablish a balance between heating and cooling.

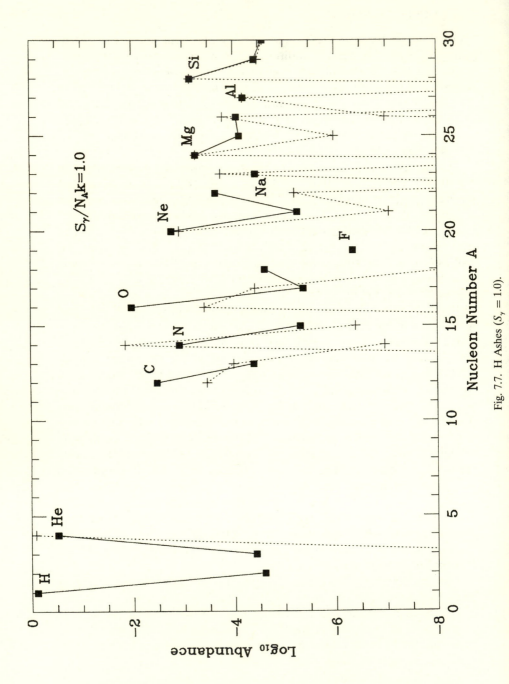

Fig. 7.7. H Ashes ($S_\gamma = 1.0$).

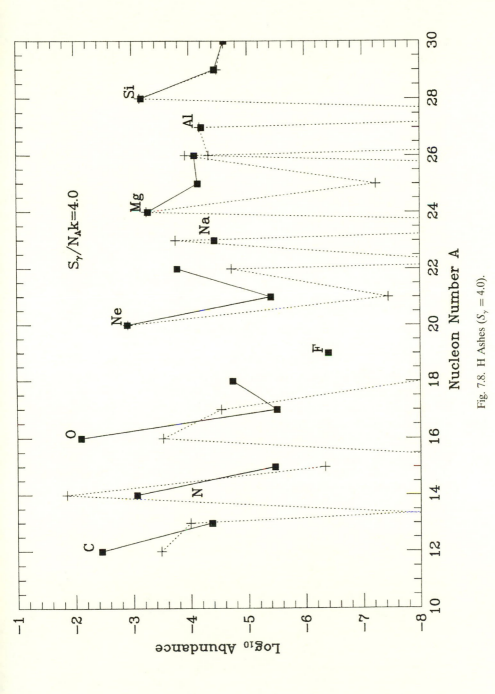

Fig. 7.8. H Ashes ($S_\gamma = 4.0$).

the products of hydrogen burning for $S_\gamma/\Re = 4.0$. This corresponds to a star of initial mass in the range 70 to $80M_\odot$, which probably does suffer extensive mass loss and becomes a Wolf-Rayet star.[7] Such massive stars that shed some of their helium core may well be ^{26}Al producers [503]. Hydrogen burning takes 1.0×10^{14}s (3.2×10^6 y). During hydrogen burning (before excited-state decay becomes important), the enhancement of ^{26}Al relative to ^{25}Mg is more than a factor of 10^3, but roughly half of the enhancement of ^{26}Mg. Another interesting site for ^{26}Al production is in the hydrogen-burning shell of an asymptotic giant branch (AGB) star [235]; see chapter 8.

For a significant yield, ^{26}Al must be ejected soon after it is produced. Here, "soon" refers not only to the evolutionary time-scale for ejection but also to the avoidance of a high temperature stage in which the decay rate could be thermally enhanced. This makes stages of rapid hydrogen burning particularly interesting as sources.

Hydrogen burning by the CNO cycle depends upon the abundance of CNO nuclei which act as catalysts. How would this description be changed if the initial abundances were deficient in these nuclei? The luminosity would be essentially unchanged, because the opacity is dominated by electron scattering. Since $\epsilon \propto X_{CNO}T^n$, where $n \approx 16.7$(see §7.5), and X_{CNO} is reduced, the temperature must increase. For Population II abundances, the reduction of CNO may be a factor of 0.01, so the temperature may increase by 50%. This allows (p, γ) reactions on higher Z nuclei. In addition to the Na enhancement found previously, a similar enhancement in Al occurs.

Sodium-oxygen abundance anticorrelations have been observed in globular cluster giants [180, 493, 187, 577, 362, 578]. Hydrogen burning in intermediate mass AGB stars [365] was applied [209] to the problem; Langer, Hoffman, and Sneden [382] give an extended discussion of the suggestion that ^{23}Na is synthesized in low-mass AGB stars and mixed to the surface. For reasons that are not entirely clear, it had been widely believed that synthesis of Na was not possible in low-mass stars, even though the Ne-Na cycle was explicitly mentioned in B[2]FH, for example [135]. Core-burning temperatures are indeed too modest in low mass stars. However, as shown above, shell burning occurs at higher temperatures. In fact, classical conditions of hydrogen shell burning for low-mass giants [600] give almost identical nucleosynthesis to that just shown for core hydrogen burning in massive stars. A possibly important

---

[7]Recent observations with COMPTEL [211] suggest that ^{26}Al originated in massive stars, presumably core collapse supernovae or Wolf-Rayet stars! It is eerie that our mathematical games can resonate so well with reality.

difference is that the shell burning is faster, giving less time for ^{26}Al to decay. This could be an efficient site for ^{26}Al production. Deep mixing, of the sort used to explain the Na-O anticorrelations, could provide a natural way to remove ^{26}Al from the flame zone. Mass loss could then distribute it to the interstellar medium.

8

Helium-Burning Stars

What is your substance, whereof are you made ...
—William Shakespeare (1564–1616), Sonnet 53

Helium burning is the last stage in which stars are cooled primarily by shining—the radiation of light. This stage poses one of the most perplexing problems of stellar nucleosynthesis—it is the stage in which both ^{12}C and ^{16}O are produced in large amounts. These are the third and fourth most abundant nuclei in the visible Universe, and crucial to the construction of those living organisms of which we know.

It would be easier to understand if helium burning produced almost all ^{12}C, or almost all ^{16}O. The production of comparable amounts implies a balance between a number of competing effects, both nuclear and stellar. Consequently, theoretical predictions have a correspondingly larger uncertainty. Various combinations of these effects can produce the same final result. There is a problem of uniqueness; knowing a possible explanation is not so satisfying as knowing the real one. However it actually comes about, a key parameter for further evolution is the ratio of ^{12}C to ^{16}O from helium burning.

The nuclear features of helium burning are curious. The original ingenious suggestion of Salpeter was that fusion of three alpha particles (^4He) to ^{12}C would occur by a nearly simultaneous collision. This rare occurrence was found to give a burning rate that was too slow for astronomical models, so Hoyle predicted the presence of just the resonance in the ^{12}C compound nucleus that was needed to speed things up. It was there! The production of ^{16}O is even stranger; it occurs almost entirely because of the uncertainty principle. The uncertainty in energy allows a subthreshold resonance to be active when classically it would have been forbidden. As oxygen breathers, we may well regard this aspect of quantum mechanics as relevant! Recent experimental work has improved our knowledge of the rate of this process, which has been the most uncertain key rate in nucleosynthesis.

Helium burning is a stage in which the *s*-process occurs; the *s* denotes neutron capture on a *slow* time-scale. A weak *s*-process occurs during core helium burning in massive stars which produces nuclei up to the $N \approx 50$ closed shell for neutrons. A stronger s-process occurs due to the interaction of separate hydrogen- and helium-burning shells in intermediate-mass stars ($1 < M/M_\odot < 8$); this promises to provide the production of essentially all the heavier s-process nuclei. This is thought to be the source of many of the isotopes of the heaviest elements, from arsenic to lead.

8.1 THERMONUCLEAR FEATURES

The triple-alpha reaction is of fundamental importance for astrophysics. Because

$$3\,^4\text{He} \rightarrow\ ^{12}\text{C} + \gamma \tag{8.1}$$

is a three-body interaction, its rate is proportional to the square of the density. Because stars have lower entropy than that found in a Big Bang cosmology, they are denser for the same temperature. This fact allows the gap at $A = 8$ to be bridged in stars but not in conventional explosive cosmologies. The nuclear nature of helium burning is summarized by the energy-level diagrams in figure 8.1; see [525].

The triple-alpha reaction is interesting from a nuclear point of view. An important aspect of the reaction is the fact that ^{8}Be is unstable against breakup into two alpha particles by only 92 KeV. The ground state has a width of 2.5 eV, so that the natural lifetime is $\tau \approx \hbar/\Delta E = 2.6 \times 10^{-16}$ sec. At helium-burning temperatures the mean time for a thermal alpha particle to move a distance equal to its diameter is less than 10^{-20} sec; this is a crude estimate of how long two alpha particles would "stick" if the ^{8}Be ground-state resonance did not exist. The ^{8}Be builds up to a tiny but significant equilibrium abundance. Salpeter [535] suggested that helium burning occurred by a two-step process:

$$2\,^4\text{He} \rightarrow\ ^8\text{Be}$$
$$^4\text{He} +\ ^8\text{Be} \rightarrow\ ^{12}\text{C} + \gamma. \tag{8.2}$$

In a tour de force of modern astrophysics, Hoyle [303] accurately predicted the position of a nuclear energy level from astronomical arguments: he pointed out that the overall reaction would not be fast enough at stellar energies unless the ^{8}Be(α, γ)^{12}C reaction was resonant. Unable to resist the challenge from a theorist, Cook, Fowler, Lauritsen, and Lauritsen [188] looked for (and were amazed to find) the 0^+ resonance in the predicted energy range: $E_0 \pm \Delta$ above the mass of ^{8}Be $+\ ^4$He. This is the 7.65 MeV 0+ state of ^{12}C shown in figure 8.1.

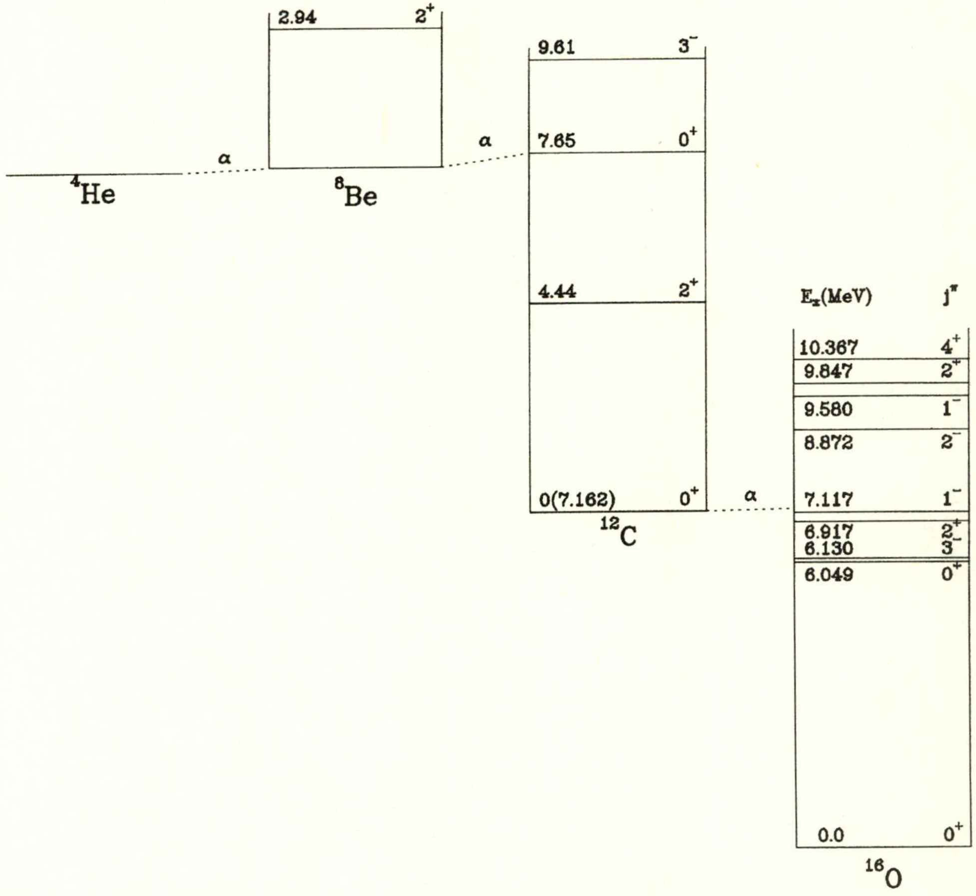

Fig. 8.1. Energy Levels of He-Burning Nuclei

The energy generation rate for the triple-alpha reaction is

$$\epsilon_{3\alpha} \approx \epsilon_0 \rho^2 \frac{X_\alpha^{\,3}}{T_8^{\,3}} f_{3\alpha}\, e^{-42.94/T_8} \tag{8.3}$$

where $f_{3\alpha}$ is the screening factor, and $\epsilon_0 = 3.9 \times 10^{11} \mathrm{erg\,g^{-1}\,s^{-1}}$. Near $T_8 = 2$,

$$\epsilon_{3\alpha} \approx \epsilon_1 \rho^2 X_\alpha^{\,3} \left(\frac{T_8}{2} \right)^{18.5}, \tag{8.4}$$

where $\epsilon_1 = 23.1 \,\mathrm{erg\,g^{-1}\,s^{-1}}$.

An interesting aspect of helium burning is the rate of the reaction $^{12}\mathrm{C}(\alpha, \gamma)^{16}\mathrm{O}$. If it is rapid, then the net result of helium burning would be $4\alpha \rightarrow {}^{16}\mathrm{O}$ (we know that $^{16}\mathrm{O}(\alpha, \gamma)^{20}\mathrm{Ne}$ is slow at these temperatures). The reaction $^{12}\mathrm{C}(\alpha, \gamma)^{16}\mathrm{O}$ would seem to be nonresonant in that no states in $^{16}\mathrm{O}$ lie near E_0 plus the mass of $^{12}\mathrm{C}$ and $^4\mathrm{He}$. The $^{12}\mathrm{C} + {}^{14}\mathrm{He}$ state lies at 7.161 MeV. Below this, at 7.115 MeV there is a 1^- state with a gamma width of only $\Gamma_\alpha = 0.066\,\mathrm{eV}$. Nevertheless, the state has a finite lifetime against radiative de-excitation, and therefore a corresponding indeterminacy in its energy. It is thought that $^{12}\mathrm{C}(\alpha, \gamma)^{16}\mathrm{O}$ can occur through the tails of this state, but the rate for this process has been difficult to measure. It is one of the oldest of the vexing problems of experimental nuclear astrophysics. The 7.162 MeV 0^+ level of $^{12}\mathrm{C} + \alpha$ and the 7.117 MeV 1^- level of $^{16}\mathrm{O}$ are shown in figure 8.1; they lie so close in energy that it is difficult to indicate on the graph.

The reduced width of the state, $\theta_\alpha^{\,2}$, is one parameterization of the desired nuclear information. Stephenson [591] estimated the reduced width to be $\theta_\alpha^{\,2} = 0.085$. For similar values of $\theta_\alpha^{\,2}$, Deinzer and Salpeter [208] had found that helium stars of most reasonable masses gave approximately equal amounts of $^{12}\mathrm{C}$ and $^{16}\mathrm{O}$ at the end of helium burning. For $\theta_\alpha^{\,2}$ larger by a factor of four, very little $^{12}\mathrm{C}$ is left. In subsequent years nuclear experimentalists suggested values that are larger by a factor of ten or more. More recent work [215, 509, 363, 472, 711, 136] shows a tendency toward convergence (see below). A direct determination requires measurement of cross sections well below the nano-barn range; for good discussions of the severe experimental difficulties see Barnes [73] and Rolfs and Rodney [525].

The production of $^{12}\mathrm{C}$ and $^{16}\mathrm{O}$ during helium-burning nucleosynthesis is illustrated in figure 8.2, as reproduced by a simple equivalent net-

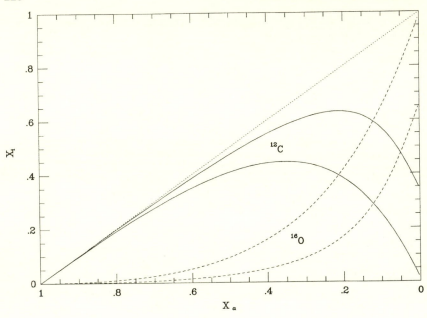

Fig. 8.2. He Burning

work. Because the $^{16}O(\alpha, \gamma)^{20}Ne$ reaction has no accessible resonances in the bombarding energies of interest for helium burning, little ^{20}Ne is produced. Therefore,

$$dY(^{12}C)/dt = \frac{1}{3}Y(^4He)^3\rho^2 N_A^2\langle\sigma v\rangle_{3\alpha}$$
$$-Y(^{12}C)Y(^4He)\rho N_A\langle\sigma v\rangle_{\alpha\gamma}, \qquad (8.5)$$

$$dY(^{16}O)/dt = Y(^{12}C)Y(^4He)\rho N_A\langle\sigma v\rangle_{\alpha\gamma}, \qquad (8.6)$$

$$dY(^4He)/dt = -Y(^4He)^3\rho^2 N_A^2\langle\sigma v\rangle_{3\alpha}$$
$$-Y(^{12}C)Y(^4He)\rho N_A\langle\sigma v\rangle_{\alpha\gamma}. \qquad (8.7)$$

Let us define

$$\Lambda = \frac{1}{3}Y(^4He)^3\rho^2 N_A^2\langle\sigma v\rangle_{3\alpha}, \qquad (8.8)$$

$$\lambda = Y(^{12}C)Y(^4He)\rho N_A\langle\sigma v\rangle_{\alpha\gamma}. \qquad (8.9)$$

By dividing the last equation into the first two, we can measure time in terms of ^{4}He consumed. Thus,

$$dY(^{12}C)/dY(^4He) = -\frac{\Lambda - \lambda}{3\Lambda + \lambda}, \qquad (8.10)$$

$$dY(^{16}O)/dY(^4He) = -\frac{\lambda}{3\Lambda + \lambda}. \qquad (8.11)$$

These last two equations are not independent if all the mass is in the form of ^{4}He, ^{12}C, or ^{16}O; we may consider only 8.10. Because $3Y(^{12}C)/Y(^4He) = X(^{12}C)/X(^4He)$,

$$dX(^{12}C)/dX(^4He) = -3\frac{(\Lambda - \lambda)}{(3\Lambda + \lambda)}. \qquad (8.12)$$

The solid lines in figure 8.2 show the behavior of ^{12}C for two different values of Λ/λ; the dashed lines represent ^{16}O. For small λ, $X(^{12}C)$ grows linearly with declining $X(^4He)$; this corresponds to a small $\langle \sigma v \rangle_{\alpha\gamma}$, and is shown as the dotted line. All solutions begin this way because λ must be small before ^{12}C is formed. Small Λ relative to λ may be obtained by a large $\langle \sigma v \rangle_{\alpha\gamma}$, by small ρ, or by small $X(^4He)$ because of the cubic dependence on abundance of helium; this gives

$$dX(^{12}C)/dX(^4He) = 3. \qquad (8.13)$$

As helium is exhausted, the rate of carbon destruction is rapid; essentially each ^{4}He destroys one ^{12}C. This implies that the final ^{12}C abundance from helium burning is sensitive to any fresh ^{4}He that is added to the convective region, especially near the end of the burning stage. In particular, numerical algorithms that give excessive mixing will result in significant underproduction of ^{12}C.

Figure 8.2 was constructed using a fixed temperature and two fixed densities, one of which was larger by a factor of four. Because of the extra density dependence of the triple-alpha reaction, it is decreased by the lower density, to give smaller Λ. Increasing the mass of the helium core increases the radiation entropy, which has the same effect. This increases the destruction of ^{12}C, so that the production of oxygen relative to carbon increases with stellar mass.

Finally, note that incomplete burning of ^{4}He gives less destruction of ^{12}C. This means that, even if complete consumption of helium in the core-burning stage produces little ^{12}C, shell burning which is terminated by mass loss or explosion—rather than by exhaustion of fuel—may still produce significant amounts of ^{12}C (see Hoyle [303]).

The uncertainty in the rate of the $^{12}C(\alpha, \gamma)^{16}O$ reaction has posed a challenge for the theorists and experimenters alike. Table 8.1 illustrates the historical range of this uncertainty, for several temperatures (and densities, using $S_\gamma/\mathfrak{R} = 1$) thought to be most relevant for nucleosynthesis. As we saw above, the abundances of ^{12}C and ^{16}O depend upon the ratio of the reaction rate for $^{12}C(\alpha, \gamma)^{16}O$ to that for the triple-alpha reaction. The values given in CFHZ85 [155] have been widely used. Also shown are values given previously (FCZI [236] and FCZII [237]), and from the fits of Thielemann, Arnould, and Truran (TAT86 [615]). The estimate by Caughlan and Fowler (CF88 [156]) includes an estimate of the uncertainty in the inferred rate, which is a factor of two up or down. It spans a range from mostly carbon to essentially no carbon produced (a small amount of ^{20}Ne is made).[1] The values for the ratio of the rates are consistent for FCZI, FCZII and CF88; the CFHZ85 and TAT86 values are significantly higher. The values labeled DA92 are those used in this book; an intermediate case seemed an appropriate example. It was chosen by examining the ^{12}C to ^{16}O ratio and other nucleosynthesis produced in core helium burning for various values of the rate, as usually done [25, 617]. The error estimate of Caughlan and Fowler (CF88 [156]) is at least marginally consistent with all these values. Strangely enough, DA92 corresponds to 1.7 times CF88, which is exactly the center of the range (1.7 ± 0.5) that Weaver and Woosley [670] have recently recommended from nucleosynthesis arguments.[2]

The experimental situation has recently become much improved. Buchmann et al. [136] have used the TRIUMF cyclotron to examine the β-delayed spectrum of ^{16}N, and infer the reduced width of the relevant state in ^{16}O. Their value is within the errors of Dyer and Barnes [215], lower than Redder et al. [509], similar to Kremer et al. [363], and larger than Ouellet et al. [472]. Zhao et al. [711], have found a value about 60% larger, and almost identical to the value of 1.7 times CF88 found from the nucleosynthesis and stellar evolution arguments just mentioned. The estimated error is mostly systematic, and about 35%. This level of modest disagreement is a major advance!

---

[1] Solutions that produce no carbon from completed helium burning, but rely upon incomplete burning, must also involve alternative sources of solar abundances of Na and Al, which are naturally produced in carbon burning. As we saw, hydrogen shell burning might produce some Na, but the present experimental situation supports carbon production in completed helium burning.

[2] This independent determination of the same value is comforting, but should not be stressed unduly. While it recommends our mathematics, it does not prove we are not both making the equivalent mistakes in physics!

TABLE 8.1

Rates for Helium Burning (He RATES)

	FCZ67	FCZ75	CFHZ85	TAT86	CF88	DA92
$T_8 = 1.6$ $\rho = 497$						
$^4\text{He}(2\alpha, \gamma)$	2.83e-12	1.60e-12	1.88e-12	1.88e-12	1.90e-12	1.88e-12
$^{12}\text{C}(\alpha, \gamma)$	6.39e-14	3.77e-14	1.18e-13	1.30e-13	4.90e-14	8.28e-14
Ratio	0.0225	0.00236	0.0627	0.0692	0.0258	0.0439
$T_8 = 1.8$ $\rho = 707$						
$^4\text{He}(2\alpha, \gamma)$	7.96e-11	4.87e-11	5.71e-11	5.71e-11	5.76e-11	5.71e-11
$^{12}\text{C}(\alpha, \gamma)$	7.16e-13	4.34e-13	1.35e-12	1.49e-12	5.60e-13	9.42e-13
Ratio	0.0090	0.00089	0.0236	0.0261	0.0097	0.0165
$T_8 = 2.0$ $\rho = 970$						
$^4\text{He}(2\alpha, \gamma)$	1.19e-09	7.74e-10	9.04e-10	9.04e-10	9.12e-10	9.04e-10
$^{12}\text{C}(\alpha, \gamma)$	5.77e-12	3.59e-12	1.10e-11	1.22e-11	4.60e-12	7.69e-12
Ratio	0.0049	0.00046	0.0122	0.0135	0.0050	0.0085

Despite this wonderful experimental progress, the question is not yet resolved. Uncertainties in the details of stellar models, such as those mentioned above, prevent us—now—from entirely disentangling the details of the nuclear from the stellar effects. The present experimental situation strongly supports the general nature of the nucleosynthesis arguments, but variations of detail are still allowed within the uncertainties of the measurements. The range of these uncertainties is still sufficient for a significant range of stellar behavior. Ultimately, further nuclear experiment will be required to place nucleosynthesis theory on a firm (and uncontested) empirical base.

8.2   IGNITION

In what follows we will examine the nature of helium stars, as an approximation to the hydrogen-exhausted cores of more massive stars of conventional composition. However, such objects are also interesting because they may be the result of the stripping of a hydrogen envelope, either by extensive mass loss or by interaction with a binary companion.

What stars can burn helium directly after core hydrogen burning? For a given temperature, the ignition mass can be estimated by equating energy release from nuclear burning to stellar luminosity. Assuming Thomson opacity, using Eq. 8.3, Eq. 6.64, Eq. 6.39, and $\beta = 1$, gives $T_8 \approx 0.88$. By logarithmic interpolation in table 6.4, this implies $M/M_c \approx 0.18$ or $M \approx 0.26 M_\odot$. Suppose this "helium star" is just the convective core of a larger star which has burned hydrogen. For $n = \frac{3}{2}$ and Kramers opacity, the fractional size of the convective core is $q_{cc} \approx 0.1$ from Eq. 7.28, so we expect this to correspond to a star of mass

$$M \geq \frac{0.26 M_\odot}{0.1} \geq 2.6 M_\odot. \tag{8.14}$$

Detailed models [323] give $M \geq 2.25 M_\odot$, which is reasonable agreement for such a simple estimate.[3] Stars more massive than this proceed directly from hydrogen exhaustion in core hydrogen burning to helium ignition.

Less massive stars would not ignite He in their cores soon after H depletion, but would instead develop isothermal cores partially supported by degeneracy pressure. These cores would slowly grow by shell burning (see chapter 7), and eventually ignite helium-burning reactions ($M_{core} \approx 0.45 M_\odot$ [318]). For such stars, over 40 percent of the hydrogen consumption occurs during this shell-burning phase.

The character of the ignition changes, depending upon whether the core is degenerate or nondegenerate. Because of the high power of temperature in the energy generation rate, it is reasonable to suppose that once begun, nuclear burning is fast compared to the diffusion time-scale. Then the first law of thermodynamics is

$$\dot{E} + P\dot{V} = \epsilon - \frac{\partial L}{\partial m} \tag{8.15}$$

$$\approx \epsilon. \tag{8.16}$$

For a hydrostatic object, we found $T \sim \rho^{1/3}$ (Eq. 6.47), or $\dot{V} = -3V\dot{T}/T$. Therefore, the energy generated is

$$\epsilon \delta t \approx \left(E_T - \frac{3V}{T}(P + E_V) \right) \delta T. \tag{8.17}$$

---

[3] A finer mesh of entries in table 6.4, and a careful treatment of electron screening and opacity, would have improved the comparison.

Now for a nondegenerate object, $E_T \approx E/T$, $E_V \approx 0$, and $3/2\, PV \approx E$, so

$$\delta T \approx -\frac{T}{E}\epsilon\delta t. \qquad (8.18)$$

An increase in energy production gives a decrease in temperature. The system has negative feedback, and is stable.

For a degenerate gas, $E_T \approx 0$ and $E_V \approx -P$, so

$$\delta T \approx +\frac{\epsilon\delta t}{\phi} \qquad (8.19)$$

where $\phi = E_T - 3V(P + E_V)/T \to 0$ as the gas becomes completely degenerate. The change in temperature is large and positive. This would increase the reaction rates in turn, and a thermal runaway would develop [558]. The system has positive feedback and is unstable. If the instability is violent enough, hydrodynamic motion can result. Such degenerate ignition of helium gives rise to a *helium flash* [355].

It is not yet completely clear how the star evolves during the flash [316, 326, 323]. The difficulties lie in the possible hydrodynamic nature of the helium flash, the uncertainty regarding mass loss, and the uncertainty regarding the nature and distribution of the final composition. Subsequent evolution is presumed to be quasi-static settling to core helium burning [618, 427]. Because such intermediate- and low-mass stars are numerous, even subtle modifications of the abundances in the ejected matter can be important for stellar nucleosynthesis. However, it appears that most of the processed matter—the helium core that flashed—remains in the white dwarf.

What are the conditions for helium ignition in more massive stars? The basic procedure is the same as it was for hydrogen burning. Because Thomson scattering dominates the opacity and the triple-alpha reaction is highly temperature sensitive, the cores will be convective. In the solar abundance pattern, the CNO isotopes comprise 75 percent of z, the total abundance by mass in elements beyond H and He. For moderate to massive stars, the abundances at helium ignition are $X(^4\mathrm{He}) = 1 - 0.75z$ and $X(^{14}\mathrm{N}) = 0.75z$, while there is little change in the nuclei for $Z \geq 10$ (see chapter 7 for details). For solar system abundances, $z \approx 0.02$. Therefore, $Y_e \approx 0.5$ and $Y \approx 0.746$. Now, $\beta = 4Y/(4Y + S_\gamma/\Re)$, so that the mass may be derived from Eddington's quartic equation (6.45). To illustrate the procedure, we follow Clayton [173] and approximate the energy generation rate near $T = 2 \times 10^8$ K by

$$\epsilon \approx 23.1\, \rho^2 X_\alpha^3 (T/2.0 \times 10^8 \mathrm{K})^{18.5}. \qquad (8.20)$$

TABLE 8.2

Main Sequence for Helium Stars (He ZAMS)

$M_\alpha/M_\odot$	β	$L/L_\odot$	$(T_8)_c$	ρ_c	R(cm)	T_{eff}(K)
0.25	0.999	9.68E+0	1.629	2.93E+5	2.80E+09	5.09E+4
0.50	0.998	7.69E+1	1.813	1.01E+5	5.02E+09	6.37E+4
0.75	0.995	2.56E+2	1.928	5.48E+4	7.06E+09	7.27E+4
1.00	0.991	5.98E+2	2.014	3.55E+4	8.98E+09	7.96E+4
1.25	0.986	1.14E+3	2.082	2.55E+4	1.08E+10	8.54E+4
1.50	0.980	1.93E+3	2.138	1.95E+4	1.25E+10	9.03E+4
2.00	0.966	4.33E+3	2.228	1.30E+4	1.58E+10	9.84E+4
3.00	0.934	1.28E+4	2.351	7.49E+3	2.17E+10	1.10E+5
4.00	0.899	2.60E+4	2.434	5.24E+3	2.70E+10	1.18E+5
6.00	0.833	6.46E+4	2.541	3.33E+3	3.59E+10	1.28E+5
8.00	0.776	1.16E+5	2.609	2.51E+3	4.34E+10	1.35E+5
12.0	0.689	2.41E+5	2.693	1.76E+3	5.60E+10	1.43E+5
16.0	0.625	3.88E+5	2.746	1.40E+3	6.64E+10	1.48E+5
30.0	0.492	9.84E+5	2.846	9.08E+2	9.47E+10	1.56E+5
45.0	0.416	1.70E+6	2.902	7.08E+2	1.18E+11	1.60E+5
60.0	0.368	2.45E+6	2.939	6.00E+2	1.37E+11	1.63E+5
100	0.293	4.57E+6	2.999	4.54E+2	1.78E+11	1.67E+5
200	0.213	1.02E+7	3.075	3.19E+2	2.53E+11	1.72E+5
500	0.138	2.79E+7	3.167	2.06E+2	3.97E+11	1.76E+5
1000	0.098	5.82E+7	3.233	1.50E+2	5.56E+11	1.79E+5

Again we impose thermal balance, and use

$$\frac{\langle \epsilon \rangle}{\epsilon_c} = \frac{3.23}{(3u + s)^{3/2}}, \tag{8.21}$$

where $s = 18.5$, $u = 3$. This is solved for temperature (and thus giving density, radius of the core, and so on). The results[4] are presented in table 8.2.

[4]For better numerical accuracy, a locally fitted power-law was used for each model, instead of using the form fitted at $T_8 = 2$. The values for the smallest mass are not very reliable because of neglect of electron screening and degeneracy; this is near the ignition mass.

8.3 CORE NUCLEOSYNTHESIS

The procedure of the previous chapter can be extended to helium burning in a convective core. We will examine nucleosynthesis in an evolutionary sequence having a constant value for radiation entropy:

$$S_\gamma/\Re = 4aT^3/3\Re\rho. \tag{8.22}$$

The corresponding mass estimates and values of β were given in table 7.4.

Because of the delicacy of the prediction of the ratio of ^{12}C to ^{16}O, a slight modification is needed. To have a "one-zone" representation of thermonuclear burning in the core, we must perform an integration over spatial variables. In this sense we use "average" values. In the previous chapter we approximated those averages as the values at the point having temperaure T_f and density ρ_f such that

$$\epsilon(T_f, \rho_f) = <\epsilon> = L/M. \tag{8.23}$$

For helium burning it is more convenient to choose the point (T_r, ρ_r) such that the ratio of the local value of ^{12}C$(\alpha, \gamma)^{16}$O/^{4}He$(2\alpha, \gamma)^{12}$C equals the same ratio for these rates integrated over the convective core. This *reaction* average point (T_r, ρ_r) is not necessarily the same as the *flame* average point (T_f, ρ_f) because of the different dependences upon temperature and density. This is approximately equivalent[5] to

$$\epsilon(T_f, \rho_f) \approx <\epsilon> = 3L/M, \tag{8.24}$$

for $S_\gamma = 1.0$ and our choice of reaction rates. This gives reasonably accurate abundances, but the times are shorter by this factor (which does not affect the discussion below).

Figure 8.3 continues the evolution of selected nuclei from figure 7.6. Even-A nuclei are represented by solid lines, and odd-A by dotted lines. Here $S_\gamma = 1.0$, corresponding to stars of initial mass $M \approx 20$ to $25 M_\odot$, or helium cores of $M_\alpha \approx 8$ to $9 M_\odot$. Actual evolutionary calculations give relatively smaller helium cores than this prescription (corresponding to $M_\alpha \approx 6$ to $8 M_\odot$); S_γ decreases slightly with evolution. At $T_9 = 0.1$ (log $T_9 = -1$), ^{26}Al and ^{18}O continue declining as discussed in §7.6. As $T_9 \to 0.125$ (log $T_9 = -0.9$), ^{13}C$(\alpha, n)^{16}$O destroys ^{18}O by ^{18}O$(n, \gamma)^{19}$F. However, ^{18}O is produced by ^{14}N$(\alpha, \gamma)^{18}$F followed by decay of ^{18}F to

---

[5]The algebra is simple, but relatively tedious. The Fowler-Hoyle approximation of Appendix C may be used for the reaction rates as well as for the energy generation rates.

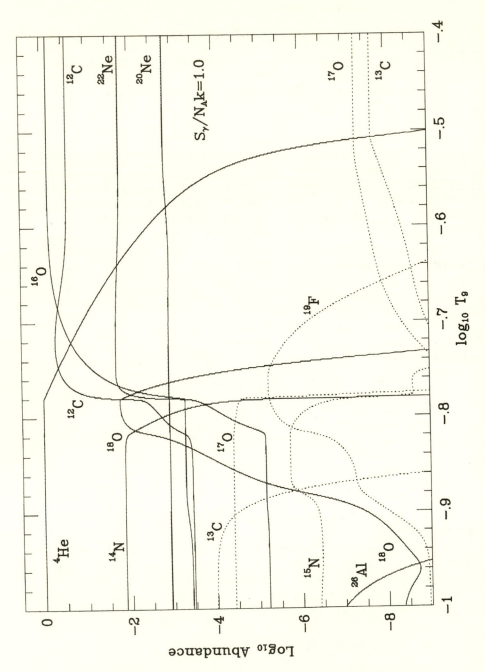

Fig. 8.3. He Burning ($S_\gamma = 1.0$)

^{18}O. This process has important implications because it converts a relatively abundant nucleus with no neutron excess (^{14}N) to a nucleus (^{18}O) having a fractional neutron excess of $\Delta\eta = 2/18 = 0.111$. For solar system abundances (Population I), this involves about $0.02 \times 2/3 = 0.013$ of the nucleon fraction, giving a total neutron excess of $\eta \approx 1.5 \times 10^{-3}$. As we shall see in chapter 9, this is roughly that needed to correctly reproduce the isotopic ratios between Ne and Ni. When $T_9 \to 0.144$ ($\log T_9 = -0.84$), ^{14}N is depleted, replaced by ^{18}O.

When $T_9 \to 0.158$ ($\log T_9 = -0.8$), ^{4}He depletion begins to be significant, and the main phase of helium burning begins. This results in production first of ^{12}C, and then ^{16}O. Simultaneously, ^{18}O$(\alpha, \gamma)^{22}$Ne occurs, as well as ^{22}Ne$(\alpha, \gamma)^{26}$Mg and ^{22}Ne$(\alpha, n)^{25}$Mg. The Mg isotopes are not shown here. Although this (α, n) is inhibited by the relatively low temperature, it does produce a significant flux of neutrons, which are captured predominantly by heavier nuclei.

Figure 8.4 shows the final abundances from helium burning (crosses), for nuclei with $A \leq 50$; the solar system abundances, normalized at ^{12}C and ^{16}O, are shown for comparison (solid squares). The ^{12}C$(\alpha, \gamma)^{16}$O rate of 1.7 times that of Caughlan and Fowler (CF88 [156]) was used (0.70 of CFHC85 [155]); this gives a ^{12}C/^{16}O ratio that is almost solar[6] for $S_\gamma = 1.0$. This normalization gives a solar value for ^{22}Ne, and a slightly deficient production of ^{21}Ne. Both ^{40}K and ^{40}Ar have the inferred value at the time of solar system formation.[7] Their production has been a puzzle since early calculations of explosive oxygen burning showed that neighboring nuclei were produced by that process, but not ^{40}K or ^{40}Ar.

The neutron-rich magnesium isotopes, ^{25}Mg and ^{26}Mg, are underproduced by about a factor of two. There are no overproductions; for example, at $A = 40$ the cross lying above the box for ^{40}Ar is actually ^{40}Ca, which is underabundant by a factor of about 100. Similarly the entry at $A = 46$ is ^{46}Ti, which is underproduced by a factor of 30, rather than ^{46}Ca, which would have been an overproduction by a factor of three.

Because of the flux of neutrons, heavy elements (especially the iron peak) are driven up to larger A by successive (n, γ) reactions. In order to examine the abundances of these nuclei, a network was used which included 297 nuclei up to ^{88}Kr; except as mentioned in the text, the rates

---

[6]Because of the uncertainties previously discussed, only a modest weight should be attached to this agreement. It is convenient for our discussion, and indicates that current estimates are in the correct range. However, a different rate—taken with different degree of mixing—could also produce a reasonable ratio. Further, no contribution from lower-mass stars is included.

[7]This is not favorable for cosmochronology. Because no other isotopes of K or Ar are produced by this process, we do not know how to construct an accurate ratio.

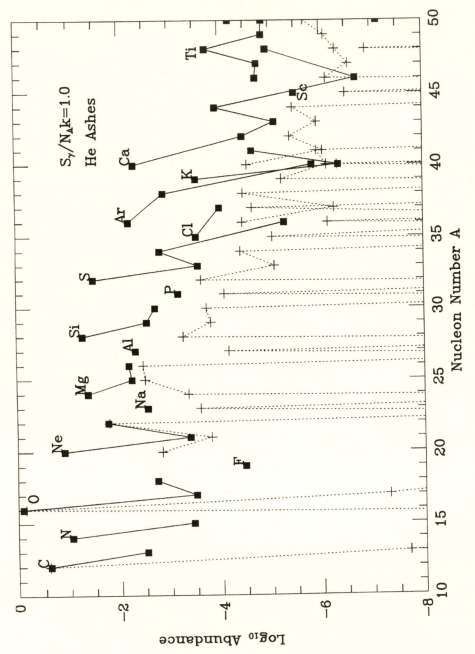

Fig. 8.4. Low-Z He Ashes ($S_\gamma = 1.0$)

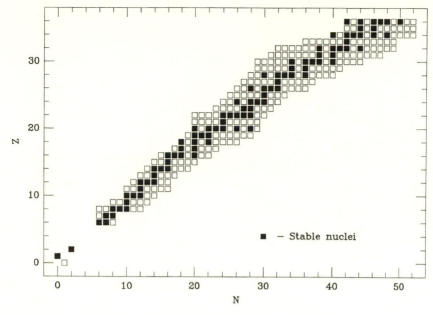

Fig. 8.5. Network

of TAT86 [615] were used. The full network is shown in figure 8.5; solid squares denote stable nuclei. This is sufficiently large to include all the charged particle reactions of importance for hydrostatic burning stages. The fragile light elements Li, Be, and B have been ignored for simplicity; they should be included for pre-main-sequence evolution, and for some high-entropy explosive processes (alpha-rich freezeout; see chapter 9).

During helium burning, heavy nuclei ($Z > 25$) are produced [492, 194]; this is a low-exposure s-process [173]. Figure 8.6 shows the final abundances from helium burning (crosses), for nuclei with $50 \leq A \leq 90$; the solar system abundances, normalized at ^{12}C and ^{16}O as in the previous figure, are shown for comparison (solid squares). A solar abundance of ^{58}Fe results. From $A = 59$ to $A = 70$, most isotopes are produced about a factor of two to four lower than the solar distribution. Above $A = 70$ the pattern continues, but the isotopes become more deficient relative to solar values. This deficiency is in agreement with that originally found [194]. A subsequent analysis [374] found similar results in general; however, in addition to the Mg isotopes, ^{40}K, and ^{58}Fe, they also found significant amounts of ^{36}S, and ^{37}Cl were produced (these effects are not evident in figure 8.6). The changes probably reflect the effect of revised reaction rates for nuclei in this region.

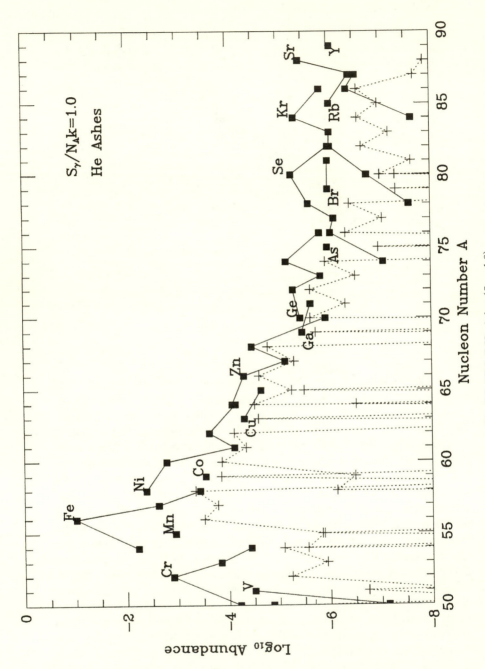

Fig. 8.6. High-Z He Ashes ($S_\gamma = 1.0$)

The strength of the neutron flux is dependent upon the rate of the reaction $^{22}Ne(\alpha, n)^{25}Mg$ relative to the rate of helium burning (for example, the triple-α rate). This may be increased by using larger values of S_γ (more massive stars). As expected, ^{12}C and ^{22}Ne are reduced, ^{25}Mg and ^{26}Mg are enhanced to essentially solar values, and the nuclei beyond $A = 70$ are enhanced. Although ^{40}K changes little, ^{58}Fe is reduced slightly. It is interesting that s-process nuclei out to $A = 90$ are produced; that is the edge of the network used here.

8.4   SHELL NUCLEOSYNTHESIS

Shell burning is an important aspect of stellar evolution, and one that is difficult to treat as simply as core burning. First, hydrogen shell burning causes the helium core to grow, which affects our previous discussion. The rate of mass processing is

$$dM/dt = L/X_f Q_f, \tag{8.25}$$

where X_f is the nucleon fraction of fuel, and Q_f is the energy released by burning a unit mass of that fuel. In massive stars, the luminosity produced by the hydrogen shell is comparable to that produced in the helium-burning core. Therefore, the relative rates of mass consumption are

$$\dot{M}_H/\dot{M}_\alpha = (L_H/L_\alpha)(X_H/X_\alpha)(Q_H/Q_\alpha), \tag{8.26}$$

$$\approx 0.1, \tag{8.27}$$

so that, due to the hydrogen-burning shell, the helium core grows in mass by (roughly) a tenth while helium burning occurs. Because the fractional size of the convective core slightly increases, the absolute size of the convective core should grow by more than a tenth. Therefore, in the last half of core helium burning, new helium should be mixed in, reducing the yield of ^{12}C. Another consequence of shell burning is that because the mass so processed must lie outside the core, it will most likely be ejected if core material is, and contribute its ashes to the yield.

In chapter 7 we saw that burning shells have characteristic luminosities, so that they also have characteristic rates of mass consumption, $\dot{M}_{shell}$. If there is only one burning shell, overlaid by a quasi-static envelope of fuel, this envelope will quasistatically settle into the shell, feeding it at whatever characteristic rate it has. Suppose that mass is being fed to the shell, at a rate $\dot{M}_{supply}$, by accretion from a binary companion, or from processing by another burning shell. It is not necessary that

$\dot{M}_{shell} = \dot{M}_{supply}$. If $\dot{M}_{shell} < \dot{M}_{supply}$, then mass will accumulate on top of the burning shell, forcing it to higher pressure, and therefore greater burning rates. Thus, $\dot{M}_{shell}$ increases, and we have at least the possibility of attaining a balance (however, overdamping is also possible). If $\dot{M}_{shell} > \dot{M}_{supply}$, then the shell will run out of fuel and must enter a quiet phase (little or no burning) while a sufficient amount of overlying fuel builds up to allow it to reignite.

From this viewpoint, it is not so surprising that Schwarzschild and Härm [559] found a hydrostatic relaxation oscillation ("thermal pulses") in the process of investigating the evolution of a star with two active burning shells (see Iben [318] for discussion and references). What was unanticipated was the mixing: the phase of active helium burning was sufficiently vigorous to drive its convective zone out to include some of the hydrogen-rich layers—beyond the position of the previously active hydrogen burning shell. The relevant point for nucleosynthesis is the production of large fluxes of neutrons. This may occur by:

- *The ^{13}C Source:* First mix ^{12}C from the helium zone into the hydrogen flame zone for $^{12}C(p, \gamma)^{13}N(\beta^+\nu)^{13}C$, and then mix ^{13}C back to higher temperatures for $^{13}C(\alpha, n)^{16}O$. Acting upon the already present seed nuclei, these neutrons can produce s-process nuclei.

- *The ^{22}Ne Source:* Alternatively, the CNOF catalyst ^{14}N may be mixed into the helium flame zone, allowing the reaction sequence $^{14}N(\alpha, \gamma)^{18}F$ $(\beta^+\nu)^{18}O\ (\alpha, \gamma)^{22}Ne(\alpha, n)^{25}Mg$ to occur. This also produces a neutron flux. The ^{14}N is made by the CNOF cycle, and may be enhanced by the mixing of ^{12}C from the helium flame zone into the hydrogen-burning region.

These two neutron sources were identified by Cameron long ago [138, 144]. The relative merits of the two proposals are the subject of some debate [636, 642]. The existence of the correct amount of mixing (or any at all) is a key difficulty with the ^{13}C source, while the chief difficulty with the ^{22}Ne source is the high temperature required to carry the $^{22}Ne(\alpha, n)^{25}Mg$ reaction to completion.

What stars might do this? The star must be sufficiently massive to ignite He shell burning, so masses above those for the He flash ($M > 2.3M_\odot$) are indicated. However, the mass should be low enough so that core carbon burning does not occur; this limits the burning to the two shells, and allows a long time for their action. This implies $M < 9M_\odot$. Such "intermediate mass" stars spend a long time in the double shell-burning phase.

Seeger, Clayton, and Fowler [564] had found that no single neutron flux provided a fit to the observed solar system abundances ascribed to the s-process, but that a distribution $\rho(\tau)$ of fluxes was required:

$$\rho(\tau) \approx G e^{-\tau/\tau_0}, \tag{8.28}$$

where the neutron exposure $\tau = v_T \int n_n(t)dt$, and the v_T is the thermal velocity (see Clayton [173]). This means that a large amount of matter sees a small flux, and a small amount of matter sees a large flux (sufficient to make Pb, the heaviest element made by slow neutron capture). Ulrich and Scalo [643, 641, 543] and Iben [319] showed that shell pulses naturally give rise to such an exponential distribution. It appears that in addition to a main component, for production of s-process nuclei in the range $90 < A < 200$, a weak component $A < 90$ and a strong component $A > 205$ are also needed [87, 641]. Low-mass asymptotic giant branch (AGB) stars have been suggested [299, 257] as the nucleosynthesis site for the main component. The site for the weak component is most likely to be core helium burning in massive stars, followed by carbon burning and neon burning (see chapter 10).

8.5 M-M_α RELATION

Nucleosynthesis yields from stars may be divided, roughly but naturally, into production by masses above, or below, about $9M_\odot$. The stars of lower mass are expected to shed their envelopes, with their cores becoming white dwarfs. Near the upper end of the range, some may ignite carbon under conditions of electron degeneracy, possibly resulting in an explosion (see chapter 11). Most of the matter returned to the interstellar medium is almost unprocessed, and acts to dilute the yields from more massive stars. Because the material is subjected to less extreme conditions, the modifications which do occur are due to the first and easiest nucleosynthesis processes—hydrogen and helium burning.

For stars above about $9M_\odot$, carbon burning ignites under nondegenerate conditions. Subsequent evolution proceeds in most cases to core collapse, forming a neutron star or black hole (see chapter 11 for an interesting possible exception, the Very Massive Objects (VMOs) which become pair-instability supernovae). These stars make the bulk of newly synthesized matter that is returned to the interstellar medium, and therefore of primary importance to our study.

As we have seen, uncertainties related to (1) mixing, (2) mass loss, (3) binary interactions, and (4) our knowledge of the microphysics (for example, the $^{12}C(\alpha, \gamma)$ reaction), plague our understanding of the evolution of these stars. To make matters worse, these stars are rare, and their spectra complex to unravel; our observations of these stars are not yet sufficiently enlightening to circumvent these theoretical difficulties.

It is prudent to pursue approximate solutions that seem to be representative. A powerful and reasonably accurate investigation may be made by use of the M-M_α relation [27, 35]. Much of the difficulty in items (1), (2), and (3) above is associated with the hydrogen-burning stage of evolution. By considering only the part of the star which has already been processed by hydrogen burning, these problems may be lumped together into the M-M_α relation, which connects the star's initial mass M to the mass M_α of a helium star whose evolution approximates that of the star's helium core. This is equivalent to considering the star to have two parts, a helium core and a hydrogen-rich envelope, and to assume that the envelope may be neglected if appropriate boundary conditions are imposed upon the core. For massive stars, the pressure and temperature at the hydrogen-burning shell are so small compared to the central values that they may be considered negligible. It is tempting to replace the core of the star by a helium star of the same mass. There are two possible problems with such a choice.

First, the hydrogen-burning shell processes matter so that the mass of the core is not constant. As we saw previously, during core helium burning, the hydrogen-burning shell moves out through a mass equal to about 0.1 of that of the core. Mass constancy fails at this level. Such a small change may be of minor importance, provided M_α is not near some critical value. Accordingly we require $M_\alpha \gg 1.45 M_\odot$ (the Chandrasekhar mass), so that $M \geq 15 M_\odot$, or so.

A second problem with this boundary-condition procedure involves the intrusion of the surface convection zone inside the position of the hydrogen-burning shell. This occurs when the envelope has expanded to the extreme red supergiant stage. Conventional treatment of the outer envelope involves nonadiabatic corrections to the mixing-length theory of convection. While this may be adequate for estimating the efficiency of heat transport by convection, it is less obvious that it will be for estimating the amount of downward penetration of the mass motions. A slight amount of extra mixing would have almost no effect upon the luminosity but would cause dramatic changes in the composition. For example, it could reduce the core mass by mixing that matter into the envelope.

Finally, both these phenomena could be modified by mass loss, and by interaction with a binary companion. However, the loss of the hydrogen envelope results in a helium star, which is precisely the approximation used. The theory of the evolution of close binaries [649] independently suggested the development of helium stars from massive cores.

Some insight as to the error involved with such a procedure may be gained by comparing the evolution of a massive star, having a helium core of mass M_{He}, and a helium star with mass $M_\alpha \approx M_{He}$. Table 8.3 shows results for a hydrogen-rich star of $25M_\odot$, and for helium stars of 6, 7, and $8M_\odot$. The helium core grows from about $6M_\odot$ at hydrogen exhaustion in the core, to $8M_\odot$ at core carbon ignition. The equivalent helium star mass M_α should lie in this range. It does, but its value depends upon which variables are considered most important. The value of the abundance of ^{12}C at the end of core helium burning is 0.32 for $M = 25M_\odot$, and varies from 0.35 for $M_\alpha = 6$ down to 0.31 for $M_\alpha = 8$. On this basis, $M_\alpha = 7.5$ would be appropriate. However, the mass of the carbon-oxygen core is another important variable; it ranges from about 3 for $M_\alpha = 6$ to $4.5M_\odot$ for $M_\alpha = 8$. Here the value for the hydrogen star is relatively small, about $3.3M_\odot$. This would give $M_\alpha = 6.5$, or so. This estimate is a measure of the uncertainty from this source (20 percent or less), but as we cannot claim that these possibilities span the range of variations in evolutionary behavior, it is not a rigorous error estimate.

Figure 8.7 gives a history of convection for a $25M_\odot$ star, from the zero-age main sequence through core carbon burning. The mass coordinate is plotted versus the logarithm (base 10) of the time in seconds left before core collapse. Convective regions are shown with solid shading; semiconvective regions appear striped. Figures of this format show at a glance the sort of processing to which any given layer is exposed, and the time of that exposure.

Following hydrogen ignition in the core, the fully convective region gradually retreats, from $M_r = 12M_\odot$ to $M_r = 5M_\odot$ at hydrogen exhaustion. Above this is a semiconvective region which grows out to $M_r = 14M_\odot$. These calculations, which used a "fast" semiconvective velocity, gave the ramp solution of Schwarzschild and Härm (see chapter 7).

TABLE 8.3

M-M_α Relation (H-He Stars)

$M_H/M_\odot$	$M_\alpha/M_\odot$	$M_{CO}/M_\odot$	$X(^{12}C)$
25	8.07	3.13-3.49	0.32
	8.0	4.3-4.63	0.31
	7.0	3.65-3.96	0.33
	6.0	2.94-3.16	0.35

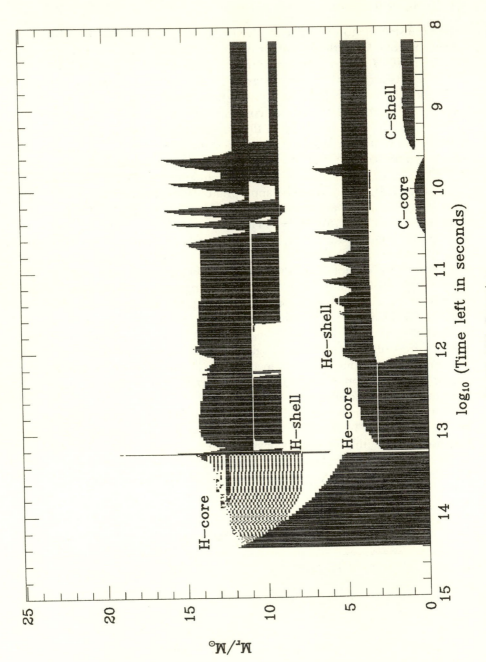

Fig. 8.7. $25 M_\odot$ Convection

At a time in seconds, $\log t = 13.22$, hydrogen burning ceases in the core. A short but eventful stage of shell hydrogen burning begins. Due to the increased luminosity, the formerly semiconvective region splits into several convective layers, breaking the ramp in composition into stairsteps. These are separated by thin regions having large gradients in composition and entropy. The flame zone of the hydrogen-burning shell has a higher helium content, and a lower opacity, than the matter just above it. Thus the flame zone can carry its luminosity by radiative diffusion while the overlying layers become convective. It is the evolution of this radiative hydrogen-burning shell that causes the helium core to grow from 6 to $8M_\odot$.

Core helium burning drives the inner region, $M_r = 3M_\odot$, to convection. This is topped by a thin region having large gradients in composition and entropy, and above that is a convective zone in what was the outer part of the old hydrogen-burning core. There is no nuclear burning in this convective zone until helium burning is finished in the core. Then the core convective zone shrinks, and shell helium burning begins (at $\log t \approx 12$). This flame zone has less helium (and more carbon and oxygen), so that it tends to be more convectively unstable than its overlying matter. The flame extends inside the mixed region (and for this calculation is contained by it). As the core contracts, it drives the shell to greater luminosity. This causes the convective instability to extend further out in mass, and mix in new fuel. This gives rise to the excursions seen in the convective helium-burning shell. At $\log t \approx 10.6$, core carbon burning begins, a subject to be dealt with in more detail later. By this stage, the thermal time-scale for the whole star is longer than the time for evolutionary change; the star is no longer a thermally static object.

The compositional structure resulting from this behavior is shown in figure 8.8. The star has just ignited carbon burning in its core. Beyond $M_r = 11M_\odot$ there is little processing, except for a slight increase in the ^{4}He abundance, and corresponding rearrangement of the CNO catalysts. From $M_r = 11M_\odot$ to $M_r = 8M_\odot$ lies most of the matter that was in the semiconvective ramp; it has been exposed to partial hydrogen burning. From $M_r = 8M_\odot$ to $M_r = 6M_\odot$ is matter in the outer part of the helium core; it has been exposed to complete hydrogen burning, but no helium burning. From $M_r = 6M_\odot$ to $M_r = 3.3M_\odot$ is matter that has also been exposed to incomplete helium burning, so that ^{12}C has been enhanced, but not oxygen. This zone (and its counterpart in stars of other masses) is interesting because of the dramatic difference in chemistry in a region with more carbon than oxygen; if ejected it will make different grains upon solidification. From $M_r = 3.3M_\odot$ to the center is matter that has

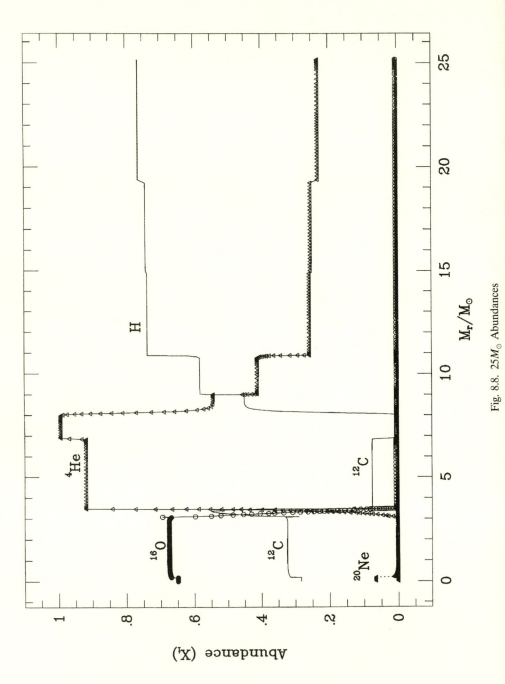

Fig. 8.8. $25M_\odot$ Abundances

undergone complete helium burning, and will proceed to more advanced burning stages. Oxygen is more abundant than carbon here.

8.6 IMPLICATIONS

After two long chapters on the two longest stellar evolutionary stages, a summary of the implications for nucleosynthesis is in order.

Massive stars have the highest luminosities, which is another way of saying that (1) they have the highest rates of thermonuclear processing, and (2) they evolve the fastest. They will be the primary sites of nucleosynthesis.

The light elements Li, Be, and B are easily destroyed in stars, and probably owe their production beyond Big Bang values to spallation by (and of) cosmic rays.

Low-mass stars produce rings of ^{3}He; whether these escape the gravitational potential well to affect galactic nucleosynthesis is unclear (conventional wisdom says no). Single low-mass stars evolve slowly and, after shedding some mass (planetary nebulae), become inert cinders (white dwarfs). Most of their nucleosynthesis is the production of ^{4}He (or if helium is ignited, ^{12}C and ^{16}O), which is buried in the white dwarf.

Hydrogen burning in massive stars produces significant amounts of ^{26}Al. This will be destroyed by accelerated decay at higher temperatures $(T > 1.0 \times 10^8$ K); however, such stars are observed to lose mass at a prodigious rate. They could be important sources for production of the ^{26}Al observed by the 1.8 MeV gamma-line from its decay to ^{26}Mg. Shell burning of hydrogen in red giants will occur at higher temperatures also, so that production of Na and Al, with O destruction, becomes possible. Similarly, ^{26}Al can be produced as well. The decay, and its acceleration at high temperature, favor hydrogen burning in objects with mechanisms for rapid mass loss as the most promising for ^{26}Al production. Intermediate-mass red giants are an interesting possibility.

CNO burning produces ^{14}N, which after helium burning leads to a neutron excess of $\eta \approx 1.5 \times 10^{-3} z/z_\odot$. This will prove important for isotopic abundances produced in subsequent burning.

The double shell-burning stage (H and He) provides an excellent environment for the s-process nucleosynthesis ($A > 90$), in stars that are likely to eject such matter. Massive stars do significant s-processing ($A < 90$) during their stage of core helium burning.

Massive stars have extensive convection zones associated with nuclear burning. This insures that large amounts of mass are processed in a homogeneous way.

Some unresolved nucleosynthesis problems which involve hydrogen and helium burning: (1) effects of binary evolution, (2) planetary nebulae formation and all other sorts of mass loss, (3) mixing, (4) novae, and (5) the core helium flash.

At this stage, the bulk yields of C and O are determined, as is most of the s-process nucleosynthesis. In helium burning, increasing mass corresponds to increasing production of O relative to C. The absolute values of the abundances of the third (^{16}O) and fourth (^{12}C) most abundant nuclei in the universe depend upon a delicate balance between production of ^{12}C by the triple-alpha reaction ^{4}He($2\alpha, \gamma$)^{12}C, and its destruction by ^{12}C(α, γ)^{16}O. This balance is as sensitive to stellar as to nuclear uncertainties, so that it is difficult to infer the correct nuclear information from astronomical observations. Recent experiments have begun to pin down the ^{12}C(α, γ)^{16}O reaction rate. While detailed uncertainty still exists in the range of astrophysical variation allowed, the empirical basis for this aspect of nucleosynthesis has been strengthened significantly.

9

Explosive Nucleosynthesis

"... for we do find
Seeds of them by our fire, and gold in them;
And can produce the species of each metal
More perfect thence, than Nature doth ..."
—Ben Jonson (1573–1637),
The Alchemist

If the role of theory in astronomy were merely to reproduce the details of observations, it would be of only modest intellectual interest. If its role is to provide the concepts for understanding existing observations, to predict new phenomena, and to guide the design of future observations, then it is a more substantial endeavor. In this light, the seminal role of simple theoretical models can be understood—they are the stuff from which our concepts are built and refined. The process often goes something like this: judicious oversimplification allows quantitative predictions of the behavior of a simpler system. Quantitative comparisons with observations, which reflect the behavior of a more complex system, suggest interpretations of data, refinement of theory, and design of new observations, until the process converges (hopefully) to what we call truth.

An example is the theory of explosive nucleosynthesis. Quantitative predictions of the nuclei synthesized in stellar explosions require (1) an adequately large nuclear reaction network, with associated reaction rates, and (2) a knowledge of the detailed behavior of the explosion. The parameterized approach to explosive nucleosynthesis consists of replacing the complexities of a real explosion with a simple model, but calculating the reaction network in detail. This approximation gave insight into the general nature of nucleosynthesis, and permitted quantitative predictions before good models of the explosion were available. The price paid is an ambiguity in the connection to explicit stellar events, which can only be removed by the construction of detailed models of the complex explosion process. While the simple model may be "more perfect" than reality, it fails when pushed to the limit. The nature of this failure is a clue to what is missing. In this case, it provides a measurement of the time taken for carbon and neon burning in stars.

9.1 PARAMETERS

Imagine a layer of unburned fuel in a star that is beginning to explode. A shock heats this matter, igniting the fuel, and accelerating it. The burned material expands and cools, shutting off the thermonuclear reactions. The challenge is to determine the composition of the ejected debris.

First, an initial composition must be specified. This is mostly the fuel for the burning process to be considered; historically this was estimated from the results expected from previous hydrostatic burning. For example, we saw in chapter 8 that hydrostatic helium burning produces primarily ^{12}C and ^{16}O, with a trace of ^{21}Ne, ^{22}Ne, ^{25}Mg, and ^{26}Mg from exposing ^{14}N to an environment rich in ^{4}He.

Because of the rapid expansion, weak interactions are generally too slow to be effective. In this case, the total neutron and proton numbers are separately conserved. The total neutron number is

$$N_n = \sum_i N_i(A_i - Z_i), \tag{9.1}$$

and for protons,

$$N_p = \sum_i N_i Z_i. \tag{9.2}$$

Their sum is the total nucleon number,

$$N_n + N_p = \sum_i N_i A_i. \tag{9.3}$$

The neutron excess per nucleon η is

$$\eta = (N_n - N_p)/(N_n + N_p). \tag{9.4}$$

Introducing mole fractions Y_i (see chapter 2),

$$\eta = \sum_i Y_i(A_i - 2Z_i). \tag{9.5}$$

By charge equality, the electron number equals the total proton number, and

$$Y_e = \sum_i Y_i Z_i. \tag{9.6}$$

By definition, $\sum_i Y_i A_i = 1$, so

$$\eta = 1 - 2Y_e. \tag{9.7}$$

The neutron excess is a useful parameter to describe the degree of neutron-proton asymmetry. We will find that often the initial abundances of many nuclei can be specified in several ways which are equivalent with respect to nucleosynthesis, providing the neutron excess η is the same.

Having specified the initial abundances, we must now design our model of the explosion. The initial density and temperature, as the initial abundance, can be estimated from stellar evolutionary models. By specifying a shock strength, the shock jump conditions may be used to infer the postshock density and temperature. This discussion relies on Landau and Lifshitz [378], especially §82. Across a simple shock, conservation of mass, momentum, and energy imply that the quantities ρv, $P + \rho v^2$, and $W + \frac{1}{2} v^2$ are conserved; $\rho = 1/V$ is the density, v is velocity relative to the shock front, P is pressure, and $W = E + PV$ is enthalpy. We translate to a coordinate system in which the preshock matter is at rest. The postshock fluid velocity v_b is related to the pressure and specific volume:

$$v_b^2 = (P_2 - P_1)(V_1 - V_2). \tag{9.8}$$

Let the subscript 2 denote the postshock condition. For simplicity in the discussion, we assume a strong shock ($P_2 \gg P_1$) and a γ-law equation of state ($W = \gamma PV/(\gamma - 1)$). Then

$$V_2/V_1 = \rho_1/\rho_2 = (\gamma - 1)/(\gamma + 1), \tag{9.9}$$

and

$$v_b^2 = \left(\frac{2}{\gamma + 1}\right) P_1 V_1 (P_2/P_1). \tag{9.10}$$

If v_b is estimated from observations (as interpreted with hydrodynamic models), both the postshock pressure P_2 and density ρ_2 are determined. The postshock temperature T_2 is then derived from the equation of state. When parameterized models of explosive nucleosynthesis were introduced, it was not understood how to infer v_b reliably from observations, and an alternative procedure had to be used.

So far the procedure has two flaws:

- It ignores the effects of nuclear burning on the hydrodynamic behavior.
- It needs some global model of the explosion.

Some insight into the first problem is directly obtained from the re-action network calculation, which gives the energy generation rate. The parameters ρ_2 and especially T_2 can be varied to approximate the effects of heating, for example.

The second flaw is dealt with as follows. First assume a relation be-tween ρ and T. A useful one is

$$\rho(t) = \rho_2 \left(\frac{T(t)}{T_2} \right)^3, \qquad (9.11)$$

which is appropriate for $\gamma = 4/3$ and no heating; it corresponds to constant photon entropy density S_γ. A real equation of state, and heating and cooling, could also be included [280]. The remaining difficulty is the time dependence. Here we assume an expansion on a time-scale τ,

$$d\rho/dt = -\rho/\tau. \qquad (9.12)$$

For many situations, the action only occurs near peak temperature (and density), so

$$\rho(t) = \rho_2 \exp(-t/\tau). \qquad (9.13)$$

However, for more complex processes (e.g., α-rich freezeout), these ap-proximations may not be adequate.

How do we estimate the expansion time-scale τ? Euler's equation in spherical symmetry is

$$du/dt = -\frac{1}{\rho} \nabla P - \frac{Gm}{r^2}. \qquad (9.14)$$

Initially the zone is in hydrostatic equilibrium, so $du/dt = 0$. For an "average" zone,

$$du/dt \approx \left(\frac{P}{\rho} - \frac{Gm}{R} \right) \frac{1}{R}. \qquad (9.15)$$

If we take our initial values (P_0, ρ_0) to equal the average ones, then

$$\frac{P_0}{\rho_0} = \frac{GM}{R}. \qquad (9.16)$$

If P/ρ is increased so

$$(1 + f) = \frac{P}{\rho} \frac{\rho_0}{P_0}, \qquad (9.17)$$

then

$$du/dt \approx f \frac{P_0}{\rho_0 R} = fGM/R^2. \tag{9.18}$$

Integrating gives an "asymptotic" velocity

$$u^2/2 \approx fGM/R, \tag{9.19}$$

where we use $u = dR/dt$ and $u(0) = 0$. The average density is $\rho = 3M/4\pi R^3$, so

$$\tau = \left(\frac{1}{\rho}\frac{d\rho}{dt}\right)^{-1},$$
$$= R/3u,$$
$$= \frac{1}{\sqrt{24\pi Gf\rho}}$$
$$= 446\,\mathrm{s}\,\chi/\sqrt{\rho}, \tag{9.20}$$

if ρ is measured in grams per cm^3 and $\chi = 1/\sqrt{f}$; Fowler and Hoyle [238] obtained this expression. Note that for $f = 1$, the explosion time is the same as the free-fall time, and that this formula is the same as used in cosmological nucleosynthesis (chapter 5).

Besides the initial composition, there are three parameters: T_2, ρ_2, and χ. They are not independent. The initial composition is related to T_1 and ρ_1 by the condition that the initial model be hydrostatically and thermally self-consistent. The shock strength then relates T_1 and ρ_1 to T_2, ρ_2, and χ.

This procedure allows us to specify an initial composition, choose T_2, ρ_2, and χ, and evolve a reaction network to predict quantitative abundances for comparison with experimental and observational data.

9.2 CARBON AND NEON

One of the simplest explosive stages of nucleosynthesis is carbon burning [54, 485]. The thermonuclear behavior is straightforward, and the hypothesized environment—unburned fuel outside the carbon-burning shell—is clearly visualized. Neon burning shares many features of carbon-burning, and produces many of the same nuclei.

The general features of nucleosynthesis during explosive carbon burning were examined by Arnett and Truran [54]; Truran and Arnett [635] considered its role in metal-poor stars, Arnett [23] discussed it in the context of galactic evolution, Hansen [280] considered the effects of coupling the energy generation to the thermodynamics, and Howard et al. [301] discussed the production of rare neutron-rich isotopes.

Following helium burning, matter is primarily in the form of ^{12}C and ^{16}O, with a trace of neutron-rich nuclei (^{21}Ne, ^{22}Ne, ^{25}Mg, and ^{26}Mg) resulting from exposing ^{14}N left from CNO cycling, to ^{4}He at higher temperature [193]. For solar system abundances, about $\frac{2}{3}$ of the "heavy" elements ($Z > 2$) are in the form of CNO isotopes. If all this is processed to ^{18}O, which has a neutron excess of $(10-8)/18 = 1/9$, the total neutron excess is

$$\eta \approx (2/27)z \tag{9.21}$$
$$\approx 1.5 \times 10^{-3}(z/z_\odot), \tag{9.22}$$

where $z_\odot \approx 0.02$ is the solar abundance of $Z > 2$ nuclei.

In order to have carbon to burn explosively, the temperature must be low enough for it to survive in the star prior to explosion. Hydrostatic carbon burning (see chapter 10) occurs at temperatures of $0.8 \leq T_9 \leq 1.0$. This implies, for a star with radiation entropy $S_\gamma/\Re \approx 1$, the density is $\rho \approx 0.6$ to 1.2×10^5 g cm^{-3}, or lower. Burning on a hydrodynamic time-scale (see equation 9.20) requires a postshock temperature of $T_9 \approx 2$.

The nucleus ^{12}C has the lowest charge Z, for this composition. Because of its lower coulomb barrier, ^{12}C $+^{12}$ C is the first signifi-cant reaction as temperature is raised. Two important channels are ^{12}C(^{12}C, α)^{20}Ne and ^{12}C(^{12}C, p)^{23}Na ; see §3.6. Above a temperature $T_9 > 1.12$, the optimum bombarding energy for the fusion of two ^{12}C nuclei exceeds the 2.60 MeV threshold for ^{12}C(^{12}C, n)^{23}Mg, so that channel will be open too. The light particles formed by these reactions, α, p, and n, will all react much faster than the basic ^{12}C $+^{12}$ C fusion.

This behavior is illustrated in figure 9.1, which shows the history of abundance of some representative nuclei as the matter cools and ex-pands from $T_9 = 2.1$ and $\rho = 1.158 \times 10^5$ g cm^{-3}. Because of shock compression, the preshock density would have been lower. The initial abundances by mass were those obtained in chapter 8 for helium burn-ing at $S_\gamma/\Re = 1.0$, except that in order to maintain consistency with the original calculations, the nucleon fractions for ^{12}C and ^{16}O were changed by a 10% switch, from 0.262 to 0.362 and 0.701 to 0.601, respectively. This is far less than the uncertainty in the ratio of ^{12}C to ^{16}O, which was discussed in chapter 8. The temperature of $T_9 = 2.1$ instead of 2.0 was used to give a similar consumption of fuel as that in [54].

First, note that as the left panel shows, the abundances of protons, neutrons, and even alphas are all small during nucleosynthesis. This sug-gests that the *steady-state* approximation of chapter 4 is valid. In fact, it

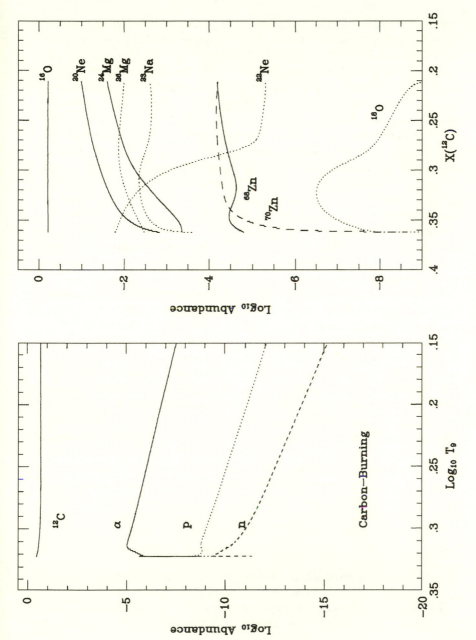

Fig. 9.1. Explosive C Burning

works well.[1] After a rapid rearrangement at the highest temperature, a steady state is established in which the evolution is driven by the consumption of ^{12}C. The production of alphas consumes the neutron-rich nuclei (mostly ^{22}Ne) by (α, n) reactions. At the highest temperatures there is a steep decrease in the abundances of neutrons and protons, and an increase in the abundance of alphas. Even with their relatively low abundances, these light particles are crucial to nucleosynthesis because of their fast reaction rates. Neutrons have no coulomb barrier; for protons and even alphas the coulomb barrier is smaller than for heavier nuclei. After a quick adjustment to the newly increased temperature, the changes are relatively slow, and fixed by the ratios of the dominant reactions *consuming* α, p, and n, to the ^{12}C $+^{12}$ C reaction which *produces* them. Notice that the neutron, proton, and alpha abundances decline exponentially with decreasing temperature; this is due to a corresponding decrease in the rate of ^{12}C fusion. After only a small fractional change in temperature, the abundance of ^{12}C stops changing—it freezes out—as do other nuclei.

The panel on the right shows the changes in abundance at these peak temperatures by plotting them against the abundance of ^{12}C. This expands the small region in temperature at which the changes occur. Protons and alphas from ^{12}C $+^{12}$ C are consumed by (α, n) reactions on ^{22}Ne and the much rarer ^{18}O. The neutrons produced eventually reside in nuclei with a neutron excess, such as ^{23}Na, ^{25}Mg, ^{26}Mg, ^{27}Al, ^{29}Si, ^{30}Si, and ^{31}P, or in much more massive nuclei beyond iron. The latter are illustrated by ^{68}Zn, and ^{70}Zn. These rearrangements occur during the first part of explosive carbon burning. Then there is a depletion of ^{12}C and a corresponding increase in the alpha-nuclei ^{20}Ne, ^{24}Mg, and to a small extent ^{28}Si.

The peak temperature and density were chosen to represent the post-shock values in the carbon-rich layer of a supernova. They correspond to a photon entropy S_γ of

$$S_\gamma/\Re = 0.1213 T_9^3/\rho_6, \qquad (9.23)$$

$$\approx 9.7,$$

an order of magnitude larger than for hydrostatic burning, but far less than the values typical of cosmological nucleosynthesis.

The final abundances are shown in figure 9.2 as pluses; also shown (as filled squares) is the solar system abundance pattern, normalized at

---

[1]In an early draft of this chapter, the author found it possible to reproduce network results to good accuracy, but the development was too complex to plague the reader with. See chapter 4 for a discussion of the steady-state approximation.

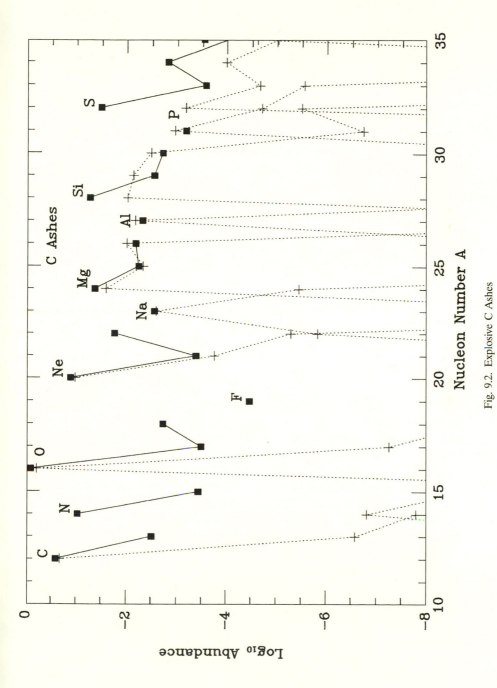

Fig. 9.2. Explosive C Ashes

^{20}Ne. This is similar to the original result of Arnett [54]. The abundances of ^{20}Ne, ^{23}Na, ^{24}Mg, ^{25}Mg, ^{26}Mg, and ^{27}Al are produced in solar system ratios. The Mg isotopic ratios are reproduced well (as well as could be expected from the accuracy of our knowledge of the reaction rates). Using a more comprehensive modern reaction rate compilation due to Thielemann, Arnould, and Truran [615] has produced some minor but interesting changes. Now ^{21}Ne, ^{29}Si, ^{30}Si, and ^{31}P are also produced in approximately solar ratios. As before, because of the incomplete burning, enough ^{12}C and ^{16}O are left to contribute significantly to the solar pattern.

The agreement with solar abundances is sensitive to the choice of parameters [54, 485]. The combination of temperature, density, and expansion time-scale should be chosen to give incomplete carbon consumption. The total abundance of neutron-rich nuclei in this range ($10 \leq Z \leq 15$) is proportional to η, the neutron excess [635, 23]. If it lies very far from 2×10^{-3}, the ratios of neutron-rich nuclei to their neighbors will be wrong.

Because of the far greater accuracy of their empirical determination, the ratios of isotopes of a given element are probably more meaningful than ratios of elements. Errors would presumably reflect faulty reaction rates (and bad theory) instead of flaws in abundance determinations as well. The structure of the Mg isotopes seen in the solar system and in figure 9.2 depend upon the treatment of neutron absorption on heavier elements ($Z > 14$). This comment applies to the ^{29}Si to ^{30}Si ratio as well.

Because of its high abundance and the larger neutron absorption cross section relative to $Z \leq 14$ nuclei, ^{56}Fe is a source of effective neutron absorbers in the solar abundance pattern (see [55] for discussion). If it is left out, the abundance of ^{26}Mg rises by about a factor of two and warps the pattern. Howard et al. [301] first discussed the nucleosynthetic consequences of these neutron absorptions; many of the rare neutron-rich isotopes from S to Se were expected to be produced in this process. We will return to this point later in this chapter.

Neon burning has important similarities to carbon burning. Figure 9.3 shows the behavior of selected nuclei as the matter cools and expands from $T_9 = 2.3$ and $\rho = 1.521 \times 10^5$ g cm^{-3}. This gives the same value for $S_\gamma/\Re$ as used for carbon burning. The initial abundances by mass were those used above for carbon burning, except that the abundance for ^{12}C was added to that of ^{20}Ne, giving it a nucleon fraction of 0.363.

The left panel shows the behavior of the abundances of protons, neutrons, alphas, and ^{20}Ne with temperature. The behavior is strikingly similar to the previous result for carbon burning (see figure 9.1). The key dif-

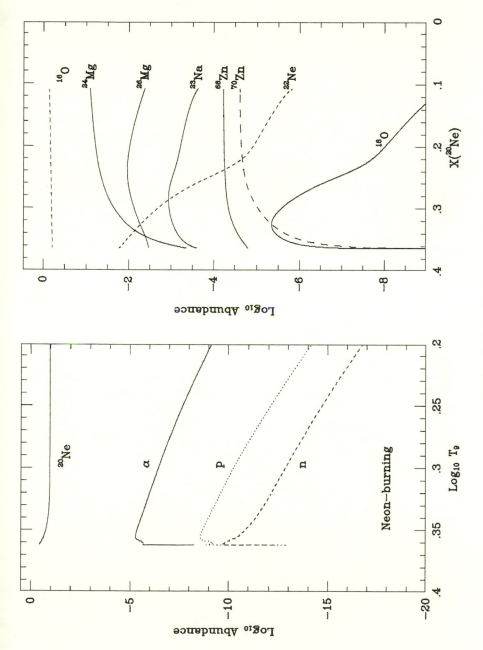

Fig. 9.3. Explosive Ne Burning

ference is that the reaction that drives the production of particles is not $^{12}C +^{12}C$ but $^{20}Ne(\gamma, \alpha)^{16}O$. The alpha particles induce the production of protons and neutrons by (α, n) and (α, p) reactions. Once produced, the protons, neutrons, and alphas do essentially the same things they did in carbon burning. There is an energetic difference: carbon burning produces about three times more energy.

The right panel resembles the previous result for carbon burning, but both ^{22}Ne and ^{23}Na are significantly reduced in abundance. Otherwise, the behavior is similar.

A comparison of figure 9.4 for neon burning with the previous result for carbon burning (figure 9.2) shows that the similarity extends to final products. Besides the underabundance of ^{22}Ne and ^{23}Na mentioned above, ^{12}C is virtually absent and ^{28}Si is almost solar. The abundances of ^{27}Al, ^{29}Si, ^{30}Si, and ^{31}P are too high, but some tuning of the parameters might improve that.

9.3 OXYGEN

While explosive carbon and neon burning are well described by the steady-state approximation, explosive oxygen burning shows quasi-equilibrium behavior. Consequently its nature is less sensitive to reaction rates, and more dependent upon binding energies and statistical factors. The higher temperatures, and more active reaction linkages, make oxygen burning nucleosynthesis less sensitive to the initial abundances that are assumed. For example, heavy elements ($Z > 28$) begin to photodissociate at these temperatures, so that the details of their initial abundances become correspondingly less relevant.

The first network calculation of oxygen burning [22], for hydrostatic burning of a pure oxygen star of $8M_\odot$, showed the quasi-equilibrium pattern of Si, S, Ar, and Ca which was later seen in the supernova remnant Casseopia A [166, 167, 357]. Following the success with explosive carbon burning, it was natural to attempt explosive oxygen burning [634]. A preliminary fit to the experimental data of Spinka and Winkler [581] was used. For densities in the range $10^5 \leq \rho \leq 10^6$ g cm^{-3}, equating the mean lifetime of ^{16}O to the expansion time gave temperatures in the range $3.5 \leq T_9 \leq 3.7$. For the standard case of $T_9 = 3.6$ and $\rho = 5 \times 10^5$ g cm^{-3}, we have $S_\gamma/\Re = 11.3$, roughly the same as for carbon burning. The network was that used previously for constant-temperature silicon burning [630]. The initial abundances were chosen to be consistent with the view that the star had equal amounts of ^{12}C and ^{16}O after helium burning, and then underwent both a carbon-burning phase [55]

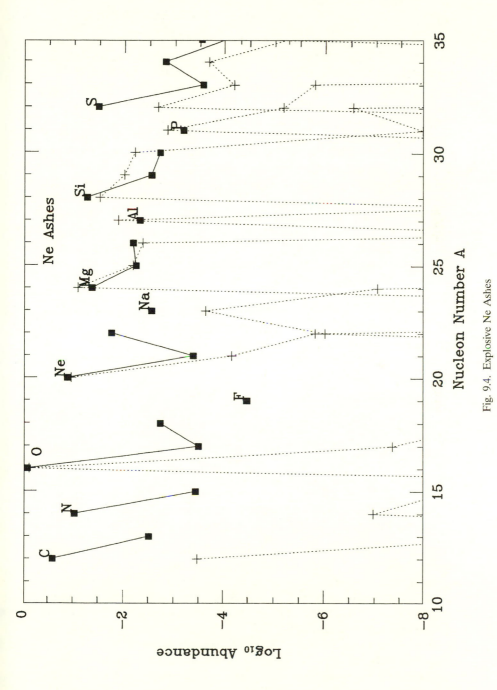

Fig. 9.4. Explosive Ne Ashes

and a neon-burning phase; it was estimated to be 54% ^{16}O, 30% ^{24}Mg, 2% ^{26}Mg, and 14% ^{28}Si by mass.

The relative abundances of the α-particle nuclei ^{28}Si, ^{32}S, and ^{40}Ca were well reproduced; ^{36}Ar agreed well with modern abundance estimates. Also reproduced were the neutron-rich isotopes of even-Z elements (^{33}S, ^{34}S, and ^{42}Ca) and isotopes of odd-Z elements (^{35}Cl, ^{37}Cl, ^{39}K, and ^{41}K). Similar results were obtained with an initial composition of ^{12}C, ^{16}O, and ^{26}Mg, providing the neutron excess η was the same (0.0015). It was found that the agreement with solar abundances was degraded if initial temperature, density, or neutron excess were varied significantly.

These features were analyzed in detail by Woosley, Arnett, and Clayton [692]. There it was found that a better treatment of the ^{12}C $+^{16}$O reaction modified the results, shifting the density for agreement with solar system abundances to a lower value ($\rho = 2 \times 10^5$ g cm^{-3}). Actual evolutionary calculations [27] of helium cores appropriate for massive stars showed that the abundances after core helium burning favored even more ^{16}O, so that the inital abundances chosen for explosive oxygen burning were too rich in the products of hydrostatic carbon burning, particularly ^{24}Mg. These two errors nearly cancel.

There is a simple procedure for defining the nucleosynthesis parameters:

1. Assuming a hydrostatic star with $S_\gamma/\Re \approx 1$, and an initial temperature less than $T_9 = 2$ so that oxygen could survive, we find preshock densities must be less than $\rho \approx 10^6$ g cm^{-3}.

2. Following [634], we equate the expansion time-scale and the mean lifetime of ^{16}O; for $\rho \approx 5 \times 10^5$ g cm^{-3}, this gives $T_9 = 3.6$.

3. Choosing the same photon entropy as previously used in carbon burning, $S_\gamma/\Re = 9.7$ gives $\rho = 5.83 \times 10^5$ g cm^{-3}. This is consistent with our previous argument, especially if we remember that there will be some compression by the shock (the initial stellar density would be $\rho \approx 10^5$ g cm^{-3}).

The question of initial abundances is straightforward, but more complicated. For a massive star, $M_\alpha = 8M_\odot$, or $S_\gamma/\Re = 1.0$, our choice of reaction rates (chapter 8) gives $X(^{12}C) \approx \frac{1}{3}$. The carbon burning will result in the production of ^{20}Ne which will dissociate in neon burning (schematically, 2^{20}Ne $\rightarrow^{24}$ Mg $+^{16}$ O) to make some ^{16}O. Suppose we assume that half of the original ^{12}C eventually becomes ^{16}O, and the remainder ^{24}Mg and ^{28}Si. Then we choose the abundances 79% ^{16}O, 13% ^{24}Mg, 2% ^{26}Mg, and 6% ^{28}Si by mass, which may be compared to the results to be presented in chapter 10. For such an oxygen-rich mixture, the ratio of ^{24}Mg to ^{28}Si is *not* crucial [692]. The choice of 2% ^{26}Mg by mass corresponds to a contribution to the neutron excess of $\delta\eta = 1.54 \times 10^{-3}$. The results are also insensitive to the abundances of nuclei with $Z \geq 15$ (phosphorus); these abundances were taken from our previous result for hydrostatic helium burning, giving $\eta = 1.58 \times 10^{-3}$.

The evolution of some important nuclei is shown in figure 9.5. The left panel shows the behavior of protons, neutrons, and alphas as a function of temperature. After a rapid adjustment near peak temperature, their abundances drop exponentially with cooling. This behavior is strikingly similar to that found previously for carbon and neon burning.

The right panel of figure 9.5 shows the evolution of abundances of some representative nuclei, as a function of ^{16}O abundance. This expands the scale of events near peak temperature. As ^{16}O $+^{16}$O releases a burst of protons, neutrons, and alphas, these react quickly with trace nuclei. Notice their tiny abundance; the scales for the right and left panels are quite different. Heavy elements ($Z > 28$) release neutrons by (γ, n) reactions at this elevated temperature. A quasiequilibrium is quickly formed with only trivial amounts of ^{16}O consumed, and further evolution proceeds through a series of quasi-equilibrium states. The ^{16}O is steadily converted into ^{28}Si, ^{32}S, ^{36}Ar, and ^{40}Ca. The neutron excess η is locked up primarily in ^{34}S, ^{38}Ar, ^{42}Ca, which are not shown, and in ^{54}Fe, which is shown as a dotted line. The element Ca has a magic proton number ($Z = 20$), so that the next elements (scandium and titanium) are weakly bound relative to the calcium isotopes. At these high temperatures they are more easily destroyed, so that they maintain a lower abundance. This inhibits flow from the Si-Ca region into the iron-peak. However, nuclei around the neutron-rich path ($A = 2Z + 2$) are more bound than those around the ($A = 2Z$) path. This enables the flow to proceed through the *Sc-Ti bottleneck* more easily for neutron-rich nuclei. The excess neutrons have a tendency to go to the iron-peak, leaving alpha-nuclei below the bottleneck.

The ratios of ^{32}S, ^{36}Ar, and ^{40}Ca to ^{28}Si are in equilibrium with respect to (α, γ) reactions, so that

$$\mu(^{32}\text{S}) = \mu(\alpha) + \mu(^{28}\text{Si}), \tag{9.24}$$

$$\mu(^{36}\text{Ar}) = \mu(\alpha) + \mu(^{32}\text{S}), \tag{9.25}$$

$$\mu(^{40}\text{Ca}) = \mu(\alpha) + \mu(^{36}\text{Ar}), \tag{9.26}$$

where the μ's denote the corresponding chemical potentials (see §4.3). Therefore,

$$Y_{40}/Y_{28} = \left(\frac{10}{7}\right)^{\frac{3}{2}} 2^{-9} Y_\alpha^3 \theta^3 \exp\frac{\Delta Q}{kT}, \tag{9.27}$$

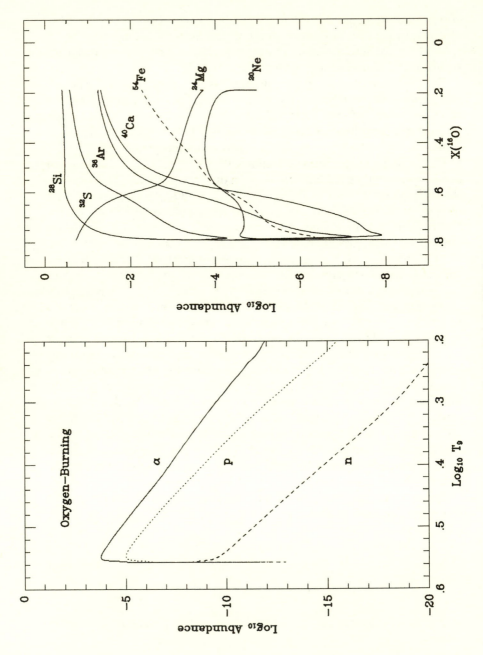

Fig. 9.5. Explosive O Burning

and

$$\theta = \rho N_A \left(\frac{2\pi\hbar^2}{M_u kT} \right)^{\frac{3}{2}} = 0.1013\rho_9/T_9^{3/2}, \tag{9.28}$$

where $\Delta Q = (M_{28} + 3M_\alpha - M_{40})c^2$ is the difference in atomic masses, M_u is the atomic mass unit, and N_A is Avagadro's number. Thus, $Y_{40}/Y_{28} \propto Y_\alpha^3$ for a given temperature and density. If the quasiequilibrium extended to ^{56}Ni, we would have $Y_{56}/Y_{28} \propto Y_\alpha^7$. This illustrates an important pattern in quasiequilibrium: more alphas cause a strongly nonlinear shift favoring the strongly bound heavy nuclei. Increasing the temperature, decreasing the density, or increasing Y_α, all have the same tendency. Note that Y_α appears with the factor ρ, so we could combine them in the number of alphas per unit volume, $N_\alpha = \rho N_A Y_\alpha$. Because the abundance of free alphas is the result of a balance between their production and their destruction, modifying either of these rates affects the quasiequilibrium.

An important nucleus which is not in the quasi-equilibrium group is ^{24}Mg. The almost exclusive product of explosive oxygen burning is ^{28}Si as long as ^{24}Mg remains undepleted. After it is depleted to small abundance, extra alphas are produced. Roughly,

$$^{16}O + {}^{16}O \rightarrow {}^{28}Si + \alpha, \tag{9.29}$$

and

$$^{24}Mg + \alpha \rightarrow {}^{28}Si + \gamma. \tag{9.30}$$

Two ^{16}O nuclei and one ^{24}Mg nucleus are converted into two ^{28}Si nuclei; without the ^{24}Mg nucleus, we have an "excess" α for each ^{16}O pair consumed. These excess alphas are captured upon existing ^{28}Si and heavier nuclei to form (primarily) ^{32}S, ^{36}Ar, and ^{40}Ca.

In figure 9.6 are shown the abundances after freezeout of explosive oxygen burning (crosses), compared with solar system abundances (filled squares), normalized at ^{28}Si. To more clearly illustrate the pattern, the most abundant nuclei for each given Z are connected, the solar system values by a solid line, and those from the explosive calculation by a dotted one. This format for comparison of observed and predicted abundances, though busy, is one of the least biased. We begin with the complete result, and proceed to the key features. At first glance, a wealth of crosses are seen lying below the solar system squares. The reaction network contains mostly *unstable* nuclei, which are usually produced in low abundance because of their lower binding energy. These contribute

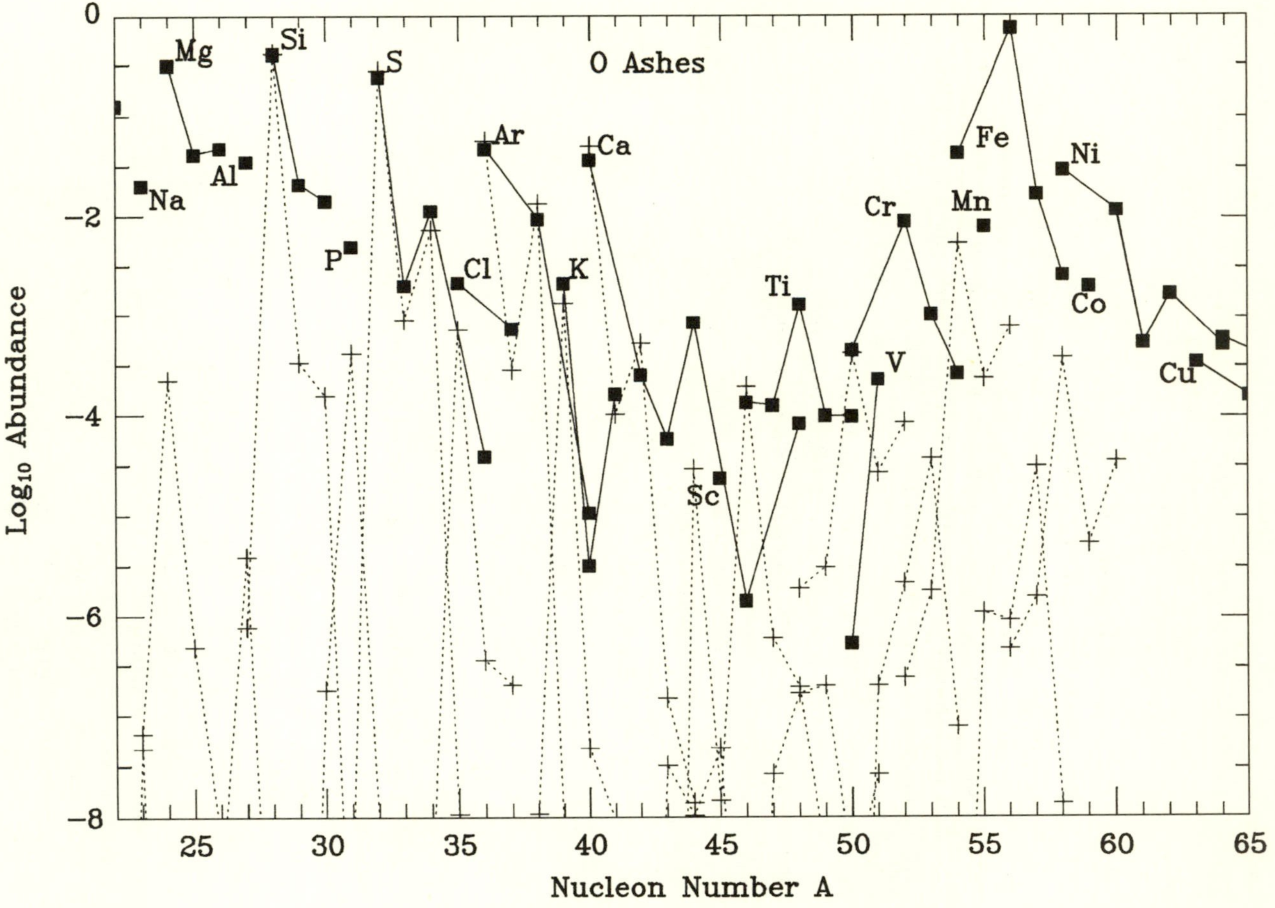

Fig. 9.6. Explosive O Ashes

little to the bulk yield, so we should focus first on the more abundant products. If we assume that a given fraction of some nucleus is made by the explosive process, we may normalize by moving the entire explosive abundance pattern up or down accordingly. This normalization defines an overabundance relative to the composition of the star at birth, which (in principle) will be predicted by theories of stellar formation, evolution, and explosion (see chapter 14). Here we use it as an adjustable degree of freedom.

All nuclei below ^{28}Si are severely underabundant in the explosive results, except for the ^{16}O that is left unburned. The nuclei ^{29}Si, ^{30}Si, and ^{31}P are underproduced; this is not a problem because they were already produced in explosive carbon (neon) burning. This interlocking of the explosive processes is encouraging. From ^{32}S to ^{42}Ca, all abundant isotopes are produced in the solar system pattern. This is a startling success; many isotopic ratios are produced strikingly well, considering reasonable estimates of the errors in our quantitative knowledge of the relevant nuclear properties. This result was not changed by updating the nuclear reaction rates. Missing are the rare isotopes ^{36}S (down by a factor of 0.12), ^{40}Ar, and ^{40}K. As we saw in the previous chapter, the latter two nuclei were produced in core helium burning. Beyond ^{42}Ca, only ^{46}Ti and ^{50}Cr are produced in significant abundance. The heavier nuclei have been destroyed.

Should the underproduction of ^{36}S be considered a major flaw or an allowable result of a simplified model? The question is an open one. Any model that reduces a complex continuum of processes to a few representative ones will be vulnerable to errors of the following sort: the results of rare processes are ignored, or poorly represented. The error of a factor of eight is not so large that it might not be made up for some different choice of parameters. It is trivial, and probably unprofitable, to add one process to reproduce the yield of one (or a few) rare nuclei. It seems better to pay particular attention to such nuclei, as we consider more and more realistic models; such flaws are clues about what we have omitted.

9.4 SILICON AND *e*-PROCESS

By the term *explosive silicon burning* we will include processes that begin with earlier burning stages—such as carbon, neon, and oxygen burning—but develop into the transformation of ^{28}Si and other S, Ar, and Ca nuclei into iron-peak nuclei by photodissassociation rearrangement (quasi-equilibrium burning). Since the pioneering paper of Hoyle [302], it has been generally agreed that the abundance peak near iron reflects the nuclear properties of the iron nuclei. Fowler and Hoyle [238] envisaged a process (now called silicon burning) which preferentially synthesized ^{56}Ni because the rapid burning time would not allow sufficient β-decays for ^{56}Fe production, but that these β-decays would subsequently occur while the matter was still at high temperatures.

The first explosive nucleosynthesis calculation [630] using an extensive reaction network was derived from the pioneering work of Truran, Hansen, Cameron, and Gilbert [632, 630] on hydrostatic silicon burning. Using results of numerical hydrodynamics [183, 16, 17] in a parameterized form, it was found that significant amounts of ^{56}Ni were synthesized in matter which the shock ejected. Bodansky, Clayton, and Fowler [109], building upon the numerical results [631, 630], were able to understand silicon burning as a quasi-equilibrium process. A more accurate numerical algorithm for solution of a network of reaction equations [55], which conserved charge and nucleon number to machine accuracy, showed definitively that ^{56}Ni was the dominant iron-group nucleus produced from neutrino-poor matter [630] (neutron excess $\eta \leq 0.003$).

Silicon burning was originally thought to produce nuclei from silicon to iron. However, oxygen burning seems likely to be a significant source of the Si-Ca region because it can also produce the rarer neutron-rich nuclei as well. The two processes are similar in that they both involve quasiequilibrium. The difference is the source of free particles (p, n, and α): for oxygen burning it is the ^{16}O $+^{16}$O reaction; for silicon burning, the thermal photodissociation of ^{28}Si. A consequence is that the nuclear energy release of oxygen burning is significantly larger than for silicon burning, making the oxygen burning matter more likely to escape the star's gravitational well.

As silicon burning becomes more extreme, it merges with the e-process and approaches nuclear statistical equilibrium (NSE). At high densities (entropy $S_\gamma/\Re \ll 10$), the number of free particles (p, n, and α) is small, so that as the matter expands and cools, the abundances follow their equilibrium values. At sufficiently low temperatures ($T \approx 3 \times 10^9$ K, but depending upon expansion rate), the nuclear reaction rates become too small to maintain NSE, and the abundance pattern "freezes out." It may be approximated by an NSE abundance at a suitably chosen freezeout temperature [277]. Such abundance distributions produce an iron-peak (for small neutron excess, dominated by ^{56}Ni), and nuclei to the low-Z side of the peak. With increasing neutrino excess η, the dominant nucleus changes [277] to ^{54}Fe ($\eta \approx 0.035$), then ^{56}Fe ($\eta \approx 0.071$), and ^{58}Fe ($\eta \approx 0.10$).

A history of the abundance during freezeout is shown in figure 9.7. In this calculation, the peak temperature was $T = 4.9 \times 10^9$ K and the

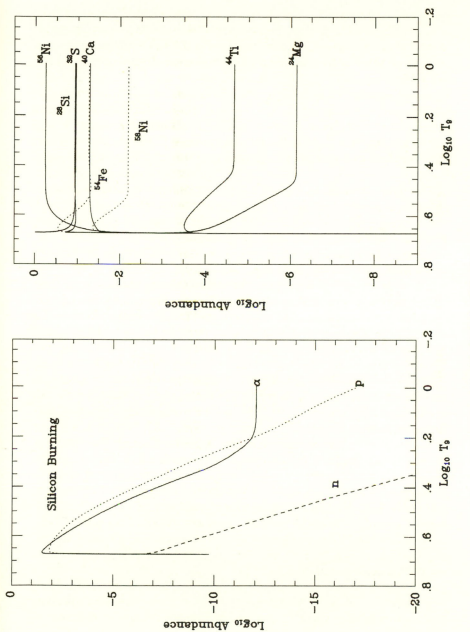

Fig. 9.7. Explosive Si Burning

radiation entropy was again $S_\gamma/\Re = 9.7$ (or density $\rho = 1.471 \times 10^6 \, \text{g cm}^{-3}$), and the neutron excess[2] was $\eta = 2.6 \times 10^{-3}$. The left panel shows the abundance of free particles; notice the higher abundances than in figure 9.5 for oxygen burning. The abundance of alphas approaches $X(^4\text{He}) \approx 0.1$, and the abundance of free protons is also in the range of a few percent. With cooling, abundances of neutrons, protons, and alphas all fall, but at $\log T_9 \approx 0.3$ ($T_9 \approx 2$) the abundance of alphas freezes out at $X(^4\text{He}) \approx 10^{-9}$. The free proton abundance is tiny ($X(^1\text{H}) \approx 10^{-16}$) at $T_9 = 1$, but because of their lower coulomb barrier, the abundance of protons is still changing well after the alphas have frozen out.

The right panel shows a rapid initial adjustment to quasiequilibrium, and then a slower evolution as temperature drops. Notice that the horizontal axis is $\log T_9$ rather than the nucleon fraction of ^{28}Si. For those heavier nuclei which are abundant, freezeout occurs at $T \approx 3 \times 10^9 \, \text{K}$. For those rarer nuclei which have low coulomb barriers, freezeout occurs slightly later, as the alpha abundance indicates. The most abundant nuclei with excess neutrons are ^{54}Fe and ^{58}Ni; the evolution of their abundances are indicated by the dotted lines. With high free-particle abundance, the nuclei that do exist tend to be neutron rich, so that ^{54}Fe and ^{58}Ni are abundant (the free protons must be balanced by neutrons in some form). As the free particles combine into nuclei, the ^{54}Fe and ^{58}Ni abundances fall, and freeze out.

The abundances after freezeout (crosses) are compared to solar system values (filled squares), in figure 9.8; the normalization is at ^{56}Fe. All of the abundant isotopes from ^{48}Ti to ^{57}Fe are produced in essentially the solar pattern. The rare ^{54}Cr is not produced, nor is there significant production of lighter or heavier nuclei, except for some ^{58}Ni. Many of the nuclei are made as proton-rich isobars which will decay to the stable nucleus found in the solar abundances. Here we have ^{48}Ti and ^{49}Ti formed as ^{48}Cr and ^{49}Cr, while ^{50}Cr is formed directly; ^{51}V is formed as ^{51}Mn. Likewise, ^{52}Cr and ^{53}Cr are formed as ^{52}Fe and ^{53}Fe, while ^{54}Fe is formed directly; ^{56}Fe and ^{57}Fe are formed as ^{56}Ni and ^{57}Ni, while ^{58}Ni is formed directly; ^{55}Mn are formed as ^{55}Co. The subsequent decay of ^{56}Ni to ^{56}Co, and then to ^{56}Fe, is the dominant source of energy that powers the brilliance of supernovae.

At lowered density and a given temperature, the number of free particles increases in NSE. As the reaction rates decrease upon expansion, the number of free particles may *exceed* the NSE value, so that the re-

---

[2] Notice that this value is larger than chosen previously. This improves the fit to the neutron-rich isotopes in the iron-peak, suggesting that such matter has increased its neutron excess by roughly a factor of 1.5 to 2 prior to explosion.

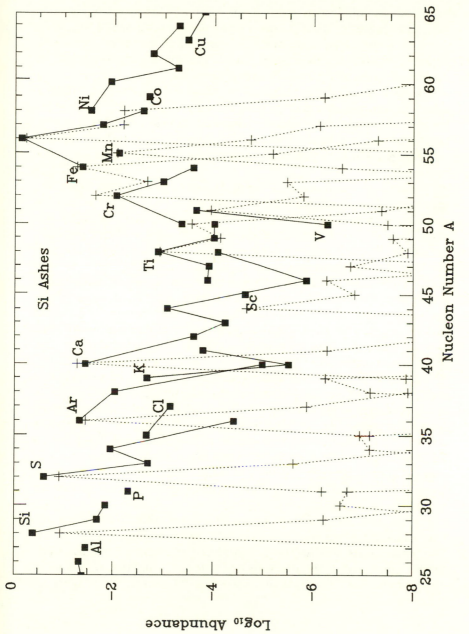

Fig. 9.8. Explosive Si Ashes

sult is a quasi-equilibrium distribution, skewed by the excess of particles. With their higher coulomb barrier, alpha particles are most likely to do this; this is called an *alpha-rich freezeout* [49]. Such abundance distributions produce an iron-peak (for small neutron excess, dominated by ^{56}Ni), and nuclei to the high-Z side of the peak.

A similar situation can occur as we slide to higher initial temperatures keeping radiation entropy constant ($\rho \propto T^3$). In NSE, the contour for constant free-particle abundance goes as a high power of temperature. Following a sufficiently high, fixed value of S_γ to higher temperature brings us to high free-particle abundance, and gives an alpha-rich freezeout.

A history of the abundance during freezeout is shown in figure 9.9. In this calculation, the peak temperature was $T = 5.2 \times 10^9$ K and the radiation entropy was again $S_\gamma/\mathfrak{R} = 9.7$, and the neutron excess was $\eta = 1.58 \times 10^{-3}$. The right panel shows the abundance of free particles; notice the different behavior from figure 9.7. Both the alphas and protons are abundant until low temperature. The left panel shows a rapid initial adjustment to quasiequilibrium, and then a slower evolution as temperature drops, with freezeout at $T \approx 3 \times 10^9$ K. Because of the large abundance of free particles, there is a shift favoring the high-A side of the $A = 56$ peak, giving a qualitatively different behavior than seen in figure 9.7. The abundance of ^{54}Fe is small compared to ^{58}Ni, rather than comparable.

Initially, ^{44}Ti drops with temperature as in silicon burning, but at $\log T_9 = 0.37$ ($T_9 = 2.34$) it begins to rise rapidly and continues to climb even at low temperatures.[3] At present this is the most plausible source for production of ^{44}Ca, to which the ^{44}Ti decays (47 years) by way of ^{44}Sc (3.93 hours). The Sc decay almost always (98.99%) gives a 1.157 MeV gamma line. As the ^{44}Ti half-life is roughly the same as the average interval between supernovae in the galaxy, an appreciable abundance could be built up [178]. This becomes an interesting nucleus for gamma-ray astronomy.

The abundances after freezeout (crosses) are compared to solar system values (filled squares) in figure 9.10; the normalization[4] is at ^{56}Fe. The pattern from ^{56}Fe to ^{62}Ni is reproduced in essentially the solar pattern (^{59}Co is low by about a factor of three). In addition to the interesting amount of ^{44}Ca made as radioactive ^{44}Ti, some ^{48}Ti is made (as radioactive ^{48}Cr) as well.

---

[3]The calculation was continued down to a temperature of $T_9 = 0.1$, at which point freezeout was complete.

[4]Strictly speaking, both this and the previous normalization for silicon burning should be adjusted to share the production of $A = 56$.

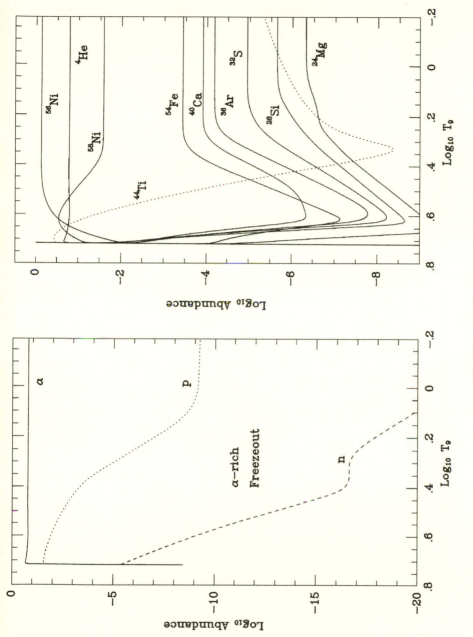

Fig. 9.9. α-rich Freezeout

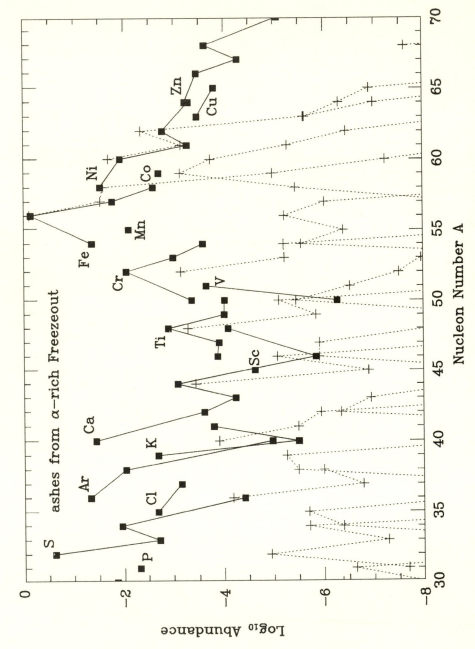

Fig. 9.10. α-rich Freezeout Ashes

Only ^{44}Ca (^{44}Ti) seems uniquely produced by the alpha-rich freezeout, although a number of the other nuclei made are weakly produced by other processes. However, this type of freezeout occurs for conditions which are commonly encountered in models of stellar explosions (that was how it was discovered [49]), so that this process may be difficult to avoid.

9.5 NEUTRON EXCESS AND GALACTIC EVOLUTION

Big Bang nucleosynthesis only makes the elements up to lithium. The subsequent element building occurs in stars. Is the nucleosynthesis from primordial stars and Population II stars different from that occurring in Population I stars? Are there differential effects in stellar nucleosynthesis as generations of stars convert pristine cosmological matter into solar system abundances? In order to pursue this question, it is useful to consider *primary* and *secondary* processes in stellar nucleosynthesis (see also chapter 14). Pristine cosmological matter is essentially just ^{1}H and ^{4}He. Nuclei produced directly from these will be called *primary*. Nuclei whose production depends upon the presence of other nuclei, which themselves require a previous epoch of stellar nucleosynthesis, will be called *secondary*. The distinction is deceptively simple. For example, consider ^{14}N. To be made in the CNO cycle, it requires the presence of C or O nuclei. If the C and O were made by helium burning in a previous generation of stars, then the N so formed would be *secondary*. If the C and O were made by helium burning in a deeper layer of the same star, and mixed into a hydrogen-burning shell to undergo the CNO cycle, the N so formed would be *primary*.

Clearly a combination of primary and secondary contributions would be complicated, and probably more realistic than only one or the other. The importance of such a distinction is related to its possible use as an observation probe of galactic evolution. In the primary case, nitrogen production is likely to be proportional to the rate of star formation (although more complicated behavior can be imagined). In the second, nitrogen production is proportional to the rate of star formation *and* to the amount of C and O in the newly formed stars. The latter is itself proportional to the rate of star formation, implying a quadratic dependence. Suppose we select stars having a variety of ages, and assume that their surface abundances reflect the interstellar abundance at their birth. Then the relative growth of the N/(C + O) ratio with abundance of C and O would show which of these ideas is correct. Apparently both [376, 479] are correct to some extent, and the unraveling of these effects is a challenge for both theory and observation.

This line of argument has relevance for explosive nucleosynthesis [23, 635], through the dependence upon neutron excess η. Explosive events, unlike hydrostatic burning, are distinguished by the absence of significant beta-decay processes. The neutron excess characteristic of the ejected matter must reflect the neutron excess of the matter prior to explosion. That, in turn, may reflect (1) the initial abundance of the matter of which the star was made, or (2) the neutron excess created during the evolution of the star prior to explosion.

All the processes discussed in this chapter have occurred in matter which had been through hydrogen burning and helium burning—at least. Hydrogen burning converts excess protons into neutrons; the primary products are ^{4}He and ^{14}N, both of which have $Z = N$. This pushes the neutron excess from negative values to zero, or slightly positive if some heavier, neutron-rich nuclei are also present. As we have seen, helium burning makes an important change in η as ^{14}N is converted to ^{18}O. This depends only upon the initial metallicity of the star (see equation 9.21). In general the neutron excess will also be increased [23, 635] by a brief phase of hydrostatic carbon burning or by significant stellar mixing (e.g., primary nitrogen production). We note that the silicon-burning results above suggest that at least some matter has had additional increases in neutron excess.

Consider two limiting cases. First (case A), suppose that there is no significant contribution to the neutron excess by hydrostatic carbon burning, primary nitrogen production, etc. Then the neutron excess is proportional to the initial metallicity of the star. Second (case B), suppose that the initial neutron excess always increases to a value of $\eta \approx 2 \times 10^{-3}$ prior to explosion, so that the neutron excess is *universal*, in a sense. We may use the explosive nucleosynthesis results to determine the dependence of abundances upon η, and predict how they should evolve in these two cases. The results are relatively simple [23, 635]. For carbon burning, the abundance of ^{23}Na and ^{27}Al relative to ^{20}Ne or ^{24}Mg is proportional to η, to a good approximation. The same is true for the neutron-rich isotopes of Mg. For oxygen burning, most of the neutron excess resides in ^{34}S and ^{38}Ar, which both have two excess neutrons. In quasiequilibrium the abundance of such nuclei, relative to abundant nuclei such as ^{28}Si, ^{32}S, and ^{36}Ar, is proportional to the square of the neutron number density N_n, so $N_n \propto \sqrt{\eta}$. For nuclei with only one excess neutron, abundances are proportional to N_n and hence $\sqrt{\eta}$. The same behavior is found for silicon burning and the alpha-rich freezeout. These simple rules fairly well describe the actual numerical results [23].

How might these abundances evolve during the evolution of a galaxy? Whether or not such simplified pictures are relevant, their predictions are clear: in case A, the neutron excess η is proportional to the metallicity z, so that the abundance of nitrogen, sodium, and aluminium behave as purely secondary, rising quadratically with metallicity. For the heavier elements made by quasiequilibrium and by equilibrium, the dependence of the abundances upon η also implies pronounced effects of odd-Z relative to even-Z. In case B, the relative abundances do not change, but march up to solar values in lockstep.

Which should we expect? In determining the parameters for explosive carbon burning, we considered the hottest matter which still contained ^{12}C. This is most likely to be found in a carbon-burning shell, so that it will have experienced incomplete carbon burning prior to the explosion (see chapter 10). For oxygen burning, it is reasonable to consider that carbon burning and neon burning have already occurred in a hydrostatic stage. In both cases η should be higher than case A, and perhaps approaching case B (how close is an important quantitative question). Carbon-burning products should show the largest effect, intermediate between case A and case B, while oxygen- and silicon-burning products should approach case B. This is roughly what the abundance data discussed in chapter 2 show (see the review by Lambert [376]). Although the effects are small—and strain our observational and theoretical abilities at present—the hard effort involved in establishing high-quality stellar abundance data is clearly beginning to pay dividends.

9.6 YIELD PUZZLE

Putting together the separate stages of explosive burning, a crude model of stellar nucleosynthesis can be built. Attributing nuclei to their appropriate process, the solar system abundances may be used to infer the relative *yields*, or production strengths, of the processes.

Suppose that ^{14}N is made by CNO cycling in stars. Some ^{4}He will be produced as well, giving a stellar addition to the cosmological helium production. If the burning goes to completion, the original ^{12}C, ^{14}N, and ^{16}O nuclei will become ^{14}N, and the rest of the matter will be almost all ^{4}He. Actually, the conversion of CNO nuclei to ^{14}N occurs before much of the hydrogen is converted to ^{4}He (see [173] and chapter 7). To produce a stellar metallicity z, we need $X(^{14}N) \geq \frac{2}{3}z\,\Delta X(^{4}He)_{stellar}$. For $z = 0.02$ and $X(^{14}N) = 1.1 \times 10^{-3}$, this implies $\Delta X(^{4}He)_{stellar} \leq 0.08$. If we subtract the cosmological nucleosynthesis from the solar value for the helium abundance, we have $X(^{4}He)_{\odot} - X(^{4}He)_{cosmo} \approx 0.045$, which is consistent. Note that $\Delta X(^{4}He)_{stellar}/z \leq 4$, so that this upper limit encloses the values suggested by observation [491, 429].

If we ascribe ^{12}C, ^{16}O, and ^{22}Ne to helium burning (presumably hydrostatic, with subsequent mass ejection), then this is about 1 percent of the solar abundance pattern, and only somewhat smaller than our estimate of the upper limit to the stellar contribution to helium production. Note that stars may make more helium and nitrogen, which is not ejected but locked up in white dwarfs, neutron stars, or black holes.

The fraction of the solar abundance pattern ascribed to explosive burning processes—carbon, oxygen, and silicon burning—is smaller, and each of the processes has a comparable contribution. The total abundance due to other processes—spallation, the s-process, and the r-process—is tiny. Thus, this gives a reasonably good, quantitative description of the noncosmological abundance pattern. The normalizations made for the explosive processes contain some interesting information concerning the yields. This may be made quantitative with a very simple model: suppose that all supernova are alike (or that a suitable average may be used), that they have a mass ejected M_{sn} and occur at a rate $\mathcal{R}_{sn}$ in the disk of the Galaxy. Let the mass of matter that participates in nucleosynthesis in the disk be M_{disk} and its age τ. Of the matter ejected from the supernova, a mass M_p is associated with a process of explosive burning, p. Then the nucleon fraction of some nucleus i is

$$X_i = \left(\mathcal{R}_{sn}\tau/M_{disk}\right)M_p X_i^p. \tag{9.31}$$

Here X_i^p is the nucleon fraction of nucleus i in the matter associated with explosive process p; it is identified with the values obtained in the calculations of explosive nucleosynthesis. We take typical values for the disk of the Galaxy, so that $M_{disk} \approx 6 \times 10^{10} M_{\odot}$, $\tau \approx 10^{10}$ years, and an ^{16}O nucleon fraction of ≈ 0.01. Let us assume that galactic disk abundances are the same as solar system abundances, as they seem to be, so $X_i \rightarrow X_i^{\odot}$. This gives a constraint on the average rate of production of oxygen from supernovae,

$$\mathcal{R}_{sn}M_O = X_O M_{disk}/\tau \tag{9.32}$$

$$\approx 0.06 \, M_{\odot} \, y^{-1}. \tag{9.33}$$

A typical estimate of the supernova rate in the disk is $\mathcal{R}_{sn} \approx 0.02 \, y^{-1}$, which would imply $M_O \approx 3 \, M_{\odot}$ per supernova. For SN1987A, $M_O \approx 1 \, M_{\odot}$ was inferred [424]. This is reasonably close: it could be that (1) the average supernova rate was higher over the lifetime of the disk, (2) the average yield of oxygen per supernova was higher than for SN1987A, or (3) the abundances of the disk stars came in part from nucleosynthesis occurring before the disk formed. A better estimate will be made in chapter 14.

The overabundance of nucleus i is defined as $\mathcal{O}_i = X_i^p / X_i^\circ$. Then,

$$\mathcal{O}_i = (M_{gal}/\mathcal{R}_{sn}\tau M_{sn})(M_{sn}/M_p). \qquad (9.34)$$

The ratio of overabundances is just the inverse ratio of mass contributions from the processes, $\mathcal{O}_j/\mathcal{O}_i = M_{p(i)}//M_{p(j)}$. These results are summarized in table 9.1, where *Factor* denotes the overabundance of the process, *Norm* gives the nucleus chosen to define the normalization, and *Ratio* gives the ratio of the mass associated with that process relative to the production of ^{16}O by helium burning.

Notice that oxygen burning, silicon burning, and alpha-rich freezeout have similar values of M_p, all of which are smaller than those for helium burning, carbon burning, and neon burning. This is a strange and interesting result: the stars must eject more of the lighter elements (unburned fuel) than heavier ones (ashes). Yet we saw in chapter 7 that the fraction of a star's core which was convective would be large, insuring that most of the fuel would be converted to ashes. We shall see that this difficulty is resolved by the effects of neutrino emission on the structure of the presupernova star.

TABLE 9.1

Yields of Processes

Nuclei	Process	Abundance	Norm	Factor	Ratio
N, some ^{4}He	CNO	0.04	^{4}He	25?	2.80
^{12}C, ^{16}O, ^{22}Ne	3α	0.0128	^{16}O	70	1.00
^{20}Ne, Na to Al	C+C	2.36×10^{-3}	^{20}Ne	67	1.05
^{24}Mg to ^{27}Al	Ne(γ, α)	0.7×10^{-3}	^{24}Mg	159	0.44
^{28}Si to ^{42}Ca	O+O	1.22×10^{-3}	^{28}Si	593	0.12
Ti to $\frac{1}{2}$Fe	*e*-process	6.5×10^{-4}	^{56}Fe/2	1390	0.051
$\frac{1}{2}$Fe to Zn	α-freeze	6.5×10^{-4}	^{56}Fe/2	1390	0.051
Remainder					
Beyond Kr	*s*-process	1.7×10^{-7}			
Beyond Kr	*r*-process	1.7×10^{-7}			
^{6}Li, Be, B	spallation	6.6×10^{-9}			

These calculations of explosive nucleosynthesis used the new reaction rate compilation of Thielemann, Arnould, and Truran [615], but differ little from the original results. Because this reaction rate data set is far more complete than the earlier ones, it is possible to extend the size of the network up to Kr, and examine the effects of neutron exposure on iron-peak elements during carbon and neon burning. For oxygen burning, and higher temperature stages, these heavy elements may be photodissociated, so that evidence of their former existence tends to be erased from the ejected abundances. Thus carbon and neon burning provide a unique and possibly important insight. Figure 9.11 is an extension of Figure 9.2 to heavier nuclei ($55 \leq A \leq 90$). The normalization has already been determined from the lighter nuclei shown previously. While many nuclei follow the solar pattern, notice that five nuclei, ^{64}Ni, ^{70}Zn, ^{76}Ge, ^{82}Se, and ^{86}Kr, are overproduced. *This represents an inconsistency in the explosive nucleosynthesis picture*; if the lighter nuclei are produced in the correct ratios and yields, we have too much of these nuclei. What does this mean?

The case is most acute for ^{70}Zn. Almost equal amounts of ^{68}Zn and ^{70}Zn are made. The yield of ^{68}Zn is nearly the value needed for it to be consistently produced with the lighter nuclei in carbon burning. However, the solar system ratio of ^{70}Zn to ^{68}Zn is 0.033; ^{70}Zn comprises only 0.62% of solar system zinc. The overproduction is a factor of 30 in an accurately known isotopic ratio!

Figure 9.12 illustrates the dominant flows for this region of the reaction network, at a time representative of this overproduction. The network is shown for $29 \leq N \leq 52$ and $26 \leq Z \leq 37$; nuclei used in the calculation are shown as squares. The stable nuclei are the filled squares. The strongest net flows are shown as solid lines connecting nuclei, and flows down by a factor of 0.1 to 0.01 from these are shown as dotted lines. Flows less than 0.01 of the major flows have been suppressed for clarity. The five nuclei that are overproduced are each enclosed by a large circle. The flows are dominated by neutron capture and intermediate between the canonical s-process and r-process (see Clayton [173]). The extent to which the nuclei of a given element move to higher N is determined *both* by beta-decay and by (γ, n) reactions. Each overproduced nucleus (Z, A) has a neighbor $(Z, A - 1)$ which is unstable to beta-decay. If these nuclei had time to decay before neutron capture, the overproductions would not occur. The rate of neutron capture is controlled by the rate of neutron production, and hence by the rate of the ^{12}C + ^{12}C reaction. *Thus we have a constraint upon the time-scale over which carbon burning occurred!* The results for neon burning are virtually identical, so that the constraint applies to neon burning as well.

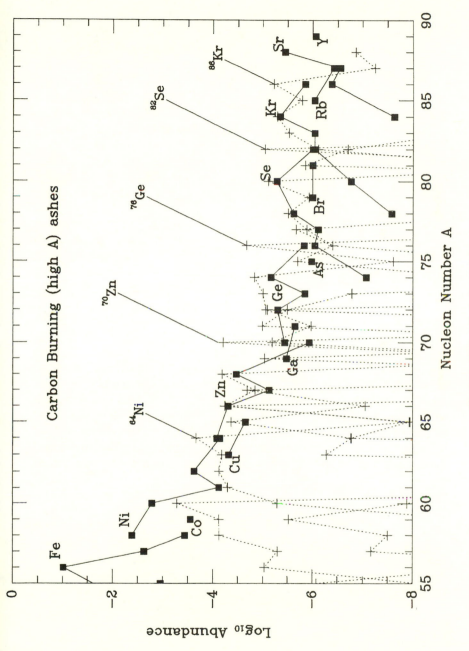

Fig. 9.11. High-A Explosive C Ashes

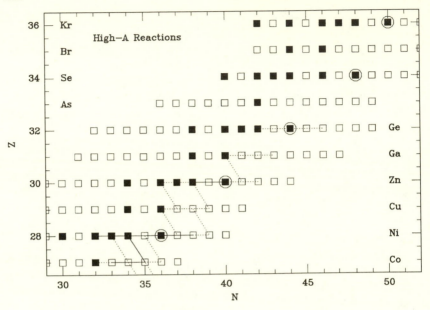

Fig. 9.12. High-A Reactions: Explosive C Burning

Table 9.2 summarizes the lifetimes of the unstable neighbors of the overproduced nuclei [388]. The lifetimes for low-lying excited states are given in parentheses. For ^{69}Zn, the ground-state decay would take about five hours to reduce the abundance by a factor of 30. This would cause a comparable reduction in the production of ^{70}Zn because the beta-decay to ^{69}Ga would be faster than ^{69}Zn$(n, \gamma)^{70}$Zn. For ^{63}Ni, a factor of four reduction would require two hundred years. The explosive time-scales are a few seconds or less. To avoid this inconsistency, the synthesized

TABLE 9.2

Carbon and Neon Time-Scales (Half-Lives)

Nucleus	Neighbor $(Z, A - 1)$	Half-life $(Z, A - 1)$
^{64}Ni	^{63}Ni	100 yr
^{70}Zn	^{69}Zn	56 m (14.0 hr)
^{76}Ge	^{75}Ge	82.8m (48 s)
^{82}Se	^{81}Se	18.5 m (57.3 m)
^{86}Kr	^{85}Kr	10.7 yr (4.48 hr)

matter must be processed by carbon and neon burning on time-scales that are *longer than those characteristic of explosions*. This suggests that we examine hydrostatic burning, at least for carbon and neon burning; the constraint does not apply to oxygen and silicon burning.

How did this effect escape detection before? The short answer is inadequate computer resources and reaction rate data sets. This restricted the networks to a size too small to include the heavy nuclei shown in figure 9.12. These nuclei were treated as a "sink" for neutrons, with an average cross section appropriate to the *s*-process. This underestimated the number density of free neutrons. The "seed" nuclei were taken initially to be in solar system abundances, rather than including the effects of neutron exposure during helium burning. This also tends to suppress the effect. Finally, without a network large enough to show the pattern, it was not clear whether overproduction was the result of inconsistencies in the fragmentary data set.[5]

While in some respects amazingly successful, this simple model of explosive nucleosynthesis is inadequate. It has an internal inconsistency regarding time-scale, and its very success raises a number of additional questions. Why do the ratios of these yields have these determined values? Explosive burning is sensitive to the parameters [54, 634, 692, 485] chosen, so how are these values selected by Nature, and how are these conditions maintained so well for such large amounts of matter? Are these processes unique? Evidently they tend to merge one into the next. Why are the yields such as to give metallicities of $z \approx 0.02$, which are typical of Population I matter? To answer these and related questions, it is necessary to relate the processes of nucleosynthesis more directly to the behavior of stars.

---

[5]This is another example of the value of the program of experimental nuclear astrophysics led so well by W. A. Fowler. It not only provided important experimental data but trained and inspired experimentalists and theorists. It provided a basis and added motivation for the effort to establish reliable semiempirical procedures to supplement direct experimental information, an approach pioneered by A. G. W. Cameron.

10

Neutrino-Cooled Stars

Combustion at the astral core—the dorsal change
Of energy—convulsive shift of sand ...
—Hart Crane (1899–1932),
The Bridge

A key to nucleosynthesis is the thermonuclear yield of massive stars. What do they make and what do they eject? This in turn is crucially dependent upon the structure of these stars before they hurl their ashes into space by mass loss or by explosion. This structure forms predominantly during the stages following helium burning; during this time the star cools by emission of neutrino-antineutrino pairs. The light radiated from its surface is a tiny part of the energy budget of the star. Four of the six major burning stages are cooled by neutrino emission; in order of occurrence they are carbon, neon, oxygen, and silicon burning. It is the detailed nature of these processes, and their effects on the star, which (1) solves the problem presented by the parameterized approach to explosive nucleosynthesis of carbon and of neon, (2) funnels the amount of mass which will become the collapsing core toward the Chandrasekhar mass, and (3) sets the stage for the explosion itself. Thus the synthesis of the elements is inextricably related to the formation of pulsars and black holes. Indeed, the structural changes caused by neutrino emission allow most stars to eject matter and form neutron stars, rather than collapse to black holes.

The use of hydrostatic and thermal balance is extended to describe core and shell burning with neutrino cooling. Detailed stellar evolutionary models show simple regularities in structure which can be captured by trajectories at constant radiation entropy, as we previously found for hydrogen and helium burning. Shell-burning-stages leave concentric layers of ashes, the mass of which is fundamental to predictions of nucleosynthesis yield. Representative burning conditions are examined quantitatively with a network of 300 nuclei.

Comparing the detailed abundance patterns from these burning stages to the solar system abundances produces two significant results. First, the matter made by carbon and by neon burning comes from hydrostatic burning stages. The difficulty of overproduction of neutron-rich isotopes of $Z \geq 28$ simply disappears. Second, the hydrostatic burning of oxygen and especially silicon makes the matter too neutron-rich; the ejected mass, which made the products of such burning, must have been faster—an explosive combustion. The same argument allows us to make a preliminary estimate of the "mass cut," that is, the region which becomes the boundary between the expelled matter, and the neutron star or black hole.

Analysis of the numerical results of detailed stellar evolutionary calculations shows several loose ends. The interaction of the convective burning shells shows an almost chaotic behavior, so that any set of numerical results may best be regarded as a sampling of a distribution of perhaps greater complexity. Fortunately, the nucleosynthetic results—when integrated over a variety of stellar masses—are relatively insensitive to this behavior. Preliminary two-dimensional hydrodynamic simulations of the convective burning of oxygen in a shell indicate that the physics is more complex than assumed: the burning and the convective flow interact. The story is not yet finished.

10.1 NEUTRINOS AND CONVECTION

Below temperatures of $T \approx 5 \times 10^8$ K, photons are the means by which a star cools. The flux of heat carried by this radiation is proportional to the gradient of the temperature, and the net loss from a unit mass of material is proportional to the divergence of that flux. The cooling rate depends upon the second derivative of the temperature with respect to space. Therefore, the cooling is *nonlocal* in the sense that it depends upon the conditions in neighboring regions. A fundamental change occurs as neutrino cooling becomes dominant. The cooling rate depends directly upon local properties of the matter, primarily the temperature, and not its spatial derivatives. This modifies the structure and evolution in ways that are crucial for understanding nucleosynthesis.

Consider an idealized but representative case—a star of pure ^{16}O [27]. Above temperatures of $T_9 \approx 0.5$, the thermal time-scale becomes a local property, $\tau \approx E/s$, where E is the internal energy per unit mass, and s is the rate of energy loss by neutrino emission in units of energy per unit mass per unit time. In a representative region of temperature and density, $E \propto T$ while $s \propto T^k/\rho$, where $k \approx 11$, so $\tau \propto T^{-10}\rho$; see figure 6.2. The hot inner regions encounter the most severe cooling.

Before thermonuclear ignition, the only energy source is gravitational contraction, but that rate of energy supply is proportional to the rate of contraction. For homologous contraction the rate of energy supply is uniform over the contracting object. The outer, cooler layers can contract almost adiabatically, but the hotter, inner regions cannot. For these regions the evolution becomes markedly nonadiabatic, and a roughly isothermal core tends to develop. The radiation entropy decreases. Because the central density rises faster than the cube of the central temperature, electron degeneracy will begin to set in. The onset of degeneracy reduces the neutrino loss rates, and the central temperature slowly rises. The star has a relatively high central density, with a relatively steep density gradient through the isothermal region. This is accompanied by a steep gradient of entropy, increasing outward to the (almost) adiabatically contracting mantle.

As the central temperature approaches the ignition temperature, nuclear reactions begin to provide another source of energy. As emphasized by Fowler and Hoyle [238], nuclear heating increases *faster* with increasing temperature than does the cooling by neutrinos. *As long as fuel remains, nuclear heating can overcome neutrino cooling.* If the rate of heating is $\epsilon \propto T^m \rho$, then $m \approx 33 > k$. In the absence of any other mechanism for energy transport, the system becomes thermally unstable with a rise time $\tau \approx E/\epsilon$. Because $m > k$, the nuclear heating dominates in the center, but the neutrino cooling dominates in the mantle. The heated region soon has a higher entropy than the surrounding cooled region, giving a convective instability. The growth of the convective core is inhibited by the entropy gradient established prior to ignition. Photon diffusion is much too slow to be an effective cooling agent. Convection can carry the excess heat from the central regions to the larger mantle, where it can be radiated away by neutrinos.

Let us explore this quantitatively, with an idealized model. The average energy generation rate is

$$\langle \epsilon \rangle = \frac{1}{M} \int_0^M \epsilon \, dm, \qquad (10.1)$$

where M is some mass, which for the moment we take to be the mass of the model, and the average neutrino cooling rate $\langle s \rangle$ is similarly defined. A thermal balance is approximately obtained if $\langle \epsilon \rangle \approx \langle s \rangle$. It is the thermal balance of the region *through which thermal communication is maintained* that is relevant. Therefore, the range of integration should be from the center to the outer edge of the convective region (0 to M_{cc}).

First let us examine the limiting case in which the thermal balance is exact. The structure will be approximated by (see Appendix C)

$$\rho = e^{-x^2},$$ (10.2)

$$T = \rho^{\frac{1}{3}},$$ (10.3)

and the rates of cooling and heating by

$$s = T^{11}/\rho,$$ (10.4)

$$\epsilon = \rho T^{33},$$ (10.5)

where we measure quantities relative to their central values. The radius scale will not explicitly enter in what follows, but $x = r/R$ where R is that scale. The dimensionless mass, neutrino luminosity, and thermonuclear luminosity are

$$m/M = \int_0^{x_1} \rho x^2 dx,$$ (10.6)

$$L_\nu = \int_0^{x_1} \rho s x^2 dx,$$ (10.7)

$$L_{nuc} = \int_0^{x_1} \rho \epsilon x^2 dx.$$ (10.8)

To be definite we take the integration to extend to $x = 2$, which is reasonably complete. The results are shown in figure 10.1. In the left panel are plotted temperature, density, neutrino cooling rate, and thermonuclear heating rate versus radius. Notice the pronounced central concentration of the cooling and heating rates. In the second panel are shown the total neutrino luminosity L_ν, the total thermonuclear luminosity L_{nuc}, and the net luminosity $L_{nuc} - L_\nu$, versus a normalized mass coordinate m/M.

The cooling and heating are normalized so that they are in exact global balance; numerical integration gives $\epsilon_c/s_c = 6.66$ for the dependences quoted above. Using the methods of Appendix C,

$$\epsilon_c/s_c \approx \left(\frac{6+33}{0+11}\right)^{\frac{3}{2}}$$ (10.9)

$$\approx 6.68,$$

which is in good agreement. The net luminosity $L = L_{nuc} - L_\nu$ has a maximum of 0.532 (normalized) at a radius $x = 0.26$ and a mass $m/M =$

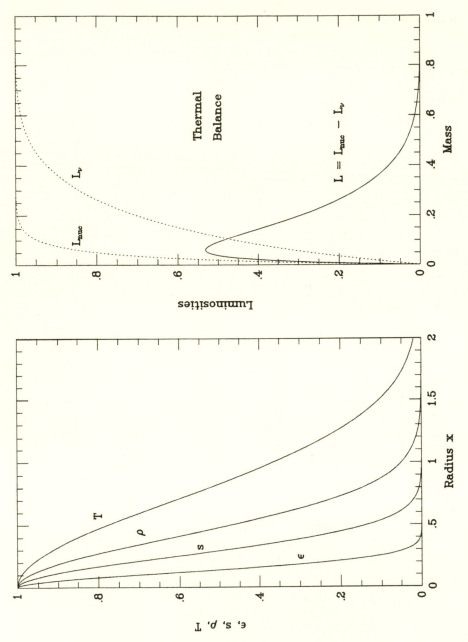

Fig. 10.1. Thermal Balance

0.0609. Interior to this radius, heating dominates; outside, cooling does. This is a small fraction of the mass of our model. The luminosity has dropped to a tenth of its maximum value at 45 percent of the mass. The ratio of central to average rates are $\epsilon_c/\langle\epsilon\rangle = 43.6$ and $s_c/\langle s\rangle = 6.55$.

Let us construct a simple approximation for later use. Suppose we define a *flame zone* which has a "typical" nuclear energy release. In such a mass, uniform burning at the maximum rate would produce the total nuclear energy release. Its fraction of the mass, m_{flame}/M, must be $\langle\epsilon\rangle/\epsilon_c = 1/43.6$, for these parameters. This flame zone is convectively mixed with a region about forty times larger. It is probably more accurate to think of convective blobs which spend about 2 percent of their time in the flame zone but otherwise reside in the cooler parts of the convective region. Notice that the part of the convective region that is outside the flame zone may not be inert with regard to nuclear burning. The flame zone nucleosynthesis must be corrected to account for reactions that occur in both regions, especially radioactive decay.

Can the thermal balance be closely attained? The thermal content of the matter is $E \approx \frac{3}{2}Y_e\Re T \approx Y_e T_9 10^{17}$ erg g^{-1}. The nuclear energy supply is $q \le X_{fuel}(5\times10^{17}$ erg g$^{-1})$, where the most energetic processes, carbon and oxygen burning, have been used. Thus,

$$q/E \approx 5X_{fuel}/Y_e T_9 \qquad (10.10)$$

$$\le 10$$

for generous estimates of the quantities. This suggests that imbalances of order 10 percent are to be expected in the global thermal balance. In such a case, the outer half of the mass may not be affected by convection because there is not time to attain exact thermal balance.

Consider a thermal balance that is not global (over the whole mass M), but only over the convective core M_{cc}. Then

$$\langle\epsilon\rangle = \frac{1}{M_{cc}}\int_0^{M_{cc}} \epsilon\, dm. \qquad (10.11)$$

Because of the steep temperature dependence of the nuclear heating, the entire contribution to the integral is soon reached, even for masses much less than M_{cc}. That is not true for the neutrino cooling, because

$$\langle s\rangle = \frac{1}{M_{cc}}\int_0^{M_{cc}} s\, dm \qquad (10.12)$$

does not reach its final value so soon, as the right panel in figure 10.1 illustrates.

Suppose the convective core is too small. Then $\langle \epsilon \rangle > \langle s \rangle$, so that the core heats, increasing its entropy, and growing. Similarly, if the core is too large, $\langle \epsilon \rangle < \langle s \rangle$, so that the core cools, decreases its entropy, and shrinks. This suggests that the system has negative feedback, and tends toward a state of thermal balance *in the convective core*.

As fuel is burned, the nuclear heating becomes less than needed, and does not quite balance the cooling. Contraction occurs, raising the central temperature, and hence the nuclear heating, so that balance is restored. In this manner the fuel is depleted in a relatively small amount of mass near the center of the star. This *thermal balance model*, while only a postulate,[1] is the model for all computations of the late stages of stellar evolution of which the author is aware. While attractive because of its simplicity, it ignores the possibility of more complex behavior, such as relaxation oscillations, excursions giving extra mixing, overshooting at the edge of the convective core, and so on.

Nothing in the discussion above necessarily referred to core burning rather than shell burning. We could equally well repeat the argument for a burning shell surrounding a core of ashes. The general features are the same.

The maximum extent of the convective core is a slowly varying function of the mass of the model, changing from $0.9M_\odot$ for a $2M_\odot$ oxygen star to $2.2M_\odot$ for $16M_\odot$. This inhibition to the growth of the convective core is to be compared to the results of chapter 7, where we would expect a much larger convective fraction. This behavior, repeated for each burning stage, reduces the central core to a smaller fraction of the star each time. If the core mass is below the ignition mass, shell burning will eventually add enough ashes to the core of the past stage to increase it until it ignites the next fuel. This is the phenomena of *core convergence*, which causes the core masses to tend toward the ignition mass of oxygen and of silicon burning, which is approximately the Chandrasekhar mass.

While the general nature of this behavior seems plausible, the exact details may be questioned. The growth of cores by convective shell evolution, and the nucleosynthesis in those convective shells, is a key part of nucleosynthesis theory. The validity of the algorithms used to describe this behavior is a long-standing question [28]. For the stages before neutrino cooling, much of the energy generation in a shell souce (the *flame* zone) does not occur inside the convection region. During a characteristic time τ, radiative energy transport will keep nearly isothermal a region of width $\Delta R \approx (\lambda c \tau)^{1/2}$, where λ is the transport mean free

---

[1] Recent two-dimensional hydrodynamic simulations support the general idea [45, 84], but not the details indicated in the one-dimensional calculations.

path. If the outer edge of such a region is convective, then isothermality insures that, in the presence of a pressure gradient, entropy decreases inward, so that most of the region is not convective. There will be a radiative region which contains most of the flame, and "pushes" the convective region before it. In a neutrino-dominated situation, the nuclear time-scale is much shorter than the local thermal time-scale as estimated from the radiative energy transport. For example, typical values for the first stage of carbon burning in a shell are $\tau < 10^8$ s, $T_9 \approx 0.9$, $\rho \approx 10^5 \mathrm{g\, cm}^{-3}$, shell radius $r \approx 1.0 \times 10^9$ cm, and the pressure scale height $h = (d \log P/dr)^{-1} \approx 2 \times 10^9$ cm. However, the width of a radiative region is $\Delta R \leq 2 \times 10^7$ cm. This is small, less than 0.01 of a pressure scale height. The problem is not just one of numerical resolution, but involves the physics of convection.

Conventional algorithms in stellar evolutionary codes give a flame zone which is inside the convective region. The shell burns until it consumes its fuel, and then "jumps" out in mass to the next region with fuel. The discontinuous behavior strongly affects the details of core convergence and nucleosynthesis in burning shells. This picture is not unique. Suppose we accept the thermal balance idea as it stands, and change only the boundary condition at the edges of the convective region. We assume that these interfaces are determined by convective overshoot rather than a thin spherically symmetric radiative layer. We note that diffusion of composition is even slower than that of radiative energy (see chapter 11). Perhaps convective blobs trade abundances weakly. Those with less fuel must sink deeper—to higher temperature—to burn as strongly. This tends to establish, statistically at least, a gradient in the abundance of fuel. If this gradient moves outward in mass in a smooth rather than discontinuous manner, the process of core convergence will be different in detail, as will the position and strength of burning shells. We have just built an alternative picture that seems as plausible. Core convergence will still happen, and the general nature of the fuel layers and burning shells will probably survive. However, if we do not understand the hydrodynamics of the flame zone, we should perhaps regard our simulations as indicative rather than definitive.

There is hope for improving this frustrating situation. A preliminary attempt [45] to directly simulate the convective motion during oxygen burning has been successful in two dimensions. Three-dimensional simulations will be required, but even these first computations provide some interesting results. The simulations demonstrate that oxygen-burning shells do approach a thermal steady state, as assumed above. The profile of the ^{16}O fuel does wander away from the steep form that the one-dimensional stellar evolutionary algorithms tend to give. Most interesting is the new view of structure of the burning and convective shells, due to the interaction of density (buoyancy) pertubations and convective flows because of the burning activity of the flame. Such behavior is not contained in conventional mixing length theory, as now used in stellar evolution. However, no light is shed on exactly how burning shells move out in mass, which remains a key uncertainty for quantitative estimates of nucleosynthetic yields, core sizes, and massive star structure.

10.2   CORE EVOLUTION

First we shall focus upon the evolution of the central regions (the core), from carbon burning up to instability. This evolution will be needed later for a discussion of core collapse. The nature and evolution of the core determines the extent and duration of convective shell burning in the overlying regions. The length of time spent at high density affects the neutron excess through electron capture, and provides a constraint on explosion models: they must not eject matter whose neutron excess is inconsistent with the solar abundance pattern. This is a statistical condition, in that rare events may be weird, but their integrated contribution should not be. Because nuclei of high neutron excess are rare, this constraint can be powerful.

In figure 10.2 are shown the trajectories in density and temperature of the centers of helium stars of 4, 8, 16, 24, and $32M_\odot$. These would correspond (roughly) to helium cores of stars of 15, 25, 50, 75, and $100M_\odot$, but the exact correspondence is affected by mass loss, convective uncertainties, and binary interactions. The locations of core burning of helium, carbon, neon, oxygen, and silicon are labeled. At the lower left-hand corner of the graph, the cores have temperatures that are too low for neutrino cooling to be effective. These cores evolve along lines having almost constant radiation entropy, so that $\rho \propto T^3$. After helium burning, the temperatures rise to the neutrino cooling regime, and the trajectories bend toward the horizontal due to cooling. The effect is largest for the lowest-mass cores, which have lower entropy and larger effects of electron degeneracy. Degeneracy pressure is affected by density, not temperature, so that cooling gives less change in pressure, hence less contraction, so less compressional heating. This results in larger amplitude excursions in ρ and T for the lower masses.

Core carbon burning occurs for temperatures $0.6 < T_9 < 1.0$. The higher masses make less ^{12}C in helium burning, and have higher neutrino energy loss rates. Higher temperatures would be required even for the same burning rate, and less fuel is available for burning, so that the

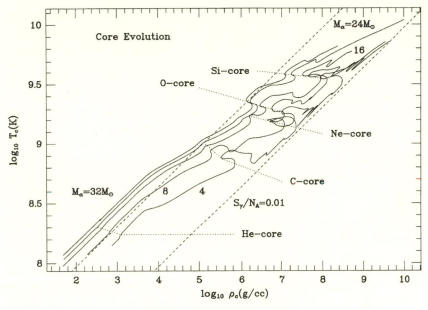

Fig. 10.2. Core Trajectories

effects of carbon burning are much more pronounced than for the lower masses. For these lower masses, electron degeneracy is not negligible when ignition occurs; there is a pronounced loop to higher entropy as the core heats up and expands. Most of the burning occurs near the labeled regions; so this is where most time is spent. After carbon is consumed, neutrino cooling is dominant. For the largest masses, this causes a continued contraction with increasingly smaller radiation entropy. The lowest masses have more complex trajectories. As carbon burning dies in the core, the entropy decreases (the core convective region shrinks). With carbon exhaustion, cooling dominates and, to the extent that degeneracy pressure is important, the temperature drops with little change in density. Shell burning begins, and the core contracts and heats. The "glitchy" behavior is due to the intermittent nature of the convective shell; as an old convective region exhausts fuel and a new one ignites, the core grows in spurts. As pointed out in the previous section, our ignorance of the physics of such convection prevents us from claiming that this behavior is the only possibility; on average, thermal balance will probably force something roughly similar, but the details could easily be different.

Eventually the core temperature rises to $T_9 \approx 1.5$, and neon burning begins. For the larger masses, which have less neon and greater cooling rates, this burning stage is relatively insignificant. For the lowest masses there is a vigorous neon flash, during which the low entropy of the core is raised significantly before steady burning sets in. At $T_9 \approx 2.0$, oxygen burning begins. This is the major stage of nuclear energy release for all masses. Even the largest masses show a significant entropy increase upon oxygen ignition. At $T_9 \approx 3.5$, silicon burning begins. By this time even the trajectories of the largest masses show wiggles that are caused by the increasing importance of electron degeneracy pressure. Notice the extent to which core convergence makes the $M_\alpha = 4M_\odot$ and $8M_\odot$ helium star trajectories similar after silicon burns. The trajectories end at the point of hydrodynamic instability to collapse; the time-scales are now so short that these stars are not so much *objects* as *events*! The trajectory for the $M_\alpha = 24M_\odot$ helium star is shown going further into collapse, to suggest the subsequent evolution. Note that because instability involves the whole stellar structure, a position of the center in temperature and density does not—by itself—define stability.

At first sight, the prospects for a simple approximation to this evolution seem hopeless. That pessimism is unwarranted. The extensive excursions do not coincide with the conditions during which the burning occurs. Figure 10.3 shows this more clearly. The trajectory of the central

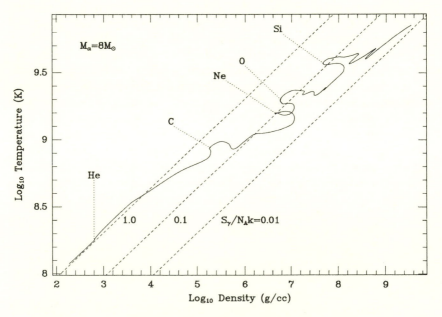

Fig. 10.3. $M_\alpha = 8M_\odot$ Core Trajectory

temperature and density is shown, as well as contours for $S_\gamma/\Re = 0.01$, 0.1, and 1.0. Carbon burning occurs at $S_\gamma/\Re \approx 0.4$, and neon, oxygen, and silicon burning occur near $0.1 < S_\gamma/\Re < 0.2$. Electron capture may occur during the time spent at higher densities (lower entropies). An attractive approximation is to use a trajectory having $S_\gamma/\Re \approx 0.2$; this will give a lower limit to the actual neutron excess because it ignores the effects of the higher density loops. Post-silicon burning evolution, which goes to considerably lower entropy, should be treated separately.

The behavior of the central values of temperature and density is an incomplete representation of what the star does. It does provide a chink in the armor of complexity, one we shall exploit.

Table 10.1 gives the behavior of the nucleon fractions of several important nuclei (^{12}C, ^{16}O, ^{20}Ne, ^{22}Ne, ^{23}Na, and ^{24}Mg) during core carbon burning, as well as temperature in the flame zone, neutron excess η, and time elapsed t_{10} in units of 10^{10} seconds. The radiation entropy is $S_\gamma/\Re = 0.4$ and the mass of the convective mass is forty times that of the flame zone. Only values of nucleon fraction above 0.001 are given. The time for burning is about three centuries.

Not only is ^{12}C consumed, but ^{16}O is also depleted. The burning is almost complete when the temperature rises above 10^9 K. The neutron excess increases by about 13 percent. At carbon depletion, ^{16}O and ^{20}Ne are the most abundant nuclei, comprising over 90 percent of the matter. The excess neutrons reside in a variety of nuclei, with ^{23}Na the most abundant; ^{22}Ne is depleted to almost a tenth of its previous value.

Core neon burning follows (table 10.2), becoming vigorous at temperatures near $T_9 = 1.5$. The radiation entropy is $S_\gamma/\Re = 0.2$ and the mass of the convective mass is forty times that of the flame zone; these values are maintained for the core burning of oxygen and of silicon described

TABLE 10.1

Core C Burning

^{12}C	^{16}O	^{20}Ne	^{22}Ne	^{23}Na	^{24}Mg	T_9	$10^3\eta$	t_{10}
0.261	0.710	0.002	0.016			0.761	2.08	0.000
0.249	0.708	0.013	0.015			0.845	2.12	0.434
0.150	0.661	0.143	0.008	0.011	0.007	0.910	2.19	0.962
0.019	0.599	0.306	0.002	0.016	0.027	1.116	2.35	1.195
0.001	0.592	0.324	0.002	0.016	0.033	1.402	2.37	1.202

TABLE 10.2

Core Ne Burning

^{20}Ne	^{16}O	^{24}Mg	^{27}Al	^{28}Si	^{29}Si	T_9	$10^3\eta$	t_7
0.324	0.592	0.043	0.006	0.002		1.402	2.37	0.01
0.311	0.598	0.041	0.006	0.003		1.471	2.34	0.67
0.278	0.613	0.058	0.006	0.005	0.001	1.501	2.30	1.14
0.250	0.627	0.072	0.006	0.008	0.002	1.514	2.29	3.55
0.189	0.655	0.096	0.007	0.016	0.003	1.534	2.28	1.42
0.112	0.693	0.117	0.010	0.031	0.008	1.564	2.29	2.52
0.022	0.743	0.115	0.015	0.064	0.013	1.643	2.36	3.21
0.001	0.758	0.104	0.016	0.078	0.014	1.742	2.41	3.46

below. As ^{20}Ne is depleted, ^{16}O increases to a value slightly higher than its value at the beginning of carbon burning. At the end of neon burning the most abundant nuclei are ^{16}O, ^{24}Mg, and ^{28}Si. The most abundant nuclei with excess neutrons are ^{27}Al and ^{29}Si. The neutron excess at first declines, and then increases to a value slightly larger than the initial. For neutron excesses near $\eta = 2 \times 10^{-3}$, both carbon and neon burning have a "funneling" effect: for smaller η, positron decay and electron captures occur to increase η, while for larger neutron excess, beta-decays occur to decrease it. This process may be partially responsible for the preference for such a neutron excess, which we found in the discussion of explosive nucleosynthesis. The duration of burning has shortened from that of carbon burning, and is now one year.

Core oxygen burning (table 10.3) is slightly faster, requiring less than eight months. Even though there is more fuel, and it has a higher energy release per unit mass, the higher temperature also increases the neutrino loss rates. Oxygen burning occurs near $T_9 = 2$. A major change occurs in the neutron excess, which more than doubles. About half of this change occurs within the first third of the burning. Such neutron-rich matter did not make major contributions to the solar abundance pattern; *presumably material processed by core oxygen burning must collapse rather than be ejected!*

The neutron excess increases still further during core silicon burning, which is illustrated in table 10.4. Silicon burning occurs over a range of temperature from $T_9 = 3$ to 4. During the last half of silicon consump-

TABLE 10.3

Core O Burning

^{16}O	^{24}Mg	^{28}Si	^{32}S	^{34}S	^{40}Ca	T_9	$10^3\eta$	t_7
0.757	0.103	0.079	0.007			1.808	2.43	0.00
0.719	0.098	0.107	0.017	0.001		1.925	2.70	0.31
0.658	0.078	0.155	0.041	0.003		1.939	3.13	0.54
0.494	0.040	0.252	0.129	0.012		1.980	4.18	1.08
0.188	0.009	0.371	0.314	0.050		2.118	5.68	1.80
0.089	0.001	0.439	0.332	0.087		2.232	6.03	1.96
0.059		0.463	0.324	0.082	0.003	2.295	6.12	2.00
0.026		0.505	0.302	0.055	0.016	2.415	6.18	2.04
0.011		0.531	0.280	0.040	0.027	2.534	6.20	2.05
0.001		0.550	0.270	0.039	0.026	2.766	6.26	2.06

tion, the neutron excess increases dramatically (a factor of five). Silicon is consumed in less than four days.

Initially, the burning consumes ^{32}S and produces ^{28}Si; oxygen burning had made more ^{32}S than quasiequilibrium allowed. The lighter neutron-rich nuclei, such as ^{34}S, are depleted early, as their heavier counterparts,

TABLE 10.4

Core Si Burning

^{28}Si	^{32}S	^{34}S	^{54}Fe	^{56}Fe	^{58}Ni	T_9	$10^3\eta$	t_5
0.550	0.270	0.039	0.006	0.005		2.765	6.27	0.
0.701	0.155	0.009	0.066	0.023	0.002	3.064	6.40	0.84
0.724	0.120	0.003	0.099	0.014	0.002	3.303	6.55	1.17
0.650	0.157	0.001	0.146	0.004	0.005	3.395	6.83	1.35
0.555	0.196		0.182	0.001	0.009	3.448	7.72	1.52
0.282	0.130		0.462	0.003	0.030	3.523	19.7	2.38
0.040	0.025		0.729	0.014	0.070	3.709	34.3	3.19
0.003	0.003		0.697	0.031	0.107	3.946	37.9	3.36
0.001	0.001		0.649	0.041	0.126	4.070	38.8	3.39

such as ^{54}Fe, ^{56}Fe, and ^{58}Ni, increase. As more mass is pushed toward the iron-peak region ($Z > 20$), the faster rates of positron decay and electron capture cause the neutron excess to change rapidly. Notice that ^{54}Fe is fifteen times more abundant than ^{56}Fe, while the solar system ratio is 0.061, a difference of a factor of 258. This matter too must be buried in a condensed remnant, and not escape the gravitational potential well.

10.3   STELLAR STRUCTURE

Much of the story of nucleosynthesis involves the ejection of previously processed layers of a massive star. To understand thermonuclear yields we must understand the structure of neutrino-cooled stars.

In chapter 8 the M-M_α approximation was introduced, in which the matter inside the hydrogen-burning shell of a star was approximated by a "helium star" of roughly the same mass. This helium star undergoes each burning stage in a way similar to that of the corresponding stellar core. Real stars have complications, such as mass loss, mixing, and inter-action with a companion star, which are not well understood. First we concentrate upon the simpler problem, hoping that it will be possible to relate it to the eventual solution.

In figure 10.4 are shown the density structure at the time of core collapse for helium stars of mass $M_\alpha/M_\odot = 4$, 5, 6, 7, 8, 10, 12, and 16. The behavior is not smooth from one mass to the next. This is not due to a flaw in computational technique or to inadequate zoning in space or in time. First, we will focus on the regularities. All central regions develop a *core* of 1.5 to $1.8 M_\odot$. As the stellar mass increases, a more pronounced *mantle* develops around the core, at increasingly higher average densities ($\rho \approx 10$ to 10^4 g cm^{-3}), or so. At the interfaces between fuels and ashes, density ramps appear, corresponding to the change in composition and entropy at burning shells. The helium-burning shell is denoted by He/C, and the neon-burning shell by Ne/O. As we go from 16 down to $8 M_\odot$ the curves change smoothly, but at $7 M_\odot$ the pattern of steadily decreasing density in the mantle is broken; here the mantle is indeed smaller, but at higher density than the $8 M_\odot$ helium star. The 7, 6, 5, and $4 M_\odot$ helium stars form a new pattern.

How might this come about? Figure 10.5 shows the history of the convective zones in the $8 M_\odot$ helium star, from the onset of core helium burning until core collapse. The vertical axis is the Lagrangian mass coordinate, and the horizontal axis is the time remaining until core collapse. The shaded regions are convective. Because the convection is driven by nuclear energy release, each such region also corresponds to a stage of burning, so that it contains the products of that stage of burning. This

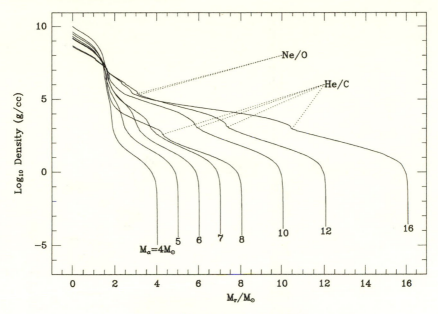

Fig. 10.4. Structure: $M_\alpha = 4$ to $16 M_\odot$

type of graph allows us to see which burning stages are most prominent in determining the nucleosynthesis yield.

The first and largest convective region is due to core helium burning; it extends to $4.3 M_\odot$. Its behavior is relatively simple. Just beyond the convective core is an unattached semiconvective region. After exhaustion of fuel in the core, the core convection dies. With contraction, a helium shell ignites (radiatively) at the outer edge of the old convective core, and drives an unattached convective shell above it. This shell mixes more opaque matter (products of helium burning, especially carbon) into material that is less opaque. Once mixed, the matter can carry less energy by photon diffusion, and tends to remain convective longer. This seems to be the cause of the pulses of shell convection seen from the end of core helium burning onward; their characteristic time is of the order of 10^9 s, so that they become frozen by the increasingly rapid evolution. They mix the products of helium burning over the outer half of the helium star.

At a time about three centuries before core collapse, carbon burning begins in the core. The size of the convective core is reduced almost a factor of ten from that in helium burning; this is due to the structural change from neutrino cooling. At the end of core burning, carbon burning ignites again in a shell. Unlike helium burning, the convective shell

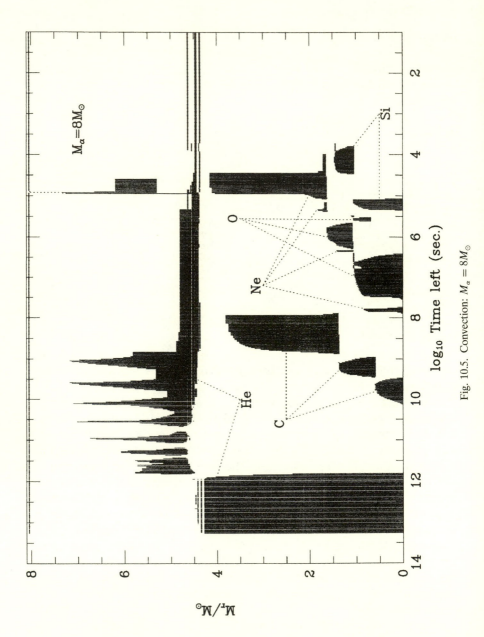

Fig. 10.5. Convection: $M_\alpha = 8M_\odot$

contains the flame zone. The shell burns itself out, to be followed by ignition of yet another shell still further out. This third stage of carbon burning occurs in a shell that contains a core of slightly less than the ignition mass for neon in this configuration. As the limiting mass is approached, the shell is forced to higher luminosity, and higher entropy. Neutrino cooling made the inner regions almost isothermal at carbon ignition, but left the overlying layers more nearly adiabatic. As the shell burning occurs further out in mass, the structure is closer to adiabatic, so that the convective region can grow more easily, and contain more mass. By the second stage of carbon burning in a shell, the convective zone is driven outward in mass, and contains about $2M_\odot$. Thus a considerable amount of matter, which lies at or beyond the nominal mass of the neutron star to be formed, is subjected to carbon burning in a vigorously burning shell. This is a common pattern; vigorous shell burning (and nucleosynthesis) happens in shells enclosing a mass approaching $1.5M_\odot$, or so. We will return to this point when we examine nucleosynthesis yields.

The pattern of the process is a general one, and not a particular feature of carbon burning. When the new shell encloses enough mass for the core to contract to ignite the ashes, a new stage of *core* burning begins. Ignition of the next fuel is delayed until some number of stages of shell burning occur, and build a core of ashes up to the ignition mass.[2]

Neon ignition in the core occurs under conditions of moderate electron degeneracy, so that there is a modest *neon flash*. Because the ignition masses of neon and oxygen are relatively close, core neon burning is followed by core oxygen burning. Neon shell burning occurs after core oxygen burning, followed by oxygen shell burning.[3] This is the same pattern, but with neon and oxygen burning occurring pairwise.

Core silicon burning follows, with a weak stage of neon burning. When silicon is exhausted in the core, a dramatic stage of shell burning occurs, in which almost all of the old carbon burning shell is processed, by high-temperature carbon burning merging into neon burning. This is a major factor in determining the yield for the $M_\alpha = 8M_\odot$ star. Core contraction continues while the shell burns, and silicon ignites in a shell, ending the episode of neon shell burning. When this is finished, the core contracts to the point of instability toward hydrodynamic collapse.

[2]This is qualitatively the same as the simpler case discussed in chapter 6, but the quantitative values for ignition masses are now affected by additional effects (for example, the finite thermal time-scale and the thermal structure of the last burning shell)

[3]After the first stage of oxygen shell burning, a weak second stage of oxygen shell burning occurs which depletes a small amount of fuel which lays interior to the flame zone of the first shell, but was too cool to ignite then.

How different is the evolution for the $M_\alpha = 7M_\odot$ star? The corresponding behavior of convective burning zones is shown in figure 10.6. At first glance, the pattern is almost the same. The excursions of convection associated with the helium shell are qualitatively similar, but not identical. The sizes (in mass coordinate) of the convective zones are slightly different, but the order of events is the same until core silicon ignition. In this star, the vigorous shell burning occurs by carbon burning, and the shell is further out in mass coordinate. An intermediate region undergoes shell neon burning. This is followed by shell oxygen burning, which is proceeding strongly as the core encounters hydrodynamic instability. This slight modification of the strength of burning shells, which occurs about a day prior to core collapse, affects the yields and the presupernova structure.

In such a situation, it is easy to see how slightly different parameters (such as initial mass of the helium star in this case) can give qualitatively different behavior. Take an idealized case, in which two neighboring masses just do (and do not) ignite a core-burning stage. The stars are qualitatively different in that one is burning ashes while the other is still burning the old fuel in a shell. Now add the complexity that the burning stages affect each other. When one shell is burning, it tends to prevent others from burning; there tends to be an alternation of burning stages. Also, an active burning stage can mix not only its own products but also those of previous burning stages. Solutions originally close can diverge as they evolve.

It should not be a surprise that such complex nonlinear dynamical systems can develop divergent behavior. This is reminiscent of the development of the theory of chaos [203, 418, 529], in which it was found that simple deterministic systems developed solutions which began from similar states but diverged dramatically. The effects seen here are hardly so spectacular, but the sensitivity to parameters should be a warning. Using a different rate for $^{12}C(\alpha, \gamma)$ will also give significant variation in the sizes and strengths of burning stages. Even if there is an element of "chaotic" behavior here, the results do not really seem chaotic so much as irregular, or "glitchy." There is a strong tendency for results to be similar for a range of masses, and then jump to a different set of results which are themselves similar. This is probably related to the strong negative feedback in these stars, which tends to force hydrostatic and thermal balance. Certain burning stages are favored to satisfy these constraints, and in a statistical sense they will be favored to contribute to the yields. We will regard our helium stars as an ensemble which reflects a possibly realistic weighting of these different burning stages; hydrogen stars evolved with no mass loss represent another ensemble, and so on.

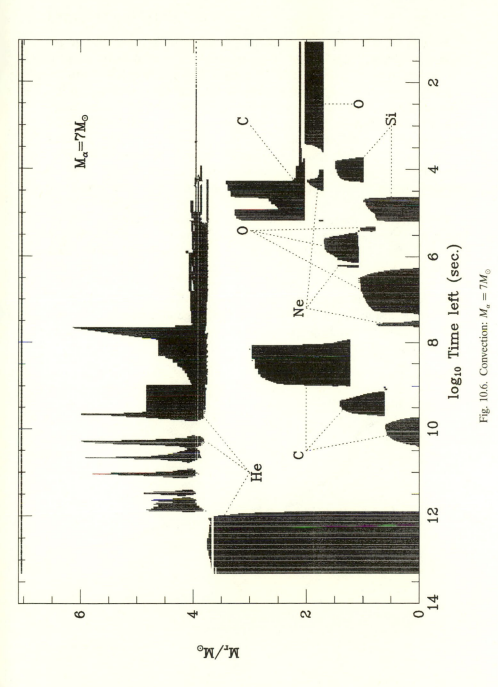

Fig. 10.6. Convection: $M_\alpha = 7M_\odot$

We will separate the nucleosynthesis problem from the stellar evolutionary problem. Examination of an ensemble of evolutionary trajectories will allow us to define a set for representative conditions for burning. These conditions will be used to explore the details of nucleosynthesis. The abundances so produced will be compared with the solar abundance pattern, thereby placing constraints upon the conditions sampled by solar system material—and to the extent that solar system matter is representative of Population I—by cosmic matter. The strategy is similar to that of explosive nucleosynthesis, but we replace a simple model of an explosion by a simple model of a shell-burning stage in a hydrostatic star.

How large are the uncertainties caused by our ignorance of the effects of mass loss, mixing, etc., in the hydrogen-rich layers? The nature of some of the error may be seen by comparing helium stars to a star of $M = 25M_\odot$, which develops a helium core of $8M_\odot$. The density versus mass coordinate for these stars is shown in figure 10.7. These stars have been evolved, using exactly the same physics, up to the point of hydrostatic instability: the cores are on the verge of collapsing. The helium stars are shown as simple lines; the $M = 25M_\odot$ star, as a heavier line with dots. The inner regions, $M_r < 1.5M_\odot$, are similar, although the

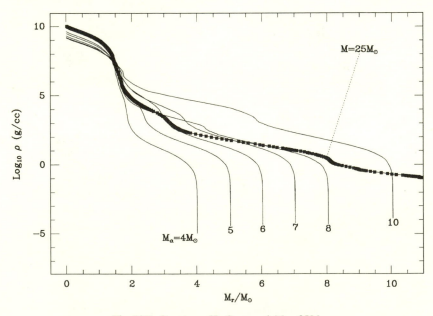

Fig. 10.7. Structure: He Stars and $M = 25M_\odot$

more massive helium stars begin to collapse at a lower central density. The $10M_\odot$ helium star has significantly higher density in the mantle, and is not a good approximation to the core of the $25M_\odot$ star. The $8M_\odot$ helium star is a reasonable approximation from $M_r = 8M_\odot$ inward to about $5M_\odot$; near $M_r = 8M_\odot$ the curves diverge because of the hydrogen-burning shell. Inside $5M_\odot$, the $25M_\odot$ tends to merge with the curve of the $6M_\odot$ helium star, but at $M_r = 2.8M_\odot$ it switches back to the $8M_\odot$ helium star.

The composition (in solar mass units) at core collapse is given in table 10.5 for $M_H = 25M_\odot$ and helium stars of 10, 8, 7, 6, and 5 $M_\odot$. The helium abundance for the $M_H = 25M_\odot$ star includes a nucleon fraction of 0.23 in the unburned hydrogen, plus some additional helium produced by the hydrogen-burning shell. Blank entries correspond to nucleon fractions less than 0.0005. The nuclei from $14 \leq Z \leq 20$ are grouped in "SiCa", and except for ^{56}Ni, those with $Z > 20$ are grouped as "Fe." Because of the increase in neutron excess during and after oxygen burning, we expect that SiCa and Fe must become a condensed remnant; to compare with gravitational masses of neutron stars, reduce this mass by 10 percent to correct for gravitational binding energy. If we consider only the yield of oxygen, the $7M_\odot$ helium star most resembles the $M_H = 25M_\odot$ star. The carbon production implies something between 8 and 10 $M_\odot$. Although the patterns are similar, no single helium star is an accurate fit to the $M_H = 25M_\odot$ star. The level of variation is roughly a factor of two. Formally, several of the stars have a nonnegligible mass

TABLE 10.5

Composition ($M_\odot$)

Species	$M_H = 25$	$M_\alpha = 10$	8	7	6	5
H	12.16	0.	0.	0.	0.	0.
He	9.148	3.864	3.418	3.086	2.807	2.462
C	0.543	0.816	0.338	0.349	0.339	0.218
O	1.040	2.497	1.626	1.056	0.757	0.430
Ne	0.357	0.614	0.302	0.358	0.240	0.141
Mg	0.177	0.197	0.292	0.225	0.129	0.087
SiCa	0.175	0.422	0.480	0.460	0.123	0.254
^{56}Ni	0.034	0.014	0.003		0.022	
Fe	1.504	1.632	1.586	1.506	1.611	1.435

of ^{56}Ni, which is freshly made by an oxygen shell as the core contracts prior to collapse; a more realistic reaction network might have produced a greater neutron excess, and given iron nuclei.

How can these abundance results be related to the evolutionary behavior? To understand we must connect the final composition to the layers of burning and convective zones which existed during the lifetime of the star. We need to introduce a vocabulary. In figure 10.8 is shown the compositional structure of the inner $8.5M_\odot$ of the $25M_\odot$ star. The abundances of the dominant nuclei are labeled; this format is the standard for this book, and will be repeated below. For the neutrino-cooled stages, there tend to be pairs of layers related to each burning process: the first is a convective region containing the flame zone, and the second is an overlying inert zone with no active burning. Several layers are labeled.

A The innermost region of the star (the core), composed of iron-group elements
B The silicon-burning zone, composed of Si to Ni nuclei, with no oxygen
C The oxygen-burning zone, composed primarily of oxygen and of Si to Ca nuclei
D The neon-burning zone, composed of Ne, Mg, and O, but no carbon
E The carbon-burning zone, composed of C and O, and carbon-burning products (mostly Ne and Mg)
F The inert carbon zone, in which helium has been exhausted but carbon is not being consumed. There were inert zones associated with the previously listed burning zones (just outside them), but not labeled because of their small mass.
G The radiative helium-burning zone, composed of helium and its products C and O, primarily. Because the helium burning is incomplete, the abundance of C may exceed that of O in much of this region
H The convective zone above the radiative helium-burning zone. This mixes the products of helium burning (mostly C) far out into the helium star
I The inert part of the old helium core, interior to the hydrogen-burning shell
J The material above the hydrogen-burning shell

The compositions used here are a shorthand; the details will be examined later. The number of electrons per nucleon, Y_e, decreases toward 0.5 at the interface between zones I and J. It has a more subtle but important decrease in zone C, so that the matter in zones A and B are too neutron rich to be compatible with massive contributions to the solar abundance pattern; zone C may or may not be. The most massive

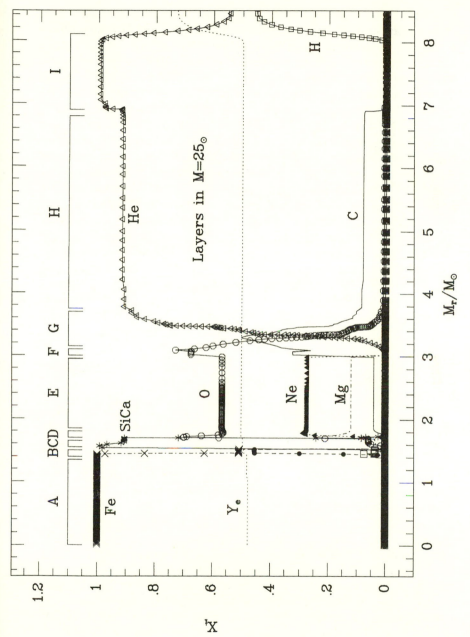

Fig. 10.8. Layers in $M = 25 M_\odot$

zones are obvious. If zones A and B become a neutron star, then most of the yield is typified by the abundances in H, I, and E. Further processing during the ejection process will be increasingly less inportant as we move outward; the innermost ejected zones will be subjected to the most extreme conditions.

How does this compositional structure change for other stars? Figure 10.9 shows this data, and that corresponding to helium stars of 6, 7, and 8 $M_\odot$. The horizontal axis ranges from 0 to 8.5 $M_\odot$ for all four cases, so they may be compared without scale changes. The width of region E steadily increases from 6 to 8 $M_\odot$, with 7 $M_\odot$ being similar to the core of the $M = 25M_\odot$ star. However, zone C is large for 7 $M_\odot$, small for 6 $M_\odot$, and almost gone for 8 $M_\odot$. It may be more accurate to say that zone E has been replaced by zone D in 8 $M_\odot$. The vigorous shell burning we discussed in regard to convective zones is the cause of the large extent of zone E (or D, if it is the neon-burning shell). The relatively large size of zone C in the 7 $M_\odot$ star is the result of a vigorous oxygen-burning shell. The net result is an ensemble of contributions, primarily from shell-burning regions for carbon, neon, and oxygen burning. To avoid the problem of too large a neutron excess in the iron-peak, this must be made from burning matter with $\eta \approx 3 \times 10^{-3}$. This in turn requires that matter at least as far out as zone C be burned, and only oxygen burning in *explosive* conditions will make the iron-peak without increasing η. Thus we are led to the suggestion that the contribution of massive stars to the solar system iron-peak nuclei must involve explosive burning, probably starting with oxygen burning.

The yield of heavy elements would be essentially the mass in zones C through G, with some contribution of carbon from zone H. This steadily increases with mass of the helium stars. Similar behavior is found in stars that retain their hydrogen. More massive stars will produce more massive mantles of heavy elements, but there are fewer massive stars. The stars shown in figure 10.9 are in the range expected to have the most important contribution.

Under what thermodynamic conditions do these stages of shell burning occur? They have relatively little variation in this range of mass, so that we simply consider the 8 $M_\odot$ helium star as a reasonably representative case. In figure 10.10 are shown the structure in temperature and density, for four epochs in the life of the star: core carbon burning, core neon burning, core oxygen burning, and the onset of core collapse. The position of the flame zone for each burning shell is shown. Also plotted, as dotted lines, are contours for radiation entropy $S_\gamma/\Re$ equal to 0.1, 1, and 10. Vigorous shell burning occurs at $S_\gamma/\Re \approx 1$ to 2 for all of these processes! It is reasonable to approximate shell burning simply

25M_⊙ and He stars

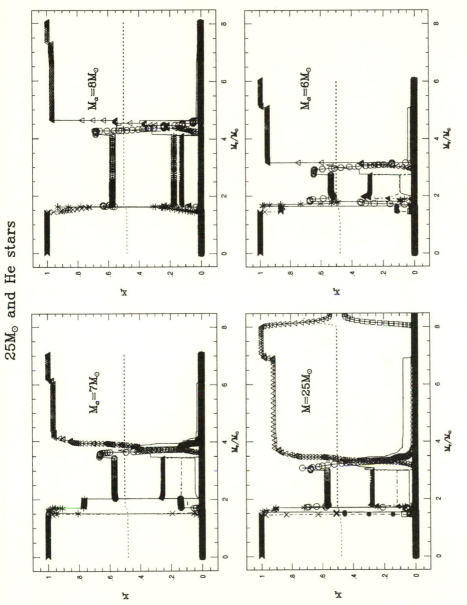

Fig. 10.9. Composition of He Stars and $M = 25M_\odot$

S_γ in Shell Burning ($M_\alpha = 8 M_\odot$)

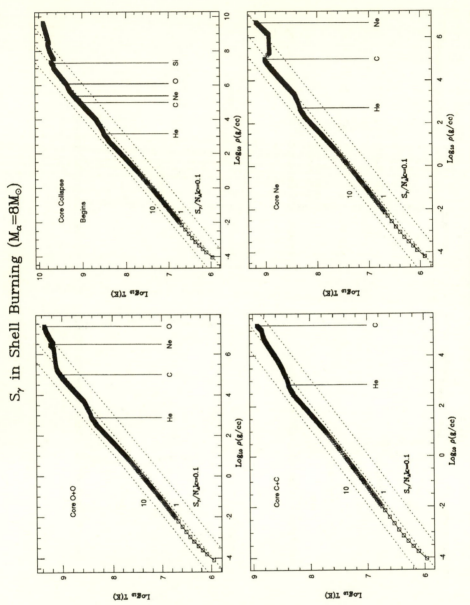

Fig. 10.10. S_γ in Shell Burning

by the value of radiation entropy $S_\gamma/\Re$, the ratio of mass in the convective zone to that in the flame zone, and the time available for burning (or fraction of fuel consumed). The choices may be guided by existing numerical evolutionary sequences, but will be more general in that they might also result from combinations of physics not yet explored. They will also provide a relatively simple set of results to aid understanding and to use as a basis for comparison.

10.4 SHELL NUCLEOSYNTHESIS

If we are to parameterize shell burning by the value of radiation entropy $S_\gamma/\Re$, the ratio of mass in the convective zone to that in the flame zone, and the time available for burning (or fraction of fuel consumed), what are appropriate values to be? In the previous section, the demonstration that constancy of radiation entropy is a reasonable approximation also supports the choice $S_\gamma/\Re = 1.5$. In Appendix C are simple approximations for the ratio of $\langle \epsilon \rangle$, the mean energy generation, to its value in the flame zone, ϵ_s. This ratio is also the ratio of the mass of the flame zone to that of the convective shell. The exact values differ for each fuel, but all lie within a factor of two of the value used previously for core burning: $\Delta M_{conv}/\Delta M_{flame} = 40$. For simplicity we will use that value throughout. Only the specification of the fraction of fuel consumed remains. For that we use the stellar evolutionary sequences for guidance in each particular case.

Before proceeding to examine the nucleosynthesis implications of this procedure, it must be tested. How well does it really work? Figure 10.11 illustrates this for carbon burning in the core, and two shell stages, for the $8M_\odot$ helium star. In the upper panel is shown the flame temperature as a function of the nucleon fraction of ^{12}C left. The stellar evolutionary sequence is shown by solid squares; the algorithm just outlined is shown as a solid line. To lessen clutter, the line for the first stage of shell burning is not plotted. The core burning is well represented by a constant radiation entropy of $S_\gamma/\Re = 0.4$ as used previously; at the very end of core helium burning the line rises above the stellar evolutionary results because of the neglect of electron degeneracy effects upon the structure. The core is beginning a swing to much lower S_γ, as was shown in Figure 10.2. The second stage of shell burning is represented by $S_\gamma/\Re = 1.5$, which follows the stellar evolutionary trajectory as far as it goes.[4]

To test the accuracy of the estimate of the ratio of convective zone to flame zone mass, the lower panel shows the same calculations but plots

---

[4]For this case, carbon is depleted down to a nucleon fraction of 11 percent.

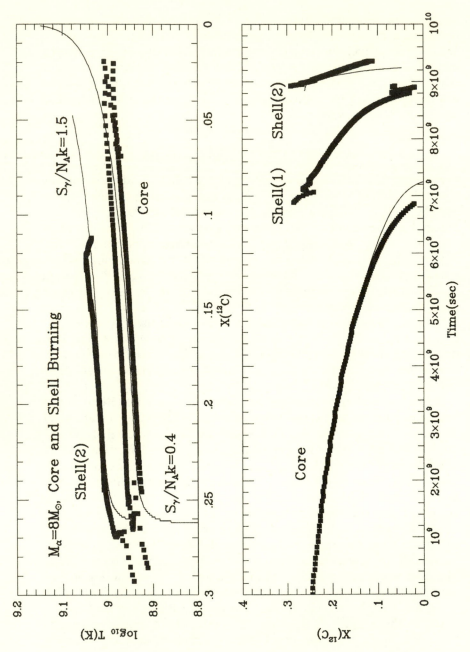

Fig. 10.11. $M_\alpha = 8M_\odot$ Core and Shell C Burning

nucleon fraction of ^{12}C as a function of time. The crude choice works well; clearly refinements would give an almost exact agreement.

Taking the ashes from core helium burning from chapter 8, we evolve the abundances to $X(^{12}\text{C}) = 0.05$, using this approximation for shell burning conditions, with our previous treatment of hydrostatic and thermal balance. The resulting abundances for $A < 50$ nuclei are shown in figure 10.12, compared to solar abundances normalized at ^{20}Ne. A significant fraction of ^{16}O is produced, and ^{22}Ne, ^{23}Na, ^{25}Mg, ^{26}Mg, ^{27}Al, ^{36}S, ^{40}Ar, ^{40}K, and ^{46}Ca are made to within about a factor of two of their solar values.

However, ^{24}Mg is underproduced by about a factor of three. In the explosive nucleosynthesis calculation shown in figure 9.2, ^{24}Mg was produced close to the solar value, as were ^{29}Si, ^{30}Si, and ^{31}P, which are underproduced by shell carbon burning. The neutron-rich nuclei ^{40}Ar, ^{40}K, and ^{46}Ca were also produced by explosive carbon burning, but ^{36}S was underproduced. The initial composition, from core helium burning (see figures 8.4 and 8.6), was deficient in ^{36}S and ^{46}Ca. On the basis of the abundances in the range $12 < A < 50$, there is no clear choice between explosive and shell carbon burning.

Having established the normalization, the abundances of the heavier nuclei are fixed. These are shown in figure 10.13. Unlike the previous results for explosive carbon and neon burning, there is no problem with overproductions. This range of nuclei does give a clear preference for shell carbon burning over explosive carbon burning (see figure 9.11). Comparison with core helium burning (figure 8.6) shows a significantly improved pattern for $A > 70$. The neutron-poor nuclei ^{64}Zn, ^{74}Se, and ^{78}Kr and the neutron-rich nuclei ^{70}Zn, ^{76}Ge, and ^{82}Se are not produced. The nuclei ^{61}Ni, ^{62}Ni, ^{63}Cu (^{63}Ni), ^{64}Ni, ^{65}Cu, ^{66}Zn, ^{67}Zn, ^{68}Zn, ^{69}Ga, ^{70}Ge, ^{71}Ga, ^{72}Ge, ^{73}Ge, ^{74}Ge, ^{75}As, ^{76}Se, ^{77}Se, and ^{78}Se are produced in solar abundances. The isotopes ^{79}Br and ^{81}Br are low by a factor of two. Near the upper edge of the network, where the results may reflect errors due to incompleteness,[5] ^{80}Se is underproduced by a factor of 30 while ^{80}Kr is overproduced by a factor of three. Finally, ^{82}Kr, ^{83}Kr, ^{84}Kr, ^{85}Br (^{85}Kr), and ^{86}Kr have solar values but lie at the network boundary.

All elements from Cu ($Z = 29$) to Kr ($Z = 36$) are reproduced; the results at Se and Kr are not exact because that corresponds to the upper edge of the reaction network that was used. It seems likely that the nuclei

---

[5]There is an additional difficulty: this region is affected by the isomeric decay of ^{79}Se. The 96 KeV isomer has a half-life of 3.8 minutes, compared to the ground-state half-life of 6.5×10^4 years. The isomeric state is easily excited at the temperatures ($T_9 \approx 1.1$) in the shell. Our flame zone approximation may give too crude a representation of the decay rate averaged over the convective region. See [663].

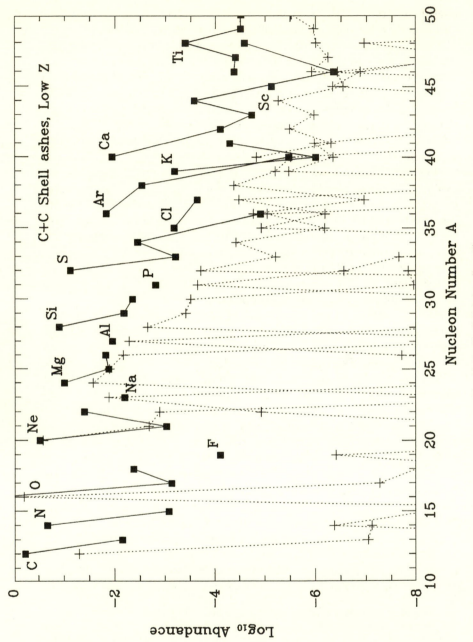

Fig. 10.12. Shell C Ashes (low Z)

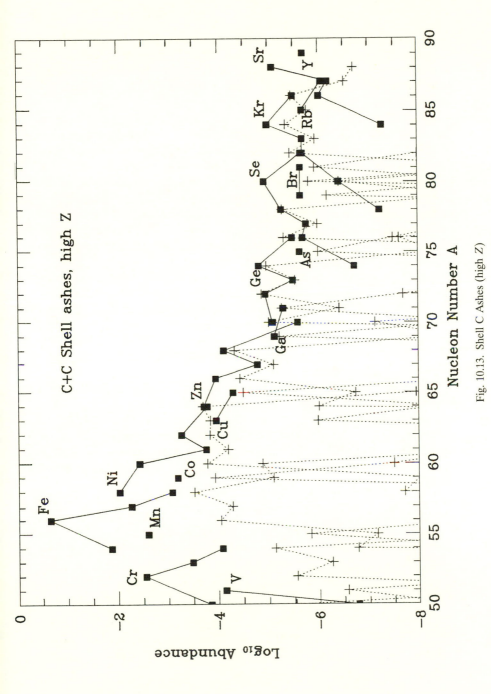

Fig. 10.13. Shell C Ashes (high Z)

to the neutron magic number ($N = 50$) will be made. *This mild s-process is the inevitable result of shell carbon burning of Population I material*; this matter should be a major component of the ejecta of massive stars. As such, it is a prime candidate for the weak component of the s-process [87, 369, 641].

Table 10.6 gives the evolution of some major constituents, from the end of core helium burning through shell carbon burning down to $X(^{12}C) = 0.05$. Most of the carbon consumption occurs for temperatures $1.0 < T_9 < 1.1$, which is significantly higher than for core burning. The variation in neutron excess is roughly proportional to the amount of carbon consumed, and rises to $\Delta\eta \approx 1.4 \times 10^{-4}$. Most of the neutron excess resides in ^{22}Ne initially; as carbon burning produces free particles, this is changed to heavier nuclei such as ^{23}Na, the neutron-rich Mg isotopes, and ^{27}Al. Some oxygen is consumed as well as carbon; ^{24}Mg increases toward the end of the burning.

Taking these ashes from shell carbon burning, we evolve the abundances to $X(^{20}Ne) = 0.10$. The resulting abundances for $A < 50$ nuclei are shown in figure 10.14, compared to solar abundances normalized at ^{24}Mg. There is the expected underproduction of ^{20}Ne, and ^{22}Ne and ^{23}Na as well. The previous problem of underproduction of ^{29}Si, ^{30}Si, and ^{31}P is now rectified. The neutron-rich nuclei ^{36}S, ^{40}Ar, ^{40}K, and ^{46}Ca are again produced in solar values.

The abundances of the heavier nuclei are very similar to those shown in figure 10.13, with a slightly larger abundance of the Kr isotopes; there is an artificial buildup at ^{86}Kr, an effect of the network boundary.

It is likely that contributions from a combination of carbon and neon shell burning will provide an excellent reproduction of the solar system abundance for the nuclei previously ascribed to explosive carbon burn-

TABLE 10.6

Shell Burning: C to Ne

^{12}C	^{16}O	^{20}Ne	^{22}Ne	^{23}Na	^{24}Mg	T_9	$\eta \times 10^3$
0.261	0.710	0.002	0.016			0.400	2.064
0.261	0.710	0.002	0.016			0.778	2.066
0.253	0.708	0.011	0.015	0.002		0.975	2.074
0.096	0.640	0.204	0.002	0.012	0.017	1.115	2.149
0.050	0.619	0.258	0.001	0.013	0.027	1.192	2.206

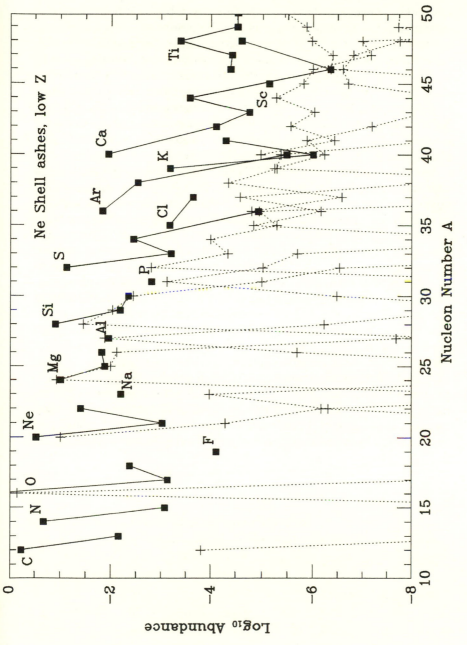

Fig. 10.14. Shell Ne Ashes (low Z)

ing, without unacceptable overproductions. As a bonus, it offers a natural explanation for the weak component of the s-process.

The evolution of some of the main constituents is shown in table 10.7, from the end of shell carbon burning through shell neon burning down to $X(^{20}Ne) = 0.1$. Burning the last 0.05 nucleon fraction of ^{12}C causes the increase in ^{20}Ne between the first and second lines; notice the change in temperature. Neon is mostly consumed between $1.6 < T_9 < 1.7$ in shell burning. The neutron excess changes little in value; it shifts toward heavier nuclei, now ^{27}Al, the neutron-rich isotopes of Si, and ^{31}P. The abundance of ^{28}Si is beginning to be significant.

What happens if we proceed? Taking these ashes from shell neon burning, we evolve the abundances to $X(^{16}O) = 0.10$. The resulting abundances for $26 < A < 86$ nuclei are shown in figure 10.15, compared to solar abundances normalized at ^{28}Si. The pattern is close to that from explosive oxygen burning shown in figure 9.6. There is an underproduction of ^{50}Cr by a factor of ten, and both ^{38}Ar and ^{42}Ca are overproduced by a factor of two or three. The neutron excess is too high for production of the Si to Ca region, but would be appropriate for the iron-peak. During the oxygen shell-burning stage, the neutron excess increases from $\eta = 2.2 \times 10^{-3}$ to 3.0×10^{-3}. The increase is less dramatic than in oxygen core burning because (1) core burning goes to completion, so that more burning occurs, and (2) the lower entropy of the core implies higher density, which assists electron capture.

At high A there is a dramatic decrease in abundance; the higher temperatures destroy these more fragile nuclei. Several p-process nuclei remain: ^{74}Se, ^{78}Kr, and ^{80}Kr. If this material is further processed upon ejection, the fragile p-process nuclei may be destroyed. Arnould [56] first explored hydrostatic oxygen burning as a site for the p-process, and

TABLE 10.7

Shell Burning: Ne to O

^{20}Ne	^{16}O	^{24}Mg	^{27}Al	^{28}Si	^{29}Si	T_9	$\eta \times 10^3$
0.261	0.618	0.027	0.005	0.002		1.192	2.210
0.284	0.612	0.054	0.007	0.005	0.001	1.583	2.254
0.161	0.669	0.104	0.010	0.021	0.006	1.645	2.235
0.098	0.701	0.117	0.013	0.035	0.009	1.676	2.259

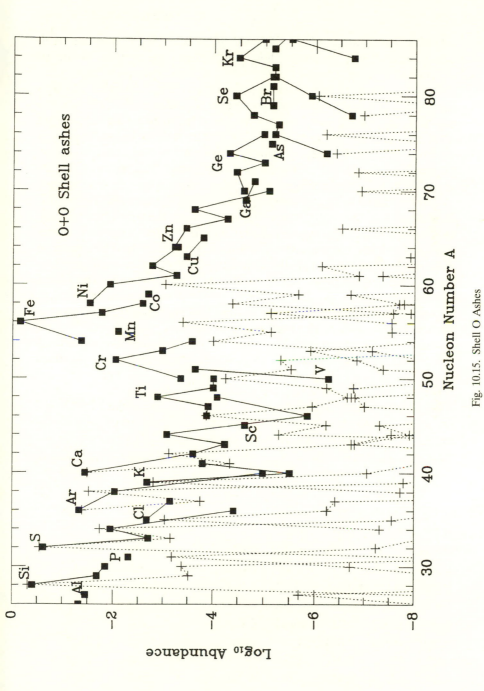

Fig. 10.15. Shell O Ashes

Woosley and Howard [696] considered the process in an exploding shell. See Lambert [377] for an excellent and wide-ranging review of the p-process.

The evolution of the major constituents is more complex for oxygen burning, as table 10.8 indicates. The abundance of ^{16}O keeps rising as neon is finally depleted; it does not begin to be depleted itself until the temperature rises above $T_9 = 2.0$. As ^{16}O burns, initially it destroys ^{24}Mg and produces mostly ^{28}Si and ^{32}S. Only near the end of the burning do the abundances of ^{36}Ar and ^{40}Ca become significant. As before, the neutron excess moves up to heavier nuclei, such as ^{34}S. Its absolute value slowly grows, so that when $X(^{16}O) = 0.1$, it is near $\eta = 3.0 \times 10^{-3}$. This is about twice the value used in figure 9.6 for explosive oxygen burning. For the point of view of nucleosynthesis yields, there are three interesting possibilities: The matter in this incompletely burned shell might be ejected as is, undergo explosive oxygen burning, or proceed still further to silicon burning. The extent to which the first is possible depends upon the mechanism of ejection. Oxygen is an easily combustible fuel in this context, so the second is certainly a possibility. The high value of neutron excess argues against its being the dominant case. In figure 9.8 for explosive silicon burning, a value of $\eta = 2.6 \times 10^{-3}$ was used, which is consistent with the third possibility.

TABLE 10.8
Shell Burning: O to Si

^{16}O	^{24}Mg	^{28}Si	^{32}S	^{34}S	^{40}Ca	T_9	$\eta \times 10^3$
0.703	0.118	0.036	0.002			1.676	2.261
0.759	0.104	0.075	0.008			1.853	2.428
0.760	0.103	0.076	0.009	0.001		1.943	2.432
0.759	0.105	0.076	0.009	0.001		2.061	2.411
0.749	0.104	0.088	0.012	0.001		2.212	2.397
0.671	0.073	0.173	0.038	0.005		2.264	2.493
0.498	0.022	0.316	0.113	0.024		2.316	2.724
0.385	0.005	0.381	0.167	0.038		2.364	2.832
0.225		0.431	0.250	0.030	0.008	2.459	2.921
0.100		0.485	0.285	0.018	0.033	2.599	2.957

The neutron excess problem is relaxed if burning during the ejection process proceeds beyond oxygen burning to silicon burning. This is simply because the ratio of the amount of mass in neutron-rich nuclei to $Z = N$ nuclei is larger for the iron-peak than for the Si to Ca region.[6] It is encouraging that the solar abundance pattern supports the increase in neutron excess found for shell burning. There is a clear hint that, unlike carbon and neon, the oxygen-burning shell will have significant explosive processing.

The next stage gives a discrepancy. Taking these ashes from shell oxygen burning, we evolve the abundances to $X(^{28}Si) = 0.10$. The resulting abundances for $26 < A < 70$ nuclei are shown in Figure 10.16, compared to solar abundances normalized at elemental Fe. There is a dramatic cliff at $A = 60$; all heavier nuclei have been destroyed by the high temperatures. While at first glance there appears the familiar iron-peak, and Si to Ca plateau, closer inspection shows that the pattern is wrong. The abundances of ^{50}Cr, ^{54}Fe, and ^{55}Mn are all a factor of ten too large. If we renormalize downward, then less than a tenth of the iron-peak can come from such material (if any at all!). During the silicon shell-burning stage, the neutron excess increases from $\eta = 3.0 \times 10^{-3}$ to 13.4×10^{-3}. Such matter cannot be the source of a large part of the solar system material. The obvious suggestion is that most matter undergoing silicon shell burning collapses to become part of a neutron star or black hole. This is an interesting point because it allows us to make a preliminary estimate of the mass of the condensed remnant from a stellar evolutionary calculation, even if we do not have an accurate model of the explosion process itself.

The evolution of the major constituents for silicon burning is indicated in table 10.8. As the last of the ^{16}O is consumed, ^{28}Si and ^{32}S are the major products. Above $T_9 = 2.8$, an episode of *sulfur burning* occurs, in which ^{32}S decreases while ^{28}Si continues to rise. This is a consequence of the particular abundances left by oxygen burning, which were higher in sulfur than the quasi-equilibrium value sought. At the same time, the neutron excess in ^{34}S shifts to the iron-peak, especially ^{54}Fe. Above $T_9 = 3.5$, a more canonical burning of silicon occurs. During this epoch, more matter is processed to the iron-peak. The new nuclei are unstable toward positron decay and

---

[6]We count ^{56}Fe as its progenitor ^{56}Ni, and so on for the other iron-peak nuclei with radioactive progenitors. Consider only ^{54}Fe, ^{56}Fe, and ^{57}Fe, formed explosively as ^{54}Fe, ^{56}Ni, and ^{57}Ni. This uses only Fe isotopic abundance rations, but ignores some other neutron-rich nuclei, ^{58}Ni in particlular, so that it is a lower limit of the value needed. The neutron excess for this combination in solar system abundances is $\eta > 2.5 \times 10^{-3}$. If ^{58}Ni is included, we have $\eta \approx 3.6 \times 10^{-3}$.

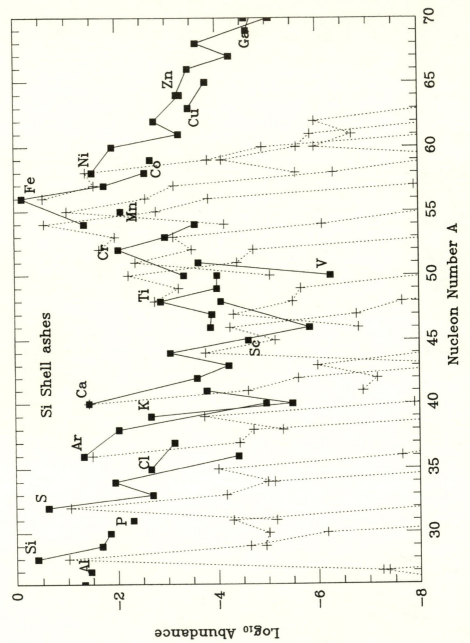

Fig. 10.16. Shell Si Ashes

TABLE 10.9
Shell Burning: Si to Fe

^{28}Si	^{32}S	^{34}S	^{54}Fe	^{56}Fe	^{58}Ni	T_9	$\eta \times 10^3$
0.486	0.285	0.018				2.599	2.957
0.517	0.309	0.017	0.003	0.001		2.815	2.982
0.577	0.281	0.005	0.043	0.004	0.002	3.096	3.014
0.612	0.253	0.001	0.069	0.001	0.003	3.238	3.037
0.562	0.252		0.080		0.005	3.672	3.505
0.097	0.087		0.261		0.040	3.951	13.39

electron capture because the valley of beta stability has shifted away from $Z = N$. The neutron excess grows to large values. If significant depletion of ^{28}Si has occurred in the shell, it will have a neutron excess too large to be consistent with a major contribution to solar system abundances.

11

Thermonuclear Explosions

Everyone knows that dragons don't exist. But
while this simplistic formulation may satisfy the
layman it does not suffice for the scientific mind.
—Stanislaw Lem (1921–   ),
The Cyberiad

Models of supernovae have often been more noted for the imagination displayed in their creation than for their intrinsic plausibility—like dragons. In the struggle to make supernovae more respectable, heavy emphasis is placed upon internal consistency, and upon unique and quantitative predictions. The conceptually simplest model for a supernova explosion is just an analog of a runaway chemical reaction that becomes explosive—a conventional bomb. The difference in energy is one of scale: both thermonuclear energies and stellar masses are much larger, so that the stellar bombs are more spectacular. As with the terrestrial case, bombs often fail, becoming duds. To avoid this, designers carefully plan the ignition and the structure of the device. Similarly, most models advanced for astrophysical bombs (supernovae) have been duds. Further constraints are that the successful astronomical device must occur naturally, and at the correct rate—there must be a convincing astronomical context.

One of the most serious uncertainties is that of the ignition itself. It is also one of the least explored aspects of the thermonuclear models of supernovae. The nature of a bomb is that it has a large amount of energy on the verge of being released. If the release is not swift, the bomb blows itself apart before it taps its entire energy reservoir. Yet, because the bomb is on the verge of releasing energy, there are often many ways that this release may be triggered. Clearly some will give more effective explosions than others. The problem is to find the one chosen by Nature.

After ignition, the explosion develops by consuming more fuel. The nature of this burning depends upon how the burning front propagates— how is the new fuel ignited? Terrestrial flames propagate by heat flowing

from hot (burned) regions to cooler (unburned) regions. Mild versions
of this process are laminar flames, in which gentle motion brings new
fuel into hot regions, where it heats up and burns (as in a candle, for example, in which gas motion brings new oxygen for combustion). A more
vigorous form is a *deflagration* wave, in which the new fuel is heated by
the violent burning in the front. A rather different mode of propagation
is the *detonation* wave, in which the burning is so violent that the burnt
fuel expands rapidly enough to drive a shock wave into the unburned
fuel, compressing and heating it to ignition. In this case the thermal energy transfer takes place not by conduction or radiative diffusion, but by
hydrodynamic motion which drives compressive heating.

In the astrophysical context, the large size of the systems introduces
additional possibilities. Stars are self-gravitating, so that hot material
would rise buoyantly, and be mixed with unburned fuel by turbulent motion, where it would heat and ignite new fuel. This mode of propagation
is usually called "deflagration" in the astrophysical literature; this terminology implies a linking with the simpler terrestrial process, which is
not correct in general. The study of such astronomical flames is in its
infancy.

11.1   THERMONUCLEAR FLAMES

Analysis of the characteristic time-scales is crucial for understanding the
physics of the problem. To begin we consider the order of magnitude of
some cross sections of interest.

For two particles of charge $Z_1 e$ and $Z_2 e$ we may define a coulomb
radius r_{coul} from $Z_1 Z_2 e^2 / 2r_{coul} = E$, where E is an appropriate energy. For example, for an atom the appropriate energy might be that
of the bound electron, so $E = p^2 / 2m_e$. Relating the momentum p to
the radius by using the de Broglie wavelength, we have a cross section
of the order $\sigma_{atom} = \pi a_0^2 Z_1 Z_2 = 0.88 \times 10^{-16} Z_1 Z_2 \, \text{cm}^2$, where a_0 is
the radius of the first Bohr orbit. However, at temperatures at which
nuclear processes can occur, atoms will be ionized. Taking $E = \frac{3}{2}kT$
gives $\sigma_{coul} = 1.01 \times 10^{-6} Z_1 Z_2 / T^2 \, \text{cm}^2$, which for higher temperature is
smaller than the previous estimate. Similarly, if the nucleus has a radius
$r = 1.4 A^{\frac{1}{3}}$ fermis, this gives a nuclear cross section of $\sigma_{nuc} = \pi r^2 = 4.5 \times 10^{-26} (A_1^{\frac{1}{3}} + A_2^{\frac{1}{3}})\, \text{cm}^2$.

Such estimates are extremely crude. These wave-mechanical systems
exhibit resonance behavior, so that the cross sections vary enormously.
With this caution, these estimates may be used to get a sense of the
magnitudes of the quantities involved.

Equating σ_{nuc} to σ_{coul} for typical nuclei of astrophysical interest gives a temperature of about 10^{10} K, higher than is usually relevant, which suggests that coulomb scattering of ions is more common than nuclear interactions. More accurate estimates generally give stronger support for this idea. Consider an extreme example: ^{12}C $+^{12}$ C at a center-of-mass energy of 3.79 MeV has a measured [420] reaction cross section of $\sigma = 3.1 \times 10^{-29}$ cm^2. This corresponds to a stellar temperature of $T = 4 \times 10^9$ K, and $\sigma_{nuc}/\sigma_{coul} = 1.5 \times 10^{-3}$. At lower temperatures the ratio is smaller. Because this number is small, we expect the ions to establish an equilibrium velocity distribution, so that we have the great simplification of dealing with the reaction rates for a thermal gas.

At high temperatures ($T > 10^8$ K), radiative diffusion is dominated by Thomson opacity. Since $\sigma_{thom} = \frac{8}{3}\pi r_e^2 = 0.665 \times 10^{-24}$ cm^2, this cross section equals σ_{coul} at $T = (\frac{3}{2}Z_1 Z_2)^{\frac{1}{2}} \times 10^9$ K. For lower temperatures, typical of burning conditions, *heat diffuses faster than composition*. The time-scale for diffusion of heat through a distance δr is $\tau_{diff} = (\delta r)^2/\lambda v$, where v is the velocity of diffusing objects and $\lambda = 1/N\sigma$ is the mean free path.[1] The form of the time-scale for diffusion of composition is identical to τ_{diff} above, but because the velocities of the nuclei are smaller and their coulomb cross sections higher, this time-scale is longer. Thus we can concentrate on the ignition of a new fuel by thermal heating rather than the action of superthermal particles, a significant simplification.

There are two separate time-scales for burning that must be considered. The first is the *ignition time-scale* of the fuel, which is defined to be the temperature e-folding time, $\tau_T = T/\dot{T} \approx C_V T/\epsilon_{nuc}$, where ϵ_{nuc} is the energy release rate of the nuclear processes, and C_V is the specific heat at constant volume. Because charged-particle reactions are heavily modified by coulomb barrier penetration, this time-scale strongly decreases with increasing temperature. The second important time-scale is the *burning time*, i.e., the time to significantly reduce the abundance of fuel, which is defined as $\tau_i = X_i/\dot{X}_i = Y_i/\dot{Y}_i$, where X_i is the nucleon fraction of the species i. In simple cases, this differs from τ_T by the ratio of the thermal energy content to the Q-value for the reaction (per unit mass burned). For example, for ^{12}C $+^{12}$ C this ratio is $0.25\,T_9$; this is of order unity at explosive conditions. For comparison, the ratio is about 0.005 for hydrostatic hydrogen burning by the CNO cycle. This ratio can also approach zero in explosive situations, if the matter is degenerate. In this case the specific heat approaches zero, so that con-

---

[1]For highly degenerate matter, electron conduction is more effective than radiative diffusion at transporting energy.

sumption of a small amount of fuel gives a large change in temperature.

Two additional time-scales must be considered that are associated with hydrodynamic motion. The first is the time for a region to react to a pressure imbalance, which is taken to be the sound travel time $\tau_{hyd} = \delta r/c_s$, where c_s is the local sound speed. Finally, there is the time for a convective element to move through the region in which convection occurs (obviously an oversimplification of a complex process). This time-scale is defined by $\tau_{conv} = \delta r_{conv}/v_{conv}$, where δr_{conv} is the width of the convective zone and v_{conv} is the typical velocity of a convective blob.

Depending on the relative size of the various time-scales, very different physical situations arise. If the nuclear time-scales τ_i are all large compared to τ_{conv}, the convective zone is approximately uniform in all abundances, which slowly evolve on the nuclear time-scales. If some of the τ_i are shorter than τ_{conv}, it is not correct to ignore the abundance gradients that this implies. In particular, these gradients may interact with the burning to modify the convective flow itself. Silicon burning is an interesting case of this: the time-scales τ_i for α, n, and p, which maintain the quasiequilibrium, are shorter than τ_{conv}, while the τ_i for ^{28}Si is longer. If the nuclear time-scales are all small compared to τ_{conv}, the problem simplifies again, and each region is loosely coupled to its neighbors. For more dynamic problems, such as pulses or explosions, the convenient fiction of steady-state convection is untenable, and the hydrodynamics must be treated as an equally important aspect of the problem.

Many different types of burning occur in astrophysical problems. The type that occurs in a particular problem is determined by the ratios of the time-scales described above. Burning can range from quiet hydrostatic burning, such as occurs in the center of most stars, to the explosive burning which can occur in supernova explosions.

The least violent form of stellar burning occurs in stars that are hydrostatic (in "radiative burning zones"). This is analogous to the familiar laminar flames of candles and campfires. In this case, the reaction rate is slow enough so that energy can be transported from the burning region by radiation or conduction as rapidly as it is generated. A variation on this theme is a subsonically convective flame region, in which heat and ashes are removed by slow convection, and new fuel brought in. In other words, $\tau_i \gg \tau_{diff}$ or $\tau_i \gg \tau_{conv}$. It is radiative burning which occurs in stars such as the Sun over most of their lifetimes. The star remains in equilibrium, quietly burning its fuel, while the energy produced is transported to the star's surface, providing its luminosity. This process will continue until the fuel is exhausted, at which time the nuclear reactions can no longer provide sufficient heat, and hence pressure support, to prevent the region from contracting. The temperature and density will continue to increase until the conditions are reached at which the reaction products of the previous burning can ignite, or the matter can support itself by degeneracy pressure.

At the other extreme, detonations are the most violent form of burning. Astrophysical detonations are thought to occur under degenerate conditions. During the initial stages of a thermonuclear runaway, before the temperature rises significantly, the reaction rate is still relatively small. In nondegenerate matter, the pressure increase produced by the reactions will cause the burning region to expand and cool, preventing the runaway from proceeding. In other words, $\tau_T > \tau_{hyd}$. On the other hand, if the matter is degenerate, the temperature increase created by the burning will not create a significant increase in pressure. Thus the temperature will continue to increase until the matter becomes nondegenerate. At this point, the energy generation rate may be too large for mild hydrodynamic motion to stop its increase, and if so, an explosion will occur. If the resulting shock is sufficiently strong to raise new fuel above the ignition temperature, a detonation wave will propagate outward from the point of ignition.

In its simplest form, detonation wave theory is easily understood (see, e.g., [197, 234]). As the shock propagates into the unburned fuel, it compresses and heats the material beyond the ignition point. Immediately behind the shock is the reaction zone, in which the fuel burns. In detonation theory the width of the reaction zone is frequently neglected, so that the detonation front is treated as a sharp discontinuity. In this case jump conditions can be derived for the change in the hydrodynamic variables across the front in much the same way as is done for a simple shock. From mass conservation one obtains the condition:

$$\rho_1 D = \rho_2(D - u_2), \qquad (11.1)$$

where D is the velocity of the detonation front, ρ is the density, u is the fluid velocity, and the subscripts 1 and 2 denote the predetonation and postdetonation states, respectively. The velocity of the material ahead of the front, u_1, is taken to be zero. A second jump condition, which expresses momentum conservation across the front, can be written as:

$$P_2 - P_1 = \rho_1 u_2 D, \qquad (11.2)$$

where P is the pressure. The third jump condition, based on energy conservation, is

$$E_2 - E_1 = P_1 V_1 + \frac{1}{2} D^2 - P_2 V_2 - \frac{1}{2}(D - u_2)^2 + q, \qquad (11.3)$$

where E is the specific internal energy, V is the specific volume, and q is the amount of energy released by the complete burning of the fuel. By eliminating the velocity from the first two jump conditions, one obtains the equation for the *Rayleigh line*,

$$\mathcal{R} = \rho_1^2 D^2 - \left(\frac{P_2 - P_1}{V_1 - V_2} \right) = 0. \qquad (11.4)$$

The equation for the *Hugoniot curve* is obtained by eliminating the fluid velocity and detonation speed from the energy jump condition, using the jump conditions for mass and momentum. The resulting equation can be written as

$$\mathcal{H} = E_2 - E_1 - \frac{1}{2}(P_2 + P_1)(V_1 - V_2) - q = 0. \qquad (11.5)$$

For a given equation of state $P = P(\rho, E)$, the intersection of the Rayleigh line and the Hugoniot curve determines the postdetonation state. Note, however, that in order to obtain the postdetonation state, *one must first choose a detonation velocity*. Unlike the case for simple shocks, the front velocity is not determined from the jump conditions. Depending on the value chosen for the detonation velocity, the Rayleigh line intersects the Hugoniot curve at 0, 1, or 2 points. If there is no intersection, no detonation wave is possible for that detonation velocity. If there are two points of intersection, there are two possible solutions. These two solutions correspond to *strong* and *weak* detonations. A strong detonation propagates at a speed slower than the postdetonation sound velocity with respect to the fluid behind the shock, i.e., $D < u_2 + c_2$, so that disturbances generated behind the front will eventually catch it. Thus this solution is unstable. Weak detonations, which propagate faster than the postdetonation sound velocity with respect to the fluid behind the shock, i.e., $D > u_2 + c_2$, are considered to be unphysical except under certain special conditions [234]. One-dimensional theory is oversimplified: real detonations are observed to have significant three-dimensional structure—the fronts are wrinkled by transverse waves. The detonation that usually occurs in Nature is the one corresponding (at least on average) to the speed at which the Rayleigh line and Hugoniot curve have only one point of intersection. This detonation speed, called the Chapman-Jouguet velocity, is equal to the sum of the postdetonation fluid velocity and sound speed, i.e.,

$$D = u_2 + c_2. \tag{11.6}$$

If we assume that the front propagates with this velocity, then the post-detonation state is completely determined. Both the pressure and the density increase across the front, while in the frame in which the shock is stationary, the fluid velocity decreases.

Consider a detonation in a rigid tube followed by a piston with a fixed velocity. Two types of detonations that have somewhat different profiles can occur. If the velocity of the piston that generates the detonation is larger than the fluid velocity of the Chapman-Jouguet state, the result is an overdriven detonation. The structure in this case consists of the shock and reaction zone, followed by a constant state corresponding to the point at which all the fuel is depleted. This is a strong detonation, and therefore, the constant state behind the front does not correspond to the Chapman-Jouguet state. For smaller piston velocities, the deto-nation is unsupported. For this type of detonation, the reaction zone and the constant state behind the front are connected by a rarefaction, called the Taylor wave. Although this type of detonation propagates at the Chapman-Jouguet velocity, the Chapman-Jouguet state does not cor-respond to the constant state behind the front, but instead is the state at the top of the rarefacton.

Although this theory provides a satisfactory explanation for simple detonations, it has significant limitations because of the assumptions that the reaction rate is infinite and that the reaction zone has zero width. A slightly more complex treatment is used in the Zel'dovich-von Neumann-Doering (ZND) model [706, 655, 212]. This theory assumes that the shock, which is taken to be infinitely thin, is followed by a reaction zone of finite width. The primary difference from the equations given above is that the term in the Hugoniot curve involving the energy generated by the reaction must now be multiplied by the extent of the reaction. Thus, each state within the reaction zone can be determined by the intersection of the Hugoniot curve for the appropriate extent of the reaction and the Rayleigh line. The final state obtained after the fuel is completely burned is exactly the same as for the simpler theory described above. Additional complications in the theory arise due to multidimensional effects, such as cellular detonations and spinning detonations [234].

Deflagrations represent a much less violent form of burning than det-onations, but in many ways are more complex (e.g., [683, 707]). They result when the *burning is unable to produce sufficient overpressure to create a shock that can ignite the fuel*. The motion of the front is usually very subsonic. Burning is initiated by the diffusive transfer of heat from the hot ashes behind the front into the cold fuel. For the case of a thin front, deflagrations must obey the same jump conditions as detonations, but the propagation velocity now depends on the rate of heat transfer. Another major difference is that the pressure and density decrease behind the deflagration front, and in the reference frame in which the front is stationary, the velocity increases. In the case of a spherically symmetric deflagration, which begins at the origin, the velocity behind the front must eventually become zero to satisfy the boundary condition at the origin. The only way in which this can happen is for the deflagration to be preceded by a compression wave that accelerates the material away from the front. The passage of the deflagration will then provide exactly the correct jump in velocity so that the material behind the front will come to rest. This can happen since the deflagration velocity is subsonic, and therefore there will be communication between the origin and the rest of the flow by sound waves.

Unlike the case of Chapman-Jouguet detonations, where it is possible to compute the exact propagation velocity, the propagation speed of deflagrations can only be crudely estimated [378, 247, 464]. For the simplest case of a laminar front that propagates as a result of radiative diffusion or conduction, it is fairly easy to obtain an order of magnitude estimate for the velocity of the wave. The width of the deflagration can be approximated by setting the heat diffusion time-scale τ_{diff} equal to the burning time-scale τ_i. Thus the width of the front is given by

$$\delta \approx \sqrt{\lambda c \tau_i}, \tag{11.7}$$

where λ and c are the mean free path of photons or electrons, and the speed of light,[2] respectively. The velocity D of the deflagration can then be estimated as

$$D \approx \delta/\tau_i \approx \sqrt{\lambda c/\tau_i}. \tag{11.8}$$

Arguing by analogy, Nomoto, Sugimoto, and Neo [462] made the ingenious suggestion that a deflagration-like flame might be propagated by turbulent convection rather than by heat diffusion alone. In this situation, two limits arise. If the scale of the turbulence is small compared to the width of the front, the deflagration may remain laminar, and the above equations could still apply if the mean free path of the photons

---

[2]We consider conditions in which the electron motion is at least partially relativistic.

or electrons is replaced by the typical length scale for the turbulence, and the velocity of the diffusing objects is replaced by the convective velocity. If, on the other hand, the thickness of the front is small compared to the length scale of the turbulence, the front will be wrinkled. The velocity of the front is increased compared to that of the diffusively propagated deflagration, both by the increase in heat transfer rate and by the increase in the surface area of the front. The upper limit to the speed of the burning front that can be obtained in an extremely wrinkled front is approximately the convective velocity. However, the interaction of the burning with the hypothesized convection is not yet understood, so estimates of this velocity are unreliable as yet.

Another complication that arises is that many astrophysical deflagration fronts may be unstable [445, 446]. Since the density decreases behind the front, when a deflagration propagates outward from the center of a star against the force of gravity, we have a hot, low-density bubble trying to push outward through a denser medium. At first sight this resembles the classical Rayleigh-Taylor instability, but the flame front will itself affect the dynamics—by the added momentum of the burned gas, for example [399, 52].

This is a difficult problem. Some progress is beginning to be made by direct simulation of the limiting cases. At the small length scale, a direct hydrodynamic simulation of the burning front in three dimensions, modeling the burning and heat conduction, may clarify the physical nature of the process and suggest what the "subgrid" velocity really should be for a large length scale computation.

In the case that the velocity of the flame front may be specified on a "subgrid" scale, algorithms have been devised that allow simulation of instabilities on larger scales [52]. Assuming this speed to be the laminar flame velocity, as calculated by Timmes and Woosley [620], the deflagration is inadequate to drive an explosion directly, although it may drive pulsations which evolve into a detonation [351, 53]. Thus, the simple analogy of Nomoto et al. does not seem to work as envisaged.

The deflagration front is unstable, as shown in figure 11.1, which is derived from two-dimensional hydrodynamics computations of Arnett and Livne [52]. Contours of energy generation rate ϵ are shown at two instants of time ($t = 0.073$ s and $t = 0.143$ s). The initial state was a spherical burning front, with relative perturbations of 0.01 in the velocities to break the symmetry; these quickly grow into a wrinkled structure, with smaller wiggles merging into larger ones. This wrinkling is inadequate to enhance the laminar flame speed (which is only of order 100 km s^{-1}) to the point of explosive energy release. Yet this is not the whole story; recent simulations by Khokhlov (in preparation), at small scales in three

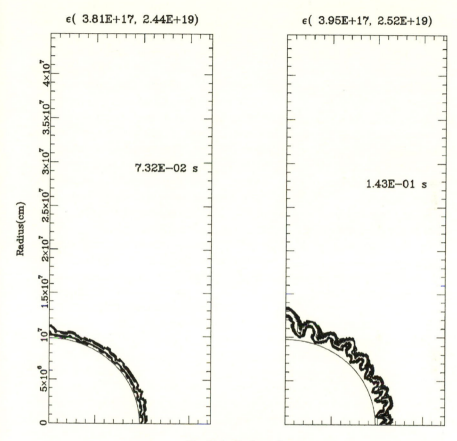

Fig. 11.1. Deflagration

dimensions, suggest that the wrinkling is more effective, and will increase the flame speed to more than its laminar value. This is a new and active area of research; the reader is advised to consult the most recent literature! It is encouraging that the underlying physics of the process can now be directly addressed.

11.2 DEGENERATE INSTABILITY

The key features of carbon ignition—in an environment of degenerate electrons—may be examined in a simple way, if at first the effects of the Urca process are ignored. These complexities will be considered in the following section.

Consider a star massive enough to have ignited helium, but having a carbon-oxygen core too small for carbon ignition. This would correspond to a star which had a zero-age main sequence mass in the range $2.3 < M/M_\odot < 9$. Mass loss may or may not have occurred; we might be considering a carbon-oxygen star of $M_{CO}/M_\odot < 1$. We consider the evolution of the central regions, which we will call the *carbon-oxygen core*.

If this core does not grow in mass, it will cool and its electrons will become highly degenerate. It is then a carbon-oxygen white dwarf. However, suppose that mass is added to the core, from its own envelope, or by accretion, and is processed by hydrogen- and helium-burning shells. If the mass addition is from a companion, then the composition becomes an issue. If the added mass has no hydrogen, there can be no hydrogen-burning shell, and similarly for helium. Some of the important parameters include the mass of the companion, its composition, and distance. The nature of the mass flow, the amount actually accreted, and the angular momentum transfer are also to be considered. With this range of possibilities, it should be no surprise that elaborate scenarios can be constructed; see Iben and Tutakov [327] for an idea of the complexities. In chapter 8 the double shell-burning stage was discussed with regard to helium-burning nucleosynthesis and the s-process.

The luminosity of a thin shell is

$$L_{sh} = 4\pi cGM_{sh}(1 - \beta)/\kappa, \qquad (11.9)$$

where we assume β is slowly varying through the shell. What would be the long-term fate [24] of such a core, if nothing catastrophic happened first? The *average* behavior of the two shells would be that they processed the matter at the same rate (see the discussion of shell burning in chapter 8). Then we have

$$dM_\alpha/dt \approx dM_H/dt = L_H/(Q_H X_H), \qquad (11.10)$$

where L_H is the luminosity of the hydrogen shell, Q_H the energy release per unit mass of hydrogen burned, and X_H the nucleon fraction of hydrogen in matter being processed by the hydrogen shell. If the opacity is dominated by Thomson scattering, the rate of increase of the core is

$$dM_c/dt \approx 2.2 \times 10^{-14}(M_c/M_\odot)(0.75/X_H)(1 - \beta)M_\odot \text{ s}^{-1}. \qquad (11.11)$$

Notice that this would correspond to an accretion rate of $6.7 \times 10^{-7} M_\odot \text{y}^{-1}$.

We have assumed that the electrons in the core have become highly degenerate, so to first order the density evolves independently of the temperature (the thermal component is only a perturbation in the equation of state). The hydrostatic condition will be adequately approximated by white dwarf models of Chandrasekhar [162] and Schatzman [544]. For these models, the density structure is determined by the mass, once the composition is defined. Therefore, the central density is a function of the mass of the core, $\rho_c = \rho_c(M_c)$.

Numerical models show that for central densities in the range $10^{7.5} \leq \rho_c \leq 10^{11}$ g cm^{-3}, this function is approximated by

$$\rho_c/\rho_0 \approx ((M_{ch} - M_0)/(M_{ch} - M_c))^{\alpha}, \tag{11.12}$$

where the coefficients are $\rho_0 = 10^{7.7} (2Y_e)^{-1}$ g cm^{-3}, $M_0 = 1.078 (2Y_e)^2 M_{\odot}$, $M_{ch} = 5.85Y_e^2 M_{\odot}$, and $\alpha = 2.533$, so that

$$\frac{d \ln \rho_c}{dM_c} \approx 2.533/(M_{ch} - M_c). \tag{11.13}$$

Now,

$$d \ln \rho_c/dt = \frac{\partial \ln \rho_c}{\partial M_c} dM_c/dt, \tag{11.14}$$

so that

$$d \ln \rho_c/dt = \frac{2.533}{M_{ch} - M_c}(2.2 \times 10^{-14} M_c) \text{ s}^{-1}, \tag{11.15}$$

where M_c is measured in solar units and $0.75(1 - \beta)/X_H$ is taken to be unity. As the core mass approaches the Chandrasekhar limit, even a tiny addition of mass causes a large increase in density. Such an enhanced rate of compression will affect the thermal behavior of the core.

The first law of thermodynamics may be written as $\dot{q} = \dot{E} + P\dot{V}$, where $\dot{q}$ is the net rate of heating minus cooling from nuclear burning and neutrino emission. The evolution of temperature is governed by

$$\dot{T} = (\dot{q} + V(P + E_V)\dot{\rho}/\rho)/E_T, \tag{11.16}$$

where, as usual, $E_T \equiv (\partial E/\partial T)_V$, and similarly for E_V.

Various components contribute to the equation of state. For a degenerate electron gas,

$$V(P + E_V)_e = \pi^2 m_e c^2 N_A(Y_e kT/xm_e c^2), \tag{11.17}$$

where N_A is Avogadro's number, Y_e the number of electrons per nucleon, and x is the electron fermi momentum in rest energy units, $x^3 = 1.0268 \times 10^{-6} Y_e \rho$. For a completely degenerate gas, $P + E_V \to 0$. For the ions (to the extent that they may be approximated by an ideal gas), we have

$$V(P + E_V)_i = \sum_i Y_i \Re T \approx 6.0 \times 10^{15} T_9, \qquad (11.18)$$

where the numerical factor is obtained by assuming a gas composed equally by mass of ^{12}C and ^{16}O. For a high degree of degeneracy, the ion contribution is larger than that of electrons.

Equation 11.16 may be integrated to determine the evolution of the thermal condition of the core. For slow changes, which are appropriate here, $\dot{T} \approx 0$, and at the center of the core,

$$\epsilon_g \approx -dq/dt \qquad (11.19)$$

$$\equiv 334.4 \, T_9 M_c / (M_{ch} - M_c) \text{ erg g}^{-1} \text{ s}^{-1}, \qquad (11.20)$$

where the core mass M_c is in solar units.

If the core mass is specified, the Chandrasekhar models give the density at the center (or elsewhere in the star). Because the compression following core growth will be nearly homologous (shape preserving), the central logarithmic compression rate, $d \ln \rho_c / dt$, also applies to other parts of the core. Then equation 11.20 may be solved for the temperature, if we know dq/dt as a function of temperature, density, and our assumed composition. This gives the steady-state solutions for the thermal behavior of the core.

For illustration, consider a pure CO core. It is a reasonable approximation for Population II stars, which because of their low abundance of heavier elements do not have effective Urca shells. The effects of nuclear Urca processes, which are important for stars of roughly solar abundance, and for Population II stars after carbon burning begins, will be discussed below.

Both ^{12}C and ^{16}O are resistant to electron capture below $\rho \approx 2 \times 10^{10}$ g cm^{-3}. The rate of energy loss due to the neutrino-antineutrino pair bremsstrahlung process of Festa and Ruderman [232] is

$$dq/dt \approx s_\nu = -1.01 \, (T_8)^6 \text{ erg g}^{-1} \text{ s}^{-1}. \qquad (11.21)$$

Using this and equation 11.20 gives a simple expression for the central temperature as a function of core mass:

$$T_8 \approx \left(33.44 M_c / (M_{ch} - M_c) \right)^{1/5}. \qquad (11.22)$$

This explicitly illustrates the tendency for the temperature to rise significantly as $M_c \to M_{Chandra}$.

A more effective cooling mechanism is the plasmon-neutrino process. In this region of temperature and density, Beaudet, Petrosian, and Salpeter [86] gave:

$$dq/dt \approx s_\nu = -8.94 \times 10^9 \, Y_e^3 \rho_9^2 \, e^{-0.5646\xi}/\xi \text{ erg g}^{-1} \text{ s}^{-1}, \qquad (11.23)$$

where $\xi = (\rho_9 Y_e)^{1/3} 60/T_8$. This is still essentially correct [332], even though the region of interest will lie near the border with the neutrino bremsstahlung process; neutral currents are included. Equating this to equation 11.20 gives the trajectory shown as a dotted line in figure 11.2; it runs diagonally from the lower right, and coincides for much of its path with the solid line labeled "stable branch."

Heating may also occur by carbon burning. At these high densities, electron screening of the colliding nuclei becomes important. As the electron fermi energy increases, the wavelength of the electrons becomes smaller, allowing them to shield the positive charges of the nuclei until closer distances of approach are attained. The coulomb barrier for fusion becomes thinner, increasing the probability of barrier penetration,

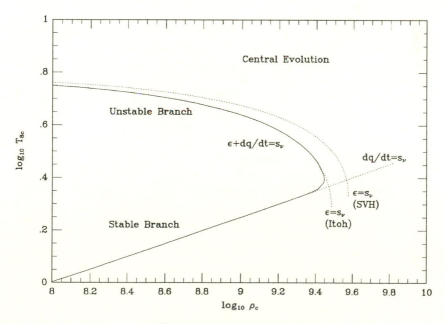

Fig. 11.2. Central Evolution

and hence the reaction rate. Salpeter and Van Horn [538] give a simple approximation to the screening factor in the strong-screening limit: $f = e^{-U}$, where $U = 37.9 \rho_9^{1/3}/T_8$ for the $^{12}C +^{12} C$ reaction. Using this correction in the energy generation rate,[3] and equating it to the plasmon-neutrino cooling rate, gives the dotted curve ($\epsilon = s_\nu$) labeled SVH in figure 11.2. It descends almost vertically at $\log \rho = 9.55$.

The use of a more recent estimate of the screening factor due to Itoh and collaborators [333, 334] gives a value which is larger by about a factor of ten. This is shown by the dotted curve labeled Itoh, that lies along the solid line labeled "unstable branch." A similar value comes from the work of Alastuey and Jancovici [3], so that the screening factor seems well determined. At these low temperatures, the optimum bombarding energies of the $^{12}C +^{12} C$ reaction lie below those that have been measured.[4] An error in the low-energy extrapolation of the $^{12}C+^{12}C$ cross section by a factor of ten would give a shift comparable to that shown between the dotted curves for the two screening values; see figure 3.4.

The steady-state solutions for the temperature are given by the more complete relation:

$$\epsilon + \epsilon_g + s_\nu = 0, \qquad (11.24)$$

which includes both nuclear and compressional heating as well as neutrino cooling by the plasmon process. This shown in figure 11.2 as a solid curve. It has high and low temperature branches, which merge in the vicinity of $T_8 = 2.5$ and $\rho_c = 3 \times 10^9$ g cm^{-3}. The low T branch is stable; the plasmon-neutrino cooling rate is more sensitive to temperature than the gravitational heating. An excess of temperature causes increased cooling, and vice versa. The high T branch is unstable; the plasmon-neutrino cooling rate is less sensitive to temperature than the nuclear heating. An excess of temperature causes increased heating, and vice versa. There is a third branch at $T = 0$ which is unstable.

If the CO core moves along the stable branch, it will lose stability at the point it merges with the upper branch. At the *ignition point* of $T_8 \approx 2.48$ and $\rho_c \approx 2.79 \times 10^9$ g cm^{-3}, we have a compressional heating rate of $\epsilon_g = 1.46 \times 10^3$ erg g^{-1} s^{-1}, a nuclear heating rate of $\epsilon = 3.37 \times 10^3$ erg g^{-1} s^{-1}, and a plasma neutrino cooling rate of $s_\nu = -4.82 \times 10^3$ erg g^{-1} s^{-1}.

---

[3]This rate is almost identical to that of Caughlan and Fowler [156]. See chapter 3.
[4]The temperatures here are even lower than used in the analysis of chapter 3 because of the strong electron screening of the nuclear coulomb potential.

What is the nature of the thermal evolution past the point of instability? First, consider the simplest case, in which the thermal behavior does not affect the structure. Then the temperature evolution is governed by

$$\dot{T} = (\epsilon + \epsilon_g + s_\nu)/\frac{3}{2}Y\Re. \qquad (11.25)$$

For slow evolution, this reduces to the thermal balance condition of equation 11.24. Evolution from the point of thermal instability to the point at which the central temperature is $T_c > 1.0 \times 10^9$ K, requires 6.8×10^{10} s, almost all of which is spent near the instability point. In this amount of time, the core would grow by $\delta M_c = \dot{M}\tau \approx 0.0022M_\odot$. This is roughly the amount of fuel consumed before the system becomes dynamic. At this point the core is within $\Delta M = 0.078M_\odot$ of the limiting mass; the change is only 2.8 percent of this.

Equation 11.25 requires that we assume the only change in specific volume be that due to compression (the term ϵ_g). More generally, heating will cause a small expansion. As the temperature rises, the matter will become convectively unstable, and the expansion of a blob of matter will give a buoyancy force, which drives motion. The matter expands more, and cools, quenching the burning. The neutrino cooling is now unbalanced, so that the blob contracts, and sinks. We have a convective overturn.

At what temperature does the burning become so vigorous that hydrodynamic motion cannot quench it? The acceleration of hot matter is proportional to the fractional pressure excess f times gravity, $a \approx fg$ (see chapter 9). The time required to move a given distance r is $t \propto \sqrt{r/a}$. For $f \approx 0.01$, a time τ of the order of ten sound travel times, $\tau \approx 446/\sqrt{f\rho} \approx 1$ s, is required. Equating this to the thermal time, $T E_T/\epsilon$, gives a *flash* temperature [54] of $T_9 \approx 1.0$. This is about a factor of four higher than the ignition temperature. There is an epoch, between ignition and runaway, lasting about two centuries, during which the nature of the explosion is affected and perhaps determined.

This epoch is difficult to compute directly. Its duration is approximately 10^{12} sound travel times. The techniques used in stellar evolutionary codes assume (1) a global spherical symmetry, and then add corrections for presumed—but uncalculated—nonspherical motion such as convection. It is further assumed that (2) convection is faster than any other process of interest, (3) connection is very subsonic, and (4) mixing is instantaneous and uniform throughout the convective region. *All of these assumptions fail between ignition and runaway.* The techniques of numerical hydrodynamics have difficulty also. The fluid has a high

Reynolds number (low viscosity), which makes inappropriate those methods having significant numerical viscosity. Implicit methods, when used for large Courant number (time steps larger than the sound travel time), gain numerical stability by damping motions occurring on shorter time-scales, that is, by increasing numerical viscosity. A challenge is to damp the unwanted solutions, but not the desired ones (presuming one can identify which are which). Explicit methods are limited by the Courant condition, so that the time step must be less than the sound travel time. This would require more than 10^{14} steps for moderate resolution.[5]

A detonation wave is a *self-consistent* solution [18, 54, 133] in this electron-degenerate environment, but it is not a *unique* one. Whether or not a detonation is initiated during this epoch depends upon the exact nature of the evolution. Immediately after ignition, the behavior is clear for this simple case in which we ignore the Urca process. The central regions begin to heat, raising their entropy. This continues until a con-vectively unstable gradient develops, and then convection begins. As the thermal runaway continues, the convective core grows, but it becomes more difficult for convection to carry sufficient energy to prevent the central regions from flashing.

In the original numerical calculations [18, 54], a single central zone flashed first. Burning of carbon, neon, oxygen, and silicon occurred rapidly, leaving the matter in nuclear statistical equilibrium with roughly equal nucleon fractions of alphas and ^{56}Ni. This occurred supersonically, so that a shock formed. The numerical method smeared the shock over several zones, reducing the temperature in the neighboring unburned zones below what it should have been. By splitting the flashed zone into several smaller zones, rezoning the neighbors likewise, and giving the burned zones the velocity implied by a detonation, it was found that a detonation propagated through the carbon-oxygen core [54], incin-erating it almost entirely [49] to ^{56}Ni. Although this large production of ^{56}Ni was recognized as a problem [49], the prediction only became seriously tested after considerable progress was made in both the col-lection and the study of supernova spectra. This abundance pattern is in conflict with the optical spectra of SNIa which are interpreted [123] as overlapping resonance-scattering profiles of OI, MgII, SiII, CaII, and CoII. There must remain a significant amount of matter in the form of oxygen, magnesium, silicon, and calcium, so that the original detonation model predicts too much burning.

---

[5]Such a direct computation was not feasible at the time of writing. Even if it were, the effect of round-off errors might be a worry.

The next proposal was the deflagration model of Nomoto, Sugimoto, and Neo [462]. It makes use of the modest[6] value of the thermal pressure: after flashing, the thermal pressure is less than 30 percent of the total [54], and is even smaller during the convective stage. In this model it is assumed that a "deflagration wave, which propagates through the core due to convective heat transport, does not grow into detonation." The velocity of this "convective deflagration wave" is taken to be an adjustable parameter. *Deflagration* and *convection* are names given to aspects of subsonic hydrodynamical behavior; it is presumed that they may be added linearly, which is not consistent with terrestrial experiment, or as we saw in the previous section, with numerical simulation. This picture grossly oversimplifies a complex phenomenon; it is an algorithm, not a theory. As such, it has been a great success! It has allowed the construction of explosion models which, by adjusting the velocity for the "deflagration," consume fractions of the available fuel, and do so after significant expansion. This is a crucial point. At lower density, electron degeneracy is raised at lower temperature. Once degeneracy pressure is not large compared to thermal pressure, a burning region can reduce its overpressure by rapid expansion, giving cooling, and quenching the flame. By burning at lower densities, significant amounts of matter survive the various intermediate burning stages.

Other models of degenerate burning which have much of the burning occurring at lower densities may well be as successful as the deflagration model. Once the total energetics are fixed, only the nucleosynthesis yield remains as a significant parameter to be specified. The *delayed detonation* model of Khokhlov [350, 351] is an example. To distinguish between such models, additional detailed information, such as abundance and velocity structure, or gamma-ray line strength and shape, is required [297, 298, 354, 448]. The helium ignition models [605, 460, 695, 398, 401], which proceed to detonation by helium ignition in a shell, also involve lower densities.

The original detonation model raised two questions. First was the perceived shortage of pulsar progenitors [276, 471]. Paczynski [474] had explicitly showed that all stars in the range of $4 \leq M/M_\odot \leq 8$ followed

---

[6]Mazurek, Meier, and Wheeler [423] used this to argue that a detonation was *impossible* for densities $\rho \geq 10^7 \, \mathrm{g\,cm^{-3}}$. Their idealized analysis was a useful step toward understanding aspects of the phenomena, but its applicability seems limited. For example, it ignores the effects of shocks running *up* density gradients, of shocks interacting, and of finite amplitude fluctuations in the convective process. After a more complete analysis, Khokhlov [351] concludes that detonation cannot be ruled out; this is consistent with the numerical simulations of Arnett and Livne [53]. The process may be more complex than we imagine, and than the terms *detonation* and *deflagration* imply.

a common evolutionary track.[7] If this continued to the formation of a $1.4M_\odot$ core of carbon and oxygen, with a loosely bound envelope, then it was attractive to assume that these were the stars that made pulsars. There was a convergent evolution, and the estimated rates of stellar death in this mass range agreed with the estimated rates of pulsar formation. The detonation model (and the deflagration model) gave complete disruption; no pulsar was left. Colgate [182] argued that the detonation calculations [54] were wrong because electron capture would overwhelm the detonation, and the core would collapse to form a neutron star. This did not survive quantitative investigation [49, 70, 127]; Bruenn found that the threshold for collapse of such detonating cores is $\rho_c \geq 2 \times 10^{10}\,\mathrm{g\,cm^{-3}}$.

The second question concerned yields. If each star in this mass range produced $1.4M_\odot$ of iron, this would be difficult to reconcile with galactic nucleosynthesis as it was understood [31, 471]. The observed $Fe/(C+O)$ ratio placed severe constraints upon either (1) the frequency of such events, or (2) the amount of iron ejected from such an event [31]. These arguments based upon nucleosynthesis theory may have been correct. Besides the observational evidence that the burning of the core is not complete [123], it also appears that the rate of Type II supernovae (producers of C and O) is higher relative to the rate of Type I supernovae (producers of Fe) than previously believed [647, 225]. The question of the pulsar rate will be dealt with later. Recent estimates are consistent with the rate of Type II supernovae, and probably could be accounted for by stars above $8M_\odot$; see Narayan [454] for a thorough discussion of the pulsar rate. Both these issures will be discussed further in chapter 14.

11.3   CONVECTION AND URCA

Given the importance of the epoch between ignition and flash for the nature of the explosion, it is necessary to include the complex effects of the Urca process in a convective environment. Paczynski [476], using the work of Tsuruta and Cameron [639] on the Urca process in pulsating stars, first suggested that convection was like pulsation for an Urca shell, and that the Urca process might halt the thermal runaway and allow these cores to collapse to form neutron stars. Let us reconsider carbon ignition.

At this point the electrons in the core are highly degenerate. The fermi energy is 5.22 MeV, and the degeneracy parameter is $u/kT = 244$. The

---

[7]The stars in the range $M/M_\odot \geq 8$ also tend to converge to a common core configuration of $M_{core} \approx 1.4M_\odot$, but this was not widely appreciated at the time.

most abundant nuclei after helium burning are ^{16}O, ^{12}C, and ^{22}Ne. As may be seen from table 6.3, they require much higher densities, $\rho_c \gg 3 \times 10^9 \, \text{g cm}^{-3}$, to capture electrons. However, an abundant nucleus is ^{25}Mg, which has a threshold for electron capture of 3.833 MeV, corresponding to a density of $\rho = 1.17 \times 10^9 \, \text{g cm}^{-3}$. Above this density, cooling and heating are modified by the Urca process on the pair ^{25}Mg-^{25}Na.

There has been much confusion about the heating and cooling effects of electron capture and beta-decay in conditions of high electron degeneracy [478, 126, 194, 321, 72]. A few comments about the physics of the process may help clarify the issues.

A completely degenerate electron gas has zero entropy and zero specific heat. Electrons fill *all* the available states below the fermi energy, and none above. If the gas is heated, this is equivalent to taking electrons from states below the fermi energy, thus making holes in the fermi "sea," and elevating them to states above the fermi energy. The additional energy required to do this is *thermal energy*, and the new equilibrium system is characterized by a finite temperature, entropy, and specific heat. Electron capture in a completely degenerate gas either removes electrons having exactly the fermi energy (at the *fermi surface*), or it makes holes in the fermi sea. Taken strictly by themselves: the first effect neither heats nor cools, but the latter effect causes heating. They do not operate in isolation however; electron capture also causes a change in nuclear composition and the emission of a neutrino.

What is the net effect of the radiation and the compositional change? Consider a heterogeneous system, for which the first law of thermodynamics may be written as

$$dE/dt + PdV/dt = -\epsilon_{emis}, \qquad (11.26)$$

where ϵ_{emis} is the rate of emission of energy by escaping[8] neutrinos (it is positive or zero), per mole of nucleons. Then, using the Gibbs relation,

$$dE/dt + PdV/dt = TdS/dt + \sum_i N_A \mu_i dY_i/dt, \qquad (11.27)$$

we have

$$TdS/dt = -\epsilon_{emis} - \sum_i N_A \mu_i dY_i/dt. \qquad (11.28)$$

If the $\sum_i N_A \mu_i dY_i/dt$ term is zero, then the entropy only *decreases*. The system is open, and the neutrino radiation connects the relatively hot

---

[8]In chapter 12 we will consider the case in which the neutrinos (and antineutrinos) do not freely escape, so that their possible absorption complicates matters.

star $(T \gg 2.7\,\mathrm{K})$ to the cold universe; the entropy of the star decreases but that of the universe increases. How might this summation term be zero? The trivial case is that in which the composition is constant, so that the time derivatives of abundances, dY_i/dt, are individually zero. A more complex case is that in which the reactions do occur, but the reactions are in "chemical" equilibrium, so that the sum is again zero. In these cases, the effect of neutrino (antineutrino) emission due to nuclear processes is cooling.

What happens if the composition is changing? The chemical potentials may be written to make the masses explicit: $\mu_i = u_i + m_i c^2$. Consider only one pair of nuclei, which have a laboratory beta-decay $(Z - 1, A) \rightarrow (Z, A) + e^- + \bar{\nu}$, and an electron capture induced by high density $(Z, A) + e^- \rightarrow (Z - 1, A) + \nu$. The energy release by beta-decay is $Q = (M_{Z-1} - M_Z - m_e)c^2$ (assuming that the mass of the neutrino is negligible); this is the end-point energy of the beta-decay. To simply illustrate the phenomenon, suppose that only electron capture is occurring. Then,

$$dY_Z/dt = dY_e/dt = -dY_{Z-1}/dt = -dY_\nu/dt, \qquad (11.29)$$

and

$$\sum_i N_A \mu_i dY_i/dt = (u_Z - u_{Z-1} + u_e - Q)dY_Z/dt. \qquad (11.30)$$

The quantity in parenthesis is the *affinity* for this reaction, and is important in the study of the thermodynamics of chemical reactions [207]. If this quantity is positive, electron captures are driven toward completion, so that dY_Z/dt is negative. Putting this back into equation 11.28 shows that it tends to increase entropy. If the affinity were negative, decay would be driven, dY_Z/dt would be positive, and again the entropy would tend to increase. The system may be thought of as transfering electron degeneracy energy into energy of radiated neutrinos, by the mediation of the weak interactions. If the rate of transfer is slow, the conversion can be done in an almost exactly reversible manner, and no heating is done in the conversion. The system simply radiates energy. If the rate of transfer is not slow, so that the conversion is no longer reversible, heat is generated, and the net effect may still be cooling, or heating, depending upon details. The transfer of degeneracy energy into neutrinos (antineutrinos) is no longer efficient; some energy is lost as heat.

So far there has been no discussion of mass motion, but as we have seen, convection is likely to accompany carbon ignition. If there is mass motion, we replace the total derivative d/dt by the operator $\partial/\partial t + \mathbf{v} \cdot \nabla$, so that

$$T\partial S/\partial t = -\epsilon_{emis} - \sum_i N_A u_i \partial Y_i/\partial t \tag{11.31}$$

$$-T(\mathbf{v} \cdot \nabla S) - \sum_i N_A u_i (\mathbf{v} \cdot \nabla Y_i). \tag{11.32}$$

The last two terms represent a current of entropy and a current of "chemical" affinity due to the mass flow at velocity $\mathbf{v}$. For slow evolution, there may be time for the convective zone to mix well toward uniformity, so that these gradients would be small. As the evolution accelerates, this will not necessarily remain true. If the currents become important, the question of heating or cooling is no longer a local question but also must involve the history of the matter and its flow. It becomes intimately connected with the hydrodynamics of the developing convection region.

The energy level diagram for an Urca pair is shown in figure 11.3. A hypothetical electron capture from the ground state of nucleus (Z,A) at energy $E = 0$ to nucleus (Z-1,A) at energy $E = Q$ is shown by a solid line. The corresponding beta-decay from (Z-1,A) back to (Z,A) is shown as a solid line parallel to the electron capture (*ec*) line. However, the fastest transitions are often through excited states, which have nuclear wave functions that are more like those of the intended final nucleus. The corresponding electron capture to the excited state in (Z-1,A) requires additional energy, E^*_{Z-1}.

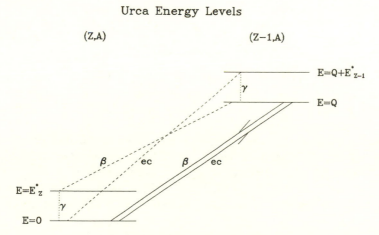

Fig. 11.3. Urca Energy Levels

What are the rates of beta-decay and electron capture, and the corresponding rates of antineutrino and neutrino emission? Integrating the weak interaction rates per particle from §3.7 over a Fermi-Dirac distribution gives the answer. For the dimensionless rate of decay, corrected for blocking of electron phase space, we have

$$W_d(q, \beta, y) = \int_0^{x_q} \frac{(q - z)^2 x^2}{1 + e^{-\beta(z-y)}} dx, \tag{11.33}$$

where $q = Q/m_e c^2$, $x = p_e/m_e c$, $z = \sqrt{x^2 + 1} - 1$, $\beta = m_e c^2/kT$, and $y = \mu_e/m_e c^2 - 1$ are dimensionless versions of the threshold energy, electron momentum, kinetic energy, inverse temperature, and chemical potential. The upper limit of integration, $x_q = \sqrt{(q+1)^2 - 1}$, is the dimensionless electron momentum for which the fermi energy equals the beta-decay end-point energy. The corresponding rate of energy emission in the form of antineutrinos is

$$\mathscr{E}_d(q, \beta, y) = \int_0^{x_q} \frac{(q - z)^3 x^2}{1 + e^{-\beta(z-y)}} dx. \tag{11.34}$$

The rate of electron capture is

$$W_c(q, \beta, y) = \int_{x_q}^{\infty} \frac{(z - q)^2 x^2}{1 + e^{\beta(z-y)}} dx, \tag{11.35}$$

and the corresponding rate of energy emission in the form of neutrinos is

$$\mathscr{E}_c(q, \beta, y) = \int_{x_q}^{\infty} \frac{(z - q)^3 x^2}{1 + e^{\beta(z-y)}} dx. \tag{11.36}$$

Notice the sign in the exponential in the denominator, which has changed from the decay case.[9] These integrals are examined in detail in Appendix B, where useful approximations are defined and compared to direct numerical integration.

For an excited state, the situation is more complex. Consider the electron capture illustrated in figure 11.3. The effective Q-value changes, $Q_d \rightarrow Q + E^*_{Z-1}$, corresponding to a neutrino energy of $\varepsilon_\nu = \varepsilon_e - Q - E^*_{Z-1}$, where ε_e is the energy of the electron. This process is shown as a dashed line. Once the excited state of (Z-1,A) is formed, it quickly decays

---

[9] A decay is reduced by filled electron states but capture is enhanced. The sign switch comes from $1 - (1 + e^a)^{-1} = (1 + e^{-a})^{-1}$.

to the ground state by emission of gamma-ray energy, E^*_{Z-1}, as shown by a dotted line. This heats the matter because, at stellar densities, the gammas are absorbed before they can escape (however, see chapter 13 for gamma-ray escape from supernovae). Similarly, beta-decay through an excited state of (Z,A) is shown as a dashed line in figure 11.3, with the corresponding γ de-excitation as a dotted line. Here the energy of the antineutrino is $\varepsilon_{\bar{\nu}} = Q + E^*_Z - \varepsilon_e$. For these processes to occur, ε_ν and $\varepsilon_{\bar{\nu}}$ must be positive. Thus, for electron capture, $\varepsilon_e > Q + E^*_{Z-1}$, and for beta-decay to be allowed, $\varepsilon_e < Q - E^*_Z$. This leaves a gap in energy of width $E^*_{Z-1} + E^*_Z$. At zero temperature, $\varepsilon_e \approx u_e \equiv \mu_e - m_e c^2$, so only decays occur for $u_e < Q - E^*_Z$, only electron captures for $u_e > Q + E^*_{Z-1}$, and nothing for the intermediate values. Finite temperatures raise the electron energies, so that both conditions on ε_e may be met, giving both decays and captures. In the limit that $kT \gg E^*_{Z-1} + E^*_Z$, these excitation energies can be ignored; unfortunately such a simplification is not often appropriate.

For temperatures sufficiently high that quasiequilibrium or nuclear statistical equilibrium are approached, the Urca *pair* is generalized to an Urca *ensemble*, and the cycle will be completed by the fastest reaction available in the ensemble of nuclei.

For brevity we focus on the simplest case, ground-state transitions. This is useful to illustrate some aspects of the physics of the Urca process, but may be inadequate in detail.

The constant K, by which these rates $\mathcal{W}$ must be multiplied to get the physical rates, may be determined directly from the experimental rate of beta-decay in a laboratory. The rate of decay is related to the half-life by $\lambda = \ln 2/t_{1/2}$. It is also given by

$$\lambda \approx \frac{g^2 |M_{ij}|^2 m^5 c^4}{2\pi^3 \hbar^7} F(Z, \infty) \int_0^q (q - y)^2 (y + 1)^2 dy, \qquad (11.37)$$

where $F(Z, \infty) = 2\pi\eta/(1 - e^{-2\pi\eta})$ and $\eta = \alpha Z \approx Z/137$. Note that use is made of the high energy of the electron relative to its rest mass, to convert $x^2 dx$ to $(y + 1)^2 dy$, where $y \approx x - 1$. The values of the coupling g^2, the matrix element $|M_{ij}|^2$, the end-point energy $E_0 = qm_e c^2$, and the decay rate λ may all be determined experimentally. Therefore, the rate of decay in a star is, if we ignore other processes,

$$dY_d/dt = -Y_d K \mathcal{W}_d, \qquad (11.38)$$

where $K = \ln 2\, F(Z, \infty)/ft_{1/2}$. For emission of antineutrinos, we have

$$\epsilon_{\bar{\nu}} = -Y_d (N_A m_e c^2) K \mathcal{E}_d, \qquad (11.39)$$

where $N_A m_e c^2 = 4.930 \times 10^{17}$ erg g^{-1}.

For a single Urca pair, the rate of change of abundance of the electron-capturing nucleus is

$$dY_c/dt = -Y_c K \mathcal{W}_c + Y_d K \mathcal{W}_d, \tag{11.40}$$

and the total energy emission by neutrinos and antineutrinos is

$$(Y_c + Y_d)\mathcal{E}_{total} = Y_c \mathcal{E}_c + Y_d \mathcal{E}_d. \tag{11.41}$$

In a steady state, $Y_c/Y_d = \mathcal{W}_d/\mathcal{W}_c$. For an Urca pair, with no other significant reactions occurring, the sum of the nucleon fractions of the Urca pair, $Y_c + Y_d = Y_p$, is constant. Figure 11.4 shows the heating and cooling from the Urca process. The numerical example has been chosen to be appropriate to the ^{25}Mg-^{25}Na pair at degenerate carbon ignition. This means $Q = 3.833$ MeV and $T = 2.7 \times 10^8$ K. The quantity shown is

$$\mathcal{E}_{net} = -dY_Z/dt \, (u_Z - u_{Z-1} + u_e - Q)/Y_p m_e c^2 - \mathcal{E}_c - \mathcal{E}_d, \tag{11.42}$$

which is related to TdS/dt above. It is plotted against $Y_c/(Y_c + Y_d)$, the fraction by number of the nuclei of the pair in the form of the nucleus that is stable terrestrially (which will *capture* electrons). In general, the composition is not equal to the steady-state value. If it is, the cooling rate is at its strongest. Three curves are shown, for low density ($\epsilon_f < Q$), threshold ($\epsilon_f = Q$), and high density ($\epsilon_f > Q$). In each case the strongest cooling occurs at the steady-state abundance; the value of abundance having the strongest cooling can be easily seen in the figure. Away from steady-state abundance, the cooling switches to heating.

Which Urca pairs are likely to be most important? The abundances after helium burning in a core of $S_\gamma/\Re = 0.2$, corresponding to a star of zero-age main sequence mass of 5 to $9M_\odot$, are shown in table 11.1, which lists all nuclei with nucleon fraction above 10^{-6}. In an extensive study, Iben [320] concluded that the most important pairs were ^{21}F-^{21}Ne, ^{23}Ne-^{23}Na, and ^{25}Na-^{21}Mg. They are 21, 12, and 10 in the list, respectively. These rankings depend upon the extent to which the original ^{14}N is converted into heavier nuclei. In this case, most resides in ^{22}Ne; the result is sensitive to the cross sections used during helium burning. A Population II composition, with its lower abundance of CNO catalysts, would have correspondingly lower abundances of these Urca pairs.

The odd-A nuclei[10] are the most effective. As mentioned above, the even-even nuclei, because of their greater binding, tend to have larger

---

[10]Here we refer to odd-Z and even-N, or even-Z and odd-N, which will both be called odd-even in this section for conciseness.

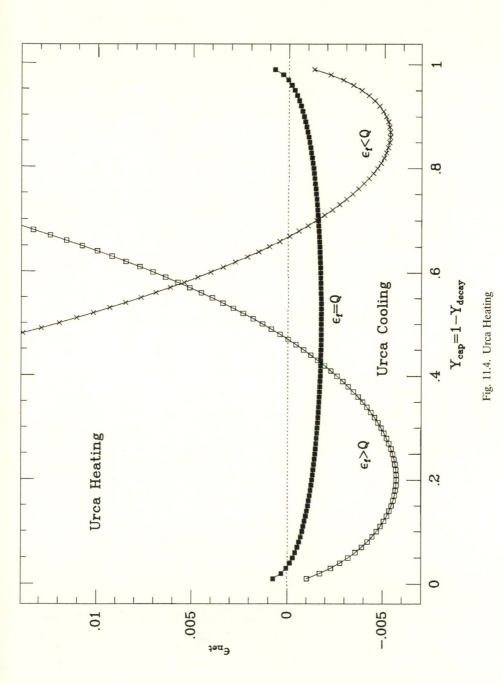

Fig. 11.4. Urca Heating

TABLE 11.1

He Ashes ($S_\gamma/\Re = 0.2$)

Nucleus	Z_i	N_i	$\log_{10}(X_i)$	Nucleus	Z_i	N_i	$\log_{10}(X_i)$
^{16}O	8	8	−0.2412	^{12}C	6	6	−0.3990
^{22}Ne	10	12	−1.6688	^{20}Ne	10	10	−2.9007
^{56}Fe	26	30	−2.9412	^{28}Si	14	14	−3.1506
^{24}Mg	12	12	−3.2682	^{26}Mg	12	14	−3.3407
^{32}S	16	16	−3.3817	^{25}Mg	12	13	−3.6844
^{57}Fe	26	31	−3.7420	^{23}Na	11	12	−3.7499
^{36}Ar	18	18	−4.0504	^{40}Ca	20	20	−4.1272
^{27}Al	13	14	−4.1542	^{54}Fe	26	28	−4.2485
^{29}Si	14	15	−4.2597	^{58}Fe	26	32	−4.3005
^{30}Si	14	16	−4.3813	^{58}Ni	28	30	−4.4888
^{21}Ne	10	11	−4.5531	^{34}S	16	18	−4.6062
^{60}Ni	28	32	−4.6317	^{52}Cr	24	28	−4.8671
^{55}Fe	26	29	−4.8680	^{31}F	15	16	−4.8701
^{33}S	16	17	−4.8976	^{59}Ni	28	31	−5.0507
^{55}Mn	25	30	−5.0686	^{59}Co	27	32	−5.2176
^{61}Ni	28	33	−5.2735	^{62}Ni	28	34	−5.2906
^{19}F	9	10	−5.4231	^{37}Cl	17	20	−5.4339
^{38}Ar	18	20	−5.4964	^{39}K	19	20	−5.5242
^{35}Cl	17	18	−5.6300	^{41}Ca	20	21	−5.6680
^{44}Ca	20	24	−5.7652	^{48}Ti	22	26	−5.7988
^{64}Zn	30	34	−5.8391	^{37}Ar	18	19	−5.8531
^{63}Cu	29	34	−5.8958	^{53}Cr	24	29	−5.9061

thresholds for electron capture (e.g., ^{12}C and ^{16}O). Even if this is not the case, the even-even nuclei are less effective than odd-even. Consider ^{24}Mg, which has a threshold of 5.51 MeV, or $2Y_e\rho = 3.16 \times 10^9$ g cm^{-3}. Its Urca partner, ^{24}Na, has a 4$^+$ ground-state spin and parity, while ^{24}Mg has 0$^+$ for its ground-state. Thus ^{24}Na decays in the laboratory primarily to the 4.1228 MeV 4$^+$ excited state of ^{24}Mg. Beta-decay directly to ground is highly forbidden, the rate is reduced by much more than is needed to make negligible both it and its inverse, electron capture (see the discussion below of how the lower abundances in Population II cores, a smaller effect, can reduce the Urca effects to unimportance). If direct electron capture from the ground state of ^{24}Mg to the ground state of ^{24}Na is too slow, what about the inverse of the laboratory decay? The temperatures are too low for the excited state of ^{24}Mg to have a thermal population sufficient to maintain a nonnegligible rate.

The Urca parameters for these pairs of nuclei are given in table 11.2. The coulomb correction factor $F(Z, \infty)$, for the deviations of the electron wavefunction from that of a plane wave, is almost constant, and fairly close to unity. The values of $\log ft_{1/2}$ refer to the ground state only.

Paczynski [476] had noted that the formal value of the convective turnover time was shorter than the effective decay time, and assumed that this implied homogeneous composition. It would also mean that the nuclear abundances were not in a steady state, so that the Urca processes might cause heating rather than cooling. This might be sensitively dependent upon the extent of convective overshooting; the coupling of the Urca process to the hydrodynamics of the convection may be important and is as yet unexplored. At these high densities, the diffusion of ions is slow, so that a convective blob will "remember" its composition, and that will affect its motion. This suggests that it is not appropriate to consider the process in a purely radial geometry, with homogeneous convective zones.

The evolution up to the development of extensive convection may be treated with greater reliability. We now put together these Urca rates with the procedure from the last section for the time-dependent evolution of the carbon-oxygen core. This trajectory in the temperature-density plane is shown in figure 11.5. The upper heavy line (labeled *Single-Star*) is the trajectory that includes Urca processes; the nearby thin line does not. The vertical dotted lines lie at the threshold for electron capture for the pairs labeled. The effect of the Urca shell is to cause cooling for electron fermi energies near threshold for electron capture.

TABLE 11.2

Urca Parameters

Pair	A	Z_d	$F(Z, \infty)$	$[\log ft_{1/2}]_0$	Q(MeV)	$K/10^{-6}\,\mathrm{s}^{-1}$
F-Ne	21	9	1.2206	5.2	5.686	5.34
Ne-Na	23	10	1.2468	5.3	4.3745	4.33
Na-Mg	25	11	1.2734	5.3	3.833	4.42

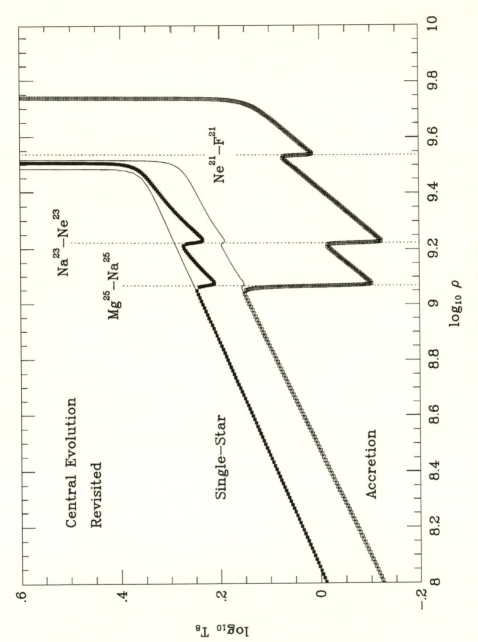

Fig. 11.5. Central Evolution Revisited

Because the evolution is driven by accretion, which is slow compared to electron capture or beta-decay rates, the abundances of the Urca pairs are near steady state, so that the result is cooling. The characteristic saw-tooth pattern is due to rapid cooling near threshold, followed by heating by compression after threshold is past. The finite thermal capacity and the low heating rate keep thermal balance from being instantaneously attained.

This figure is almost identical to figure 1 of Iben [322]; his graph uses linear rather than logarithmic variables, so it first appears more different than it is. A real difference occurs at a temperature of $T_8 \geq 2.5 \approx \log T_8 = 0.4$. In Iben's computation, convective cooling halted the thermal runaway at this point. He assumed that the convective regions would be homogeneous in composition, and paid special attention to the energetics of such convective regions. He found behavior of great complexity, involving three Urca shells and thermal oscillations, and concluded that the Urca process only delayed[11] the thermal runaway. Core densities were about a factor of 1.5 larger than they would have been if the Urca process had been ignored.

The thick curve labeled *Accretion* is the trajectory with Urca effects included, but an accretion rate smaller by a factor of 40, in order to correspond to Iben's white dwarf accretion (his figure 5). Again, up to ignition the agreement is excellent. This curve is shifted downward from the *Single-Star* curve because of the smaller rate of gravitational heating, which results from the smaller rate of core growth. This smaller growth rate results in longer time-scales, so that the Urca cooling has more time to act and causes larger amplitude changes in temperature. The lower trajectory, being cooler for a given density, has ignition at a higher density. For this particular choice of screening, burning rate, and accretion rate, the ignition is postponed until after the ^{21}Ne-^{21}F Urca shell has become operative. Runaway happens for a central density of $\rho_9 \approx 6$. This is well below the value $\rho_9 \geq 20$ needed for electron capture to be fast enough to prevent a detonation (if formed) from giving an explosion [126]. For a variety of accretion rates, a variety of ignition temperatures are expected. Does this give a variation in the nature of the ensuing evolution? More specifically, would it predict variations in supernovae of Type Ia, which have been attributed to such objects? Answers to these questions require an understanding of how the core evolves after carbon ignition; at present this is incomplete.

---

[11]This provides a much tighter limit than the previous one of Couch and Arnett [194, 195], who attempted to show that even for a set of assumptions favoring collapse, the Urca process allowed explosion.

The thin solid line, which coincides in part with the *Accretion* line, is a calculation which is identical, but with Population II abundances (the Urca pairs were taken to be 100 times *less* abundant). It is almost identical to the accretion case with no Urca rates. A calculation, with the accretion rate reduced by 4,000, required a time of about 10^{10} y to arrive at instability; this occurred at a central density of $\rho_c \approx 1.5 \times 10^{10}\,\mathrm{g\,cm^{-3}}$, close to the boundary for electron capture to cause a collapse.

Time-scales are crucial for understanding the convective Urca process. Figure 11.6 shows the thermal runaway of the *single-star* core case. It gives a measure of the time-scale associated with each temperature. Central temperature (in units of 10^8 K) is plotted, as solid squares, against time (in seconds) remaining until the *flash* temperature ($T_8 \approx 10$); both variables are in logarithms. The temperature $T_8 \approx 2.5$, at which Iben found convection to halt the runaway, is shown as a dotted line. At this point in the evolution, the characteristic times are about 10^{10} s. The time to relax the abundances to an Urca steady state is about 10^7 s. The half-lives for free decay of ^{21}F, ^{23}Ne, and ^{25}Na are shown; they are changed because the reaction is slowed by phase space inhibition. The convective turnover time estimated by Iben [322] from mixing-length theory is shown as the short line labeled ML. The fastest turnover time is esti-

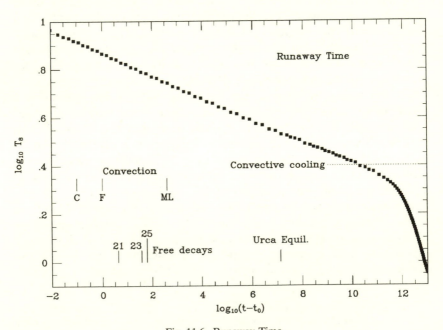

Fig. 11.6. Runaway Time

mated above, and shown labeled as F. The time labeled C is the sound travel time through the core. Because the relaxation time to Urca steady state is longer than that for convective turnover, the core is not in Urca equilibrium, and to the cooling is added some heating, because the affinity is not zero. As the core grows, the matter samples regions of greater differences in density, so that the affinity also grows, giving more heating. Initially these effects happen on a time-scale that is faster than the runaway time of 10^{10} s, which is shorter than the carbon fusion time. Thus they may modify the convective flow. This might be the physical basis for Iben's thermal oscillations, filtered through his assumptions of spherical symmetry and mixing-length convection. In any case, the nature of the convective Urca process is still not well understood. At present it seems plausible to assume that a thermal runaway follows carbon ignition, perhaps after some delay; there seems no particular reason to suppose that the core somehow manages to collapse.

11.4 YIELDS FROM DEGENERATE INSTABILITY

Despite considerable effort, we seem to have no very complete picture of what happens after the ignition of $^{12}C + ^{12}C$ under conditions of high degeneracy of electrons. It seems likely that the result is a thermal runaway at densities less than $10^{10}\,\mathrm{g\,cm^{-3}}$, probably $\rho_9 \approx 3$ to 4 for single-star evolution. The behavior during the time from ignition to runaway may determine the process of burning, and hence the explosion mechanism.

A consequence of the uncertainty regarding this epoch is that our predictions of the detailed nucleosynthesis have an unknown degree of reliability. While the abundances may be accurately determined for a given model of the explosion, this model may be flawed. Nevertheless, there are some aspects of such nucleosynthesis that are broadly relevant. One is the possibility that the high densities cause so much electron capture that it is not possible to make iron-peak isotopes in the solar system ratios.

Suppose a blob, composed of ^{12}C and ^{16}O, burns rapidly at density so high that the electrons are highly degenerate. What is its new state? For simplicity we will first assume that the transition is so fast that only the change in temperature is important. Therefore, density is constant and there is no flux out of the blob. For example, if a nucleon fraction $X(^{12}C) = 0.4$ is burned to equal ^{20}Ne and ^{24}Mg by mass, then about $2 \times 10^{17}\,\mathrm{erg\,g^{-1}}$ is released. This change in energy defines a new temperature, which is shown in figure 11.7 as a dashed line labeled $C + C$. For reference are shown the single-star and accretion trajectories discussed in the previous section. The increase in temperature is large, the

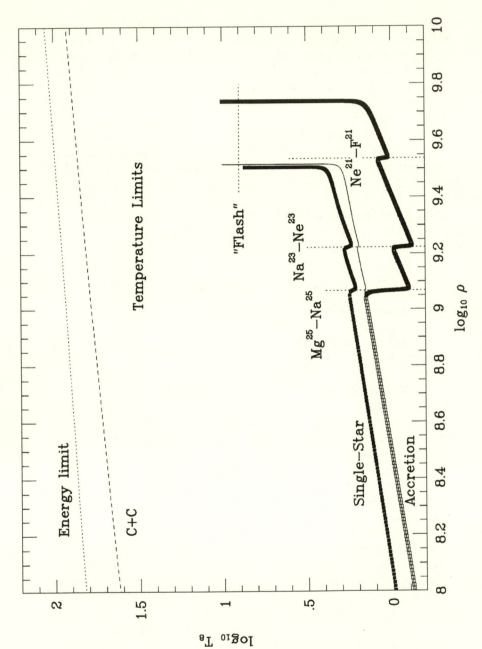

Fig. 11.7. T Limits

final value being over twenty times that at ignition. The energy release is large compared to the thermal energy capacity of the highly degenerate gas. Only a moderate amount of burning would bring the temperature to the "flash" point, at which the nuclear burning time equals the hydrodynamic time-scale. If the carbon burns, the new temperature is above the flash temperature for neon, oxygen, and silicon burning, and is large enough for photodisintegration to occur [18, 54]. The new state at high temperature will be in nuclear statistical equilibrium (NSE), because the time needed to approach NSE is short compared to hydrodynamic timescales. The burning forms iron-peak nuclei, which are more tightly bound than ^{12}C and ^{16}O; the extra energy is used to increase the temperature and to free nucleons and alphas. This is shown as the dotted line labeled *Energy limit*, which is an important limiting case for discussion of explosion models. For brevity we will call such material *carbon-flash matter*.

Some of the characteristics of carbon-flash matter are given in table 11.3. For each initial density ($\rho_9 = \rho/10^9 \, \text{g} \, \text{cm}^{-3}$) there is a corresponding temperature. Because of the photodisintegration, the adiabatic exponent Γ is less than 4/3. The electron chemical potential $u_e = \mu_e - m_e c^2$ is given in MeV, as is the difference in chemical potentials for neutrons and protons, $\hat{u} = u_n - u_p = \ln(Y_n/Y_p)$. The entropy per nucleon for the ensemble of particles and radiation is given in dimensionless form, $S/\mathfrak{R}$. The mole number of free protons Y_p is slowly varying, and roughly 2%. The nucleon fractions of ^{4}He (X_α) and of all heavier nuclei (X_{ZA}) are also insensitive to density. The fractional increase in pressure, $\Delta P/P$, is near unity only at the lowest densities, but is much larger than its value at ignition.

At these high densities, electron capture occurs rapidly on free protons and the abundant ^{56}Ni. Table 11.4 gives an estimate of the change in neutron excess η; the scale is chosen so that one unit is a change of $\Delta\eta = 10^{-3}$. The temperature and density entries are the same as in table 11.3. The rate of decline in the nucleon fraction of electrons, due *only* to electron capture on free protons, is $\Lambda_p = \rho Y_e Y_p N_A \langle\sigma v\rangle$. The units are inverse seconds. The rate of decline in the nucleon fraction of electrons, due *only* to electron capture on ^{56}Ni (denoted by subscript Z), is $\Lambda_Z = \rho Y_e Y_Z N_A \langle\sigma v\rangle$. The rates of capture on ^{56}Ni are comparable[12] to those for free protons. Electron capture is

---

[12]The effective $ft_{1/2}$ values of Fuller, Fowler, and Newman [254] are comparable to those for free protons, and the phase space factor is larger due to the nuclear energy difference. This suggests that the capture rate of ^{56}Ni is the larger. However, the abundance by *number* of ^{56}Ni is smaller, giving rates that are almost equal.

TABLE 11.3

C-Flash Matter

T_9	ρ_9	u_e	$\hat{u}$	$S/\Re$	$\Delta P/P$	Y_p	X_α	X_{ZA}
6.37	0.10	0.940	−4.568	4.695	0.850	0.019	0.093	0.887
6.88	0.20	1.451	−4.351	4.635	0.599	0.021	0.121	0.858
7.24	0.32	1.863	−4.233	4.568	0.482	0.022	0.135	0.843
7.42	0.40	2.094	−4.179	4.529	0.434	0.022	0.141	0.836
7.61	0.50	2.342	−4.128	4.487	0.391	0.023	0.147	0.830
7.80	0.63	2.609	−4.079	4.443	0.354	0.023	0.152	0.825
8.00	0.79	2.897	−4.032	4.396	0.321	0.024	0.157	0.820
8.20	1.00	3.207	−3.986	4.347	0.291	0.024	0.161	0.815
8.41	1.26	3.541	−3.941	4.297	0.265	0.024	0.164	0.811
8.64	1.58	3.900	−3.897	4.246	0.242	0.024	0.168	0.808
8.86	1.99	4.287	−3.853	4.195	0.220	0.025	0.170	0.805
9.11	2.51	4.704	−3.809	4.141	0.201	0.025	0.173	0.802
9.36	3.16	5.153	−3.766	4.087	0.184	0.025	0.175	0.800
9.62	3.98	5.637	−3.722	4.034	0.169	0.025	0.177	0.798
9.90	5.01	6.160	−3.678	3.979	0.155	0.026	0.178	0.796
10.19	6.31	6.724	−3.634	3.924	0.142	0.026	0.179	0.795
10.49	7.94	7.337	−3.590	3.866	0.130	0.026	0.180	0.793
10.81	10.00	8.005	−3.545	3.804	0.120	0.027	0.180	0.793

turned off by expansion, which may be driven by an overpressure, or by buoyancy forces. An accurate estimate of the increase in neutron excess requires knowledge of hydrodynamics which we do not have; it depends upon the model. An estimate, which is only illustrative, comes from multiplying the rate of electron capture, $\Lambda_p + \Lambda_Z$, by a hydrodynamic time $(d \ln \rho/dt)^{-1}$ which has been corrected for the overpressure not being unity (see §9.1). *The precise value of the expansion time depends upon the detailed nature of the hydrodynamics of the explosion.*

In chapter 9 we found that to produce the solar system ratios of isotopes in the iron-peak required a neutron excess of $\eta \approx 3 \times 10^{-3}$. From table 11.4, the *increase* in neutron excess is larger than this for densities above $8 \times 10^8 \, \mathrm{g\,cm^{-3}}$, which is considerably lower than the ignition

TABLE 11.4

C-Flash e^- capture

T_9	ρ_9	Λ_p	Λ_Z	$10^3 \, \Delta\eta$	$(d \ln \rho/dt)^{-1}$
6.374	0.100	0.002	0.000	0.118	0.048
6.888	0.200	0.007	0.002	0.348	0.041
7.239	0.316	0.014	0.006	0.734	0.036
7.420	0.398	0.021	0.011	1.078	0.034
7.607	0.501	0.030	0.018	1.543	0.032
7.798	0.631	0.044	0.028	2.155	0.030
7.996	0.794	0.065	0.043	3.014	0.028
8.201	1.000	0.095	0.067	4.223	0.026
8.414	1.259	0.140	0.103	5.929	0.024
8.635	1.585	0.206	0.161	8.346	0.023
8.865	1.995	0.303	0.251	11.780	0.021
9.106	2.512	0.448	0.387	16.552	0.020
9.357	3.162	0.663	0.588	23.085	0.018
9.621	3.981	0.981	0.898	32.292	0.017
9.897	5.012	1.455	1.379	45.328	0.016
10.187	6.310	2.162	2.136	63.900	0.015
10.491	7.943	3.221	3.336	90.572	0.014
10.812	10.000	4.819	4.944	125.158	0.013

density. While this is not the complete story, it is clear that this is an important constraint if much of the solar system matter has been synthesized in such events. The effect of increasing neutron excess upon the abundances after freezeout is simple. For a particle-poor freezeout (see the discussion of *silicon burning* and the *e-process* in chapter 9), the final abundances are similar [277] to those in nuclear statistical equilibrium (NSE) at a temperature $T_9 \approx 3$.

The dominant nucleus is that with the largest binding energy *and* satisfies $Z/A \approx Y_e$ (charge conservation). This may be seen in a striking way in figure 4 of Hainebach et al. [277]. For $Y_e \approx 0.5$, the dominant nucleus is ^{56}Ni, for which $Z/A = 28/56 = 0.5$. As the number of electrons per nucleon (Y_e) drops, ^{54}Fe, with $Z/A = 26/54 = 0.481$, and ^{58}Ni, with $Z/A = 28/58 = 0.483$, become the dominant nuclei. Their abundances reach a maximum near a neutron excess $\eta = 1 - 2Y_e \approx 0.036$, or $Y_e \approx 0.482$, as this simple argument suggests. Further decrease in Y_e brings the dominance of ^{56}Fe, which peaks at $Y_e \approx 26/56 = 0.464$. This is followed by a peak in ^{58}Fe at $Y_e \approx 26/58 = 0.448$. In the solar system abundance pattern, the nucleon fraction of ^{54}Fe is only 0.0612 of that of ^{56}Fe, which ^{56}Ni eventually becomes (the percentages by number are 5.8 and 91.7). As Clayton and Woosley [178] have argued, the solar system abundance of the iron-peak nuclei is dominated by production of ^{56}Fe as ^{56}Ni. However, some modest production of ^{56}Fe as itself is not disastrous. On the other hand, any significant contribution of NSE matter dominated by ^{54}Fe *must* be only about 6 percent or less of that making ^{56}Fe or ^{56}Ni. This provides a stringent constraint on the amount of matter with $Y_e \approx 26/54 = 0.481$.

For an alpha-rich freezeout (see chapter 9), the charge conservation contraint is also effective, but the relevant nucleus is shifted from ^{54}Fe to ^{58}Ni, for example. Thielemann, Nomoto, and Yokoi [616] found that while the alpha-rich freezeout reduced the problem with ^{54}Fe, there arose a corresponding problem with ^{58}Ni and ^{62}Ni. The constraint survived.

A quantitative calculation of the increase in neutron excess may be easily illustrated. We use the electron capture rates given in table 11.4. We choose the expansion time to be $\tau = 446\chi/\sqrt{\rho_c}$ s. Here the central density is used for the hydrodynamic time-scale. However, we use the structure of an $n = 3$ polytrope to relate density for electron capture to the mass at that density. We extrapolate the neutron enrichment by simply multiplying the two, at different layers in the star. For a central density of $\rho_c = 3.16 \times 10^9$ g cm^{-3}, the resulting values of Y_e are shown in figure 11.8, plotted against fractional mass coordinate. For the pressure excesses in table 11.3, $\chi = 2$ is an extremely fast expansion; $\chi = 4$ and 9 are more realistic. Clearly much more than 6 percent of the degenerate core is in the "forbidden" region. The actual values of Y_e which show a big decrease are only qualitatively correct. Detailed integration gives smaller values if the changes are large, because the abundances of ^{56}Ni and free protons decrease, making our estimates of the electron capture too high for integrated rates. The precise shape of the curves depends upon the hydrodynamics assumed, but the original[13] detonation model [49] (see their table 2) and the deflagration model [619] (see their

---

[13]The estimate of neutron excess was flawed by an error in the conversion of energy from electron rest mass units to MeV, resulting in the "low-η" case being a factor of four too low. Correction gives the low estimate only a factor of five below the high estimate, which is not drastically different from the results here.

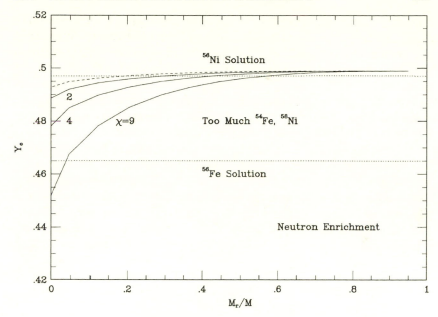

Fig. 11.8. Neutron Enrichment

figure 13) give similar behavior. Both have the excess neutron region more concentrated toward the center. For the detonation model, the expansion times are shorter for matter farther from the center [49]. For the deflagration model, there is significant expansion of the outer layers before they burn. The delayed detonation model of Khokhlov [350] also shows this concentration toward the center of the excess neutron region (see his figure 13) for his preferred parameters.

The dashed line in figure 11.8 represents a case with a relatively slow expansion ($\chi = 9$), but a central density of only $\rho_c = 10^9 \, \text{g cm}^{-3}$. It clearly is far less problematic. If the convective stage that follows ignition can cause such an expansion of the core, then the neutron excess problem would be reduced. Similarly, if the burning is by deflagration, but is weak, the CO core may be expanded. If this core now detonates, most of the burning will occur at the lower density, again reducing the neutron excess problem [350, 53].

Such a decreased density would also remove the difficulty regarding the observed significant abundance of intermediate mass elements (O, Mg, Si, and Ca; see [123]) in Type I supernovae. The thermal runaway is quenched when the increase in temperature causes a significant increase in pressure, which causes expansion and adiabatic cooling. If this maximum temperature is lower than the flash temperature for a fuel, that fuel does not burn significantly. To preserve these elements, the temperature should be less than that for explosive silicon burning (for O, Si, and Ca), or explosive oxygen burning (for Mg). Consider a cold gas ($T_8 \approx 2$) of half ^{12}C and half ^{16}O by mass. Assume that carbon burning and neon burning occur, but oxygen burning does not go to completion (so that Mg still survives). This increases the energy of the gas by about $X(^{12}\mathrm{C})Q \approx 2.5 \times 10^{17}$ erg g^{-1}.

The temperature and density combinations resulting from this procedure are shown in table 11.5. Also shown is the mean time for oxygen destruction by the ^{16}O+^{16}O reaction, $\tau_O = -Y_O/\dot{Y}_O$, the electron chemical potential divided by kT (which indicates the degree of degeneracy), and the increase in pressure due to the heating, $\Delta P/P$. At what point will oxygen burning quench? Suppose that $\dot{Y}_O \propto T^\alpha$ and $\rho \propto T^3$, where $\alpha \approx 30$. Then to reduce the oxygen abundance by a factor of e requires burning

TABLE 11.5

O-Flash

T_9	ρ	τ_O	u_e/kT	$\Delta P/P$
2.441	3.16E+06	7.55E+02	0.0738	2.0166
2.555	3.98E+06	1.52E+02	0.2363	1.7867
2.672	5.01E+06	3.22E+01	0.4030	1.5981
2.794	6.31E+06	7.03E+00	0.5723	1.4315
2.921	7.94E+06	1.58E+00	0.7448	1.2847
3.053	1.00E+07	3.68E-01	0.9205	1.1551
3.190	1.26E+07	8.80E-02	1.0996	1.0406
3.332	1.58E+07	2.17E-02	1.2823	0.9394
3.480	2.00E+07	5.53E-03	1.4693	0.8498
3.632	2.51E+07	1.45E-03	1.6612	0.7706
3.790	3.16E+07	3.93E-04	1.8580	0.7002
3.954	3.98E+07	1.10E-04	2.0599	0.6375
4.296	6.31E+07	9.37E-06	2.4806	0.5316
4.660	1.00E+08	9.00E-07	2.9283	0.4466
5.246	2.00E+08	3.30E-08	3.6538	0.3480
5.665	3.16E+08	4.23E-09	4.1813	0.2966

for a time τ_O at constant conditions. If the matter expands and cools, less burning occurs. This must be compensated for by a shorter τ_O (more vigorous burning). Using $d \ln Y_O/dt \approx (\alpha/3)d \ln \rho/dt$, we estimate that this occurs for $\tau_O \approx 0.1(d \ln \rho/dt)^{-1}$. Hydrodynamic times of the order of $0.1\,s$ are typical of such explosions [49, 619]; $\tau_O \approx 0.01\,s$ occurs for $\rho = 1.2 \times 10^7\,\mathrm{g\,cm}^{-3}$. This is close to the value found by direct hydrodynamic calculations of the deflagration model (see figure 5 in [619]). In a detonation, the transition occurs at a similar density [331, 349]. Both the deflagration model and the delayed detonation model utilize the same physical effect to preserve a layer of intermediate mass elements at the outer edge of the exploding object. These arguments are also relevant to detonation in sub-Chandrasekhar mass white dwarfs (see §11.5).

For an $n = 3$ polytropic structure, 80 percent of the mass lies inside a radius at which the density is 6 percent of the central value. If we want the outer 20 percent of the degenerate object to avoid oxygen burning, this implies a central density of, roughly, $\rho_c \le 2 \times 10^8\,\mathrm{g\,cm}^{-3}$. The gravitational binding energy B of a white dwarf of mass M is

$$B/M \approx (B/M)_1 \left(1 - (\rho_1/Y_e\rho)^{1/3}\right)\left(M/(2Y_e)1.42M_\odot\right)^{1/3}, \qquad (11.43)$$

for $\rho_c \ge 10^9\,\mathrm{g\,cm}^{-3}$, where $(B/M)_1 = 2.46 \times 10^{17}\,\mathrm{erg\,g}^{-1}$ and $\rho_1 = 2.7 \times 10^7\,\mathrm{g\,cm}^{-3}$. For smaller central densities, direct numerical evaluations are required. To reduce the central density from the value at ignition ($\rho_c \approx 3 \times 10^9\,\mathrm{g\,cm}^{-3}$) to the value that would allow an outer layer of intermediate mass elements, ($\rho_c \approx 2 \times 10^8\,\mathrm{g\,cm}^{-3}$), requires about 1.5×10^{50} erg. This could be supplied by burning $0.15M_\odot$ of $^{12}\mathrm{C}$ to Ne and Mg; more energy would be needed if Urca or other losses were important. If this could happen during the epoch between ignition and flashing, it would provide another path to an acceptable explosion.

Suppose that a parcel of matter does burn to form carbon-flash matter. How will it interact in an environment of unburned matter? An attempt to answer this question leads to a discussion of laminar flames [690, 622, 352, 399, 53]. Suppose the matter is in pressure equilibrium, and initially at rest. The hot matter will heat its surroundings, primarily by electron conduction of thermal energy. As the new fuel is heated it begins to burn, heating still more. If this new heat can be balanced by loss by conduction, then a stable flame front is at least possible. Consider a layer of width Δr, area A, and mass $\Delta m = A\rho\Delta r$. The rate of energy release by burning is $\Delta m\epsilon = A\rho\Delta r\epsilon$, while the cooling rate is FA, where the heat flux is $F = -\frac{\lambda c}{3}daT^4/dr$. The temperature of the burned matter is much larger than for unburned matter, so $F \approx \frac{\lambda c}{3}aT^4_{Cflash}/\Delta r$. Notice the different dependence of heating and cooling upon the width of the zone. Reducing Δr enhances cooling over heating. High heating rates demand small widths for balanced cooling. For $\kappa = 10^{-3}\,\mathrm{cm}^2\,\mathrm{g}$, $\rho = 1.6 \times 10^9\,\mathrm{g\,cm}^{-3}$, $T_{Cflash} = 7 \times 10^9\,\mathrm{K}$, and $\epsilon = 1.2 \times 10^{30}\,\mathrm{erg\,g}^{-1}$, this implies $\Delta r \approx 5 \times 10^{-6}\mathrm{cm}$ is the zoning needed to resolve the burning front. Direct numerical calculation of such a front, using the physics in an old paper of Arnett [21], gives a propagation velocity of $v_{flame} \approx 270\,\mathrm{km\,s}^{-1}$. This is less than a factor of three above the recent value of Timmes and Woosley [620], and is similar to that of Khokhlov [352]. Evidently such values are not controversial. For comparison, the sound speed is about $8 \times 10^3\,\mathrm{km\,s}^{-1}$; the flame motion is quite subsonic.

The flame is a surface which grows to enclose more volume. A sphere is that geometrical shape which has a minimal surface area relative to its volume. The rate of increase in volume is proportional to the velocity of the flame and to the area of the surface. This will be increased if the surface is distorted, resulting in a faster rate of burning of matter. Numerical experiments by Müller and Arnett [445, 446] indicated that a spherical flame would be subject to instabilites with large length scales. Woosley [690] developed an ingenious scenario in which the crinkling of the flame front is assumed to be of a fractal nature, and gives a slow onset of burning which accelerates into an explosion. This ameliorates the electron capture problem by a stage of subsonic expansion prior to most of the burning. Thus we return to the flame problem discussed earlier.

Could it be that the presence of intermediate mass elements in SNIa is not a clue about the nature of the burning front, but about the geometry of the initial star? Steinmetz, Müller, and Hillebrandt [589] conducted a set of numerical experiments to investigate whether rotational flattening of a white dwarf could "tame" a detonation by allowing incompletely burned matter to be ejected. This is particularly relevant to scenarios in which the exploding white dwarf is in a binary. Their conclusion was that it is impossible to reproduce the desired $^{56}\mathrm{Ni}$ production in detonation models, whether they involve rapid rotation or off-center ignition. Thus we are still left with the question of how the ignition evolves to an explosion.

11.5 HE DETONATION

We have explored the idea that some supernova explosions involve carbon ignition under conditions of extreme electron degeneracy, with the burning initiated by the approach to the Chandrasekhar limiting mass.

Are there variations of these concepts which avoid some of the problems encountered above?

What options are available for the fuel to be burned? Hoyle and Fowler [305] concluded that the light nuclei, especially ^{12}C and ^{16}O, were the most likely explosive agents. As they pointed out, hydrogen burning is weakly explosive at best, because of the slowness of the necessary weak interactions. Even in the case of CNO catalyzed burning, the cycle time is about 300 seconds. This is longer than the hydrodynamic time-scale for a stellar core, even for a star on the main sequence, and much longer still for more evolved stars. The energy release is limited by that from (p, γ) reactions on the catalyst nuclei (which are used up after one or two captures), and is too small to power a supernova explosion, although it is important for novae [587]. Heavier nuclei than ^{12}C require more extreme conditions for ignition, and have correspondingly greater difficulties with problems of neutron excess due to electron capture. What about ^4He as a fuel? Hoyle and Fowler originally excluded this possibility because the intermediate nucleus, ^8Be, would never be very abundant. However, the low abundance is compensated for by a large reaction rate (see [236] and §8.1), so that ^4He is an interesting possibility for an explosive fuel. It has some advantages. First, the extra energy released by converting the ^4He to the fuels for the next stage, ^{12}C and ^{16}O, roughly doubles the net energy release from complete combustion. This makes a detonation more likely, giving a more robust type of explosion. Second, the ignition temperature is lower than for ^{12}C burning. This allows burning to commence at lower densities, which reduces the problem of an excessive amount of electron capture.

Where and how might the explosive ignition of ^4He occur? As was seen in chapter 8, helium burning in single stars seems to be placid. Only for stellar masses below about $2.25 M_\odot$ does the helium flash occur, and that is thought to be comparatively mild. Above this mass, helium burning makes a core composed of ^{12}C and ^{16}O. A better possibility is a composite system, in which accretion from a binary companion adds matter to a white dwarf, building a degenerate layer of helium around a CO core [461, 605, 328]. This conception is vastly more difficult to evaluate quantitatively. First, there are many more parameters necessary to specify the binary system and its evolution. Second, there are unresolved points of principle about that evolution. We do not yet have a real understanding of the hydrodynamics of accretion and mass loss from stars, despite significant progress. The change in angular momentum is crucial for the evolution of a binary; this is directly related to the way mass is lost or gained by the system. In the much simpler case of mass loss from a single star, the mechanism of wind formation, and its MHD computation in a full three-dimensional geometry with heating and cooling processes, is formidable. Thus we are reduced to scenario building, in hopes that despite our ignorance, we may still capture some of the basic features of the phenomena. The best justification of scenario building may be that it allows the construction of more incisive observational programs.

If matter is added to the white dwarf so slowly that it cools effectively, a degenerate layer is built up. If the companion is a helium star [328], the accreted matter is already mostly ^{4}He. If the companion contributes hydrogen, then this must be burned to helium without too much mass ejection, so that the layer actually grows. Ignition, as we have seen, depends upon a complex balance of heating and cooling, which determines that path to ignition temperature. Thus, in addition to the uncertainties related to binary evolution, the evolution will also be sensitive to details of this cooling and heating. For example, helium ignition may occur not by the triple-alpha reaction, but by

$$^{14}N(e^-, \nu)^{14}C(\alpha, \gamma)^{18}O. \tag{11.44}$$

Hashimoto et al. [285] show ignition curves for this reaction which are analogous to those above for carbon ignition. For accretion rates in the range $\dot{M} = 10^{-8}$, 10^{-9}, and 10^{-10} $M_\odot$ y^{-1}, ignition occurs at about $\rho = 10^6$, 2×10^7, and 2×10^8 g cm^{-3}.

The helium layer can ignite before the CO core [605], in which case ignition probably occurs at a point in the helium shell. This gives symmetry about a line through the center of the CO core and the ignition point. This two-dimensional geometry is an essential feature [401], a fact that cannot be dealt with in one-dimensional simulations of such an event (see [700] and references therein).

This ignition is followed by development of a detonation wave in the helium layer. Livne [398] pointed out that the detonation drove a compressional wave into the CO core which could cause a *carbon* detonation there. If the mass of the degenerate regions—He shell and CO core—is not large, the central density may easily be low enough to avoid excessive electron capture. For example, we found $\rho_c \leq 2 \times 10^8$ g cm^{-3} to allow an outer layer of intermediate-mass elements and to provide little electron capture; this corresponds to a mass of about $1.2 M_\odot$, or less (see figure 6.4). On the other hand, the densities must be large enough to avoid the production of too little Ni relative to Si and Ca. In addition, observations [122] suggest that in supernovae of Type Ia, the velocity of silicon-rich matter is $v_{Si} > 9 \times 10^3$ km s^{-1}. This may be difficult[14] to obtain if the Si is not in a relatively thin shell [53], so that much of the core must burn past Si to Ni.

Recently [400], the hydrodynamics of detonations of white dwarfs, having a carbon-oxygen core and a helium envelope, have been examined in a two-dimensional geometry. Various combinations of the masses of the CO core and the overlying He layer were examined. Figure 11.9 shows such a simulation during the early development of the detonation in the He layer. Model 4 from Livne and Arnett [400] is shown; it had a CO core of $0.7M_\odot$ with a He layer of $0.17M_\odot$. Notice that the front in the He moves faster than the pressure wave in the CO core, so that the hot region wraps around the core. This eventually results in the converging wave on the opposite side of the core which was discussed by Livne [398].

Nuclear burning was directly computed with an alpha-network from ^{12}C to ^{56}Ni as a part of the hydrodynamic calculation. The nucleosynthesis yield was similar to that needed for Type Ia supernovae to explain the variation in alpha-elements relative to iron in Population II stars (see chapter 14). Figure 11.10 shows the resulting yields, as a function of mass coordinate (upper panel), and of expansion velocity (lower panel). The two-dimensional simulations were taken well beyond the time burning had ceased, and spherical symmetry was being approached. Then they were projected into a spherical, one-dimensional geometry and continued until the expansion converted all the internal energy into kinetic energy of expansion. Significant amounts of radioactive ^{44}Ti were produced. In the outer parts of the helium layer, incomplete burning to ^{56}Ni means an alpha-rich freezeout. In the CO core, the lower density allows a higher entropy, enough for alpha-rich freezeout to occur, at least for white dwarfs of such small total mass. Sandwiched between these regions are two layers. The overlying one, just inside the original He layer, results from carbon burning which is not followed by oxygen burning; the major ashes are ^{16}O and ^{24}Mg. Below this, both carbon and oxygen burning occur, but not silicon burning; the ashes are Si, S, Ar, and Ca, with increasing amounts of ^{56}Ni. These yields are similar to those predicted from one-dimensional computations, despite the very nonspherical hydrodynamic behavior. This curious behavior results from the fact that the detonation is sensitive to density and composition, and the fact that

---

[14]The observed lines are SiII (singly ionized silicon). If there were Si at lower velocity, but it was in a different ionization state, then the lack of SiII lines at low velocity would not be a problem. Reliable quantitative models of the ionization and excitation in such events are needed to remove this ambiguity in interpretation.

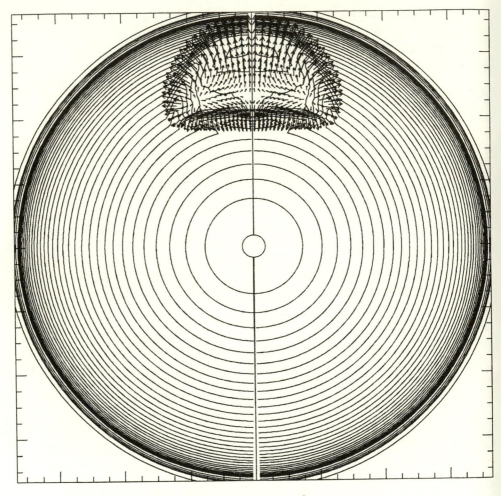

Fig. 11.9. He Detonation

the initial structure is changed little for the *unburned* matter prior to the arrival of the front.

Light curves from these models will be compared to observations of Type Ia supernovae in chapter 13. What had been a perplexing problem, the observed variation in peak absolute luminosity and in postmaximum decline rate [495, 419, 279], is reproduced in a natural way.

A rough comparison of predicted velocities with the well-observed [678] Type Ia, SN1989B in NGC 3627, of CaII H&K, MgII λ4481 and SiII λ6355 is striking. Models of this sort have the interesting property

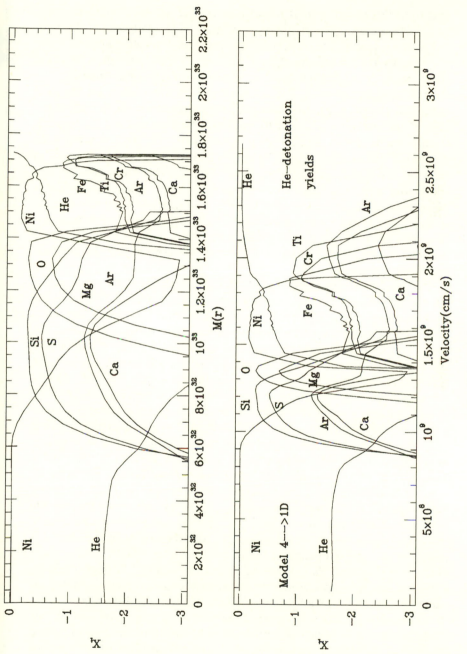

Fig. 11.10. He Detonation Yields

that their nucleosynthesis is robustly connected to their expansion rate, so that their absolute brightness is related to the velocity of the various burning layers. The maximum brightness is determined by the mass, the velocity of expansion, and the amount of ^{56}Ni produced. Because the burning is by detonation, it is sensitive only to initial density. For electron-degenerate stars, the initial density is determined by the total mass. The density for the He detonation depends upon the density at the base of the helium layer prior to ignition, which should not vary greatly from event to event. The He detonation ejects the He layer, but does not affect the density in the CO core significantly. Thus the density in the CO core is also near its initial value. The density determines how hot the matter gets, and this maximum temperature is a fine discriminator of burning stages. The degree of burning determines the energy release, and sets the velocity scale. Finally, the homology of expansion connects this velocity scale with the edges of the burning zones. Thus, in principle, by reading the velocities of the layers, we are reading the absolute luminosity of these events.

This is shown explicitly in figure 11.11, in which the minimum velocities of Mg, Si, and Ca are plotted for a variety of sub-Chandrasekhar masses, versus the maximum luminosity. The solid horizontal lines are a crude estimate of the corresponding minimum velocity taken from the MgII, SiII, and CaII features observed in SN1989B [678]. All three features imply a similar maximum luminosity for this event, which is an interesting consistency check. For all but the most massive white dwarfs, these velocities increase rapidly with luminosity at maximum light.

Detonations in sub-Chandrasekhar white dwarfs seem to be a promising explanation for most Type Ia supernovae. They appear to have a plausible set of astronomical progenitors, and occur at the appropriate rate [328, 514, 449, 346]. They provide a range of diversity in light curves that agrees well with that observed. For the well-observed Type Ia, SN1989B in NGC 3627, the correspondance of observed and predicted velocities of Ca, Mg, and Si is striking, and seems promising for spectral synthesis. Such events could not only provide the excess Fe needed to explain the observed abundance changes with metallicity in Population II stars but also the old puzzle of the paucity, relative to solar system abundances, of ^{44}Ti and ^{48}Cr produced in massive stars. Although their explosions are dramatically nonspherical, they evolve toward homologous spherical expansion, with different layers characterized by specific compositions and velocities. Because of the relative simplicity of the burning, and the ease with which signatures of the variations may be identified in their spectra, these supernova models offer considerable promise for the problem of extragalactic distance determination.

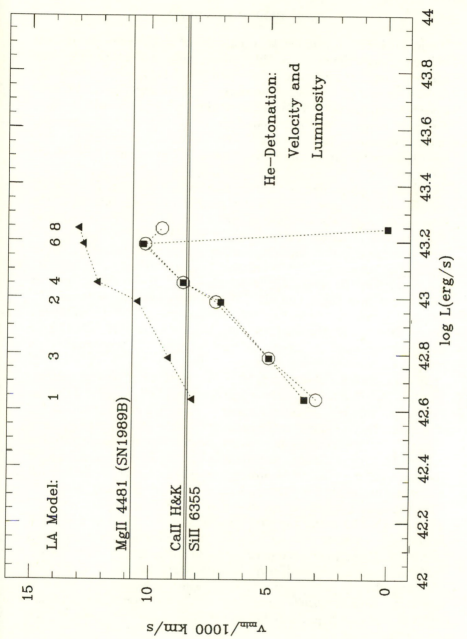

Fig. 11.11. Velocity and Luminosity

11.6 PAIR INSTABILITY

It is ironic that the *pair-instability* mechanism of stellar explosion, which is in many respects the most plausible and well analyzed, may be rare in occurence. The discussion here follows Bond et al. [114], who defined *Very Massive Objects* (VMOs) as those stars which are unstable during their oxygen-burning stage due to the formation of electron-positron pairs. The mass range of VMOs extends from $\approx 10^2$ to $\approx 10^5 M_\odot$, depending upon composition. There seem to be few at the present epoch, but they have been proposed as candidates for the first generation (Population III) stars.

The upper mass limit represents a boundary between the pair instability and the general relativistic instability. Above this mass, the designation *Super-Massive Objects*, or SMOs, is used. Zel'dovich and Novikov [708] estimated this boundary at $M = 6.4 \times 10^4 M_\odot$ if no nuclear burning were effective. Including nuclear effects, the boundary moves up to $M = 5.5 \times 10^5 M_\odot$ for Population I and II, and $M = 1.7 \times 10^5 M_\odot$ for Population III; see [114]. Fricke [245, 246] found slightly smaller values, and that Population III SMOs always form black holes. For helium burning, the boundary above which the relativistic instability wins is $M \approx 4 \times 10^4 M_\odot$; see [245, 114]. More massive cores will already be collapsing when they begin oxygen burning, and are expected to become black holes.

Massive stars are observed to be uncommon, more so with increasing mass (see chapter 14). Larson and Starrfield [386] suggested that an upper limit for stars arises as a result of processes which occur during formation. They assume that the collapse of gas leads to a strong central condensation; this embryo star accretes matter from the protostellar envelope. The time-scale for accretion may be estimated by assuming that the envelope of a given mass protostar begins collapsing at the free-fall rate when its density has risen to the point at which it is just marginally Jeans unstable (roughly, when its self-gravity exceeds its pressure gradient force). The cooler the cloud, the longer the accretion time for a region of given mass, because a smaller density is needed for the Jeans instability. These arguments make plausible the paucity of massive stars, but may be circumvented. The problem of star formation is more complex than this simple picture (see §7.1), which is heavily influenced by experience with spherically symmetric calculations. For example, clumping, rotation, and magnetic fields cannot be properly considered in such an approximation.

The observational situation is made still more difficult by the facts that (1) massive stars have high rates of mass loss, so that interpretation of their atmospheric conditions is complicated, (2) they tend to be in regions of active star formation, and hence of great physical complexity, and (3) it is difficult to distinguish between a close group of single massive stars and one very massive one. Discussion of stellar objects of $M \approx 100 M_\odot$ is not considered outlandish, but at $M \approx 200 M_\odot$ the critical observer will probably demand evidence. Given the uncertain observational and theoretical situation, it is prudent to consider what very massive stars might be like, and how they might evolve.

Upon ignition of hydrogen, VMOs become pulsationally unstable [389] if their mass is above

$$M_p \approx (108 + 20X_\alpha)(1 - 5X_\alpha/8)^2 M_\odot; \qquad (11.45)$$

this expression is a fit to Population II calculations [593]. If primordial helium is $X_\alpha \approx 0.23$, then $M_p \approx 83 M_\odot$. Higher metallicities give higher pulsation masses. These values are larger than the often quoted value of $60 M_\odot$ found by Schwarzschild and Härm [557] for models with only electron-scattering opacity. Linear stability analysis predicts that the amplitude of the fundamental radial pulsation grows with an e-folding time of about one-thousand years. Although the development of an abundance gradient acts to increase this time, if $M \gg M_p$, there are so many e-folding times that the surface velocities would quickly exceed the escape velocity (in the linear case). Large mass loss would drive M below M_p on time-scales small compared to the nuclear burning time, as Schwarzschild and Härm [557] suggested. As stars seem to exist which have $M > M_p$, they suggested that the inclusion of nonlinear terms would aid in damping, and giving a finite amplitude of pulsation. This was borne out by numerical computations [15, 606, 712, 484]. The actual mass loss rate is difficult to compute, because (1) the pulsational periods are short ($\approx \frac{1}{2}$ day) compared to the e-folding times, and (2) the effects of radiative transfer in the outer layers are complex.

Even if VMOs are found to be rare today, they are of considerable theoretical interest. There is also the suspicion (worry?) that they may have been more common in the past. After decoupling, the cosmic background would have been hotter than now, and if this translated into a higher thermal pressure in the gas, larger masses would be required for gravitational contraction. The distribution in mass of newly formed stars (the *initial mass function*; see chapter 14) might be biased in favor of larger masses. Testing this possibility requires a quantitative knowledge of how VMOs might behave.

Operation of the CNO cycle requires the presence of the catalyst nuclei; these were are not made in significant amounts in the Big Bang.

Stars generate their own. The weakness of the pp-chains allows the stars to evolve to higher central temperatures, until the 3α reaction produces ^{12}C. This continues until the abundance of catalysts is high enough for the CNO cycle to provide the stellar luminosity; as this occurs, a further increase in the catalysts allows the luminosity to be produced at lower central temperatures, decreasing the 3α rate. This occurs at a nucleon fraction $Y_{CNO} \approx 10^{-9}$; see [114] for a simple quantitative description.

From chapter 7, we see that for a hydrogen-rich VMO of mass $M > 150 M_\odot$, we have $\beta < 0.5$, so that the fractional size of the convective core is $q > 0.89$. These objects may convert most of their hydrogen to helium, making them energetic, and possible sources of noncosmological helium production. The stellar layers which have processed helium mixed in with their cosmological allotment could be lost by a stellar wind, perhaps driven by nuclear-energized pulsations [609]. If the mass loss is too slow, the star will proceed to burn the helium to heavier nuclei. If the mass loss is faster than the nuclear burning rate, the ejected matter will reflect the abundance of the initial, unprocessed matter. The maximum helium production occurs if the time-scale for steady mass loss is comparable to that for burning hydrogen. Similarly, for episodic mass loss, the event should occur after the helium is made, but before it is destroyed. This would correspond to times between the later stages of hydrogen burning and the onset of helium burning.

Suppose the VMO manages to rid itself of hydrogen. Bare helium cores are also pulsationally unstable if their mass exceeds

$$M_{\alpha p} \approx (13 + 120z)(Y_T/0.75)^2 M_\odot. \tag{11.46}$$

This is a fit to the numerical results of Stothers and Simon [593], which holds for metallicity z from zero to Population I values. A pulsating core might drive mass loss, and it might build an extended envelope which would damp the pulsations. Convective cores tend to be large in massive helium stars ($q > 0.9$), so that mass loss in this phase might soon reflect the products of incomplete helium burning in such stars (^{12}C, ^{16}O, and some ^{20}Ne).

Suppose that the pulsational instability is associated with extensive mass loss. Then, because the helium cores are more unstable than hydrogen stars of similar mass, it is possible that initially more massive stars suffer so much mass loss that they end life as relatively low-mass objects, with masses well below the initial mass of stars which suffer little mass loss. To be explicit, a $70 M_\odot$ star might lose its hydrogen, become a Wolf-Rayet star, have extensive mass loss during its helium-burning phase, and die with less than $10 M_\odot$ left; see [383].

11.7 OXYGEN BURNING AND BEYOND

If the helium cores manage to finish helium burning, the subsequent evolution will be so rapid that further mass loss will not be relevant. After helium burning the core is mostly ^{16}O, with some ^{12}C and ^{20}Ne. The low ^{12}C abundance and the small energy release from burning ^{20}Ne insure that the burning of these fuels will have only minor effects. The stars are essentially oxygen cores. For example, a $64M_\odot$ helium star produced a $62.3M_\odot$ core of 83% ^{16}O, 10% ^{12}C, and 7% ^{20}Ne for a $^{12}C(\alpha, \gamma)^{16}O$ rate which was 0.7 of the CFHZ [155] value. This model explodes in a manner similar to that of the $60M_\odot$ oxygen star described below (and is probably very close to a $57M_\odot$ oxygen core). It is useful to consider the simpler question of the evolution of oxygen cores, remembering that a more realistic situation will involve some carbon and neon, which will have minor effects on the evolution but significant yields of these elements and their ashes.

The effect of electron-positron pairs upon the equation of state is shown in figure 11.12, in which the adiabatic exponent, $\Gamma_1 = (\partial \ln P / \partial \ln \rho)_S$, is plotted versus temperature, for radiation entropy $S_\gamma / \Re = 0.1$, 1, and 10. The trajectory for constant total entropy is more complicated but similar to this, as is the true trajectory, which includes nonadiabatic effects. For low entropy (low mass), Γ_1 does not drop below $4/3$, and the core does not lose stability, until $\log T > 9.5$. This is above the temperature at which oxygen burning occurs on a dynamic time-scale, so the matter will be made of more complex nuclei. The figure was constructed assuming the matter was ^{56}Ni, and the decrease in Γ_1 is due to dissociation of ^{56}Ni into alphas, and then a second dip is due to the dissociation of alphas into nucleons. As entropy increases to $S_\gamma = 1.0$ (larger mass), the effect of the pairs is more pronounced and causes a loss of stability at $\log T \approx 1$. For a temperature $T_9 \approx 3$, Γ_1 rises to $4/3$, but stability is lost again at still higher temperature. At an entropy of $S_\gamma = 10$, the dip due to pairs is pronounced. Sufficiently large masses will overshoot the oxygen-burning temperatures and collapse due to nuclear dissociation.

The original numerical experiments concentrated on the lower mass range, since it might be expected that a few of these might be exploding in the present epoch. Fraley [240] evolved cores of 45, 52, and 60 $M_\odot$. The first did not explode but oscillated, slowly consuming oxygen and expelling only a few solar masses; the implication was that a black hole would be the final state. The other two stars exploded, with no remnant. Barkat, Rakavy, and Sack [71] evolved 30 and 40 $M_\odot$ oxygen cores. The smaller did relaxation oscillations driven by burning, and did not ex-

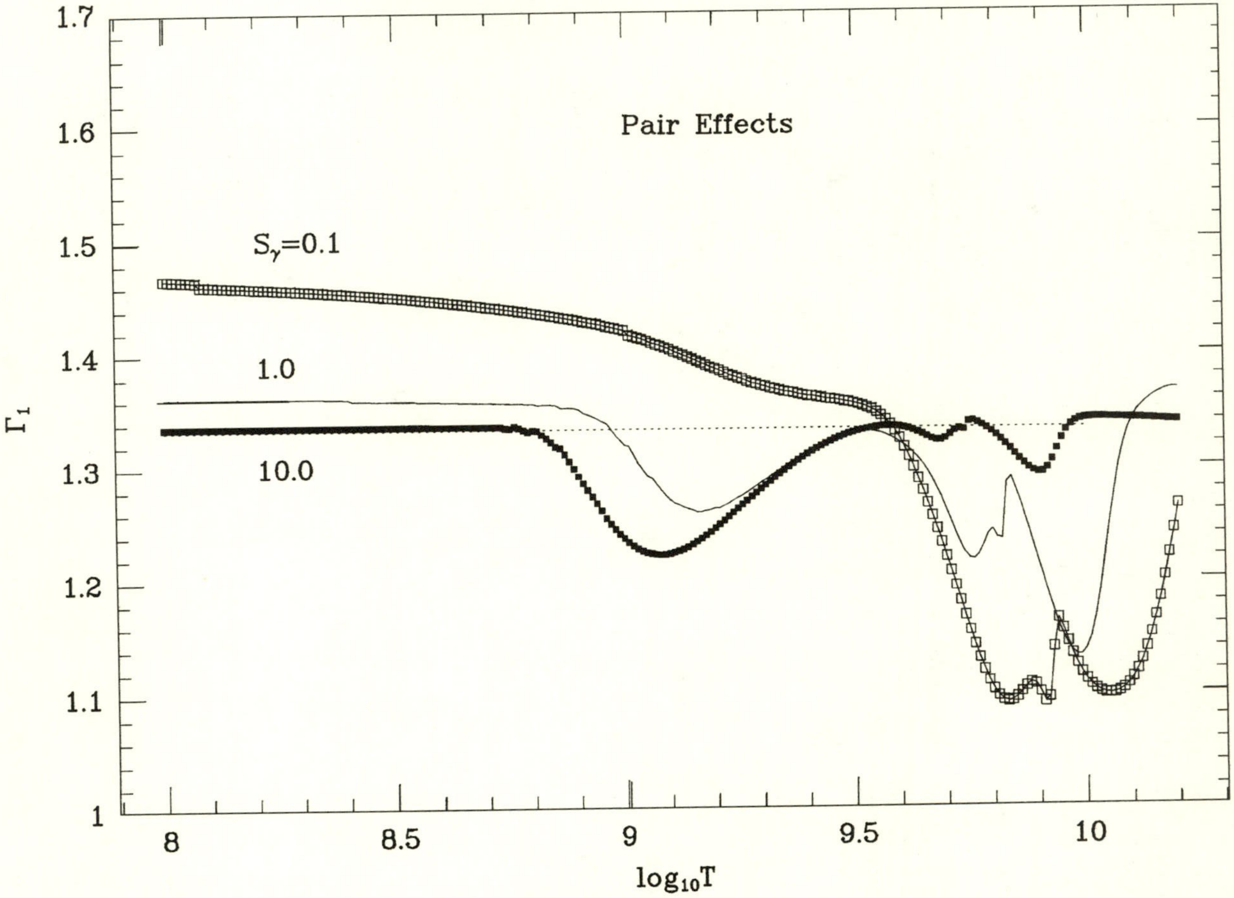

Fig. 11.12. Pair Effects on Γ_1

Pair Effects

$S_\gamma = 0.1$

1.0

10.0

Γ_1

$\log_{10} T$

plode; the larger did explode. Arnett [29] evolved 64 and 100 $M_\odot$ helium cores (corresponding to 58 and 93 $M_\odot$ oxygen cores). The smaller exploded, leaving a 2.2 $M_\odot$ silicon remnant whose fate was not determined. The larger exploded, leaving no remnant. Wheeler [679] evolved 10^3 and $10^4 M_\odot$ oxygen cores; both cases collapsed with no mass ejection. The experiments agree upon general trends: low masses oscillate without shedding much, if any, mass; intermediate masses completely disrupt, creating very energetic supernova explosions; high masses form black holes without ejection of matter. This, taken with more recent work by Woosley and Weaver [698], and by Ober, El Eid, and Fricke [465, 466, 220], provides a body of numerical results, which is well explained by analytic work [113, 114].

Figure 11.13 is a snapshot of the structure, in density and temperature, of oxygen stars of 40 and $50 M_\odot$. The region in which the adiabatic exponent Γ_1 is less than 4/3 is enclosed by a line with diamonds. If enough of the mass of the star lies inside this region, it becomes unstable. Both the oxygen stars of 40 and $50 M_\odot$ do become unstable, and contract to sufficiently high density and temperature to ignite oxygen burning. Notice that the curves of structure bend at the top to be more nearly horizontal; this is the result of neutrino cooling. The structure, and the evolution, is not really adiabatic. This curved structure implies that the central entropy is smaller, and the entropy in the outer regions larger, than *average*. The analytic work [113, 114] uses the approximation of polytropic structure. This means that, because the inner values are more important for the equation of state and for the fuel consumption, the entropy of the polytropic structure corresponds to a lower mass than the more realistic model. In the more realistic model, the outer regions have a higher entropy; they can support more mass. Therefore the masses obtained with the analytic approach will be systematically lower than with the numerical results. When this shift is accounted for, the results should be correct.

What is the magnitude of the shift in mass? The analytic models give little help for masses that are near the minimum which explode, and the numerical results show why. The $40 M_\odot$ model undergoes a collapse, and an oxygen flash. This drives an expansion (central density decreases by a factor of ten) which does not give significant mass ejection. The star then undergoes oscillations driven by oxygen burning. They appear to damp; if this is true, such a star may evolve onward to core collapse.

This is a nasty problem. Because the average adiabatic exponent is near 4/3, there is almost no "restoring force," so small perturbations give rise to large amplitude motions. Also, pulsational periods are much longer than the sound travel times. A computation limited to steps less

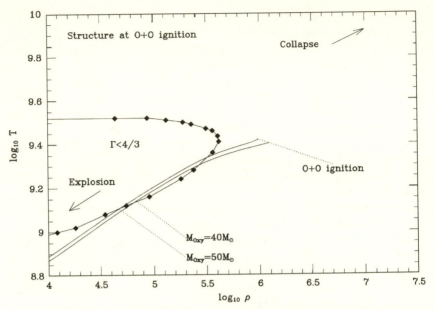

Fig. 11.13. $M_{Oxy} = 40$ and $50M_\odot$ at O+O ignition

than the sound travel time over a zone requires many steps to evolve to interesting times; many steps may allow errors to build. At first it might appear that the solution would be to use an *implicit* method, in which the solution involves undetermined values at the new time; this allows time steps larger than the sound travel time. However, if the time steps exceed the sound travel time, the price for the larger time step is *damping. Implicit calculations underestimate the possibility of instability and explosion.* It is very difficult to decide how the marginal masses actually behave. The $50M_\odot$ oxygen star is above this difficult region; it clearly ignites sufficient oxygen to blow itself apart.

What about the upper boundary in mass, above which collapse continues? The behavior for $M_{Oxy} = 120$ and $150M_\odot$ is shown in figure 11.14, in terms of the trajectories of central density and temperature with time. At the onset of instability (where global Γ_1 falls below 4/3), the core conditions for 120 and $150M_\odot$ lie inside the boundary for local $\Gamma_1 = 4/3$. This is necessary if the pressure average of Γ_1 is to be less than 4/3.

The $120M_\odot$ oxygen star contracts to oxygen-burning temperatures ($T > 3 \times 10^9$ K); this is apparent from the "wiggle" in the trajectory, which corresponds with an increase in entropy. It continues through silicon burning (with little change in entropy), and up to a temperature

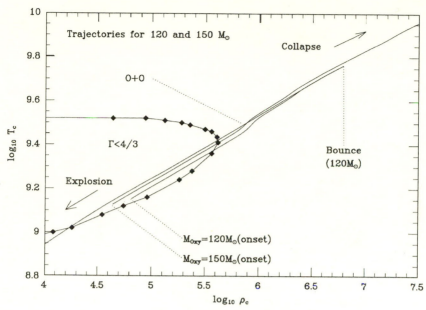

Fig. 11.14. T_c and ρ_c for 120 and $150M_\odot$

of $T_c = 5.76 \times 10^9$ K. At this point contraction ceases and bounce occurs. This gave a vigorous explosion, with no remnant left.

For $150M_\odot$, the contraction phase is almost the same, but it extends too far into the nuclear dissociation region for bounce to occur, and collapse ensues. The boundary mass M_{Ocrit} above which collapse occurs lies in the range $120 < M_{Ocrit}/M_\odot < 150$. The analytic estimate is $100M_\odot$ so that the mass shift is a factor of about $4/3$.

The yields from oxygen stars of 40 to $150M_\odot$ are given in table 11.6. The mass of ^{16}O ejected, $\Delta M(^{16}O)$, is almost constant from core masses of 50 to $120M_\odot$. The mass of the ashes of oxygen burning which is ejected rises rapidly with mass, and then flattens. The yield of ^{56}Ni is negligible for the lower masses, but rises to almost $40M_\odot$ for the $120M_\odot$ core. These larger cores would give exceptional supernovae. For example, the bolometric light curve of the $100M_\odot$ core had a peak at $\log_{10}(L/L_\odot) \approx 10.6$ which lasted about one hundred days, with a nominal temperature of 10^4 K. It would be interesting to see what observational constraints could be placed upon such bright, rare events, especially at large redshift. The explosion energies rise to almost 10^{53} erg for $M_{core} = 120M_\odot$. The central temperatures at bounce rise with increasing

TABLE 11.6

VMO Yields

Mass (Core)	ΔM (^{16}O)	ΔM (Si-Ca)	ΔM (^{56}Ni)	Mass (rem)	E_{explos} (10^{51} erg)	$T_c/10^9 K$ (bounce)
32						
40				40(?)	0	2.83
50	45.3	4.7	2×10^{-5}	0	2.2	>3.1
60	46.7	13.3	0.086	0	8.1	3.6
80	45.6	30.1	4.3	0	28.2	4.36
100	49	37.3	13.7	0	48.7	<4.7
120	44.4	37.1	38.5	0	90	5.76
150				150(BH)		

core mass; the limits in this column are the result of coarse sampling in these two computations, so that the precise values were not saved.

The cores of $M_{core} \leq 40 M_\odot$ did not explode by this mechanism. These cores may be so large that the mechanism by which supernova SN1987A exploded (presumably related to neutron star formation; see chapter 12) is not effective [17]. If this is true, then these objects would collapse to form black holes of mass $M_{bh} \approx M_{core}$, with little ejection of matter. For $M_{core} > 120 M_\odot$, the pair-instability mechanism does not work, and collapse occurs. Here black hole formation would also be expected, with $M_{bh} \approx M_{core}$.

12

Gravitational Collapse

The brilliant Cerebron, attacking the problem
analytically, discovered three distinct kinds of
dragon: the mythical, the chimerical, and the purely
hypothetical. They were all, one might say, nonexistent,
but each nonexisted in an entirely different way.
—Stanislaw Lem (1921–), *The Cyberiad*

One of the most intriguing of scientific puzzles of our time is gravitational collapse. A facet of this problem—the question of how a star collapses and gives rise (at least in many cases) to a supernova—has been one of the most frustrating to the discerning astrophysicist. For three decades there has been a progression of ideas which were each proclaimed to be *the* solution to the problem. The failure rate may have equaled this impressive rate of production.

Yet there has been great progress and spectacular confirmation of some of the basic aspects of the theoretical development. The detection of neutrinos from SN1987A was incredible luck. The neutrinos and photons from the supernova arrived at a time preceded by intensive research on the problem, so that there was a body of ideas to test. Because of previous interest by particle physicists in proton decay and grand unification theories, neutrino detectors of sufficient size and sensitivity had been built, and were in a mature state. These detectors happened to be on line and working during the crucial twelve seconds. Even so, the Large Magellanic Cloud was barely close enough for success!

The problem abounds with the extreme and the exotic. Neutrinos, which eluded direct experimental detection for decades because of the extreme weakness of their interaction, are trapped in the collapse. The collapse is so fast that a mass, more than half that of the sun, shrinks in upon itself at velocities approaching the speed of light. Einstein's general relativity theory becomes necessary to describe the event, not merely a subtle correction to the gravitational theory of Newton. Perhaps nowhere else in the Universe does it matter so much whether neutrinos obey Pauli's exclusion principle.

The reaction to a discussion of the gravitational collapse of stellar cores is itself remarkably bimodal. For those with a detailed understanding of nuclear and particle physics, the response can be a kind of gleeful exhilaration at seeing familiar concepts applied successfully to an amazingly exotic problem. For others, numbed by the soporific weight of unfamiliar detail, the response is boredom. The following exposition is an attempt to spread the excitement more fairly. Pundits must pardon the intentional suppression of detail—hopefully the attempt to focus strongly on key issues will make some amends.

12.1 HISTORICAL OVERVIEW

With the discovery of a maximum possible mass for white dwarfs by Chandrasekhar [161], it became clear that a star of larger mass could not simply pass to the white dwarf stage, and one was "left speculating on other possibilities." The next clear reference is to Baade and Zwicky [63], who proposed both the idea of neutron stars, and that these objects were produced in supernova explosions. Oppenheimer and Volkoff [470] first calculated neutron star models, assuming a dense gas of free neutrons. Oppenheimer and Snyder [469] first examined the continued collapse of larger mass objects to black holes.[1] Gamow and Schönberg [259] identified the importance of neutrino emission and nuclear weak interactions for supernovae and core collapse. Hoyle [302] identified the key role of nuclear statistical equilibrium, and examined in some detail the possibility of the formation of the elements in supernovae from matter ejected by the collapsing core. Cameron [142] constructed neutron star models with an equation of state which included the effects of the nucleon-nucleon interaction, and found a maximum neutron star mass which was larger than the Chandrasekhar limit. With the success of the conserved-vector-current theory of the weak interactions [233], Fowler and Hoyle [238] and Chiu [170] explored the enhanced cooling implied for massive stars, and its implications for collapse. This was followed by numerical experiments [506, 22] which showed that such cooling changed the structure of the star during these stages, giving a convergence toward a small, partially degenerate core.

Colgate et al. [183, 184] suggested that the transfer of energy, from the gravitational potential energy released by the imploding core to its exploding mantle, took place by emission and deposition of neutrinos,

---

[1]It is striking in retrospect that this work by Oppenheimer and his students was so far ahead of its observational confirmation, that its full import could not be properly recognized in Oppenheimer's lifetime.

and presented one of the early published examples of numerical hydro-dynamics. To understand this and subsequent work, it is useful to define a "neutrinosphere," in analogy to a photosphere. This is a surface at a depth of one mean free path for a typical neutrino[2] $\tau_\nu \approx 1$. Below this surface, the neutrinos may be thought of as diffusing; we will call a description of neutrino transport in this region the "inner problem," for which $\tau_\nu \geq 1$. Above this surface, the neutrinos stream toward in-finity, many escaping with no further interaction. The primary process is attenuation, with some reemission. The flow of these neutrinos is the "outer problem," for which $\tau_\nu \leq 1$. The history of the subject is littered with theories that failed because one or the other of these aspects were ignored.

Colgate and White [184] argued that previous work had ignored "the dynamic effect of the rarefaction wave created by the implo-sion itself." Their collapses were violent; in the range of density $10^{10} \leq \rho \leq 10^{11}$ g cm^{-3}, their pressure was only 0.25 of the value needed for hydrostatic equilibrium. Such strong collapses gave small cores that contracted quickly, more or less as a unit,[3] to supernuclear density. The overlying matter crashed down, setting up a strong shock wave; this was beyond the capabilities of the numerical resolution they used, so that they imposed an approximation. It was assumed that in the thick regime of the "inner problem," neutrino diffusion was fast enough to supply flux for the "outer problem" on the hydrodynamic timescale. They simply took half the thermal energy in the core and put it in the surrounding mantle. Of course this exploded violently. With this algorithm, they obtained violent explosion for all core masses they discussed, 1.5, 2.0, and 10.0$M_\odot$. No black holes were formed, just neutron stars of 0.87, 0.98, and 1.8$M_\odot$. The binding energy of such cores is about 0.1 of their rest mass energy, so the explosion energies were enormous, from 80 to 160 foes[4] (10^{51} erg). These values are a factor of 50 to 100 times larger than the explosion energies of actual supernovae (see chapter 13).

The vigor of the collapse was overestimated. The Colgate-White equa-tion of state was simply incorrect [19] above densities of $\rho \approx 10^{10}$ g cm^{-3}; the error was in the sense of a deficit in pressure, which exaggerated the violence of the collapse (and the vigor of the rarefaction). This was ex-acerbated by an overestimate of the rates of electron capture, due to

---

[2]A note of caution: this surface is not sharp, because the neutrino mean free path varies with neutrino energy. The thickness of the neutrinosphere may be significant in comparison to its radius. Therefore the concept of a typical neutrino should be used with some caution.

[3]Such regions are now called "homologous cores."

[4]We will use Gerry Brown's convenient and apt energy unit, the "foe," which is ten to the *fifty one ergs*.

several causes: an overestimate of the free proton density, a neglect of the threshold for electron capture, and a poor representation of neutrino diffusion. They found that more than 85 percent of the neutrinos in the homologous core escaped before bounce. They also concluded that muon neutrinos were not important because the time for formation was too long. As we shall see, these conclusions were not correct.

Although flawed, the Colgate work was both imaginative and influential, and did inspire a serious effort by many theorists to establish the physics of stellar core collapse. An explicit calculation of neutrino diffusion [16, 17] gave attention to the inner problem. Even without knowledge of neutral current interactions, the neutrinos diffused out so slowly that the catastrophic cooling of Colgate and White did not occur; the heat from compression was trapped for seconds—as SN1987A indicated [47], not lost in microseconds. The other types (flavor) of neutrino then known—the muon neutrino and antineutrino—were found to be important coolants. Larger masses, which formed denser cores, were more opaque to neutrinos, and collapsed to black holes.

Numerical techniques were improved as well. Schwartz [554] and May and White [417] added general relativity to the dynamics. Then, in a burst of improvement, Wilson [684, 387, 685] added integration over frequency group and over angle to the treatment of neutrino transport, rotation, and magnetic fields with the first two-dimensional simulations, and the transport of the tau neutrino family as well (see [121]). Mazurek [422] and Sato [541] argued that not only would the thermal energy be trapped, but lepton number as well. A realistic initial model, evolved from well-explored stages of stellar life, was followed directly into collapse [34]. Careful attention to the reaction kinetics of electron and neutrino emission, absorption and scattering, including neutral current processes [242, 243], showed that the loss of lepton number and energy were small during collapse, and that the mean neutrino energies were distributed in a broad peak around 8 MeV. Bethe, Brown, Applegate, and Lattimer (BBAL) [96], in a beautiful analysis, showed that this was a general result, implying an enormous simplification: entropy changes little during collapse. Lamb, Lattimer, Pethick, and Ravenhall (LLPR) [371, 372] derived a finite temperature equation of state from the compressible liquid-drop model of Baym, Bethe, and Pethick [83]. Using this, Bethe et al. [96] showed that collapse would continue until the core reached supernuclear densities.

Having sketched some of the early history, we now examine the main features of the process of core collapse. With the advantage of hindsight, the discussion can be much simpler than the actual historical development.

12.2 NEUTRONIZATION AND DISSOCIATION

The small mass of the electron—relative to that of the neutron or the proton—is of fundamental importance to evolution beyond silicon burning. The de Broglie wavelength is $\lambda = h/p \approx h/mv$. To fit an electron into the same size box as a proton (that is, for it to have the same wavelength) requires that the smaller mass is balanced by a larger velocity, and hence a larger kinetic energy. Neutrino cooling insures that in the late stages of evolution of massive stars, this energy is not thermal energy but quantum degeneracy energy. Ordinary nuclei have the luxury of storing their electrons in the much larger volume of the electron orbits rather than in the crowded nuclear region. For densities $Y_e\rho \gg 10^6 \text{g cm}^{-3}$, the electrons are relativistic, and the fermi energy is

$$\epsilon_f \approx 11.0\,\text{MeV}\,(Y_e\rho_{10})^{1/3}. \tag{12.1}$$

Such high energies make electron capture favorable, even though it produces nuclei that are unstable to beta-decay in the laboratory.

For nuclei in the iron-peak, the threshold Q for electron capture is roughly

$$Q \approx 190\,\text{MeV}\,(0.45 - Y_e). \tag{12.2}$$

Notice that this increases to large values as Y_e drops below 0.45; for Y_e above this value, this $-Q$ corresponds to the energy available for beta-decay, but the expression is not accurate for $Y_e > 0.45$.

In chapter 10, we left core silicon burning at a temperature of $T_9 \approx 4$, which for $S_\gamma/\Re = 0.2$ corresponds to a density of $\rho \approx 3.9 \times 10^7$ g cm^{-3}. However, in figure 10.2 for example, the subsequent evolution is shown, in which the density increases until the radiation entropy approaches $S_\gamma/\Re \approx 0.01$, and the evolution then follows a path of roughly constant S_γ. This is an old result, taking the trajectory into the vicinity of $T = 10^{10}$ K and $\rho = 10^{10}$ g cm^{-3}. Collapse does not actually become hydrodynamic until $\rho \approx 5 \times 10^9$ g cm^{-3} however, so that there is an extended period between the termination of core silicon burning and the onset of true collapse. During this stage several important quantities are determined: the initial entropy, the initial electron fraction Y_e, and the size of the "iron" core.

After core silicon burning, the neutron excess was $\eta = 0.0388$ (see table 10.4), or $Y_e = 0.480$. Electron capture will occur until the threshold rises to $Q = \epsilon_f$. This removes pressure, causing the core to tend to contract, increasing the tendency to capture electrons. Heating or cooling may occur, depending upon the hydrodynamic state of the matter (see

§11.3). What are typical conditions prior to collapse? As an example, we take a core which has increased its central density to $\rho \approx 2.3 \times 10^9$ g cm^{-3}. For the fermi energy to be at threshold, as is the case when electron captures continue until the threshold rises to shut them off, we may solve to obtain $Y_e \approx 0.42$ and $\epsilon_f \approx 5.0$ MeV. The actual value of Y_e is slightly higher because the rate of contraction, while slow, is not negligible. Figures 10.5 and 10.6 indicate that the entire stage, from the end of the last active stage of silicon shell burning to hydrodynamic collapse, requires about 10^4 s. The effect of back reactions, decays in this case, also inhibits the approach to equilibrium. For a proton magic number of $Z = 28$, this implies that $A \approx Z/Y_e \approx 66$, so we should expect nuclei such as ^{66}Ni to be abundant; it is a known but highly radioactive nucleus, with a beta-decay half-life of 54.8 hours.

The central temperature is $T_9 \approx 7.5$, or $kT = 0.086T_9 \approx 0.645$ MeV (see §10.2). This gives $\epsilon_f/kT = 8.5$. The pressure is essentially that of degenerate electrons. Photodissociation of nuclei begins to have a modest effect; the abundance of free nucleons is $Y_p = 1.0 \times 10^{-5}$ and $Y_n = 1.1 \times 10^{-3}$, while that of alpha particles is $X_\alpha = 4.2 \times 10^{-3}$.

As a prelude to later discussion, let us examine the entropy of this matter. The entropy per electron is

$$S_e/\Re \approx Y_e \pi^2 (kT/\epsilon_f), \tag{12.3}$$

where ϵ_f is the electron fermi energy. This gives $S_e/\Re = 0.50$; the accurate value is 0.525. The radiation entropy is $S_\gamma/\Re = 0.022$, a small value. For nuclei and nucleons, the entropy is

$$S_i/\Re = Y_i(5/2 - u_i/kT) \tag{12.4}$$

(see Appendix B), and

$$u_i/kT = \ln(Y\theta/zA^{\frac{3}{2}}), \tag{12.5}$$

where z is the partition function (just $2J + 1$ for nucleons or ground-state nuclei), and $\theta \approx 0.1013 \, \rho_9/T_9^{3/2}$. We find $S_i/\Re$ to be 0.0160, 1.94×10^{-4}, and 0.0167 for neutrons, protons, and alphas, respectively. For the nuclei, $S_i/\Re = 0.347$, giving a total entropy of $S/\Re = 0.93$. While these values are respresentative, the determination of the entropy at this epoch may be changed with more careful treatment of the effects of the Urca process and its interaction with hydrodynamics. If these effects produce no major surprises, the entropy will have a small value such as this.

Using the methods of Appendix B (see §B.6), the rate of electron capture may be written as

$$dY_e/dt = -Y_t Y_e \rho N_A \sigma_0 c I(\epsilon_f, Q)/(m_e c^2)^2, \qquad (12.6)$$

where

$$I(a, b) = 3a^2/5 - 3ab/2 + b^2 - b^5/10a^3. \qquad (12.7)$$

The constant[5] $\sigma_0 \approx 2 \times 10^{-44}$ cm^2 is defined in §3.7. Using the time elapsed between the end of core silicon burning and the start of core collapse as 10^4 s, and equating this to $(d \ln Y_e/dt)^{-1}$ gives an estimate of the degree to which the fermi energy ϵ_f approaches the threshold Q. They agree to 0.03 MeV, or better than 1 percent, but this ignores decays, which will spoil the agreement somewhat. Still, at the onset of collapse, the fermi energy is close to the threshold for electron capture. A zero temperature white dwarf with this Y_e has a mass of only $1.03M_\odot$. This is a useful measure of the effective "size of the core."

A simple model indicates how the core collapses. Consider the contraction of an initially hydrostatic core, at a rate controlled by the loss of electrons. We approximate this using the "hydrodynamic" time-scale,

$$\tau_{hd} = 446 \text{ s} / \sqrt{\rho}, \qquad (12.8)$$

and

$$d \ln \rho/dt = b(Y_e(0) - Y_e)^{2/3}/\tau_{hd}. \qquad (12.9)$$

The functional form and the constant $b \approx 5$ were obtained from study of 1D hydrodynamic simulations; b is adjusted so that the contraction to density $\rho > 10^{13}$g cm^{-3} takes about 0.8 s. By integrating $d\rho/dt$ and dY_e/dt ahead in time, we obtain an estimate of their behavior along the collapse trajectory. Figure 12.1 shows the evolution of Y_e, *assuming neutrinos escape freely at the speed of light*, as the lower dotted line in the bottom panel (the higher dotted line shows the effects of neutrino trapping on Y_e, which will be discussed below). The value of Y_e stays near 0.42 until the density approaches $\rho = 10^{10.5}$g cm^{-3}, and then slowly declines. The lower solid line denotes the mole fraction of leptons, $Y_l = Y_e + Y_\nu$; neutrinos are produced so rapidly that a significant density builds up, even though they escape at the speed of light,[6] This simple model

---

[5]The actual value to be used would be an average over an ensemble of nuclei in nuclear statistical equilibrium, and over all effective states in these nuclei [250, 254].

[6]This is not so strange as it seems at first. It may help to note that the velocity of the other lepton in the reaction—the electron—is near the speed of light as well ($0.996c$).

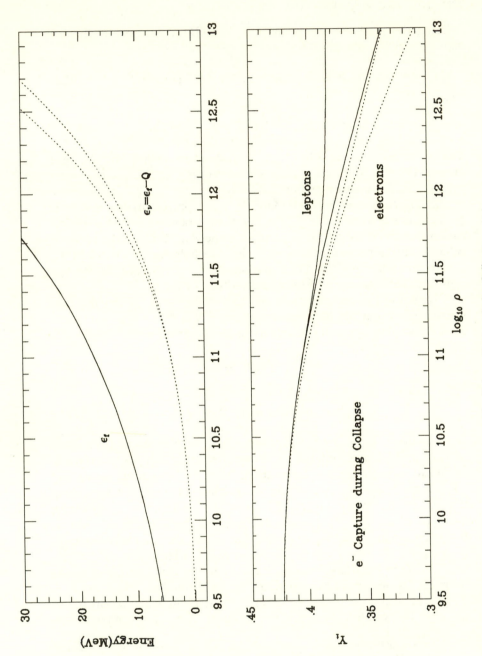

Fig. 12.1. e^- Capture during Collapse

ignores some of the feedback effects of the captures on the collapse, but is in reasonable agreement with most one-dimensional hydrodynamic simulations.

The slow change in Y_e is due to (1) the weakness of the weak inter-actions, and (2) the inhibition due to the threshold energy Q. In the upper panel, the growth of the electron fermi energy ϵ_f is shown as a solid line. The energy of the neutrino produced by the electron capture is (on average) $\epsilon_\nu = \epsilon_f - Q$; this is shown as the lower dotted line. It quickly rises, surpassing 10 MeV at $\rho = 10^{11.7} \mathrm{g}\ \mathrm{cm}^{-3}$. At $\rho = 10^{11} \mathrm{g}\ \mathrm{cm}^{-3}$, we have $Y_e \approx 0.4$. At this point our estimate of Q is still based upon known nuclei (for example, ^{50}Ca), but for higher density Q is simply an extrapolation. A high value of Q inhibits electron capture directly, but also indirectly. Nuclei with high Q quickly gobble up free protons un-der conditions of nuclear statistical equilibrium (NSE), reducing electron capture on free protons.

Early models [184] were predicated on the idea that electron pressure was quickly removed; this is clearly not the case, even if we assume that *every* neutrino escapes as fast as is possible. This mistake was partially due to the notion that the entropy was much higher than seems to be the case, so that electron capture could occur on abundant free protons (for which there is no threshold: $Q = 0$).

12.3 NEUTRINO TRAPPING

An early impediment to the understanding of core collapse was the notion that neutrinos, once made, did not interact. They could travel through "light years of lead." We review some simple estimates, of the sort that were common long before SN1987A made the subject fashion-able. First, the conventional estimate: the mean free path is $\lambda = 1/N\sigma$, where the number density of target particles is $N = YN_A\rho$, and the cross section is $\sigma = \sigma_0 \varepsilon^2$, with $\sigma_0 \approx 2 \times 10^{-44}$ cm^2 and ε is the relative energy in units of electron rest mass. Lead has a density of 11.34 g cm^{-3}, and atomic mass $A \approx 208$. Consider a neutrino energy of about 1 MeV, or $\varepsilon = 2$. Then $\lambda \approx 3.8 \times 10^{20}$ cm, or about 380 light years.

However, the situation is different in a collapsing stellar core. Suppose the average density in the newly formed core to be about $\rho \approx 4 \times 10^{14}$ g cm^{-3}, which is near nuclear density. The appropriate[7] neutrino energy may be of order 150 MeV, or $\varepsilon = 300$. The number of targets will

---

[7]For $Y_e \approx 0.3$, we have a fermi energy of 250 MeV for electrons. In weak interaction equilibrium with nucleons, an average energy in the degenerate neutrino sea of 3/5 of this is not implausible. Notice the trend in table 12.1 below.

be approximately the number of nucleons, or $Y \approx 1$. Then, $\lambda \approx 2.2$ cm. For a radius of 10 km, the diffusion time is

$$\tau_{dif} \approx 3R^2/\pi^2 \lambda c \approx 5 \text{ s,} \qquad (12.10)$$

which is slightly larger than that subsequently observed for the neutrinos emitted from SN1987A [295, 99, 4, 65]. The coefficient $3/\pi^2$ is for a uniform density sphere, but its value is insensitive to the radial distribution of $\lambda = \rho \kappa$ (Appendix D). While this estimate of the diffusion time gives the correct order of magnitude, a more careful treatment would involve the effects of neutral currents, phase space blocking by degenerate electrons, nucleons, and neutrinos, the presence of nuclei and their phase transition to nonsymmetric nuclear matter, as well as the effects of hydrodynamic transport on the loss of neutrinos for SN1987A, and so on. The hydrodynamic time-scale of such a core is roughly $\tau_{hd} \approx 446 \text{ s}/\rho^{1/2} \approx 2 \times 10^{-5}$ s, considerably shorter than the neutrino escape time. The newly collapsed stellar core is a "neutrino star," at least briefly, in which diffusing neutrinos play the role of diffusing photons in more normal stars.

Let us consider the "inner problem" of neutrino transport more carefully. The choice above for the mean free path gives a value which is numerically similar to that for neutral current scattering of nuclei, and is appropriate for the diffusion of neutrinos. The rate of change of neutrino abundance is

$$dY_\nu/dt = -(dY_e/dt) - Y_\nu/\tau_{escape}, \qquad (12.11)$$

where $\tau_{escape} \approx \tau_{dif} + R/c$. The extra term R/c limits the escape rate to that implied by the finite velocity of light, and was used to construct the results discussed in the previous section.

Figure 12.1 shows the modifications due to neutrino diffusion. In the lower panel, the values for Y_e and $Y_l = Y_e + Y_\nu$ lie above the corresponding ones for free escape. Notice that the curves follow nearly the same track until the density approaches $\rho = 10^{10.5}$g cm^{-3}, above which they diverge. The lepton abundance varies slowly above this density; although the electrons continue to be converted to neutrinos, the neutrinos do not escape but build up their own fermi sea. Because neutrinos are fermions, their behavior is in many ways similar to highly relativistic electrons, except that their helicity implies half as many per state. Thus, $Y_e \approx 2Y_\nu$ is approached for $\mu_e \gg Q$.

In the upper panel, the electron fermi energy is shown as the solid line; it is almost identical to the case of freely escaping neutrinos. However, the neutrino energy, shown by the higher dotted line, is increased because the number of neutrinos builds up and requires new neutrinos to be put into still higher energy states—the neutrino fermi sea is being filled. This discussion is based upon the assumptions of lepton conservation and no mixing of neutrino flavors [255, 256]; a fascinating aspect of this problem is its dependence upon the exact nature of neutrinos.

The evolution during neutrino trapping is shown more quantitatively in table 12.1. As the newly formed neutrinos must have a higher energy, they are less likely to escape. Once it begins, the trapping becomes more effective. For this set of parameters, the lepton loss is only $\Delta Y_l = -0.047$, which is comparable to the results of complex numerical simulations (for example, see [132, 451, 431, 432, 433]), which may range down to $\Delta Y_l = -0.1$ for some parameter choices. The quantity $\Delta \mu = \mu_e - Q - \mu_\nu$ is a measure of the degree to which weak interaction equilibrium is approached; at a density of $\rho \approx 2 \times 10^{13}$ g cm^{-3}, $\Delta \mu \approx 4.36$, which is much smaller than any of its separate terms. The diffusion time has risen to 35 msec while the collapse time has decreased to 0.12 ms at this epoch.

An interesting point of nuclear physics arises here [250]. As the collapse proceeds, the material becomes more neutron rich. Eventually the

TABLE 12.1

Neutrino Trapping

time(s)	$\log \rho$	Y_e	Y_ν	Y_l	μ_e	μ_ν	Q	τ_{dif}
0.01278	9.48	0.422	5.73E-08	0.422	5.53	0.04	5.32	1.8E-09
0.74213	9.85	0.421	3.26E-05	0.421	7.34	0.39	5.52	2.9E-07
0.79391	10.25	0.417	2.55E-04	0.417	9.99	1.07	6.32	2.9E-06
0.80775	10.67	0.409	9.13E-04	0.410	13.64	2.25	7.84	1.7E-05
0.81237	11.03	0.399	2.14E-03	0.401	17.92	3.95	9.75	6.7E-05
0.81419	11.33	0.389	3.93E-03	0.393	22.28	6.07	11.62	1.9E-04
0.81515	11.59	0.379	6.75E-03	0.386	27.10	8.91	13.45	4.7E-04
0.81575	11.86	0.370	1.12E-02	0.381	32.84	12.89	15.23	1.1E-03
0.81615	12.12	0.361	1.74E-02	0.378	39.95	18.31	16.97	2.5E-03
0.81641	12.39	0.352	2.48E-02	0.377	48.60	25.29	18.66	5.2E-03
0.81659	12.65	0.343	3.28E-02	0.376	58.84	33.88	20.32	1.0E-02
0.81671	12.90	0.335	4.09E-02	0.375	70.82	44.28	21.93	1.7E-02
0.81683	13.32	0.321	5.46E-02	0.375	94.65	65.71	24.58	3.5E-02

mean nucleus will have all the neutron fp-shell orbits filled, with valence neutrons beginning to fill the gd-shell orbits. The valence protons will be filling the fp-shell. In this situation all allowed neutron hole orbits—available via the Gamow-Teller operator to protons capturing electrons—are filled, or in other words, "blocked." Electron capture will proceed more slowly. However, the nuclei represent a thermodynamic ensemble in nuclear statistical equilibrium; at finite temperature there will be some "unblocked" nuclei. The exact rate of electron capture will depend upon the nuclear details. If blocking on nuclei is effective, most electron captures will occur on free protons, whose abundance also is sensitive to the nature of these neutron-rich nuclei, as well as to the entropy of the matter. A very clear discussion of these issues, with a variety of "one-zone" results to be compared to table 12.1, has been presented by Fuller [250]; see also [190, 343].

12.4   COLLAPSE

The dynamics of collapse naturally separates the core into two dynamical regions. Consider the characteristic time for contraction to be

$$\tau_{con} = (d \ln \rho / dt)^{-1}. \qquad (12.12)$$

There is some radius r to which a sound wave can propagate and return in a time τ_{con},

$$2r = v_s \tau_{con}, \qquad (12.13)$$

where the sound speed is $v_s = (\gamma P / \rho)^{\frac{1}{2}}$. Matter contained inside this radius can fall as a unit, maintaining some structural integrity. This material is called the *homologous core*, and represents the first dynamical region. To preserve its shape, the velocity is proportional to the distance from the center, $v \propto r$, in this material.

In contrast, matter outside this radius cannot respond to the contraction of underlying matter until the information that the core has collapsed actually reaches it. Consequently it lags behind. This region has behavior that is more akin to accretion. As a rarefaction wave moves out at the local sound speed, this matter is accelerated inward by change in the pressure gradient, which no longer balances the force of gravity. Here, the velocity approaches the form of the free-fall velocity, so that $v \propto 1/\sqrt{r}$. Goldreich and Weber [268] discovered solutions for homologous collapse of the core; see also the discussions of Yahil and Lattimer [704, 703].

Suppose the contraction time has its maximum rate, that of free-fall, so that

$$\tau_{con}^{-1} = v_{ff}/r = (2Gm(r)/r^3)^{\frac{1}{2}}. \qquad (12.14)$$

Then at the edge of the homologous core, we have

$$v_{ff}/r = (2Gm(r)/r^3)^{\frac{1}{2}} \approx 2v_s/r, \qquad (12.15)$$

or

$$\gamma P/\rho \approx Gm(r)/r, \qquad (12.16)$$

which is approximately the condition for hydrostatic equilibrium (see chapter 6 and Appendix C). Equivalently,

$$\gamma P/\rho^{\frac{4}{3}} = D_n G(m_{hc})^{\frac{2}{3}}, \qquad (12.17)$$

where m_{hc} is the mass of the homologous core.

Figure 12.2 illustrates the two regions. The infall velocity (filled squares) and the sound speed (thin line) are plotted versus the logarithm of the radial coordinate, at the time when the central density has reached $\rho_c = 3 \times 10^{13}$ g cm^{-3}. In the innermost regions, the sound speed far exceeds the infall velocity, so that the motion is subsonic. The curves cross at the inner sonic point (in this case, at a radius $r \approx 10^7$ cm); this is essentially the edge of the homologous core. Beyond this is the infall region, which extends out to another sonic point, at a radius $r \approx 10^8$ cm, which lies just inside the flame of silicon burning. There is a kink in the infall velocity curves at this flame, as the silicon burning has flashed through the matter lying just inside this layer, and affected its motion. Just above the silicon shell, at a radius of $r \approx 2.5 \times 10^8$ cm is the oxygen-burning flame. This corresponds to a discontinuity in neutron excess; abundance arguments (see §10.4) suggest that matter interior to this level should not escape collapse. These outer regions have not yet been affected by the collapse of the core.

The pressure is predominately that of relativistic leptons: electrons and neutrinos. The pressure is $P_i \approx K(Y_i\rho)^{\frac{4}{3}}/g^{\frac{1}{3}}$, where $g_e = 2$, $g_\nu = 1$, and $K = hcN_A^{\frac{4}{3}}(3/4\pi)^{\frac{1}{3}}/4$. Hydrostatic equilibrium is closely maintained, initially. We saw that the nucleon fraction for electrons is $Y_e = 0.421$. Upon contraction and neutrino trapping, we found $Y_e = 0.321$ and $Y_\nu = 0.055$. The degeneracy pressure scales as density to the 4/3 power, so

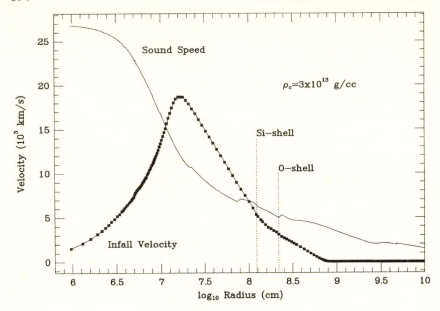

Fig. 12.2. Infall and Sound Speed

that the effective pressure-deficit factor D is

$$D = \quad (P/P_0)(\rho_0/\rho)^{\frac{4}{3}},$$
$$= (Y_e/Y_{e0})^{\frac{4}{3}} + 2^{\frac{1}{3}}(Y_\nu/Y_{e0})^{\frac{4}{3}},$$
$$\approx \quad 0.780. \tag{12.18}$$

Goldreich and Weber [268] found that for their simple solutions, a deficit factor of $D > 0.97$ would allow the whole core to collapse homologously, but for smaller values, the two dynamical regions develop. The corresponding homologous core is a fraction,

$$m_{hc}/m_c = D^{\frac{3}{2}} \approx 0.690, \tag{12.19}$$

of the precollapse core m_c. The effective mass of this core depends upon structure and temperature; using our previous estimate of the core mass ($m_c \approx 1.0 M_\odot$ for a central value of $Y_e \approx 0.42$), we have $m_{hc} \approx 0.69 M_\odot$, which is in reasonable agreement with numerical computations (for example, see [130, 131]).

Notice that more lepton loss (lower $Y_l \equiv Y_e + Y_\nu$) gives a smaller homologous core; this will prove to be a difficult point for the prompt

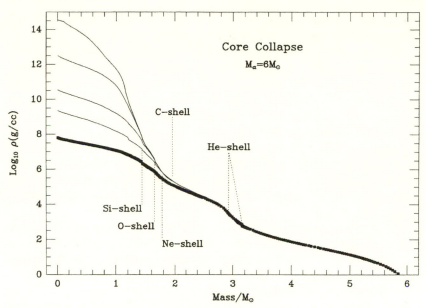

Fig. 12.3. Collapse: Density

shock mechanism discussed below. For example, suppose $Y_e = 0.28$, and $Y_\nu = 0.04$, so $Y_l = 0.32$. This corresponds to $\Delta Y_l = -0.1$, and we have $D^{3/2} = 0.509$, so $m_{hc} = 0.5 M_\odot$ (see Bruenn [132]).

Figure 12.3 summarizes the changes in density during core collapse, up to the time at which the central regions reach nuclear density, for the center of a $6M_\odot$ helium star. The results are typical of stars having helium cores of this size, such as $4 \leq M_\alpha/M_\odot \leq 8$ and $15 \leq M/M_\odot \leq 25$ for typical evolutionary histories. The heavy line shows the structure at the end of core silicon burning; here $Y_e \approx 0.48$. The next higher solid line corresponds to the end of the hydrostatic stage of electron capture; $Y_e \approx 0.42$. Notice that the outer regions are little affected by the collapse. The helium shell is unaffected, and the changes in density at the carbon and neon shells are small. The oxygen shell suffers some modification. The silicon shell is vigorously burning during the collapse.

12.5 BOUNCE

As a consequence of the low entropy increase and small lepton loss upon collapse [34], BBAL [96] showed that core collapse would continue until nuclear densities were reached. Below nuclear density, the pressure is primarily due to relativistic electrons and neutrinos, so that the effective adiabatic exponent is close to 4/3. The increase in pressure scales with compression the same way that the increase in gravity does.[8]

Although the nucleon-nucleon strong interaction is attractive at longer range, with closer encounters it has a repulsive component (originally, the "hard core"; [112], see p. 245). It is this component that prevents nuclei from collapsing due to purely nuclear forces, and is the reason for the small variation in the density inside nuclei. As the stellar core approaches nuclear density, the nuclear surfaces disappear, and the nucleons merge into an enormous nucleus. Further compression is resisted by the repulsive component of the strong interaction, increasing the adiabatic exponent. The pressure rises faster than the gravitational force, and the homologous core stops contracting. At this point, the epoch of "core bounce" begins.

The following discussion is based upon two sequences of calculations that were chosen for illustrative purposes. One calculation, which we shall call "case A," involved complete trapping of neutrinos and a hard equation of state. This is clearly a limiting case in which the effect of neutrino transport is a minimum; it gave a relatively strong bounce shock, and illustrates that physics. Figure 12.3 was constructed from the case A sequence. The second calculation, "case B," was more complicated. Neutrinos were transported by flux-limited diffusion [121]. The neutrino distribution function was taken to be that of a Fermi-Dirac gas, so that it was defined by two Lagrange multipliers (here neutrino temperature and neutrino chemical potential). These were determined by differential equations for number conservation and for energy conservation.[9] Enforcing momentum conservation gave the acceleration of matter by the neutrinos. Absorption and coherent scattering by nucleons and nuclei, as well as incoherent scattering by electrons and positrons, gave the coupling to the matter. Inverse processes were included in a thermodynamically self-consistent way. These results illustrate the behavior of state-of-the-art transport calculations for the collapse and early accretion stages (for example, see [132]). However, in our calculation the effects of antineutrinos, as well as mu and tau neutrino families, were intentionally suppressed, so that the effect of lepton trapping alone could be examined. The equation of state above nuclear density was taken to

[8]General relativistic corrections to gravity are small at densities less than nuclear. However, these corrections may rise to 25 percent in the inner core, so that newtonian treatment is inappropriate for a serious simulation of later evolution.

[9]Although different in detail, this approach was developed in response to work by Imshennik and Nadyozhin [330, 329], and by Bludman and Van Riper [107], for the purpose of analyzing multigroup computations [34, 39]. See also [191, 158, 157] for more recent efforts.

be of the simple functional form assumed by Baron, Cooperstein, and Kahana [74], with a compression modulus $K_0 = 120$ MeV (see [94]) and a maximum neutron star gravitational mass of $M_g = 1.55 M_\odot$ ($M_A = 1.86 M_\odot$ for the nucleon mass contained). This is comfortably above that measured for the binary pulsar PSR 1913+16 [677].

Figure 12.4 shows, in more detail, snapshots of density versus mass coordinate, from the beginning of collapse until the accretion shock contains $1.55 M_\odot$ (which occurs 0.2 s after bounce); this was constructed from case B. The first models represent the last part of the neutrino-trapping epoch discussed in the previous section. Until the epoch of core bounce the two sequences are similar, except that case B has a smaller homologous core (0.5 versus $0.65 M_\odot$) because (1) some neutrinos escape, leaving a smaller pressure-deficit factor, and (2) the neutrino transport rearranges the pressure distribution to slow the outer parts of the homologous core, decreasing its mass. At one millisecond after trapping, the effects of the nucleon repulsion on the equation of state are becoming evident. A small region of the core is resisting further compression and maintaining an almost uniform density. This is indicated by the dotted line, with the label "Bounce shock begins." As the collapse proceeds, the jump in density at the edge of this uniform core grows in size, as is shown. At the second dotted line from the label, we see that the density at which this jump begins has suddenly started to decrease. In this simulation, the homologous core had a mass of about $0.5 M_\odot$, a value not uncommon for simulations including neutrino transport [39, 129, 450]. Beyond this homologous core, the sound speed is inadequate to communicate to the infalling matter that the inner region has stopped. A shock develops, and begins to propagate outward, down the density gradient. The sequence takes only a few milliseconds.

While this shows the processing of matter by the shock, it does not indicate whether the shock is moving outward in radial coordinate (an explosion shock), or remaining relatively stationary (an accretion shock). This is illustrated in figure 12.5, which gives snapshots of velocity for the corresponding models in the previous figure. As the core halts, strong pressure waves ripple through the homologous region. As the pressure waves build up to a front at a mass coordinate of $0.6 M_\odot$, this front becomes a shockwave. After it propagates beyond $0.7 M_\odot$, the velocity becomes significantly positive in the core: then an explosion shock is born. However, by $1.1 M_\odot$, it has weakened to the point of having little significant positive velocity. The explosion shock has become an accretion shock, after propagating through only $0.5 M_\odot$ of matter. This is still well inside the silicon- and oxygen-burning shells, so that the explosion has failed.

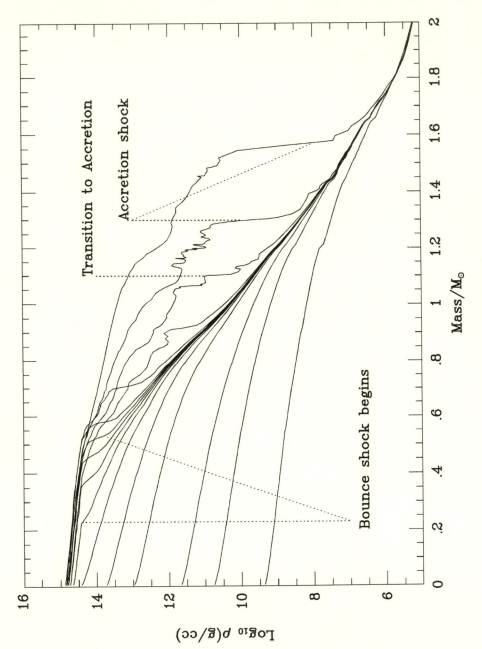

Fig. 12.4. Bounce Shock: Density

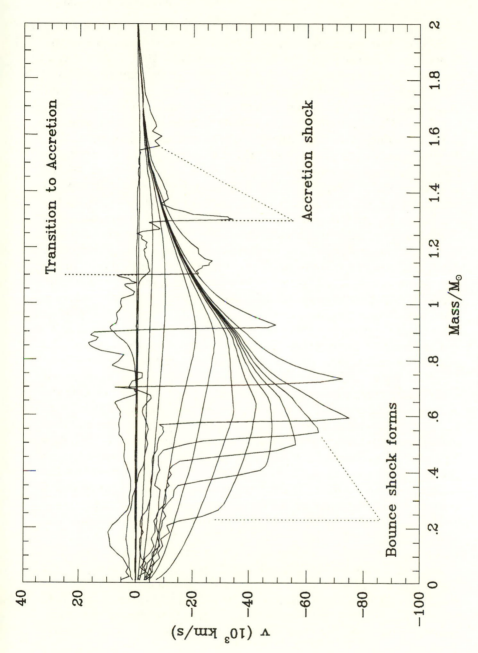

Fig. 12.5. Bounce Shock: Velocity

Why does this happen? At the point of bounce, the homologous core is close to hydrostatic equilibrium, so that the energy available to the bounce shock is essentially the kinetic energy [124, 704]. Even for case A, this is 11.7 foe. As the shock moves through the matter, it leaves behind a hot, almost hydrostatic configuration; the excess energy goes into a relatively small amount of matter which moves outward, driving the shock. The shock heats the material to above the nuclear dissociation temperature. This requires about 8 MeV per nucleon, or 16 foe/$M_{\odot}$. To propagate to the oxygen shell, where we expect the mass cut to be formed, requires the processing of $1.6 - 0.7 = 0.9M_{\odot}$ of matter, or 14.4 foe of energy, which is 2.7 foes more than is available to this mechanism, even in the favorable case A. The shock might barely reach the silicon-burning shell near $1.4M_{\odot}$. Even this is a gross overestimate because it ignores other important losses. As the shock moves outward, it moves to lower density. The probability of neutrino escape increases. Electron capture on free protons is not inhibited by a large threshold energy, so the neutrinos are freely produced and freely escape, robbing the shock of its excess pressure. It stalls, becoming an accretion shock [39]. Thus case B gives the same qualitative result. The supernova explosion has not yet happened.

The transition to accretion shock has already occurred in figure 12.5 at only 5.3 milliseconds after core bounce. As we have seen, the time for neutrinos to diffuse out of the core is several seconds, as discussed years before confirmation by SN1987A [295, 99, 4, 65]. At this time the accretion shock encloses only $1.1M_{\odot}$, leaving a mass of about 0.4 to $0.5M_{\odot}$ outside which has an embarrassing neutron excess. The observed abundance pattern suggests that such matter could be ejected in less than 1 percent of the events. These two problems have a single solution: more accretion. This suggests that the explosion mechanism is connected to the release of neutrinos from the core. Is it the loss of trapped lepton excess (electrons and electron-type neutrinos), or the release of gravitational energy as mediated by all flavors of neutrinos and antineutrinos, and other means of energy transport?

If it is the loss of lepton excess that is crucial, the case B should indicate this. Janka and Hillebrandt [339] show that local emission and absorption in shocked matter dominates the neutrino flux from at least several tenths of a second after core bounce. The failure of case B to explode suggests that this short-term effect is not the cause of the explosions in nature, and focuses attention on the longer-term loss of energy from the core.

The last three snapshots in figures 12.4 and 12.5 correspond to 5.33, 24.0, and 209 milliseconds after bounce, respectively, for case B. At these times the accretion shock encloses 1.10, 1.30, and $1.55M_\odot$, and the pre-shock density is $\log \rho = 10.86$, 9.87, and 7.37, in cgs units. As the evolution proceeds, matter is being fed to the accretion shock by the expansion of surrounding matter inward toward the new neutron star. The rate is roughly $dM/dt \approx 4\pi r^2 \rho v_s f$, where ρ is the local density and v_s the local sound speed, and f is a factor of order unity. To a fairly good approximation, the density scales inversely with the cube of the radius, $\rho \propto r^{-3}$. The sound speed satisfies $v_s^2 = \gamma P/\rho$, so for $\gamma \approx 4/3$, $dM/dt \propto r^{-3/2} \propto \rho^{1/2}$. The rate of mass infall decreases as the accretion continues.

The accretion shock moves out in radius, so the preshock density decreases and the mass accretion rate decreases as well. This is the "thinning" stage of Bethe [94]. Why does the radius increase? The matter accreted through the shock has a finite pressure and supports a mantle around the original homologous core. As matter piles on, the mantle extends to lower and lower density, and larger and larger radii. The pressure is finite because of the competition between heating and cooling. The neutrinosphere is deep inside the mantle; it cannot extend to such low density. The escaping neutrinos are hotter than the overlying matter, so they heat it. The heating is proportional to the matter available to heat, and thus to the density. However, cooling by emission of neutrinos is often by collision processes, which are proportional to the number of collision pairs, that is, to the square of the density. Thus we have the possibility of a thermal runaway; as the density decreases, the heating becomes more effective. With decreasing density and increasing temperature we form an "entropy bubble" between the core and the Si and O shells. The steep density gradient thus formed is what Bethe [94] called the "cliff" in reference to Wilson's computation [686].

This can be understood in terms of an example that is closer to home. The surface of the Sun exhibits similar characteristics for similar reasons. As radiation streams out from a photosphere, or neutrinosphere, it maintains the energy (temperature) characteristic of that surface. It tends to bring the atmosphere lying above the $\tau \approx 2/3$ surface to this temperature. In pressure equilibrium, or an approximation thereto, a flatter temperature gradient is compensated for by a steeper density gradient. This analogy is a rich one; the Sun has a surface convection zone, and a solar wind which is driven by high entropy gas above the photosphere. As we shall see, the collapse problem also exhibits overturn instabilities, and a "neutrino star wind" might be involved in the explosion mechanism. However, an important difference is that while the solar surface is in a quasi-stationary state, the core collapse has rapid secular changes, and is consequently much more dynamic in nature.

Figure 12.6 shows the compositional structure of the helium core at
0.2 s after core bounce. The inner region has a low entropy; most of the
matter was in the form of nuclei before reaching nuclear density and
merging into a gigantic nucleus. This is labeled "Heavy." The high value
of electron abundance Y_e is due to lepton trapping. As we move out
beyond $0.5M_\odot$, the entropy increases due to the previous effects of the
shock. As we move further outward, at first the abundance of alpha particles rises, but then at still higher entropy it decreases. The abundance
of free nucleons rises to almost 100 percent, and due to lepton loss,
shifts toward lower Y_e and more neutrons. Much of the former Si shell
has burned and then dissociated; the accretion shock is at $1.55M_\odot$. The
oxygen shell is beginning to burn explosively at $1.65M_\odot$. Beyond this, the
composition is the same as it was at the onset of collapse.

This is approaching an optimum time for explosion, from the point of
view of nucleosynthesis yields. The steep density gradient in figure 12.4
coincides with the outer edge of the neutron-dominated region. This
is a natural point for a bifurcation into ejecta and neutron star matter. It would avoid the ever-present problem of excess production of
neutron-rich ejecta. In general, simulations will not succeed in avoiding
this problem, which promises to be a keen test of their plausibility.

Even though the bounce shock is long dead at this epoch, the reserves
of energy have hardly been tapped. The inner core is still lepton rich; the
total fermi energy of these leptons alone is about 50 foe. The thermal
energy of the neutron-rich mantle is about 30 foe. The gravitational
binding energy is as yet only 62 foe; a neutron star of $1.55M_\odot$ would
have a gravitational binding of 280 foe, so that over 200 foe remain to
be released. It is interesting to note that this $1.55M_\odot$ of matter would
have a gravitational mass of about $1.40M_\odot$ as a neutron star, which is
fascinatingly close to the binary pulsar PSR 1913+16 [677] values.

Figure 12.7 shows the energy per nucleon in the form of neutrinos,
versus mass coordinate, from the center out to the He/H interface. The
neutrino-burst from the collapse and bounce shock stage is traveling out
at the speed of light and has reached a mass coordinate of $3.3M_\odot$. The
accretion shock is at $1.55M_\odot$; there is no indication of it in this variable. However, the neutrinosphere is evident at $1.25M_\odot$. The dominant
feature is the large energy in the neutrinos trapped in what was the
homologous core, $M < 0.8M_\odot$.

The thermal energy of nucleons is also significant. Figure 12.8 shows
the temperature versus mass, for the same sequence of snapshots of case
B as previously shown. While the beginning of the bounce shock shows
little that is new, as the transition to accretion is approached there is a
marked spike in temperature just behind the shock. This is characteris-

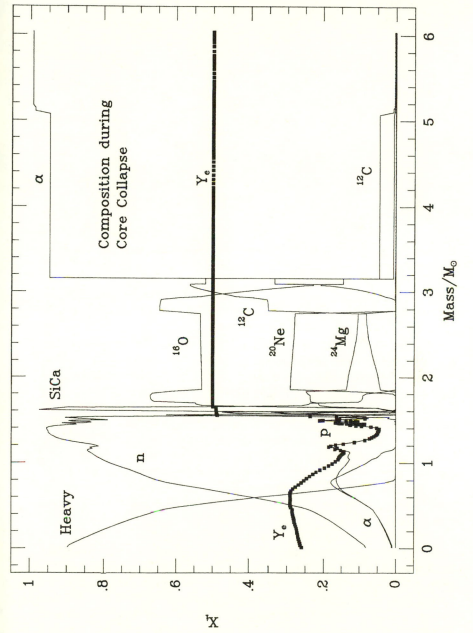

Fig. 12.6. Composition

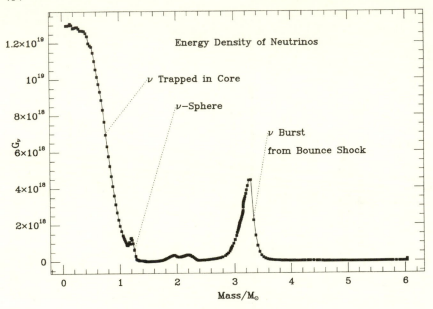

Fig. 12.7. Neutrino Energy Density

tic of a radiating shock. Compressional heating gives a high temperature which is then reduced by the enhanced rate of energy loss by radiation. It seems that an instability develops, which will probably require multidimensional simulations to represent properly. Notice that $kT \approx 10$ MeV for most of the matter inside the shock. Comparing this figure with the previous one, it is clear that while most of the neutrino energy lies within $M < 0.8M_\odot$, the thermal energy extends further, even to $M \approx 1.5M_\odot$ in the last snapshot. Neglect of the electron antineutrinos, and of mu and tau neutrino-antineutrino pairs, is clearly incorrect at this point. The trapping of neutrino heat seems to have become more important than trapping of lepton number, and the additional neutrino species are important for its conduction.

Another interesting facet of this object is illustrated in figure 12.9, which shows masses for cold neutron stars with this equation of state. Such a neutron star will have a small abundance of protons and electrons, here illustrated by $Y_e = 0.03$. The gravitational mass is shown as a dashed line; it peaks at $1.55M_\odot$, slightly above the masses of the binary pulsar PSR 1913+16 [677], the largest of which is shown as a dotted line. The mass of the nucleons required to build such a neutron star, denoted M_A, is larger by the gravitational binding energy, and peaks at $1.81M_\odot$. However, if lepton number is trapped, the neutrinos are accompanied

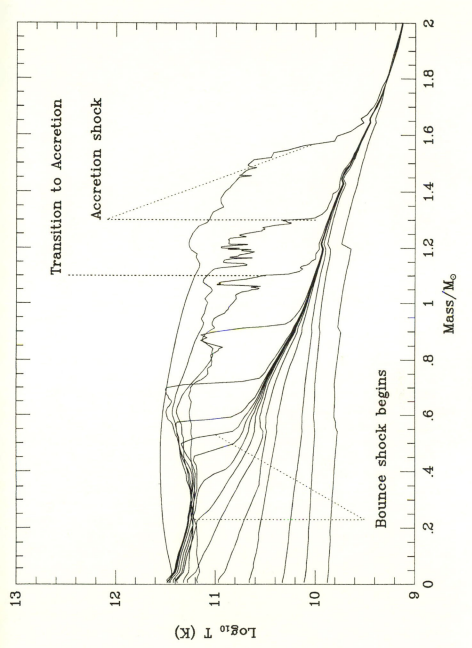

Fig. 12.8. Bounce Shock: Temperature

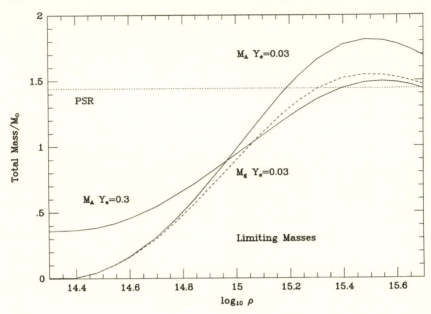

Fig. 12.9. Limiting Masses

by roughly twice the number of electrons, and therefore twice this number of protons. This means that the nucleons fill two fermi seas—both neutron and proton. In the simple case of a noninteracting gas, the corresponding fermi energies are less, and the fermi pressures are less. Let us suppose that the same effect occurs in an interacting nucleon gas; in particular we scale the pressure and energy with isospin (neutron excess) as in the noninteracting case. While there is an added pressure due to the leptons, it is dwarfed by the loss of nucleon pressure. This is indicated by the solid curve labeled M_A $Y_e = 0.3$, in which the lepton component is approximated simply by a relativistic electron gas with this mole fraction. Notice that the maximum mass now corresponds to $M_a = 1.49 M_\odot$, which is less than the matter inside the accretion shock at 0.2s. Such an object avoids collapse because of its thermal pressure. As lepton number decreases due to neutrino escape, the equation of state stiffens, and the future of the core is more secure—provided further accretion does not overwhelm it. Thus the question of collapse to a black hole depends upon the initial core mass, the amount of accreted mass, the degree of lepton trapping, and the thermal energy distribution (and possibly rotation).

12.6 EJECTION OF MATTER

The questions of collapse to form a black hole, and of ejection of nu-clearly processed matter, are inextricably intertwined. It may be that the formation of a neutron star, followed by a failed explosion, results in the whole star falling into its Schwarzschild radius. In some events ejection might occur, followed by collapse of a slightly too massive core as it cools and contracts. While both black hole formation and explosive ejection of matter seem to occur in Nature, a deep understanding of the causes and origins of these two modes still eludes us.

Thermonuclear

The first quantitative ideas about the mechanism of explosion are due to Hoyle and Fowler [302, 303, 305, 238]. In this mechanism, the collapse is used as a trigger for the release of thermonuclear energy, not as the primary energy source for the explosion. The onset of instability gave a rapid contraction of the mantle regions, made of explosive fuels (espe-cially ^{16}O), which then burned rapidly. The core was spinning, so that centrifugal forces then slowed the collapse of the core, while the mantle exploded away, ejecting the newly synthesized matter as well as the outer parts of the star. To properly evaluate this concept, plausible evolution-ary sequences of rapidly rotating presupernovae are needed. None yet exist. Colgate and White [184] were the first to simulate the nonrotating case, and found that the core collapsed before the mantle exploded; as we have seen, the race is much closer than they imagined.

With SN1987A we may estimate the amount of thermonuclear energy released. Burning matter in the oxygen shell to ^{56}Ni releases about 1.5 foe/$M_\odot$; for the $0.075 M_\odot$ of ^{56}Ni implied by the observations (see chap-ter 13), this implies an energy release of about 0.1 foe. Because of the sensitivity of explosive burning to the peak temperature in the shock which ejects the matter from the star, this is likely to be accompanied by explosive oxygen burning which does not proceed to ^{56}Ni formation (chapter 9). This releases about 1.2 foe/$M_\odot$, and the mass so processed is likely to lie in the range $0.1 \leq \Delta M/M_\odot \leq 0.3$. This gives a thermonu-clear energy release[10] of $0.2 \leq \Delta E_{tnuc} \leq 0.4$ foe for SN1987A, which is about 20 percent of the explosion energy required in chapter 13 for the light curve and expansion observed. This suggests that the core collapse does supply most of the explosion energy, but that thermonuclear energy

---

[10]In §9.6 we saw that carbon and neon burning were likely to be hydrostatic, so that their energy release would not be part of the explosive event.

release might be important for defining the interface between ejecta and collapsed remnant.

Prompt shock

The idea that the energy of collapse gives rise to ejection of matter is an old one, going back at least to Baade and Zwicky [63]. Colgate and Johnson [181] emphasized that a bounce shock would form, and eject matter, but the "prompt shock" mechanism has been most strongly advocated by Brown [125]; see also [74, 76, 191]. To enhance the chances of a successful explosion, one may try to increase the energy available and to decrease the losses. A bounce shock of larger energy results if the homologous core is closer to the limiting gravitational mass [651, 652, 38]; a gentler rebound occurs which puts more energy into ordered motion of expansion, and the larger homologous core has more energy available. For this mechanism to have the best chance at success, one needs (1) a soft equation of state, so that the limiting mass is as close as it can be to the homologous core mass at bounce, (2) a steep density gradient in the presupernova, so that a minimum amount of mass must be dissociated by the shock, and (3) a small loss of energy by neutrino radiation. Notice the difficulty here: a large core is desired, close to the limiting mass, yet a steep density gradient is also wanted. The presupernova models give the steep density gradient with the small cores.

While a softer equation of state has been argued (see Bethe [94] for a review), it is difficult [267, 125] to have both a very soft equation of state and yet have neutron star masses as large as $1.444M_\odot$ and $1.384M_\odot$, as measured [677] for the binary pulsar PSR 1913+16. These are gravitational masses, so they are reduced by a binding energy of order 0.1 of the rest mass [119]. Thus the total mass that would collapse to form such a neutron star would be about $1.6M_\odot$; therefore a reasonable estimate of the minimum precollapse mass for formation of a black hole is probably above $1.7M_\odot$.

Massive stars which develop cores near the white dwarf limit have steeper density gradients (see figures 10.4 and 10.7); these are the "lower-mass" massive stars. By including coulomb corrections and modifying their convective algorithm, Woosley and Weaver [699] obtained lower mass cores (in better agreement with [32], which had both). Nomoto and Hashimoto [458] obtained a presupernova core of $1.18M_\odot$, with a very steep density gradient, for a star of $13M_\odot$.

While initial results for this model seemed promising [74], it was soon shown that simulations with more accurate treatments of the neutrino radiation [129, 108, 451, 452] gave greater energy losses by neutrino escape, and gave no explosions by this mechanism. Bruenn [130, 131] estimates that the core mass must be less than $1.1M_\odot$ for this mechanism to succeed.

Rotation

Hoyle [302] considered rotation to be a key ingredient in supernova explosions, and since the discovery of pulsars, rotation has often been related to the explosion mechanism. Intuitively the general idea is appealing. The dramatic compression, with conservation of angular momentum, causes a rapid spin-up of the core, which then flings off the overlying matter. In the latter regard, magnetic fields have often been invoked. However, there are serious problems of physics with this picture. It is notoriously difficult to trap magnetic fields in plasma; they are buoyant and unstable [486] (it is no accident that magnetically contained fusion has not been an easy success). Prior to collapse, presupernovae spend long periods in a relatively quiescent state, during which such buoyant expulsion could occur. This may be a critical process for angular momentum redistribution in the star, and the reason that we have been relatively successful with nonrotating models! Prior to collapse, it would seem that the magnetic field pressure is small compared to the pressure of matter. In this picture, the magnetic field would be carried along with the collapsing matter, and have little effect on the dynamics. Eventually, perhaps assisted by dynamo action, the fields would emerge from the surface of the new neutron star and begin the complex activity that we observe in pulsars. But what about the rapidity of pulsar spins? Actually, except for millisecond pulsars which are thought to be spun up by accretion, pulsar periods are so slow that they correspond to slow rotation. The Crab pulsar, with a period of 0.033s, probably has a rotational velocity of only 2,000 km/s, while its escape velocity is a significant fraction of the speed of light. See Michel [437] for a comprehensive and entertaining discussion.

Given the severe difficulties encountered even in the one-dimensional (spherical) case, relatively few simulations have been done in the two-dimensional (axisymmetric) case [387, 442, 623, 443, 110, 601, 441]. The added degree of freedom allows more complex solutions; only a few of these complexities will be mentioned.

First, consider the bounce. In the spherical case, because the adiabatic exponent remains near 4/3, collapse continues to nuclear density. With even a moderate centrifugal force, smaller than either gravity or the pressure gradient force, this delicate balance can be destroyed. The bounce can occur at lower densities, and with the spherical symmetry broken [442, 601, 441]. This will tend to lower the neutrino energies, and while it reduces their diffusion time, the release of gravitational energy is now determined by the time-scale for angular momentum transfer. SN1987A does not seem to have been such an event. The quantitative nature of these events would be sensitive to the exact amount of rotation present in the presupernova, and might be highly varied [223].

Second, consider energy transport. If the amount of rotation is insufficient to halt the collapse, it may still have a significant role because of its breaking of spherical symmetry. This gives rise to global circulation currents in the newly formed core, which advect neutrinos and heat from deep within to above the neutrino-trapping surface. This modifies the "inner problem," enhancing the transport of energy, and providing the "outer problem" with increased luminosity, possibly aiding the explosion mechanism.

Early ν transport

The suggestion, that neutrino transport on the collapse time-scale had difficulty in providing an explosion [16, 17, 19], has been confirmed by many independent simulations [684, 34, 120, 129, 450, 295]. The problem may be simply stated: the mechanism does not work if the time to transport energy from the core to the mantle is longer than the time to accrete enough mass to form a black hole. This difficulty is part of the "inner problem." If the core mass at bounce is close to this limit, there will be little time for energy transfer by this mechanism. If the neutrino opacities are larger, the diffusion time is longer, so that pure diffusion will be less adequate. A larger core, and a higher density, make the diffusion time longer. Even with the small opacities used for neutrino transport prior to the discovery of the weak neutral currents [16], massive cores still collapsed while the lowest-mass cores exploded. This is a simple consequence of the relative sizes of these time-scales. The qualitative idea of the "neutrino transport explosion mechanism" is physically correct; whether it works in Nature is a quantitative question. Unfortunately for theorists, the best estimates for the correct physical parameters seem to give a failure to explode, but by a narrow margin. Definitive results require extremely accurate simulations; for example, see [431, 432, 433].

Late ν transport

Due to computer resource limitations, collapse simulations are not continued indefinitely. Jim Wilson [686, 97, 687] allowed one simulation to run far past the time required for core bounce and shock stagnation (out to 0.8 s after bounce) and was pleasantly surprised by an explosion! This is now called the "late mechanism" or the "Wilson mechanism." It is produced not by the outburst of neutrinos from the bounce shock, but from the longer-term loss from the core. Unfortunately, the strength of the explosion is inadequate to explain the light curve and expansion observed in SN1987A. Further, the mechanism does not always work [132]. Nevertheless, the result is an important step: it shows that, given the appropriate conditions for the "inner problem," the "outer problem" can give an explosion. If the infalling matter above the neutrinosphere is properly exposed to neutrino luminosity, it can cause an explosion, ejecting overlying matter.

First, some qualitative thermodynamic considerations provide a guide, as we indicated in the previous section. Neutrinos emerge from the neutrinosphere with a distribution function typical of the electrons from which they are made and with which they interact. They may be characterized by a temperature of $kT \approx 1.5$ to 2 MeV, and a chemical potential $u_\nu \approx 8$. As they escape their density decreases, so that the effective chemical potential decreases, but the energy distribution remains fixed, with mean energy $< \epsilon_\nu > \approx 3kT \approx 5$ MeV. The matter temperature above the neutrinosphere will be lower, so that as the neutrinos escape, any interactions will tend to heat the matter. The heating rate per unit volume will be proportional to the density ρ; it depends upon how much matter is available to interact with the neutrinos. The cooling rate per unit volume will be proportional to ρ^2, because pairs of particles are required. Therefore, at low density, heating is favored over cooling, and for sufficiently low ρ the temperature will rise. This will give a region of high entropy and pressure, called the "high entropy bubble." If energy is transferred into this region with sufficient speed, the pressure can be high enough—in principle at least—to explode the presupernova.

Overturn

Can we "supercharge" this weak but promising mechanism of late ν transport? It is tempting to focus on processes that might enhance the rate of energy release from the denser regions. The analogy introduced in the previous section, between Bethe's "cliff" and the solar surface immediately suggests a possibility: hydrodynamic overturn. In the outer layers of the Sun, there is too much opacity to carry the luminosity, which gives rise to steeper temperature gradients that drive convection. Although the collapse case is more complex, with entropy and chemical potential gradients that are sensitive to the details of the coupled effects of hydrodynamics, compositional change, and transport, this seems a promising possibility.

The first serious examination of the idea that convective overturn should be important in core collapse, and could be the cause of explosion, is due to Epstein [222]. Early numerical simulations [128, 397] and arguments [185], based on the idea of a catastrophic overturn of the whole core, mixing the unshocked inner core and releasing its leptons, were shown to be overly enthusiastic by more accurate simulations of Smarr et al. [574]. However, it was suggested [574] that outer core overturn would be a generic and important feature of core collapse.

This milder conclusion has been amply confirmed by a new generation of multidimensional simulations of the collapse hydrodynamics [291, 136, 340]. At the time of writing, the treatment of neutrino transport has been overly simplified. Further, there are differences in interpretation of the results; see [89] and the references above. Part of the difficulty is that overturn not only transports neutrinos but also energy by gravity modes, sound waves, and shocks [84]. Despite these complications, the outlook for an answer to the question of how core collapses become supernova explosions is brighter than it has been in years.[11]

Black holes

If there is no explosion, accretion continues until the mass of the black hole grows to that of the whole star. Thus, in stellar collapse, the black hole formed may have a mass much larger than that of a neutron star. Suppose a massive star has such a large core when it collapses, that there is not enough time for explosion energy to be transported to the mantle. Such a massive star would be expected to have suffered extensive mass loss, perhaps losing all of its hydrogen envelope. As we shall see in §14.4, the mass of the iron core at the onset of collapse increases from 1.6 to above $2.0M_\odot$ as the helium core mass changes from $M_\alpha = 10 \to 12M_\odot$. This might be a likely mass for a black hole which results from the failed explosion of a single star. The case for binary evolution is more complex, and could easily produce other masses as well.

Weak interaction

The collapse of a stellar core is perhaps the only place in the Universe as we know it, where it is important that neutrinos are fermions—so they fill

---

[11]The reader may test this optimism by examining the current contents of the annual indexes of the journals referred to above.

a fermi sea—rather than bosons. It matters whether neutrinos are their own antiparticles or not. While the previous discussion only considered the simplest electro-weak theory of these interactions, this choice was simply for convenience in writing the story. This environment is one in which the exact nature of the weak interaction is crucial. Advances in our understanding of weak interaction will have direct implications for the gravitational collapse problem.

13

Supernovae

> . . . He saw; and blasted with excess of light,
> Closed his eyes in endless night.
> —Thomas Gray (1716–1771),
> *The Progress of Poesy*

In November 1572 a new star, or *nova*, appeared in the constellation
Cassiopeia; in modern terminology it was a supernova in our own Milky
Way galaxy. From his own careful observations, a 26-year-old, eccentric
Danish nobleman, Tycho Brahe, became convinced that this new object
(now called Tycho's supernova) was in the supposedly immutable celes-
tial regions. The beautiful and comforting medieval cosmology began to
crumble. On October 9, 1604, another new star appeared in the sky, less
than three years after Tycho's death. This supernova was named for his
student and assistant, Johannes Kepler, who was 33 at the time. Kepler's
use of Tycho's planetary observations and his own theoretical genius led
to his laws of planetary motion, which supplanted Ptolemy's model of
the solar system. The supernova also interested a 40-year-old professor
at Padua, Galileo Galilei; this was his first serious astronomical observation. He concluded that the new star had no parallax, and that it must
be among the immutable fixed stars. The response was that since the
supernova was obviously not immutable, it could not be among the fixed
stars, and therefore Galileo must be wrong. This was an unwise assertion
for supporters of the status quo. In the summer of 1609, when he heard
of the invention of the "eye-glass," he may have had something to prove.
He proceeded to build one, and used it to invent telescopic astronomy.
Near the end of his life, banished to his villa in Arcetri and becoming
blind, he smuggled his last book, *Two New Sciences*, to Holland for publication. This established the beginnings of the science of motion, and
laid the foundations for the Newtonian revolution. Medieval cosmology
died.

Since then Nature has been parsimonious with her Galactic supernovae (although many supernovae are observed each year in other galaxies). Perhaps she feels that modern astrophysicists do not deserve one. In
1885 Hartwig discovered a supernova in our companion galaxy, M31. But

the most spectacular display since Kepler's supernova began on February 23, 1987, when a blue supergiant star, Sanduleak -69° 202 , underwent core collapse and exploded, in a satellite galaxy of our own Milky Way (the Large Magellanic Cloud, or LMC for short). Only its neutrinos were observed in the Northern Hemisphere, after they had passed through the earth. Almost as if to even the score for northern observatories, SN1993J appeared in the nearby galaxy M81; it was the second brightest supernova since Kepler's.

What are supernovae? Much of the story is the decay of ^{56}Ni and ^{56}Co heralding the synthesis of iron-peak nuclei, "writ large in the heavens." Supernovae are probably the most brilliant *events* that we observe. As such they provide a potential probe for the long ago and far away—if we can understand them well enough to interpret what is seen.

13.1   AN OVERVIEW

The discussion below focuses on the physical nature of the supernova events; a more complete mathematical treatment is given in Appendix D.

For weeks a bright supernova can compare in luminosity to its parent galaxy. How does a single star manage to become so brilliant? Because Tycho and Kepler discovered their supernovae by naked-eye observations, the most cautious assumption[1] is that most of the light comes in the visible wavelengths. An object that is 10^{10} times the brightness of the Sun, and has roughly the same temperature, must have an area that is 10^{10} times larger. Supernovae are observed to have temperatures near maximum brightness that are hotter than the Sun; a factor of two hotter allows an area smaller by a factor of 1/16. Suppose the supernova is roughly spherical; then its radius is $R_{sn} \approx 0.25 \times 10^5 R_\odot \approx 2 \times 10^{15}$ cm, which is about twenty times larger than the largest stars known (red supergiants). *This is the first rule for making supernovae: they must have large surface areas.*

An object radiating at $10^{10} L_\odot$, for a few weeks (say 2×10^6 s), emits an energy $\mathscr{E}_{rad} \approx 8 \times 10^{49}$ erg. This is a large energy, about twenty times the rest mass of the earth. From the Doppler-shifted emission and absorption lines, we know that supernovae are explosions having mean expansion velocities of order 2,000 to 10,000 km s^{-1}. Any hydrodynamically plausible model seems to imply kinetic energies that are much larger than $\mathscr{E}_{rad}$: $\mathscr{E}_{ke} \approx 10^{51}$ erg. Detection of neutrinos from SN1987A indicate that for this type of supernova, the energy release is larger still,

---

[1]In this case, caution seems correct.

$\mathscr{E}_{\nu\bar{\nu}} \approx 0.15 M_\odot c^2 = 2.7 \times 10^{53}$ erg. *Our second rule for making supernovae: they require large energy release.*

There is a consistency in this: if the outer edge of the explosion moves at 10^4 km s^{-1}, it will take several weeks to go a distance of 2×10^{15} cm, which is a time-scale roughly typical of observed supernovae.

The simplest model of a supernova is to imagine that a large amount of energy is released in a star. This generates a shock wave which heats and accelerates the matter, so that it expands and radiates. Upon expansion, the internal energy is converted into kinetic energy as the pressure does work upon the expanding matter, further accelerating it. Thus, the average fluid velocity tends toward

$$v \approx 10^9 \text{ cm s}^{-1} \left(\frac{\mathscr{E}_{sn}}{10^{51} \text{ erg}} \frac{M_\odot}{M} \right)^{\frac{1}{2}}. \tag{13.1}$$

The fluid velocity at the surface will be higher (see Appendix D).

This conversion of internal energy into kinetic energy poses a quandry for the construction of models of bright supernovae: any such conversion reduces the amount of thermal energy which may escape to be seen! In fact, as we shall see, simple explosions of initially dense stars will be ineffective producers of luminosity. By the time they expand to become big, they have been cooled by just that expansion. Radioactive heating is a solution to this problem. The energy of nuclear decay is unaffected by the expansion of the ejecta, and for a lifetime in the appropriate range, will be released when the radius is of supernova dimensions.

Let us consider these two possibilities for the ejection of radioactive debris: some or none. Similarly, let us consider two possibilities for the size of the progenitor: a red supergiant, and smaller. These will determine the nature of the luminosity. We may then construct a logical table, shown in table 13.1.

Case A will have contributions to the light curve both from radioactivity and from shock heating; most Type II supernovae fit here. Case B will

TABLE 13.1

Type II Parameters (Supernova Cases)

	Large Progenitor	Small Progenitor
Radioactive debris	A	B
Nonradioactive debris	C	D

be dominated by radioactive heating; Type I supernovae fit here. How-ever, the observational classification places the presence of hydrogen in the spectra as of paramount importance—Type II's have it and Type I's do not. Thus we might expect some objects of case B which still have hydrogen to cause problems with the classification, just as SN1987A did. Are there objects of case A which have lost their hydrogen? It seems that SN1993J nearly did so. Because of its greater transparency, helium tends to prevent envelopes from expanding to red supergiant dimensions. It is interesting to speculate as to whether other compositions might do so. Objects of case C might be classified as Type II supernovae if they showed hydrogen lines. For case D, the luminosity of the event would be low, and its evolution fast, so that it might not even be classified as a supernova.

For reasons that are perplexing to the author, the idea that hydrogen-rich events could be either case A or case B met with a considerable lack of enthusiasm, until the advent of SN1987A. The idea that case B could correspond to both core collapse and thermonuclear events has received a similar response, although SN1993J now supports it. Such ideas were discussed well before these supernovae were discovered (e.g., [33, 38]). This is not a trivial matter, because the history of astronomy suggests that we tend to see only what we look for, and that observational genius lies in noticing the unexpected. If our conceptual framework is limited, it often causes us to delay discoveries, which probably happened here. If we include additional attributes, such as explosion mechanisms (core collapse or thermonuclear?), and surface compositions (hydrogen or not?), then the logic table grows. We now have four sets of binary decisions, or sixteen possibilities, even in this very simplified scheme. It might be hoped that the present supernova classification schemes will evolve into something more meaningful and fundamental.

The evolution of the supernova luminosity—its *light curve*—contains information about the explosion. There are five aspects of the event which together dominate the nature of that light curve: shock emer-gence, expansion, radiative diffusion, heating (by radioactivity, or possi-bly a pulsar?), and recombination. Each will be examined in turn, and then judicious combinations will be used to analyze the data on a variety of well-observed events, including SN1987A and SN1993J.

13.2 SHOCK EMERGENCE

Consider a star that is the progenitor of a supernova. In hydrostatic equi-librium, the gravitational binding energy of the star is $B/M \approx GM/R$, (see §6.5) and the mean pressure and density satisfy $P/\rho \approx GM/R$ as well (see §6.2). The sound speed s is $s^2 = \gamma P/\rho$. The explosion energy of the supernova, $\mathscr{E}_{sn} \approx v^2/2$, is much larger than the binding energy B, which implies

$$v \gg s, \tag{13.2}$$

so the expansion velocities are supersonic. Without an exceedingly slow acceleration, which is not a feature of most mechanisms, the explosion must cause a shock wave. The emergence of the shock wave at the surface of the progenitor is the first visual indication of the explosion (neutrinos and gravitational waves can escape sooner).

The ratio of radiation pressure to gas pressure is

$$(1 - \beta)/\beta = aT^3/3\Re Y\rho \tag{13.3}$$

(see chapter 6). If the average density is $\rho = 3M/4\pi R^3 \approx 0.5 \times 10^{-12} m/R_{15}^3$, where $R_{15} \equiv R/(10^{15} \text{ cm})$ and $m \equiv M/M_\odot$, and the supernova thermal energy is $\mathscr{E}_{sn} \approx (1/2)aT^4 4\pi R^3/3$, then

$$T \approx 6.3 \times 10^4 \text{ K } (\mathscr{E}_{sn}/R_{15}^3)^{1/4}. \tag{13.4}$$

This gives

$$(1 - \beta)/\beta \approx 1.6 \times 10^4 (R_{15} \mathscr{E}_{sn}/10^{51} \text{ erg})^{1/4}/m, \tag{13.5}$$

so that radiation pressure should always dominate, as anticipated. This is a significant simplification for the relevant mathematics; except for radioactive and recombination energy, the internal energy of the matter may be ignored. A supernova is basically a ball of light.

This has an interesting implication. Since the radiation entropy is

$$S_\gamma/\Re = 4Y(1 - \beta)/\beta, \tag{13.6}$$

we have

$$S_\gamma/\Re \approx 6.4 \times 10^4 (R_{15}\mathscr{E}_{sn}/10^{51} \text{ ergs})^{3/4}/m. \tag{13.7}$$

Previously in explosive burning we have found $S_\gamma/\Re \approx 10$, which is clearly smaller. To get such high entropies requires heating at lower temperature than that of explosive burning; this is an indication of radioactive heating after the explosion, as we shall see.

For a strong shock, the internal energy density is $E = aT^4V \approx v^2/2 = 5 \times 10^{17} \text{ erg g}^{-1}(v/10^9 \text{ cm s}^{-1})^2$. Typical observed velocities are 5,000 km/s for Type II and 10,000 km/s for Type I supernovae. Upon expansion the internal energy is converted into kinetic energy. At the time of shock emergence t_0, $E_0 \approx v_a^2/4$, where v_a is the asymptotic fluid velocity. Let $R_{14} = R/10^{14}$ cm and $v_9 = v/10^9$ cm/s. If at t_0, the radius is $R_0 \leq 10^{14}$ cm, which is true for almost all stars, the hydrodynamic expansion time,

$$\tau_h \equiv 10^5 \, \text{s} \, (R_{14}/v_9), \tag{13.8}$$

is short, a day or less. The subsequent coasting phase will be roughly homologous,[2] so we take the time derivative of the specific volume to be $dV/dt \approx 3v_a V/R$ where R is the surface radius, and $R \approx R_0 + v_a t$.

First, let us consider the nature of the epoch of shock emergence. In figure 13.1 are shown the luminosity and effective temperature in a model of SN1987A [41, 42], during the first 180 minutes after shock emergence. Notice that the shock takes about 53 minutes after core collapse to arrive at the surface ($R = 2 \times 10^{12}$ cm), corresponding to an average velocity of $< v > = 6{,}250$ km/s. The light travel time is 67 seconds, which happens to be comparable to the time it takes the surface to heat. Consider a region δR below the surface. The hydrodynamic time is $\delta R/v$ while the radiative diffusion time is $3(\delta R)^2/\lambda c$, where $\lambda = 1/\rho\kappa$ is the effective mean free path for radiative heat flow. For sufficiently small δR these times are equal; this occurs at an optical depth $\delta R/\lambda = c/3v \approx 15$. The mass contained in δR is just

$$\delta M = 4\pi R^2 (\delta R/\lambda)/\kappa \approx 10^{-6} M_\odot. \tag{13.9}$$

Thus almost all of the mass of the progenitor is exposed to a shock wave in which the radiation is well-coupled to the matter.

As the shock moves out through the star, it rearranges the density structure. The steep density gradient between the inner and outer regions is reduced by the great expansion of those inner regions. See, for example, figure 1 in [57], which shows the evolution of density structure from core collapse to mantle ejection. We may estimate the peak temperature to be expected from the escaping radiation by assuming uniform density, and using $\mathscr{E}_{sn} = 2$ foe, $R_{15} = 0.002$ in the relations above, we find $T \approx 4.2 \times 10^5$K. The actual value found in the numerical simulation was $T \approx 5.0 \times 10^5$K, with the difference probably due to shock steepening in the density gradient near the stellar surface, which our estimate did not take into account.

---

[2]Explosion from a point (a small volume) generates homologous expansion, in which $v \sim r$; this is similar to the cosmological Hubble flow discussed in chapter 5.

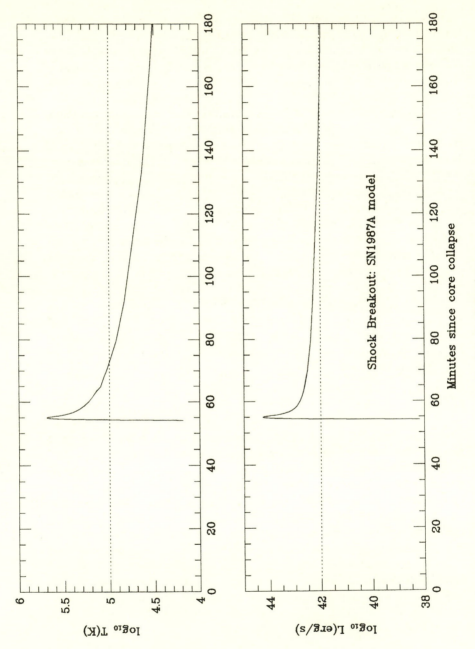

Fig. 13.1. Shock Breakout

There are other reasons to suspect that the radiation temperature is actually higher than the numerical estimate. At the low optical depths encountered at shock breakout, one might wonder if the radiation field really has the same temperature as the matter. If much of the opacity is supplied by electron scattering, the two temperatures might deviate, in the sense that the matter would be hotter. Perhaps the radiation gas is dilute, with a higher color temperature than its equilibrium value [661, 249, 292]; see also [221]. This is relevant to the question of how much ionizing radiation is produced in the burst; Fransson and Lundqvist [241] found that modeling the state of ionization of circumstellar matter required a peak radiation temperature of $(3 - 6) \times 10^5$ K for the shock breakout. This seems consistent with the earlier numerical simulations [41, 42, 570, 691]; for example, the model shown above radiates about 1.5×10^{46} ergs while the effective temperature is above 1.0×10^5 K. The proper treatment of the decoupling of the radiation from the matter remains a formidable challenge [221], but would allow a more reliable comparison of theory and observation.

13.3 EXPANSION AND RADIATIVE DIFFUSION

Figure 13.1 showed the effective temperature and luminosity for the first three hours after core collapse for the same SN1987A model. The short duration of the burst was evident. While the effective temperature continues to decrease after the burst, the luminosity approaches an almost constant value. This behavior will be described below by relatively simple analysis.

Figure 13.2 shows snapshots of the density structure, at 54 m, 100 m, 10.4 h, and 1.85 d after the explosion began [42]. Most of the ejected matter resides in a central mass of surprisingly uniform density, topped by a power law distribution ($\rho \propto r^{-\alpha}$, with $\alpha \approx 7$ to 10), which is produced in the first hour after the shock reaches the surface (this happens between 54 m and 100 m in the figure). Also shown, as a large open circle, is the position of the photosphere. After 10.4 h, it begins to move in through the matter—but out in radius. Thus there will be an increasing amount of transparent matter overlying the photosphere, and at first this matter will have a power-law density distribution in radius [165].

After the shock has emerged, the expansion is soon homologous. Further evolution is governed by the first law of thermodynamics:

$$\dot{E} + P\dot{V} = \epsilon - \partial L / \partial m. \qquad (13.10)$$

This can be simplified by use of a "one-zone" approximation (which, as shown in Appendix D, is equivalent to separation of space and time

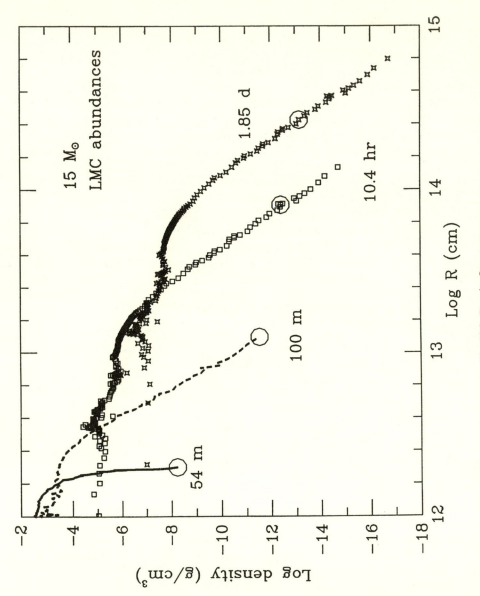

Fig. 13.2. Density Structure

variables). Therefore,

$$\partial L/\partial m \approx L/M = E/\tau_{diff}, \qquad (13.11)$$

where M is the total mass, $\tau_{diff} = \kappa M/\beta cR$ is the time-scale for radiative diffusion, and here β is just a numerical constant which is $\beta \approx 13.8$ for a variety of distributions (see §D.1). Note that $\tau_{diff} \approx 3R^2/\pi^2\lambda c$, the value for a homogeneous sphere. For solar composition, $\kappa \approx 0.4$ cm^2/s, which gives

$$\tau_{diff} \approx 2 \times 10^7 \text{ s } (m/R_{14}), \qquad (13.12)$$

where m is the mass in solar units, so that $\tau_{diff} \gg \tau_h$. However, while the expansion time-scale τ_h increases with time, the diffusion time-scale decreases. Equating them gives $R \approx 1.4 \times 10^{15}$ cm $(m/v_9)^{1/2}$, which roughly coincides with the duration of high luminosity in this simple model.

For a zeroth-order solution, suppose $\mathscr{E}_{rad} \ll \mathscr{E}_{ke}$, as suggested above. Then we have adiabatic expansion of a radiation-dominated gas, $\dot{E} + P\dot{V} \approx 0$, or $\dot{T}/T \approx -\dot{R}/R = -v_a/R$. Thus, $T \propto R^{-1}$ and $E \propto T^4R^3 = R^{-1}$. *Expansion leads to cooling of the radiation gas, before it can escape to power the light curve!* We estimated that $\mathscr{E}_{rad}$ was a few percent of $\mathscr{E}_{ke}$; dumping an energy $\mathscr{E}_{ke}$ in a red supergiant of initial radius $R_0 \approx 10^{14}$ cm would require an expansion of roughly a factor of 14 to maximum light, and a reduction of the thermal energy to roughly this amount (about 7 percent). For a progenitor of smaller radius, such as that of SN1987A, the fraction of energy in thermal form when it escapes freely is even less, implying a lower luminosity.

Suppose that there is no heating ($\epsilon = 0$). Then Eq. 13.10 may be written as

$$d\ln(T^4R^4)/dt = -(\tau_h + t)/\tau_h\tau_d(0), \qquad (13.13)$$

where $\tau_h \equiv R_0/v_a$ and $\tau_d(0) \equiv \kappa M/\beta cR_0$. The opacity κ is taken to be constant for the moment. The solution is

$$(TR/T_0R_0)^4 = e^{-[(\tau_h t + t^2/2)/\tau_h\tau_d(0)]}. \qquad (13.14)$$

From Eq. 13.11,

$$L = ME/\tau_d = (4\pi\beta ac/3)T^4R^4/\kappa M, \qquad (13.15)$$

so that

$$L = L_0 e^{(\tau_h t + t^2/2)/\tau_h\tau_d(0)]}, \qquad (13.16)$$

where $L_0 \equiv (4\pi\beta ac/3)T_0^4 R_0^4/\kappa M$. If τ_h is small, the logarithm of luminosity falls quadratically. The time constant is determined by the product $\tau = (2\tau_h\tau_d(0))^{1/2} \approx 2 \times 10^6$ s $(m/v_9)^{1/2}$. This is roughly the correct time-scale for supernovae; better estimates will be discussed below. The important general point is that objects which are less massive, less opaque, and expand faster also have light curves which evolve faster. For the SN1987A model, this time-scale is much longer than the span of time shown in figure 13.1. Therefore the luminosity should be roughly constant, while the radius increases, giving a decreasing effective temperature, as is indeed the case.

What is the value of this luminosity? If $\mathscr{E}_{rad} \approx \frac{1}{2}\mathscr{E}_{sn}$ just after shock breakout, and we take this to evaluate $(\mathscr{E}_{rad})_0 = (4\pi/3)R_0^3 aT_0^4$, then

$$L_0 = \frac{1}{2}(\mathscr{E}_{sn}/M)R_0\frac{\beta c}{\kappa}, \tag{13.17}$$

or $L_0 \approx 2.5 \times 10^{43}$ erg/s $(\mathscr{E}R_{14}/m)$, where $\mathscr{E}$ is the explosion energy in units of 10^{51} erg (foes). Note that R_0 is (essentially) the initial radius of the progenitor, a constant, not the time-dependent radius of the outer edge of the ejected mass. Many Type II supernovae have luminosities of several times 10^{42} erg/s , which is consistent with $\mathscr{E}/m \approx 0.1$ and $R_{14} \approx 1$.

Several general properties of light curves may be seen from these results, even though they refer strictly to cases in which there is no heating. The luminosity is inversely proportional to the opacity κ; lower opacity simply means shorter times for radiative diffusion. This trend will survive even if the opacity is not constant in time or space. The luminosity depends upon the explosion energy per unit mass, so that for the same $\mathscr{E}_{sn}$, less massive stars are brighter. The luminosity is proportional to the initial radius. Large (tenuous) stars will be brighter than small (dense) stars. This is because small (dense) stars must expand more, and suffer more adiabatic degradation of thermal energy, before radiation easily escapes. For very small (dense) stars, some other source of energy besides the thermal energy of the shock is required for them to be bright enough to be supernovae.

13.4 RADIOACTIVE HEATING

As we have seen, the doubly magic nucleus ^{56}Ni is likely to be produced in explosive nucleosynthesis, whether in a detonating/deflagrating white dwarf, or the mantle ejected from the collapsed core of a massive star. Consequently, an important component missing from the previous discussion is the effect of radioactive heating. The energy of radioactive decay cannot be lost by adiabatic cooling until after the decay occurs. If the lifetime for decay is of the order of weeks or months, this energy will be released at an epoch at which it can readily escape, and thereby influence the light curve.[3]

The radioactive energy does not necessarily appear as heat; it originally appears by the production of gamma rays and positrons, which have to be thermalized to provide thermal luminosity (UVOIR, which is short for *ultraviolet*, *optical*, and *infrared*) from the supernova. Thus, the effective heating rate is not generally the same as the simple radioactive decay, but modified by the probability of thermalization in the radiating gas.

Following the suggestion of Nadyoshin [453], a simplified decay scheme for $^{56}Ni \rightarrow ^{56}Co$ is shown in figure 13.3. The terrestrial decay is by electron capture from a K-shell electron, with a half-life of 6.10 days, and populates the 1.720 MeV excited state of ^{56}Co. By the time decays would occur, the temperature of the supernova ejecta is so low that the K-shell of Ni will be again filled by recombination, so that the supernova decay occurs this way too. Figure 13.3 shows the most important energy levels in ^{56}Co, and the cascade of gamma transitions (the energy of the line in MeV, and the branching fraction, are given).

The cross section for the interaction of gamma rays with electrons has a minimum around twice the electron rest mass energy, $2m_ec^2$. At higher energy, electron-positron pairs are readily produced. At lower energy, the dominant process is Klein-Nishina scattering, with an average cross section that is about one-third the value for Thomson scattering [231]. The scattering accelerates the electron and reduces the energy of the gamma ray. When, after several scatters, the energy is reduced to the point at which K-shell electrons begin to resonate with the gamma ray, absorption occurs readily. For example, when the scattering reduces the gamma energy W_γ to 100 KeV (about two scatters on average; see [287], p. 208), photoelectron absorption on Ni, Co, or Fe becomes dominant, and as it increases as W_γ^{-3}, further thermalization proceeds rapidly. Note that the absorption energy depends upon the binding energy of the K-shell electron, and therefore upon the composition of the gas. Thus the question of thermalization reduces to this: do the gammas escape before they scatter down to absorption energies? The gammas begin to escape without depositing their energy when the mean number of scat-

---

[3]At the time of writing, there is no clear evidence for other sources of heat in SN1987A, even though the neutrino-burst clearly indicates the formation of a condensed remnant, and evidence for a new pulsar is still awaited. The discussion here can be readily generalized [51] to include other such sources of heating.

Simplified ⁵⁶Ni Decay

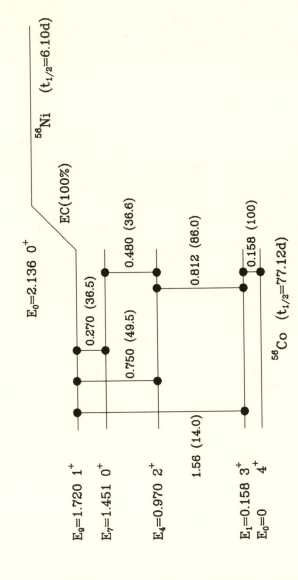

Fig. 13.3. Simplified ⁵⁶Ni Decay

ters n drops. The ratio of radius to Klein-Nishina mean free path λ is $R/\lambda = 3\kappa_{KN}M/4\pi R^2 \approx 32m/r_{15}^2$. At $n \approx (R/\lambda)^2 \approx 1$ the gammas begin to escape freely; this occurs at $R_{15} \approx 5.6\sqrt{m}$. Using the velocity of expansion, these radii can be converted into times since the explosion began. Notice that for larger masses and slower explosion velocities, the epoch of gamma escape occurs later. Thus the best targets for gamma-ray astronomy are supernovae which are (1) nearby, (2) ejecting large amounts of radioactive matter, (3) of low mass, and (4) expanding rapidly. The last three conditions favor supernovae of Type Ia.

Figure 13.4 gives the corresponding simplified decay scheme for ^{56}Co. Here the decay also occurs by positron emission (about 19 percent of the time). The format is the same as in the previous figure. The product, ^{56}Fe, is stable. The positron channel will be interesting for our discussion of light curves. The positron scatters readily with electrons, thermalizing its kinetic energy. Then, when the positron is slowly moving, annihilation is more probable, giving two gammas of $m_e c^2 = 0.511$ MeV. These gammas may or may not escape, but the positron kinetic energy is very likely to be thermalized, thus giving a minimum rate for radioactive heating.

We recall the first law of thermodynamics, and consider a sphere of uniform density, so that

$$dE/dt + PdV/dt = \epsilon - E/\tau_d(t), \tag{13.18}$$

where $\partial L/\partial M \approx L/M = E/\tau_d(t)$, and $\tau_d(t) = 3R^2/\pi^2\lambda c \approx \kappa M/4\pi cR$ is the radiative diffusion time-scale. This may be written as

$$4(\dot{T}/T + \dot{R}/R) = 1/\tau_{heat} - 1/\tau_d(t), \tag{13.19}$$

where $\tau_{heat} \equiv E/\epsilon$. Notice that this combination on the left-hand side suggests the transformation

$$T^4 R^4 = T_0^4 R_0^4 \phi(t)\Psi(x), \tag{13.20}$$

for separation of space and time variables (see Appendix D and below). The luminosity is

$$L = ME/\tau_d(t) = (16\pi^4/27)acR^4 T^4/\kappa M. \tag{13.21}$$

The maximum of the (bolometric) light curve occurs when $dL/dt = 0$. Now,

$$dL/dt = L\left(4(\dot{T}/T + \dot{R}/R) - \dot{\kappa}/\kappa\right), \tag{13.22}$$

Simplified ^{56}Co Decay

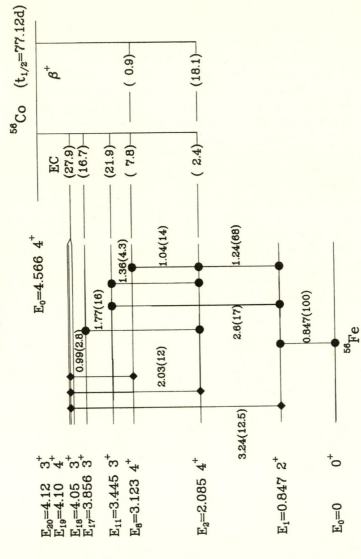

Fig. 13.4. Simplified ^{56}Co Decay

so this implies that, *at maximum bolometric luminosity*,

$$1/\tau_{heat} - 1/\tau_d(t) = \dot{\kappa}/\kappa. \qquad (13.23)$$

For the special case of constant opacity κ, the instantaneous rate of heating equals that of cooling by radiative diffusion, at maximum luminosity. In other words, at that moment, the luminosity equals the total radioactive heating [36, 39]. If the effective opacity decreases with time, as would be expected from both the Karp[4] effect [344] or from recombination, then the luminosity is still increasing at this point (at $\tau_{heat} = \tau_d(t)$) because $dL/dt > 0$, and the maximum luminosity is shifted to a later time. The time of maximum bolometric luminosity is therefore an important diagnostic for light curves.

13.5 RECOMBINATION

As the supernova expands and cools, ionized matter may recombine. This will reduce the opacity (the matter becomes almost transparent to its own thermal radiation), and will release the energy of ionization. A recombination wave develops, which moves inward in mass until all the ejected matter has become transparent, and the supernova has changed from a superstar to a supernebula.

Much of the opacity is due to Thomson scattering of the thermal photons by ionized electrons; this component of the opacity is easy to calculate for a gas in thermal equilibrium. Figure 13.5 shows the result for some representative cases. Mixtures of pure H and He are shown; the solid lines represent $X_H = 0.6$ and $X_{He} = 0.4$, while a solid line represents pure He, $X_{He} = 1.0$. A typical photospheric density for a solar-like star is about 10^{-8} g cm^{-3}, which is much larger than the values estimated above for a supernova, which we represent by 10^{-11} g cm^{-3} and 10^{-12} g cm^{-3}. The H/He mixture shows several ledges, as different ionized states recombine (He^{++}, He$^+$, and H$^+$ as the temperature decreases). This occurs at lower temperature as the density considered is reduced. For supernova densities the opacity drops drastically for temperatures below 6,000 K.

This general behavior should be compared to figure 6.1, which includes the opacity contributions from absorption processes. Below 10^4 K these contributions tend to increase the opacity, so that it is almost a

---

[4]The rate of leakage of photons by diffusion is reduced, relative to static stars, because Doppler shifting of absorption lines closes "holes" in frequency through which stellar photons would escape.

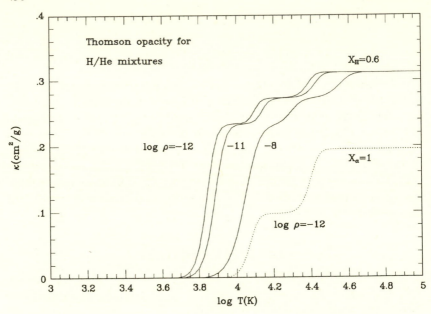

Fig. 13.5. H/He Opacity

single-step function, disappearing below a cutoff temperature. This temperature is lower than before, but of the same order.

For pure He, the steps for He^{++} and He^+ are more pronounced, and the drastic decrease in opacity now occurs at hotter temperatures, about 10 to 12 thousand degrees kelvin.

A supernova with overlying hydrogen would have a photospheric temperature of about 6,000 K, or more. A supernova with overlying helium would have 10,000 to 12,000 K, or more. These values are slightly above those observed for Type II (with hydrogen), and Type I (no hydrogen lines). In the latter case, other elements might be in high abundance in the overlying layers as well.

The recombination wave essentially sets a minimum effective temperature, but does it affect the shape of the light curve? Suppose the recombination region to be thin, so that the wave is almost a discontinuity. As this wave moves inward in mass, the rate at which energy sweeps through the front is

$$L_{ion} = -4\pi R_{ion}^2 v_{ion}\left(aT_{ion}^4 + \rho_{ion}Q_{ion}\right), \tag{13.24}$$

where R_{ion} is the radius of the recombination front, T_{ion} and ρ_{ion} the temperature and density there, Q_{ion} the energy per unit mass released by recombination, and v_{ion} the velocity of the front (positive outward) relative to the matter. Here the photosphere and the recombination front are taken to be essentially coincident; they are expected to be close together. The first term is the loss of radiation energy due to "advection," that is, to the motion of the photosphere inward. This increases the luminosity because the radiation escapes sooner than it would for a photosphere fixed at the outer boundary of the ejected mass. The radiation energy density is reduced by expansion, so that this is similar to an increased leakage rate. The second term is different; it involves the atomic binding energy of the matter, and is not affected by expansion. The relative importance of this term is measured by the ratio $E_{ion}/(E_{ion} + E_{rad}) = \rho Q_{ion}f_i/(aT^4 + \rho Q_{ion}f_i)$, where f_i is the fractional ionization.

Figure 13.6 shows this energy ratio for the same H/He mixtures in the previous figure. The largest effect is for the hydrogen-rich case, at the highest (that is, stellar) density, for which the ionization energy is much larger than the radiation energy density for a wide range of temperature.[5] For the lower densities that are expected in supernovae, the ionization energy is significant if hydrogen is present, but negligible for the pure helium case (shown as a dotted line, which hugs the lower axis). This is because the helium recombination temperatures are higher, and the energy *per unit mass* does not show the large increase that the energy per atom does.

For hydrogen, $Q = 1.31 \times 10^{13}$ erg/g so that the energy released by recombination of hydrogen is $MX_H Q_H = 2.61 \times 10^{46}$ erg mX_H. If this energy is lost over a time of, say, one month, the average luminosity is 10^{40} erg/s mX_H. This is relatively dim, unless mX_H, the mass of hydrogen, is large. Thus we should expect only the hydrogen-rich supernovae (Type II) to have their light curves modified by the release of recombination energy itself.

The inward motion of the recombination wave affects the temperature structure. This will cause an additional modification of the luminosity. If the wave moves *slowly*, radiative diffusion will reestablish the form of $\Psi(x) = \sin(\pi x/x_i)/(\pi x/x_i)$, the spatial part of the solution for a uniform sphere, where $x_i = r_i/R$ and r_i is the radius of the wave (see Appendix D). Consequently we can simply use $d\Psi/dt = (\partial\Psi/\partial x_i)(dx_i/dt)$ to find the additional change in luminosity due to loss of thermal energy.

If the motion is *fast*, the shape of the thermal distribution will be almost unchanged inside the recombination wave, but the external radi-

---

[5]The ionization energy may have important implications for the dynamics of the solar convection zone [508].

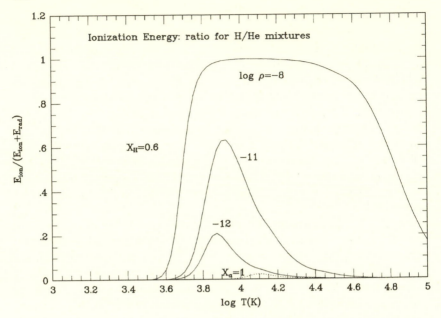

Fig. 13.6. Ionization Energy

ation gas will be "sliced off." For the uniform sphere, this would imply
$\Psi(x_i) = \sin(\pi x)/\pi x$, for $x \le x_i$. These two limiting examples are illustrated schematically in figure 13.7. For the simple and interesting case of
uniform density, these two extremes span the range of possibilities and
allow us to obtain an internal measure of the accuracy of our computations.

First let us examine the "fast" case. There is an additional energy loss
between times t_1 and t_2 in figure 13.7. For the fast case, this may be
estimated from the change in thermal energy which results from slicing
off a fraction δx of the ejecta. The total thermal energy $\mathscr{E}_{th}$ inside radius
r_i is

$$\int_0^{r_i} 4\pi r^2 aT^4 dr = 4\pi R_0^3 aT_0^4 (R_0/R(t))\phi(t) \int_0^{x_i} \Psi(x)x^2 dx, \qquad (13.25)$$

so that

$$\partial \mathscr{E}_{th}/\partial x_i = 4\pi R_0^3 aT_0^4 (R_0/R(t))\phi(t)\Psi(x_i)x_i^2. \qquad (13.26)$$

If the recombination front is assumed to be thin, we may also regard this
as a contribution to an "advection luminosity"; it is generally larger than

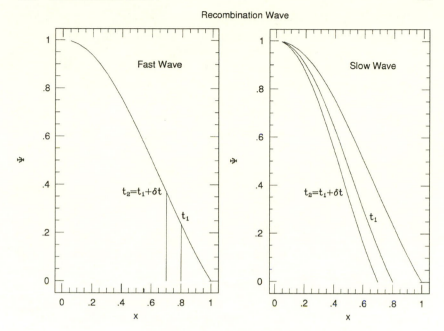

Fig. 13.7. Recombination Wave

the contribution referred to above.[6] Thus,

$$L_{advec} \approx -v_i \partial \mathcal{E}_{th}/\partial x_i, \qquad (13.27)$$

where $v_i = dx_i/dt$ is the velocity of the recombination front with respect to the matter.

The luminosity due to radiative diffusion changes with the motion of the recombination front as well, because it depends upon the slope of the temperature distribution. At x_i the luminosity due to diffusion from deep inside is

$$L_{diff} \approx (\mathcal{E}_{th}(0)/\tau_{diff}(0))\phi(t)[-x^2\partial\Psi(x)/\partial x]_{x_i}, \qquad (13.28)$$

and

$$-x^2\partial\Psi(x)/\partial x = x\cos\pi x - (1/\pi)\sin\pi x, \qquad (13.29)$$

for this choice of $\Psi(x)$.

---

[6]This is because some of the lost radiation is hotter than the photospheric temperature, and this energy had to be reemitted as several photons at that photospheric temperature.

If the photospheric temperature is large compared to the recombination temperature, the recombination front lies in the transparent region, and we may simply ignore it and take the luminosity to be the diffusion value, $L = L_{diff}$, and $x_i = 1$. As the supernova cools, this will no longer be the case, and the recombination front must be included. The luminosity emerging from the photosphere, if it is assumed to be near the recombination front, is $L = 4\pi r_i^2 \sigma T_e^4$. For a plane-parallel atmosphere, Eddington's approximation is $T^4 = \frac{3}{4} T_e^4 (\frac{2}{3} + \tau)$, where τ is the optical depth. Taking the recombination front to be at the $\tau = 0$ boundary, we would have

$$T_i^4 = T_e^4 / 2. \tag{13.30}$$

This suggests a simple approach: for $T_i^4 < T_e^4 / 2$, the recombination front will be ignored; otherwise it will be included. We will use this approximation in what follows. It will fail as the ejecta make the transition to the supernebula stage because of the assumption of planar symmetry, making our light curves less reliable there. However, at such a time, the light curves are beginning to follow the instantaneous rate of energy deposited, so that the actual errors tend to be more modest than would otherwise be the case.

Energy must be conserved through the front, so we have the constraint equation:

$$L = L_{ion} + L_{advec} + L_{diff}. \tag{13.31}$$

Using the expressions just derived for the fast case gives a relation between the velocity of the recombination front, $v_i = dx_i / dt$ and the other evolution variables:

$$3x_i^2 dx_i / dt = \left[\pi^2 I(x_i) \phi / \tau_{diff}(0) - (\pi^2 c / 3R_0)(T_i^4 / 2T_0^4)(R/R_0)^4 x_i^2 \right]$$
$$/ \left[MQ_{ion} / \mathscr{E}_{th}(0) + \phi(R_0 / R)(\pi^2 \Psi(x_i) / 3) \right]. \tag{13.32}$$

Using our standard transform for separation of space and time variables (Appendix D),

$$T^4 R^4 = T_0^4 R_0^4 \phi(t) \Psi(x), \tag{13.33}$$

and the first law of thermodynamics, we have

$$d\phi(t)/dt = \left[(M < \epsilon > + L_{psr}) / (\mathscr{E}_{th}(0) \pi^2 I(x_i)) - \phi / \tau_{diff}(0) \right] (R/R_0), \tag{13.34}$$

where L_{psr} is any energy input from the newly formed neutron star, $M < \epsilon >$ is the energy deposited in the gas by radioactive decay, and finally $\pi^2 I(x_i) = -x_i \cos \pi x_i + \frac{1}{\pi} \sin \pi x_i$. Note that for this choice of Ψ, $[-x^2 d\Psi/dx]_{x_i} = \pi^2 \int_0^{x_i} \Psi(x) x^2 dx$. There has been a cancellation of the terms L_{ion} and L_{advec} with the corresponding changes in thermal energy; this must occur because, in this approximation, the thermal energy inside r_i is unaffected by the motion of the recombination front.

Let us now repeat the analysis for the "slow" case. The total thermal energy is now

$$\mathcal{E}(t) = 4\pi R_0^4 a T_0^4 (R_0/R)\phi(t)(x_i^3/\pi^2), \tag{13.35}$$

because

$$\int_0^{x_i} [\sin(\pi x/x_i)/(\pi x/x_i)] x^2 dx = x_i^3/\pi^2. \tag{13.36}$$

Therefore, taking the partial derivative with respect to x_i, we find

$$L_{advec} \approx -3x_i^2 v_i \mathcal{E}_{th}(t). \tag{13.37}$$

The recombination luminosity is the same as in the "fast" case, but the diffusion luminosity is now

$$L_{diff} \approx (\mathcal{E}_{th}(0)/\tau_d(0))\phi(t)x_i. \tag{13.38}$$

Notice that the faster leakage time is overwhelmed by the decrease in the number of photons due to the decrease in volume. The first law of thermodynamics implies

$$d\phi(t)/dt = [(M < \epsilon > +L_{psr})/(\mathcal{E}_{th}(0)x_i^3) - \phi/x_i^2 \tau_{diff}(0)](R/R_0), \tag{13.39}$$

and the conservation of energy constraint on the luminosity components gives

$$3x_i^2 dx_i/dt =$$
$$[\phi/\tau_{diff}(0) - (\pi^2 c/3R_0)(T_i^4/2T_0^4)(R/R_0)^4]$$
$$/[MQ_{ion}/\mathcal{E}_{th}(0) + \phi(R_0/R)]. \tag{13.40}$$

The coupled solution of these two equations gives the evolution of the light curve and the recombination front.

13.6 SN1987A

On February 23, 1987, a supernova (SN1987A) was detected in the Large Magellanic Cloud (LMC). It was the brightest supernova detected since the invention of the telescope, and became one of the most extensively observed objects in the history of astronomy. It was observed in all wavelengths from gamma rays to radio, and for the first time, neutrinos were detected from the collapse of a stellar core [99, 295]. The literature on the event is extensive; for a beginning, see [47, 424] and references therein.

Because of its small initial radius, the development of this supernova was greeted by some as a great puzzle, even though many aspects of its behavior were previously discussed (e.g., see [33]). Rather than treat it as anomalous, we will take SN1987A to be the standard for our discussion of supernovae.

Several parameters are assigned fixed, plausible values: for the opacity to gamma rays, $\kappa_\gamma = 0.07$ cm^2/g, and the initial thermal energy was chosen to be half of the explosion energy. The parameters which were varied are listed in table 13.2, for the Type II supernova models. The pertinent observational criterion for the classification is the presence of hydrogen lines in the spectra. Stars can lose their outer layers of hydrogen by mass loss, or by binary interaction. Table 13.2 gives the distances used for converting the observational data for comparison with the theoretical models. Also listed are distances determined by the expanding photospheres method (EPM) [547, 216, 546], the Tully-Fisher method (TF) [496], and Cepheids (Cep) [244]. Our adjustment of the distance to obtain the desired apparent brightness for a given time behavior of luminosity, effective temperature, and velocity is independent of these distance estimates, so that the general agreement is encouraging.

In order to determine a good choice of parameters, we fit both the UVOIR light curve, and the effective temperature. The UVOIR light curve is shown in figure 13.8. The pluses represent the observational data of the South African group [154, 681]; see also [102, 278, 117, 118]. The results from Cerro Tololo International Observatory, and from European Southern Observatory, are in excellent agreement until the last, difficult epochs. These observational data sets are completely unprecedented in quality and coverage. The thin solid line is the theoretical curve for the slow approximation; the fast approximation will be shown below. Over the first 800 days, the observed UVOIR luminosity, which varies through a factor of about 3,000, is nicely described. At 139 days the Ginga/Mir experiments detected x-rays ([213], see discussion and detailed references in [47]). The first detection of gamma-ray lines (from ^{56}Co) was by the

TABLE 13.2

Type II Supernova Parameters

Supernova	1987A	1969L	1980K	1993J
Type	SNII-S	SNII-P	SNII-L	SNII/I
$M_{ej}/M_\odot$	15	17	2.2	3.26
E_{51}	1.7	1.7	1.0	1.7
$R_0/10^{12}$ cm	3	15	22	0.7[a]
$M(^{56}Ni)/M_\odot$	0.075	0.075	0.075	0.075
$\kappa(cm^2/g)$	0.2	0.1	0.2	0.075
$T_{ion}(K)$	4,500	4,200	3,000	6,200
q_{ion}(ev/nucleon)	13.6	13.6	13.6	13.6
R_{Ni}/R	0.65	0.65	0.65	0.65
$E(B-V)$	0.15	0.06	0.41	1.00
Galaxy	LMC	NGC1058	NGC6946	M81
D(Mpc)	0.050	10.8	5.7	3.6
D(Mpc)EPM	0.049 ± 0.006	10.6 ± 1.5	5.7 ± 0.7	2.6 ± 0.4
D(Mpc)TF		9.2 ± 1.3	5.5 ± 1.0	
D(Mpc)Cep	0.049 ± 0.004			3.63 ± 0.34

[a]This does not include the outer, hydrogen-rich layer.

Solar Maximum Mission (SMM) at 178 days [416], with subsequent detections by many balloon experiments [189, 263, 409, 539, 613] (see [47]; Caltech: CIT, Lockheed-Marshall: LM, Florida-Goddard: FG are shown in figure 13.8). The model does not accurately account for mixing due to hydrodynamic instabilities in the explosion [48], which will cause the gamma lines to emerge earlier than shown, agreeing better with the data. The average loss is reasonably well reproduced if the ^{56}Co extends out to $R_{Ni}/R = 0.65$ for this model.[7] We will keep this value of R_{Ni}/R for all subsequent Type II supernovae; the UVOIR light curves are not sensitive to it. The ^{56}Ni, which was produced explosively, decays to ^{56}Co. The first part of the exponential tail determines this mass, $M(^{56}Ni)$. The

[7]This corresponds to a velocity of 3,000 km/s, so that the bulk of the radioactive Ni(Co) should be at this or lower velocity. This agrees well with the observed $v_{max} \leq 3,000$ km/s in the ^{56}Co lines [640].

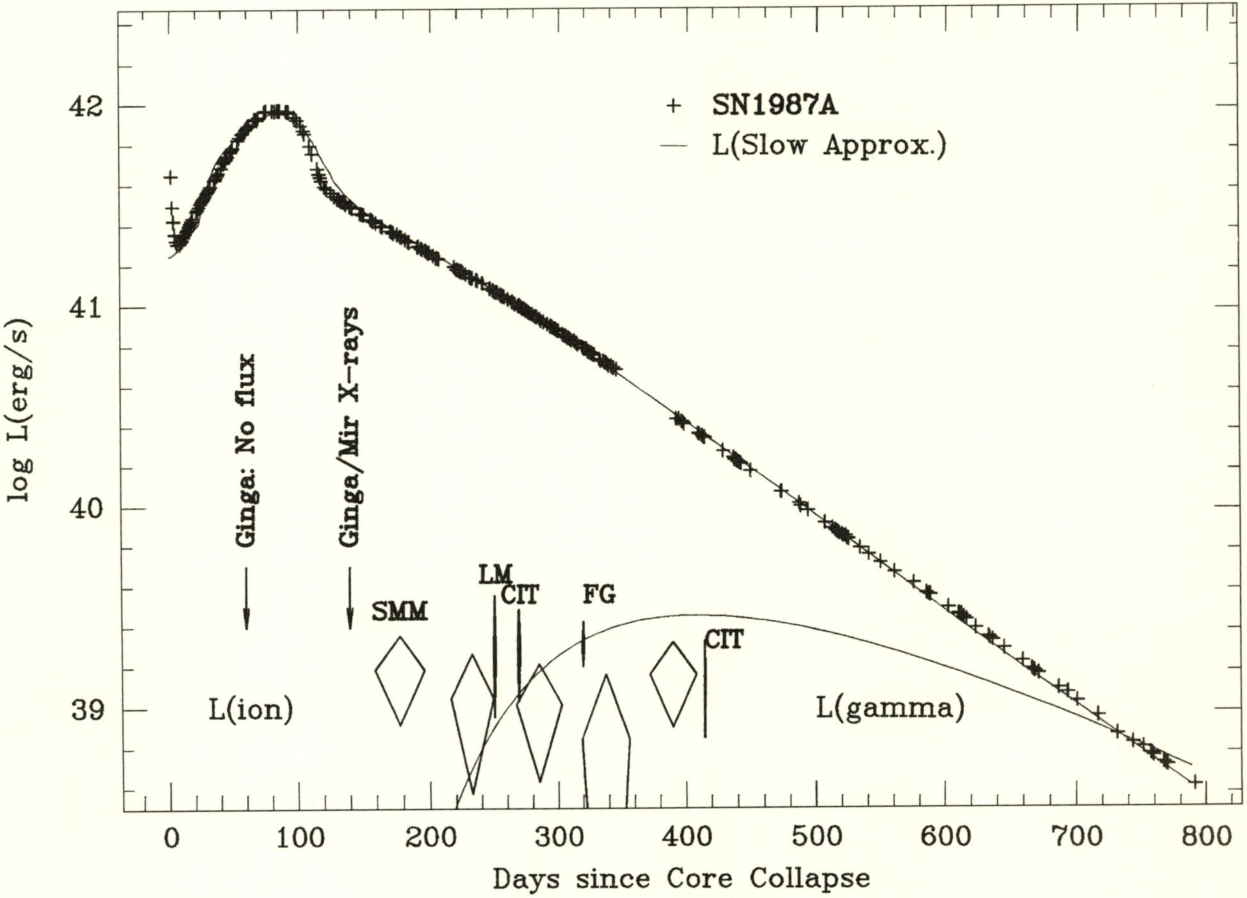

Fig. 13.8. SN1987A

escape of gammas causes the light curve to sag below an exponential decay having the lifetime of ^{56}Co. This constrains some of the parameters. The fraction of decay energy lost to gamma-ray escape is $1 - e^{-\tau} \approx R\rho\kappa_\gamma$. At late times, $R \approx vt$, so that $R\rho\kappa_\gamma \approx 3\kappa_\gamma M/4\pi(vt)^2$. Knowing the time t, we can fix the combination $\kappa_\gamma M/v^2$. Since $\mathscr{E}_{sn} \propto Mv^2$, and κ_γ varies weakly, light curves with the same $M^2/\mathscr{E}_{sn}$ have the same curvature in the quasi-exponential tail. If we simply choose $\mathscr{E}_{sn} = 1.7 \times 10^{51}$ erg as a plausible value, the corresponding mass of ejected matter is $15M_\odot$.

Let us now concentrate on the earlier part of the light curve, shown in figure 13.9. The pluses again represent the observational data of the South African group [154, 681]. We need to examine the thermal opacity, the initial radius, and the ionization energy. From the previous discussion we have seen that the value of the ionization energy is relatively unimportant, so we take the value for hydrogen in what follows. This slightly overestimates the effect when it is at all significant. The opacity is probably dominated by electron scattering. It depends strongly on the number of free electrons. We will use the observed effective temperatures to specify the ionization temperature, T_{ion}. Having done this, there remains the weak dependence on Y_e, the total number of electrons per nucleon; we take the constant (the maximum opacity) to be $\kappa = 0.2$ cm^2/g for Type II supernovae. This choice has a small effect on the light curve; the recombination effects are much larger. The only effective parameter left is the "initial" radius; we choose it to be $R_0 = 3 \times 10^{12}$ cm, which is comparable to the radius of the progenitor. There should not be an exact correspondence because, as we saw, the shockwave drastically readjusts the structure of the progenitor. This "initial" radius corresponds to the value after the passage of the shockwave; we expect it to be comparable to or larger than the radius of the presupernova.

But what about the two approximations we developed? The thin solid line is the theoretical curve for the slow approximation; the fast approximation is shown as a thicker line. The kink at 30 days in both theoretical curves is the point in time at which the photosphere begins to recede into the model. Because the density distribution was that of a homogeneous sphere, this effect begins abruptly; a power-law density distribution at the outer layers of the sphere would be more realistic, less abrupt, and agree better with the observational data. This kink is the price paid for simplicity in our discussion; it clearly is a small one in this case. At earlier times the fast approximation should be better, but the true solution should approach the slow approximation as expansion reduces the diffusion time. Both curves fit the data fairly well. Mild adjustment of the parameters would improve the fast model at the expense of the slow model. An appropriate combination of the two would be better, as it

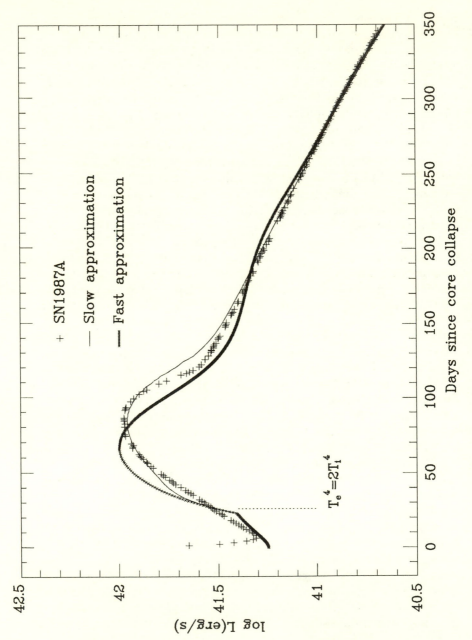

Fig. 13.9. Approximations for SN1987A

should. We will take the deviations between the two theoretical curves
to be a measure of the uncertainty.

Figure 13.10 indicates how the various effects contribute to the luminosity, for the "fast approximation" light curve of SN1987A. The observations are again plotted as pluses. The theoretical light curve, L_t, is
composed of two major contributions, the diffusion luminosity L_d, and
the advection luminosity, L_a. During the first 40 days, the luminosity due
to radiative diffusion is the most important component. A minor contribution due to the release of ionization energy by recombination, L_i, may
be seen at the lower part of the graph, from 20 to 100 days. At the top
of the figure, the degree of recombination is indicated by several vertical
lines, denoting the time that various fractions of the ejected matter are
opaque. For example, by day 43, 80 percent of the matter is still opaque,
but by day 100 only 10 percent is. During this epoch the advection contribution exceeds that of radiative diffusion. Thus, the peak at maximum
light is due to the release of trapped radiation by the inward motion of
the recombination front. At day 160 in the model, the radiative diffusion
luminosity regains dominance, but at this point the ejecta are mostly
transparent, so that the model is breaking down. Beyond this time the
luminosity is really due to reradiation of heat deposited by gamma-ray
interaction with the matter, so the sum of L_d and L_a should be thought
of as this.

Figure 13.11 shows the evolution of the effective temperature and the
photospheric radius for SN1987A, for the first 200 days after core collapse. Both the fast (solid line) and slow (dotted line) approximations
are shown. The observational data is from [278]. At about day 105, the
transition to supernebula is essentially complete, so that the idea of a
photosphere is becoming inaccurate. The fraction of the ejected mass
that is still opaque is indicated in the upper panel. After the initial burst,
to day 15, the models have a higher temperature and smaller radius than
the observations. After day 20, until day 105, the opposite is true. Some
of the error in radius compensates for a slight underestimate of the effective temperature. It is this flat part of the effective temperature curve,
from day 20 to day 100, that we use to fix the ionization temperature.
The value used, $T_{ion} = 4,500$ K, is slightly below what we estimated earlier. The evolution of these variables is sensitive to inadequacies in the
atmospheric/nebular radiation models; both the interpretation of the observational data and the boundary condition for the theoretical models
are affected. This evolution is also dependent upon the density structure,
which is more complex than we have assumed. Despite these problems
of detail, the very simple model reproduces the character and some of
the quantitative detail inferred from the observations.

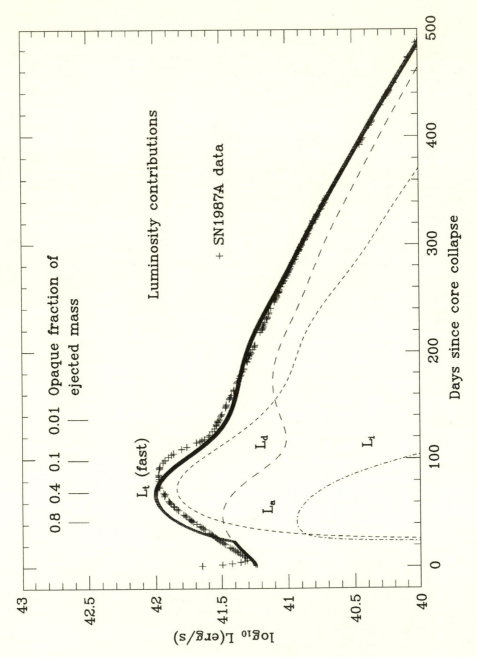

Fig. 13.10. Luminosity Contributions

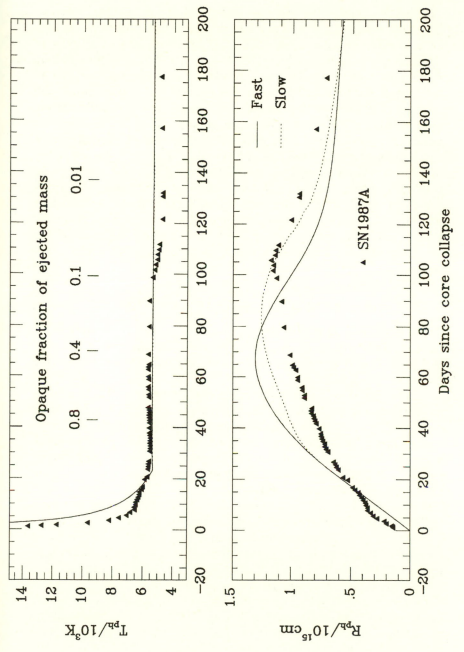

Fig. 13.11. T and R for SN1987A

The study of SN1987A promises to remain active for years to come [424]. With the explosion, the previous epochs of mass loss from the progenitor were illuminated [664, 666]. Mass loss from massive stars is an important and difficult problem [168, 214], so that this provides a unique opportunity to probe the historical development of the mass loss in one star. One of the first surprises was the discovery of a ring around the supernova, followed by the discovery of two fainter rings with Hubble Space Telescope [338, 482]. These can be understood as the results of interaction of a slow, red supergiant (RSG) wind being overtaken by a faster blue supergiant (BSG) wind [105, 412]; see also [403, 498, 665]. Extreme asymmetry is required, probably in the RSG wind, which would be almost disklike. The cause of this asymmetry is not yet known. However, given this asymmetry, Martin [412] has shown the interacting winds model to give good quantitative agreement for the emission line image [482], as well as the continuum light echo image [202]. When the ejecta from the explosion collide with the rings, SN1987A will enter a new epoch of development [424].

13.7   TYPE II SUPERNOVAE AND SN1993J

Having developed an understanding of SN1987A, it is interesting to attempt to extend this to other events. We begin by considering the morphological changes in the light curve as we vary the parameters from the values used for SN1987A.

The effect of different initial radii is shown in figure 13.12. The heavy line corresponds to the SN1987A parameters. The initial radius R_0 is varied from 10^{12} cm to 10^{14} cm. With increasing R_0, the early light curve rises, making a "plateau" structure. For larger radii, the effect of shock emergence becomes more important. This is ignored in our approximation, so that real supernovae are expected to have an initial spike in luminosity, above the curves shown in figure 13.12.

The canonical Type II supernovae are divided into two groups [68], depending upon the shape of the light curves: plateau (P) and linear (L). SN1969L, a Type II-P, was historically important for the development of the theory of supernova light curves. It was the first to be quantitatively understood [50, 669].

What change in the parameter set for SN1987A will make a light curve that resembles SN1969L? Figure 13.13 makes this obvious; increase the initial radius. In table 13.2, the following differences are indicated: the ejected mass is increased from 15 to 17 $M_\odot$, the initial radius is increased to $R_0 = 1.5 \times 10^{13}$ cm, and the distance is taken to be 10.8 Mpc. The effective opacity is smaller. The explosion energy, the Ni mass, the re-

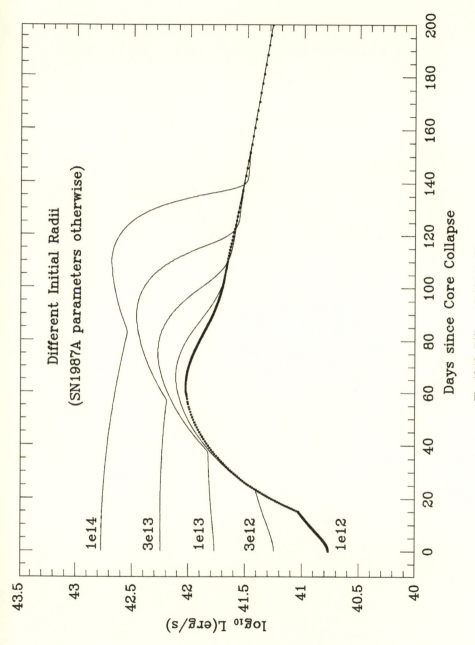

Fig. 13.12. Different Initial Radii

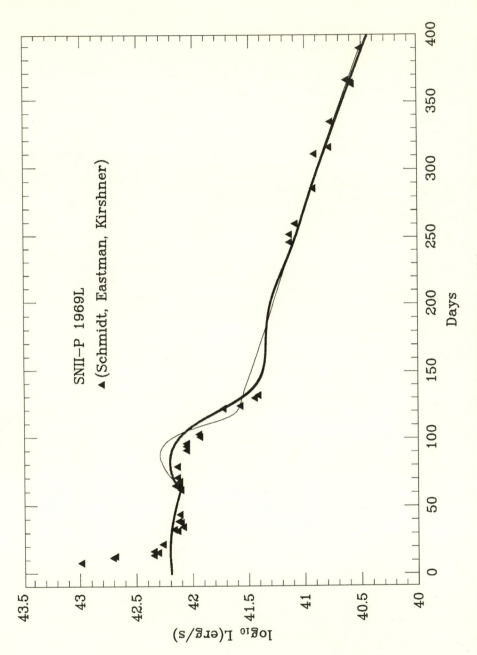

Fig. 13.13. SN1969L, a Type II-P

combination temperature, and all the other parameters are the same. Essentially the progenitor is approximated by almost the same object, except that it is a red supergiant. The resulting light curve is shown in figure 13.13, as well as the inferred UVOIR luminosity [548]. The slow approximation gives the thin line, while the fast approximation gives the heavy line. Because of the larger initial radius, the early light curve is strongly affected by the shock emergence, which is not treated in the analytic model. The lack of an outer power law in the density structure also is unrealistic. Hydrodynamic models do this phase better [50, 669]; this kind of supernova is probably the worst case for the simple analytic models we use. Still, the height and duration of the plateau are reproduced. The exponential tail corresponds to the supernebular stage, while the plateau corresponds to the last of the superstar stage. Consequently the inferred luminosity is determined differently, so that there may be some error in the relative levels.[8] The agreement is sufficiently good to suggest that such supernovae are core collapse events like SN1987A, but occurring when the progenitor was a red supergiant.

The degree of mass loss is another parameter that we need to examine. Figure 13.14 shows how the light curves for a red supergiant model change with increasing amounts of mass lost. The initial radii are all $R_0 = 3 \times 10^{13}$ cm. The mass, which remains to be ejected by the supernova explosion, ranges from $15M_\odot$ down to $2.5M_\odot$. All other parameters are the same as for SN1987A. The light curve shape collapses toward an initial hump, and an exponential tail. Again, for these large radii, we expect a spike in luminosity at early times, which will enhance this initial hump.

What is a Type II-L? In table 13.2, we see that to fit SN1980K, the ejected mass is reduced to $2.2\ M_\odot$, the initial radius is $R_0 = 2.2 \times 10^{13}$ cm, and the distance is taken to be 5.7 Mpc. The energy of explosion is reduced to 1 foe. The recombination temperature is reduced to 3,000 K; this might reflect an increased abundance of easily ionized elements, or a complication in the ionization-recombination physics. The other parameters are kept the same. The progenitor has a red supergiant envelope of much lower mass than we have used above. Presumably this would result from a star having extensive mass loss. The resulting light curve is shown in figure 13.15, as well as the inferred UVOIR absolute luminosity from [548]. The heavy line is the fast approximation; the thin line, the slow. Given the sensitivity of the shape to these variations in the recombination, the agreement is good. Again, at the earliest times, we see flaws due to the lack of shock emergence in the analytic models.

---

[8]Private communication from Brian Schmidt.

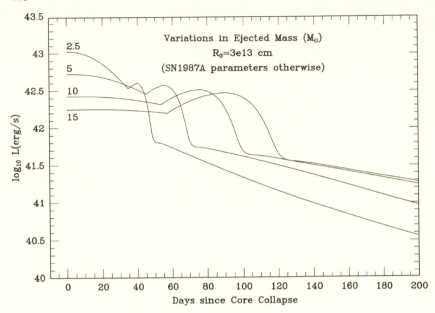

Fig. 13.14. Variations in Ejected Mass

It is interesting that the Ni mass inferred for SN1987A seems to work reasonably well for both SN1969L and SN1980K, suggesting that many of the collapsing cores may have similar properties.

Supernova 1993J, in the nearby spiral galaxy M81, was the brightest supernova seen in the Northern Hemisphere since the invention of the telescope. Like SN1987A, it provided evidence for new variations in supernova behavior. The first spectra showed a blue continuum with weak lines of hydrogen; however, the spectrum evolved in an unprecedented way, with these lines weakening, and transforming itself into that of a Type Ib supernova. The progenitor seems to have been a G or K supergiant [9]. Based upon the early light curve, the event was interpreted as the explosion of a star having initial mass $\approx 15 M_{\odot}$, but having lost all but ≈ 0.1 to $0.9 M_{\odot}$ of its hydrogen prior to core collapse [459, 507, 499, 77, 79, 645, 693]. The mass loss probably occurred by Roche lobe overflow in a binary system. In table 13.2 are found the parameters needed to explain the light curve, which is shown in figure 13.16. The data from several groups, [392, 521, 547], are shown, with their different estimates of the extinction A_v. The solid line denotes the fast approximation; the dotted line, the slow.

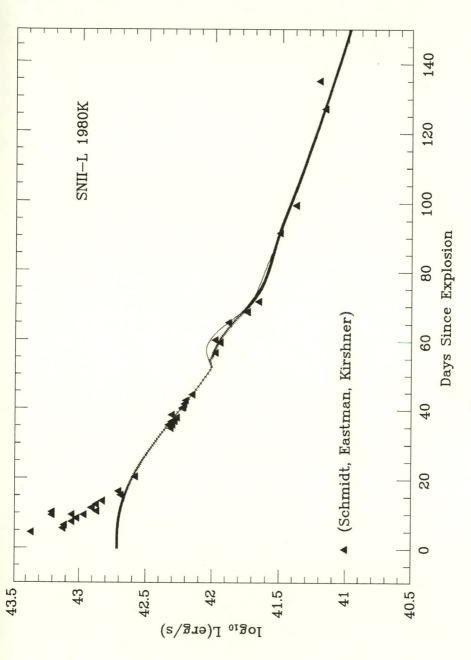

Fig. 13.15. SN1980K, a Type II-L

The figure shows data plotted as log₁₀ L(erg/s) versus Days Since Explosion for SNII-L 1980K. The plotted points are labeled ▲ (Schmidt, Eastman, Kirshner).

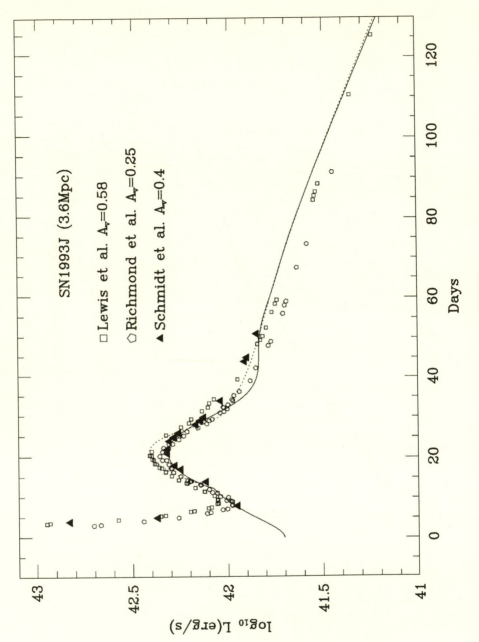

Fig. 13.16. SN1993J

Notice that we have used exactly the core properties of SN1987A, including the mass of Ni ejected, and the Cepheid distance to M81 [244]. The agreement in the level of luminosity is nontrivial: it illustrates the consistency[9] of Cepheid and supernova luminosity estimates, and hence distance scales. Note that because this is a core collapse event, and produces less ^{56}Ni than a thermonuclear event, the peak luminosity is less than that of a typical Type I supernova. Unlike SN1980K, the mass of the hydrogen envelope was not sufficient to spread the shock outburst over to the recombination peak, and produce a Type II-L.

13.8   TYPE I SUPERNOVAE

The behavior of SN1993J provides an intermediate step toward Type I supernovae. In fact, simply removing the hydrogen entirely gives what is generally regarded as a Type Ib. Despite many years of being used as "standard candles," Type I supernovae are now beginning to be recognized as a heterogeneous class. We will examine both core collapse and thermonuclear models.

Figure 13.17 shows light curves produced by a SN1987A-like core collapse, which ejects a hydrogen-poor mantle ranging from 6 down to 1 $M_\odot$. All initial radii are small. The low mass and higher recombination temperature ($T = 7.5 \times 10^3$ K) give a faster light curve, and a higher initial (^{56}Ni) peak. This is an important effect; a light curve computed with constant opacity will be dimmer and more extended in time. The exponential tails vary due to differing transparency to gamma rays. The effect is much more pronounced in these lower mass, faster expanding events, than in SN1987A. With no H lines and such light curve shapes, these events would be classified as "Type I."

It is difficult for core collapse events to produce bright Type I supernovae. The stars in the lower end of the mass range having core collapse, also have steep density gradients due to core convergence (see chapter 10). The reason is simple. High density forces electron capture, so that the neutron excess becomes too large for ^{56}Ni to be abundantly produced. Low density implies a specific heat so large that peak temper-

---

[9]The expanding photospheres method (EPM) was used on this supernova. The calculations of [546] used a grid of models which did not include the large mass loss indicated for this event, and obtained a smaller distance (2.6 ± 0.4 Mpc). However, better agreement is found by Baron et al., [78], with a grid of models which do include mass loss. Refinement of the EPM method may allow extreme cases to be accurately determined as well, according to Eastman and Pinto [217].

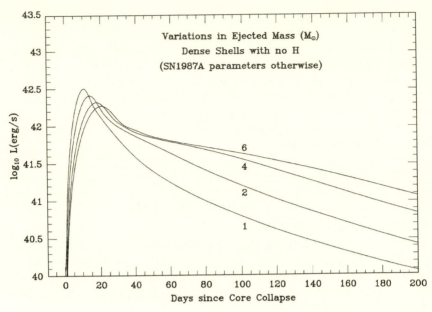

Fig. 13.17. Non-H Ejecta

atures are too low to allow burning all the way to ^{56}Ni. A steep gradient means little mass between these extremes to make ^{56}Ni. Larger-mass stars might make more ^{56}Ni (they have shallower density gradients; see chapter 10), but their larger cores may be more likely to become black holes [17], which could affect the ejection of ^{56}Ni. Also, larger mass in the ejected mantle tends to flatten the light curve too much (see the $6M_\odot$ curve in figure 13.17). It seems plausible that at least most Type I supernovae which are due to core collapse will be relatively dim ($\log_{10} L_{UVOIR} \leq 42.3$), probably comparable to SN1993J after 10 days, when the hydrogen layer became unimportant for the light curve.

While the core collapse models have no simple, tightly defined relationship[10] between the important parameters of explosion energy, mass ejected, and ^{56}Ni ejected, this is often not true for thermonuclear models. The production of ^{56}Ni is usually related to the same burning that releases the explosion energy. Both are usually related to the total mass, which is also the mass ejected. These models rely on an electron-degenerate configuration for the thermal instability to start the explosion [305], so that as much as $1.4M_\odot$, a Chandrasekhar mass of

[10]At least, none is presently known.

^{56}Ni, might be produced. Thus, bright Type I supernovae are likely to be thermonuclear events rather than core collapse.

What energies are available for thermonuclear events? Table 13.3 lists Q-values, both in units of erg per gram and of 10^{51} erg (foe) per solar mass burned, for several likely combinations of fuel and ashes. A typical supernova energy of 1.5×10^{51} erg per Chandrasekhar mass is about 1 foe per solar mass. All entries exceed this, suggesting that the burning may not go to completion, or that the masses may be below the limiting value. Burning from ^{4}He is significantly more energetic than burning from ^{12}C and ^{16}O.

With the discovery of accurate analytic representations for the light curves of Type I supernovae [36, 39], it became clear that such events could well be heterogeneous and still maintain their characteristic light curve shapes, and temperature evolution. SN1970J and SN1972E were used as examples, and indicated a variation in explosion energy of a factor of two. Unfortunately, different observational techniques and conditions could give spurious variations. Significant new observations [495, 419, 279] of high quality suggest that an intrinsic inhomogeneity among SNIa does indeed exist. An intrinsic dispersion in the peak absolute magnitudes of SNIa of $\sim$ 0.8 mag in M_B and $\sim$ 0.5 mag in M_V was inferred from a sample of close objects. Even when "peculiar" subluminous objects like SN1991bg are eliminated, the dispersion reduces to $\sim$ 0.3 to 0.5 mag [279]. This range is consistent with that suggested from SN1970J and SN1972E [39]. Moreover, the new data confirm the findings by Phillips [495] that the absolute magnitudes are correlated

TABLE 13.3

Q-values for Type I Supernovae

Fuel	Ashes	Erg/g	foe/$M_\odot$
^{4}He	^{56}Ni	1.57(18)	3.007
^{4}He	^{28}Si $+^{32}$S	1.40(18)	2.677
^{4}He	^{12}C $+^{16}$O	7.28(17)	1.446
^{12}C	^{24}Mg	5.60(17)	1.113
^{12}C $+^{16}$O	^{56}Co	8.22(17)	1.645
^{12}C $+^{16}$O	^{56}Ni	7.86(17)	1.561
^{12}C $+^{16}$O	^{28}Si $+^{32}$S	6.20(17)	1.231

with the decline rate of the B light curve. This suggests that the progenitors of SNIa have a range of parameters. In chapter 11 we saw at least two interesting possibilities for the parameter which varies: (1) the mass at ignition in the helium detonation of sub-Chandrasekhar white dwarfs [401, 700, 400], and (2) the density at which the transition to detonation occurs in delayed detonation events [350, 351, 399, 52, 53]. Understanding what the variation implies will require high-quality, quantitative simulations of the supernova spectra, with careful treatment of time dependence and geometric effects. At maximum light, Type I supernovae have already begun the transition from superstar to supernebula.

Figure 13.18 compares light curves from models of helium detonation of sub-Chandrasekhar white dwarfs (solid lines), and of delayed detonation events in Chandrasekhar-mass white dwarfs (dotted lines). For purposes of comparison, a fairly consistent set of models was used [53, 400]. Notice that the brightest of both classes of models are similar in luminosity at maximum ($L_{max} \leq 2 \times 10^{43}$ erg s^{-1}) and in light curve shape. While both classes show a range of brightness at maximum light, the range is larger for the sub-Chandrasekhar models; these also show a more pronounced change in slope of the postmaximum luminosity. The Chandrasekhar-mass white dwarfs have a later occurrence of maximum light for the dimmer models. At present there seems to be no clear reason for ruling out either possibility, so that an attempt to determine the relative frequency of occurrence may be in order.

Recognizing its many weaknesses, we shall see how far we may push our simple light curve model. Table 13.4 shows parameters needed to fit the Suntzeff [599] data on 1991T, 1989B, 1992A, and 1991bg (which were kindly provided in machine-readable form). The Suntzeff sample is on a consistent $H_0 \approx 85$ kms^{-1}Mpc^{-1} distance scale [336]. The fits to SN1989B, SN1992A, and 1991bg are not unique, but the values given are representative. For this distance scale, we find that for the dim supernovae SN1991bg, the mass of radioactive Ni is $0.075 M_\odot$, similar to that for SN1987A. For the bright supernova SN1991T, we find the much larger value of $0.7 M_\odot$. This light curve is reasonably well represented by models r and q in [53]. While SN1991T is one of the brightest observed supernovae, these theoretical models were not the brightest computed; they lie just above the middle of the range shown in figure 13.18 (see the dotted light curves).

As an experiment, we shall deal with this inconsistency by shifting the Suntzeff sample of SNIae by a factor of 1.3 in distance, to bring them to a consistent "supernova" distance scale. Figure 13.19 shows these data, and the data on the SNIc supernova 1994I by Brian Schmidt [549].

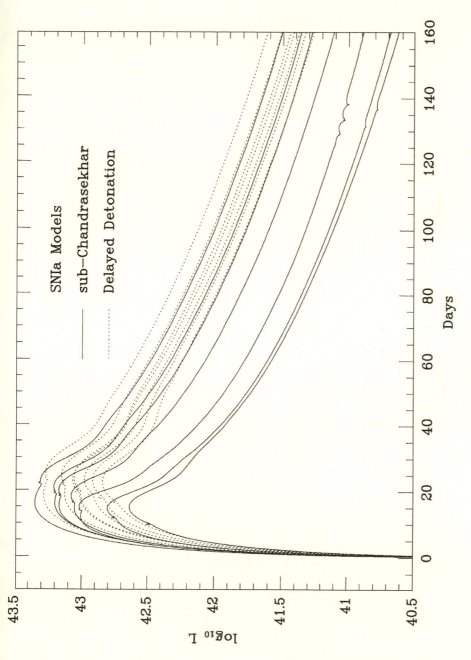

Fig. 13.18. SNIa Models

TABLE 13.4
Type Ia Supernova Parameters (1)

Supernova[a]	1991T	1989B	1992A	1991bg
Type	SNIa	SNIa	SNIa	SNIa
model	r, q [53]			
$M_{ej}/M_\odot$	1.33	1.1	1.1	1.1
E_{51}	1.27	1.7	1.7	1.7
$R_0/10^{12}$ cm	0.001	0.001	0.001	0.001
$M(^{56}Ni)/M_\odot$	0.692	0.2	0.15	0.075
$\kappa(cm^2/g)$	0.15	0.15	0.15	0.15
$T_{ion}(K)$	7,500	7,500	7,500	7,500
q_{ion}(ev/nucleon)	13.6	13.6	13.6	13.6
R_{Ni}/R	0.727	0.65	0.65	0.65
Galaxy	NGC4527	NGC3627	NGC1380	NGC4374

[a]These fits are based on the short distance scale [336].

Also plotted, as solid lines, are the light curves predicted by the sub-Chandrasekhar mass models [400], and by a putative core collapse model for SN1991I as the dotted line.

The parameters are summarized in table 13.5. An attractive feature is apparent: these models reproduce both the range of observed luminosity and the variation postpeak decline rate with luminosity at maximum [495]. The SNIc, 1994I, lies well below the Ia curves, as we would expect for a core collapse event. Its implied mass of radioactive Ni is about two-thirds that of SN1987A, which does not seem implausible. Because of the convergence of the light curves for bright events which was pointed out above, the delayed detonation models would give the same result for these most luminous supernovae, but might have difficulty reproducing the dimmest SNIa events like SN1991bg.

Taken with high-quality observations, made primarily since SN1987A, this simple theory of the light curves gives considerable insight into the systematics of supernova events. The consistent difficulty in making supernovae dim enough for the shorter distance scale is suggestive. As we saw in chapter 11, a preliminary comparison of velocities with observed spectral features (SiII, CaII, and MgII) is consistent with the longer dis-

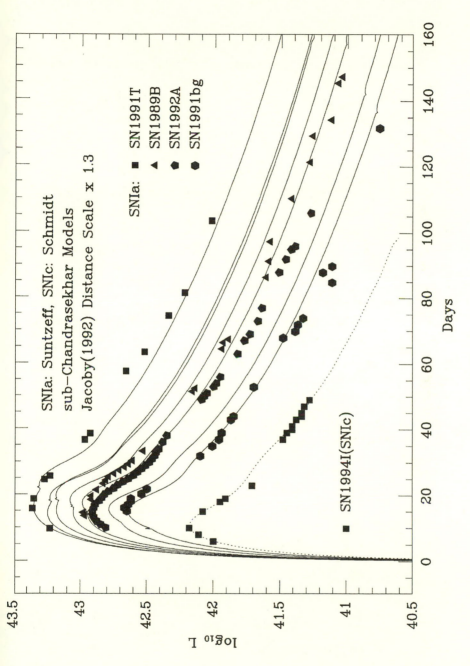

Fig. 13.19. SNI Models

TABLE 13.5

Type I Supernova Parameters (2)

Supernova	1991T	1989B	1992A	1991bg	1994I
Type	SNIa	SNIa	SNIa	SNIa	SNIc
LA model	9	4	2	1	fit
$M_{ej}/M_\odot$	1.1	0.87	0.8	0.7	1.2
E_{51}	1.48	1.18	1.0	0.69	1.7
$R_0/10^{12}$ cm	0.001	0.001	0.001	0.001	0.001
$M(^{56}Ni)/M_\odot$	1.0	0.466	0.358	0.140	0.045
$\kappa(cm^2/g)$	0.15	0.15	0.15	0.15	0.15
$T_{ion}(K)$	7,500	7,500	7,500	7,500	7,500
q_{ion}(ev/nucleon)	13.6	13.6	13.6	13.6	13.6
R_{Ni}/R	0.7	0.7	0.7	0.7	0.65
D(Mpc)[a]					7
Galaxy	NGC4527	NGC3627	NGC1380	NGC4374	M51

[a]The Suntzeff sample of 1991T, 1989B, 1992A, and 1991bg have been shifted (increased distance by a factor of 1.3) from a consistent $H_0 \approx 85$ kms^{-1}Mpc^{-1} distance scale, to one corresponding to $H_0 \approx 65$ kms^{-1}Mpc^{-1}.

tance scale. However, no definite conclusion should be made until physically consistent simulations of the spectra allow us to fully exploit the wonderful data now becoming available. An understanding of supernova light curves and spectra will automatically imply the establishment of an independent astronomical distance scale.

14

Galactic Evolution

The end is where we start from.
—T. S. Eliot (1888–1965),
Four Quartets

How do these ideas combine to explain the abundance patterns with which we began this book? Rather than a completed answer we find new questions taking shape—how do galaxies come into being, and how do they evolve?

The Standard Model of cosmological nucleosynthesis deals with a single event, but stellar nucleosynthesis deals with an ensemble of events at different places and times. It occurs within the context of a galaxy, that is, a complex which we observe mostly as stars, interstellar gas, and dust. The abundances that ultimately result depend upon the integrated effects of all these different stellar events, with their individual, and perhaps different contributions. Given this possible complication, it is astounding that there should be a "cosmic abundance pattern" at all. How does it happen?

Parts of the answer can be pieced together, but there remain areas of incompleteness. For example, a quantitative theory of star formation eludes us. Galaxies as we observe them are ensembles of stars and interstellar gas. The gas radiates, cools, and contracts by self-gravitation. Somehow new stars are formed. The stars evolve, ejecting matter back into the interstellar medium, either by steady mass loss or by more dramatic events (such as formation of planetary nebulae, novae, or supernovae). Progress has been made by using those regularities observed in star formation, and inferred from the supposed evolution in nearby systems, but as we lack a real understanding, extrapolation to other regions of space and time is dangerous. To complete our picture of the evolution of matter, we must somehow fill this gap. What happened after the Big Bang? What was the nature of the first generation of stars (Population III)? What were/are protogalaxies? How do galaxies evolve?

A first step is to consider the implications of what we have learned, for the best known sample of a galaxy we have—the solar neighborhood. We have seen that the relative yields of stars give plausible explanations of the details of the solar abundance pattern, and to less certainty, the Population I pattern. What are the absolute yields? Do these ideas provide an explanation of the rates of supernovae, pulsar formation, or white dwarfs? The answers are encouraging, and may provide a basis for attacking the questions raised above.

We may be bolder. Suppose the first generation of stars was simply related to those we now observe to be forming nearby. The first stellar nucleosynthesis would be the yields of these massive and fast-evolving stars. This simple idea seems to hold some promise when compared to the abundances of Population II ("metal-poor") stars. Further, the added contributions of thermonuclear supernovae, and of intermediate mass stars, seem to provide what is needed to flesh out the full solar system abundance pattern.

14.1 GALACTIC EVOLUTION EQUATIONS

The interstellar medium is a complex plasma. To be treated as a gas, a minimum requirement is that the matter have a sufficiently large rate of collisions. The collision cross section for two hydrogen atoms is $\sigma \approx 10^{-16}$ cm^2, so that the mean free path is $\lambda \approx 10^{16}/n$ cm, where n is the number density of hydrogen atoms. For $n > 1$, $\lambda < 10^{16}$ cm. In describing interstellar space, the characteristic distances are of order 10^{18} cm, the distance between stars, which is much larger than stellar radii, of order 10^{14} cm or less, or the collisional mean free path λ. This gives some support to the approximation of the interstellar gas (on larger scales) as a hydrodynamic fluid. For the present we will suppress discussion of the magnetic field, but emphasize that in a more complete picture it plays an important role (see [486], ch. 22).

The continuity equation expresses the conservation of nucleon number:

$$\partial \rho / \partial t + \nabla \cdot \rho \mathbf{v} = 0. \tag{14.1}$$

Consider a domain consisting of interstellar space, but not including the space *inside* stars. The continuity equation is modified, so that integrating the divergence term over the stellar surfaces gives rise to source and sink terms located at the stellar positions:

$$\partial \rho / \partial t + \nabla \cdot \rho \mathbf{v} = s_{mass}. \tag{14.2}$$

We may also write a formally similar equation for each species i, where i might be, say ^{4}He or ^{16}O,

$$\partial \rho^i / \partial t + \nabla \cdot \rho^i \mathbf{v} = s^i_{mass}, \tag{14.3}$$

where $\rho^i = X_i\rho$. Similarly the hydrodynamic equations for momentum conservation,

$$\partial(\rho\mathbf{v})/\partial t + \nabla(\rho(\mathbf{vv} + \varphi) + P) = s_{mom} \qquad (14.4)$$

and energy convervation,

$$\partial(\rho(E + v^2/2))/\partial t + \nabla \cdot \mathbf{v}(\rho(v^2/2 + E + \varphi) + P) = s_{energy} \qquad (14.5)$$

will be modified in the same way to include analogous source terms.

To derive the corresponding set of equations to describe the dynamics of the stars, we perform a corresponding integration over the domain which includes *only* the volume inside the stars. To the extent that the stellar radii are small, the stars may be approximated by points of mass. This results in usual equations of stellar dynamics [99], but with three additional terms for each star, accounting for its exchange of mass, momentum, and energy with the interstellar medium. Consider a volume over which the interstellar gas density ρ is defined. The quantity s_{mass}, introduced above, represents the sum of mass exchange terms for all stars in this volume. Usually in stellar dynamics, a simpler case is considered, in which the stellar masses are taken to be constant $(dm/dt = 0)$. Now for the j stars in the volume,

$$s_{mass} = -\sum_j dm^j/dt, \qquad (14.6)$$

describes the mass interchange. Similarly, momentum and energy may be exchanged between *each* star and the interstellar medium. To correctly specify these exchange terms requires knowledge of stellar evolution, and of the interaction of the stars with the interstellar medium. Because stars are born—and die—the number of mass points is not conserved.

These coupled equations for gas and star dynamics are complex, so that further simplifications are almost always made. The most popular and extreme is the *one-zone model*, in which the details of structure are either ignored or lumped into boundary conditions. Less extreme approximations may still be flawed with respect to *resolution*; a "fluid element" for a hydrodynamic description of the interstellar medium may be far from homogeneous, so that its average properties would behave in a complicated way.

Let $\mathcal{G}$ be the mass in the form of interstellar gas,[1] and $\mathcal{S}$ be the mass in the form of stars . Consider a spatial domain D, with external surface Σ. Assume $\mathcal{G}$ and $\mathcal{S}$ to be smoothly varying quantities, in a statistical sense, on subvolumes d of D. Then,

$$\mathcal{G} = \int_d \rho d^3 x. \tag{14.7}$$

Here $\mathcal{G}$ is the total mass of interstellar matter in the subvolume d. Using this choice, and integrating 14.2 gives

$$\partial \mathcal{G}/\partial t + \nabla \cdot \mathcal{G} \mathbf{v} = \int_d s_{mass} d^3 x + \mathcal{I}. \tag{14.8}$$

Here $\mathcal{I} = -\int_\Sigma \rho \mathbf{v} \cdot d\Sigma$ is the net mass flowing in through the external boundary; this expression is often called *infall*. The integral on the right-hand side is the net amount of mass added by stars. In the *one-zone model*, it is further assumed that the system is homogeneous over D, so that $\nabla \cdot \mathcal{G}\mathbf{v} = 0$, as are all spatial derivatives, simplifying Eq. 14.8. This reduces the partial differential equation to an ordinary differential equation; the left-hand side becomes simply $d\mathcal{G}/dt$.

The quantity $\int_D s_{mass} d^3 x = -\mathcal{F} + \mathcal{E}$ is of fundamental importance to the theory of galactic evolution. We have changed the summation to an integration over m, assuming a statistical treatment of the stars to be adequate, as above. The first term on the right-hand side, $\mathcal{F}$, is the rate at which interstellar mass is *formed* into stars of all masses m,

$$\mathcal{F} = \int_0^\infty \mathcal{B}(m, t) \, dm, \tag{14.9}$$

where $\mathcal{B}(m, t)$ is called the *birthrate function*, and is the rate at which mass is converted at time t into stars of mass m. Here $\mathcal{E}$ is the rate of mass *ejection* from stars of all masses during their evolution and death. How do we relate this $\mathcal{E}$ to $\mathcal{B}(m, t)$? The mass of matter that went into making stars of mass m to $m + dm$, and of age a to $a + da$, is $\mathcal{B}(m, t - a) \, dm \, da$, which is just the birthrate at time $t - a$. Therefore

---

[1]It would be more realistic to consider the interstellar medium to be heterogeneous, consisting of several types of "gas" (for example, phases dominated by molecules, by atoms, or by ions). This would necessitate extra subscripts on $\mathcal{G}$, and a set of coupled differential equations, representing each component of the interstellar medium and their interactions. We will suppress this complexity in the present discussion.

we have the double integral,

$$\mathscr{E}(t) = \int_0^\infty \int_0^t \mathscr{B}(m, t - a)\mathscr{R}(a, m) \, dm \, da, \tag{14.10}$$

where we integrate over all possible masses, and over all possible ages. The return function $\mathscr{R}(a, m)$ depends upon the stellar age a and the mass m; it tells how the initial matter of each star is (or is not) returned to the interstellar medium. In fact, $\mathscr{R}(a, m)$ probably depends upon other variables such as initial composition, presence of a companion, etc., in which case there are additional integrations over each of these variables as well. Many stars spend most of their lives burning hydrogen without ejecting much matter. The most prominent exceptions are massive stars which have lifetimes short compared to most timescales of interest for galactic evolution. While not strictly true for all cases, a useful and often-used approximation is that the star returns whatever matter it ejects at the end of its life, at age $a = \tau_m$. Then the return function becomes

$$\mathscr{R}(a, m) \approx \delta(a - \tau_m) \, \mathscr{R}(m), \tag{14.11}$$

so that

$$\mathscr{E}(t) \approx \int_0^\infty \mathscr{B}(m, t - \tau_m)\mathscr{R}(m) \, dm. \tag{14.12}$$

A related approximation is that of *instantaneous recycling*. If the rate of change of the birthrate $\mathscr{B}(m, t)$ is slow compared to the lifetime of the stars of interest, then

$$\mathscr{B}(m, t - \tau_m) \approx \mathscr{B}(m, t). \tag{14.13}$$

This will be invalid for lower-mass stars (e.g., $m < 1$), but will be true for the more massive stars which do almost all of the nucleosynthesis. We use it as an approximation that is both illustrative and simple, but warn of its limitations. A further step is to assume that $\mathscr{B}(m, t)$ is separable in the variables m and t, so

$$\mathscr{B}(m, t) = \Lambda(t)\xi(m). \tag{14.14}$$

The function $\xi(m)$ is the *initial mass function*, whose properties are discussed below. The rate of mass *ejection* from stars of all masses and ages,

$\mathscr{E}$, is related to the rate at which interstellar mass is devoured by star formation, $\Lambda(t)$, by $\mathscr{E}(t) = f\Lambda(t)$, where

$$f = \int_0^\infty \mathscr{R}(m)\xi(m)\,dm, \tag{14.15}$$

is the return fraction, a constant in time in this approximation.[2]

In order for the one-zone model to be valid (or even the less restrictive models which use coarse-grained hydrodynamics), the processes that homogenize the gas (in the reference volume) must have time to operate. Suppose that mass, composition, momentum, and energy are all transported at some velocity v. Then if the volume has a length scale ℓ, it requires a time larger (pehaps much larger) than $t = \ell/v$ to establish homogeneity. Suppose that the mixing velocity is sufficiently small that it does not ionize the interstellar gas. To do so might inhibit star formation; in any case we can take a typical velocity observed in the interstellar medium. Thus, if $v \le 10\ \mathrm{km\,s^{-1}}$, then to mix on a kiloparsec scale takes a time $t \ge 3 \times 10^{15}\ \mathrm{s} \approx 10^8\mathrm{y}$. This is comparable to the dynamical time for stars at the solar position in our galaxy, and a factor of ten larger than the time of evolution of massive stars which do most nucleosynthesis. A random-walk estimate of the mixing time, which might be more reasonable, would be still more restrictive. *The one-zone equations are not valid for the initial phases of galactic evolution,* although they have often been so used.

To understand the source terms, s_{mass}^i, s_{mom}, and s_{energy}, some idea of their magnitude and nature are needed. Such estimates are the key to plausible models of galactic systems, but depend in complicated ways upon the interaction of stars with the interstellar medium. This short discussion is intended to illustrate some aspects of an unsolved problem; it is certainly not complete.

First, consider the transfer of mass between stars and the interstellar medium. The mass of a star or protostar might be changed by accretion, by mass loss during evolution, or by an explosive event. The last two topics will be explored in more detail in later sections.

A gravitating, point mass m in a gaseous medium undergoes Bondi-Hoyle accretion [115]. If the relative velocity is v_{rel}, then gas that is less than a distance r_g from the point will be strongly influenced by

---

[2]If the age of stars of mass m and below is sufficiently large that they return little to the interstellar medium, then $\mathscr{R}(m) \approx 0$ and the integral is truncated. For ages of the order of the Hubble time (say 10 Gy), stars with masses just above this limit are thought to become white dwarfs, with ejection of matter. Thus f is not really constant, but can be treated as slowly varying.

gravity. This gravitational radius is defined by $r_g = 2Gm/v_{rel}^2$. At low relative velocity v, this expression is singular. Then $r_g = 2Gm/v_s^2$, where v_s is the sound speed; this is essentially the virial theorem. Thus, in the sense of an interpolation formula, we take v to be the larger of v_{rel} and v_s. Suppose $m = 2 \times 10^{33}$ g, $v = 10^6$ cm s^{-1}, and the gas density is $\rho_{ism} = 1.6 \times 10^{-24} n$ g cm^{-3}, where n is the interstellar nucleon density by number. Then the area swept out of the gas by the gravitating point is πr_g^2, so the mass accretion rate[3] is

$$dM/dt = \rho_{ism} 4\pi (Gm)^2/v^3,$$
$$\approx 5 \times 10^{11} n \text{ g s}^{-1}$$
$$\approx 8 \times 10^{-15} n \text{ M}_\odot \text{ yr}^{-1}. \tag{14.16}$$

Consider a typical interstellar density, $n \approx 1$ cm^{-3}. Then the time to accrete a solar mass on a solar mass object is $\tau_{acc} \approx 10^{14}$ yr, much longer than the Hubble time or the age of the galaxy. However, in a giant molecular cloud, $n \approx 10^4$ cm^{-3}, and for a temperature $T \approx 25$ K, the sound speed is $v_s \approx 5 \times 10^4$ cm s^{-1}. This gives $\tau_{acc} \approx 10^6$ yr, which is shorter than the lifetime of the even the most massive stars. Star formation is observed to be associated with such giant molecular clouds, so we anticipate that mass accretion will be important for protostars (which might move this slowly relative to the gas in such regions), but not for ordinary stars in regions of lower interstellar gas density.

Similarly, we can esimate the momentum exchange from Bondi-Hoyle accretion. The rate of change of momentum is $d(mv)/dt \approx v \, dm/dt$, if the velocity of the star is slowly changing. Then the time for significant momentum change is $\tau_{mom} = mv/[d(mv)/dt] \approx \tau_{acc}$ given above. This effect also would be most important for protostars in giant molecular clouds, and probably unimportant for ordinary stars.

Stellar winds are another source of momentum; they exert a pressure on the interstellar medium. The ram pressure of the wind is $P_w = \rho_w v_w^2$, where ρ_w is the wind density and v_w its velocity. The gas pressure of the interstellar medium is

$$P_{ism} \approx N_A k \rho_{ism} T_{ism} \tag{14.17}$$
$$\approx 10^{-16} n_{ism} T_{ism} \text{ dyne cm}^{-2}. \tag{14.18}$$

---

[3]Three-dimensional numerical simulations of Bondi-Hoyle accretion [530] indicate that the scaling laws so derived are correct on average but that the flow shows a chaotic behavior, and is much more complex in detail.

For interstellar number density of $n_{ism} = 1$ cm^{-1} and temperature of $T_{ism} = 10^4$ K, $P_{ism} \approx 10^{-12}$ dyne cm^{-2}, and magnetic pressure ($P_{mag} = B^2/8\pi$) is expected to be comparable. The mass loss rate from the star is $dm/dt = 4\pi r^2 \rho_w v_w$. Taking typical values to be $dm/dt \approx 10^{-6}$ M$_\odot$ yr^{-1} and $v_w = 100$ km s^{-1}, $P_w = 5 \times 10^{26}/r^2$ dyne cm^{-2}. For $P_w = P_{ism}$, this implies

$$r \approx 2 \times 10^{19} \text{ cm}, \tag{14.19}$$

or about 7 pc. The scaling with mass loss rate implies that this boundary radius would approximate a typical interstar spacing (1 pc) at a modest mass loss rate of $dm/dt \approx 2 \times 10^{-8}$ M$_\odot$ yr^{-1}. This suggests that the source s_{mom} is not necessarily negligible in regions having stars with appreciable mass loss (e.g., OB associations, giant molecular clouds). It also implies that treating the gas as a homogeneous quantity (or assuming that it has a structure independent of the presence of these stars) is not valid.

The transfer of energy between stars and the interstellar medium is an important effect which is difficult to evaluate precisely. The kinetic energy from a stellar wind is $K_w = \Delta m v_w^2/2$, where v_w should be the average wind velocity and Δm the total mass loss over the time of interest. For a massive star, we take $v_w \approx 200$ km s^{-1} and $\Delta m \approx 0.5m$. Then $K_w \approx 10^{47}m$ erg $\leq 10^{49}$ erg, where m is the mass in solar units.

However, the energy radiated by a star is

$$L\tau = \int_0^\tau L \, dt = \phi M Q \approx 5 \times 10^{51} m \text{ erg}, \tag{14.20}$$

where ϕ is the fraction of the mass of the star that is burned, and Q is the energy released by that burning. While this energy $L\tau$ is much larger than that from a stellar wind, its efficiency at heating gas is not high (we do *see* stars, which means their light generally escapes through the gas in their vicinity). The largest cross section for interaction is from the resonance lines. For hydrogen atoms, the ionization limit signals the onset of large opacity in the continuum; we require that the typical photon have such an energy $3kT \geq 13.6$ ev, or $T \geq 45,000$ K. This would correspond to O stars, and $m \geq 15$. Thus, these relatively rare stars are the most effective at heating the surrounding gas, and their effectiveness depends upon the distribution of gas in their neighborhood.

Supernova explosions deliver about (1 to 2) 10^{51} erg (or 1 to 2 foe) directly as kinetic energy of ejected mass, as we saw in chapter 13. The integrated photon luminosity of supernovae is less by a factor of 0.01 or so. Many of the supernova progenitors would have been O stars during some part of their evolution, which adds to the energy input. Larger star-forming regions should display violent gas motions that could affect their nature, and the star-formation process itself. On the other hand, the ejected matter will create a shock wave that may be radiative. This causes much of the supernova kinetic energy to be reradiated in the form of photons that escape the interstellar medium. In the limit of a nonadiabatic shock wave, the supernova ejecta may be thought of as having an inelastic collision with a part of the interstellar medium, $m_{sn}v_{sn} \approx m_{sw}v_{sw}$, where the mass and velocity of the swept-up gas are m_{sw} and v_{sw}. If we take values appropriate for a Type II supernova, $m_{sn} = 10$ in solar units, and $v_{sn} \approx 3 \times 10^8$ cm s^{-1}, and identify v_{sw} with a typical turbulent velocity in the interstellar medium (15 km s^{-1}), we have $m_{sw} \approx 2 \times 10^3$ M$_\odot$. Thus the kinetic energy left is $K \approx K_{sn}v_{sw}/v_{sn}$, a reduction down to about 0.005 of its initial value.

These estimates, while crude, do suggest that the traditional practice of ignoring all source terms—except those for mass ejected from stars—is probably not adequate, particularly in regions of active star formation.

14.2 INITIAL MASS FUNCTIONS

Salpeter [536] introduced the notion of an *initial mass function*, or IMF, $\xi(m)$. Suppose that stars are being formed over some interval of time t_0, and that t_0 is shorter than the evolutionary time of any of the stars. Let the least massive star considered have mass m_{min} and the most massive m_{max} (not necessarily very different). The total number ΔN of stars formed in this region of space, of all masses m in the range $\Delta m = m_{max} - m_{min}$, is

$$\Delta N = \int_0^{t_0} (\mathscr{B}(m, t)/m)dt \; \Delta m, \qquad (14.21)$$

where the birthrate *by mass*, $\mathscr{B}(m, t)$, is the rate at which the mass of interstellar gas is converted into stars of mass m at time t. The birthrate *by number* of stars is $\mathscr{B}(m, t)/m$. Salpeter assumed that $\mathscr{B}(m, t)$ is separable in the variables m and t, so

$$\mathscr{B}(m, t) = \Lambda(t)\xi(m), \qquad (14.22)$$

where $\Lambda(t)$ is the rate at which mass of gas is converted into stars, and $\xi(m)$ is the IMF. Thus, the IMF is the function that describes the distribution of newly formed stars in terms of their most crucial parameter, the initial mass. For example, it gives the ratio of mass tied up in stars of 1 $M_\odot$ to that of stars of 10 $M_\odot$, from the process of star formation.

One interpretation of the IMF is that it is a universal function, the same for all bursts of star formation at all places and times. A less extreme interpretation is that it is the appropriate average of a variety of different episodes of star formation (see [220, 542] for discussion), and is an average over both space and time. In the latter case, the presumed constancy (or approximate constancy) of the IMF is surprising, and must reflect a statistical regularity in the star-formation process which is not yet understood.

Volumetric IMF

In order to appreciate some of the issues involved in deriving an initial mass function, it is instructive to repeat the Salpeter analysis (his mass function is still a reasonable approximation for the mass range he considered!). We will use the same basic observational data that Salpeter used, but with modern bolometric corrections and stellar evolutionary theory, and supplement the data for stars of very large and very small luminosities. The original discussion, in analogy with Boltzmann statistical mechanics, dealt with volumetric densities, so that N was the number of stars in a unit *volume* near the Sun.

The original observational data consist of counts of stars having absolute visual magnitudes in a given range, giving the total luminosity function, $\phi_t(M_v)$. This implies a determination of the stellar distances, and the associated errors. These data are corrected further, first to eliminate stars not on the main sequence,[4] and then to convert visual luminosities to bolometric luminosities,[5] giving the bolometric luminosity function for main sequence stars, $\phi(M_b)$.

Because

$$dN/dM_v = (dN/dM_b)\,(dM_b/dM_v), \qquad (14.23)$$

we have $\phi_v = \phi_b\,dM_b/dM_v$. Thus, a knowledge of the derivative of the bolometric correction is needed to convert from M_v to bolometric luminosity functions. Table 14.1 gives a relatively smooth version of the bolometric correction for main sequence stars; see[191, 92, 289].

We now generate a theoretical estimate of $\phi(M_b)$. Let $l = L/L_\odot$ and $m = M/M_\odot$. In chapter 7 we derived the mass-luminosity relation for

---

[4]This is done by spectral analysis, primarily on the basis of post-main-sequence stars having lower densities, and therefore higher ionization at the same temperature. All types of stars could be considered if their evolutionary state could be reliably determined.

[5]Although bolometric corrections for red stars do not seem to have been made in the original work, the inclusion of modern values gives the same final *form* for most of the mass function.

TABLE 14.1

M_v and M_b

M_v	M_b	M_v	M_b	M_v	M_b
-6.0	-11.40	3.0	3.00	12.0	9.70
-5.0	-8.30	4.0	4.01	13.0	10.30
-4.0	-6.80	5.0	4.90	14.0	10.90
-3.0	-5.20	6.0	5.80	15.0	11.45
-2.0	-3.90	7.0	6.52	16.0	11.99
-1.0	-2.31	8.0	7.03	17.0	12.30
0.0	-0.90	9.0	7.72	18.0	12.70
1.0	0.49	10.0	8.44	19.0	13.26
2.0	1.71	11.0	9.08		

main sequence stars from simple arguments. Our purpose here is to illustrate the method rather than provide an embellished product. Rather than fit data from detailed stellar evolutionary sequences (for example, [311, 312, 313, 314, 315, 474]), we use the method of chapter 7, that is, Eddington's Standard Model constrained by the Fowler-Hoyle approximation to total nuclear heating. While less accurate, this approach provides a clearer view of the underlying physics. Stars brighten while on the main sequence; the luminosity here should correspond to the average luminosity of the star during its time on the main sequence, and will be larger than the luminosity on the zero-age main sequence. This brightening, which is less pronounced for massive stars, can be approximated by considering the effective opacity in the Eddington model to be reduced (by the conversion of hydrogen to helium, which is more transparent).[6] In particular, the constant 5.88 in equation 6.66 is lowered by a factor of 0.6. For a star of one solar mass, this gives a luminosity—averaged over the main sequence lifetime—of $\log(L/L_\odot) = -0.06$, which is close to the solar value (or $M_{bol} = 4.8$ versus 4.65). Iben [323] finds the average main sequence luminosity to be $\log(L/L_\odot) \approx +0.15$. Our main sequence lifetime is $\tau_{ms} = 12.2$ Gy instead of 10.1. For a star of $20M_\odot$, this gives $\log(L/L_\odot) = 4.85$ instead of 4.79 [82], and a lifetime of $\tau_{ms} = 6.9$ My instead of 7.7. Our solar mass star is slightly subluminous, and our $20M_\odot$

---

[6]Another important effect is the loss of pressure per nucleon, as hydrogen is converted to helium.

star is slightly superluminous. Errors of this magnitude will not change the general nature of our results.

In chapter 7 it was shown that such models did not reproduce the observed mass-luminosity relation below $m \approx 0.5$ (due to our neglect of envelope convection). The values for the lowest masses are set to approximate the empirical mass-luminosity relation from Henry and McCarthy [289]. Specifically, for $m \leq 0.45$,

$$l \propto m^2, \tag{14.24}$$

with the constant of proportionality chosen to insure continuity at $m = 0.45$.

The bolometric magnitude was assumed to be given by

$$M_b = -2.5 \log_{10}(l) + 4.65. \tag{14.25}$$

For stars having masses from $m = 0.3$ to $m = 10$, which is approximately the range appropriate to the data Salpeter used, the luminosities vary by a factor of about 10^7. For a given apparent brightness, this corresponds to a difference in distance of a factor of 3×10^3. Low-mass stars are dim; suppose we can count them out to 20 parsecs; see [337, 545]. Massive stars are rare; we need to count many to obtain statistical accuracy. Because they are so bright, we could count massive stars out to a distance of 30 kiloparsecs, which includes the whole galaxy, if there were no other observational problems. Actually, the massive stars are usually associated with star-forming regions, which have high obscuration. The volumes of space in which we count low-mass and high-mass stars are dramatically different. We must convert the observational data to a standard reference volume. In order to do so, we need additional information to estimate what volumes the stellar orbits sample. All the corrections introduce new sources of possible error.

First we consider a volume defined by the counts of low-mass stars: a region within 20 pc of the Sun [337], and estimate the equivalent density of higher-mass stars from counts in larger volumes.

If the star burns a fraction $f \approx 0.1$ of its mass on the main sequence [560], the energy liberated is fMX_HQ_H, where the hydrogen nucleon fraction is $X_H \approx 0.7$ and $Q_H \approx 8 \times 10^{18}$ erg/g is the energy released per gram of hydrogen burned. The lifetime [173], for this stage of main sequence burning, is then

$$\tau(m)_{ms} = \tau_\odot m/l, \tag{14.26}$$

where $\tau_{\odot} = 108f$ Gy. For $1 \le M/M_{\odot} \le 9$, f varies slowly from about 0.1 to 0.14. Above $M = 9M_{\odot}$, the increasing importance of radiation pressure causes the convective cores to grow beyond $f = 0.14$ during main sequence burning (roughly as $0.14 + 0.1(m - 9)/11$ for $m > 9$). This gives slightly longer lifetimes on the main sequence than without such increased core size.

Suppose interstellar gas is converted into stars during a time τ_0, at a rate Λ (which we take to be constant for the moment). Then the number of stars formed of mass m is equal to the total mass converted into those stars, $\tau_0 \Lambda \xi(m)$, divided by the mass per star, m, or

$$dN/dm = \tau_0 \Lambda \xi(m)/m. \tag{14.27}$$

The bolometric luminosity function is defined by

$$\phi(M_b) = \int_{M_b - \frac{1}{2}}^{M_b + \frac{1}{2}} (dN/dM_b)\, dM_b. \tag{14.28}$$

Now, using the expressions above, we have

$$dM_b/d\log_{10} m = -2.5(d\ln l/d\ln m), \tag{14.29}$$

and derive

$$dN/dM_b = (dN/dm)/(dM_b/dm) \tag{14.30}$$

$$= -[\ln(10)/2.5\chi]t\Lambda\xi(m), \tag{14.31}$$

where $\chi = d\ln l/d\ln m$, and t is the smallest of $\tau(m)_{ms}$ and τ_0. Therefore,

$$\phi(M_b) = \int_{m_+}^{m_-} [\ln(10)/2.5\chi]t\Lambda\xi(m)\, dm, \tag{14.32}$$

where m_- is the mass having a bolometric luminosity $M_b - \frac{1}{2}$, and m_+ similarly for $M_b + \frac{1}{2}$. This is our theoretical expression for the bolometric luminosity function.

Star formation occurs over time τ_0, the age of the disk (or of that part of the galaxy represented by the sample of stars). If a mass is considered for which the stellar age is less than this, $\tau(m) < \tau_0$, then only stars formed in the interval between $\tau_0 - \tau(m)$ and τ_0 still survive on the main sequence. By adjusting the parameters in $\xi(m)$ and the value of τ_0 we can adjust the theoretical luminosity function to fit the observed one. If the star-formation rate is not constant in time, the product Λt becomes

$$\int_0^{\tau_0} \Lambda(t')dt'. \tag{14.33}$$

For stars that have lifetimes $\tau(m)$ less than τ_0, we have

$$\int_{\tau_0-\tau(m)}^{\tau_0} \Lambda(t')dt'. \tag{14.34}$$

Dividing this integral by $\tau(m)$ gives the average rate of star formation over the time interval $\tau_0 - \tau(m)$ to τ_0. For presently observable stars, the low-mass ones reflect the net effect of star formation over all of galactic history, while the high-mass ones represent only the results of recent star formation.

Figure 14.1 shows the bolometric luminosity function per unit volume, $\phi(M_b)$, based on the data used by Salpeter but with an updated set of bolometric corrections. Also shown (as a solid line) is a theoretical luminosity function constructed from Eq. 14.31 above. Our toy model assumes that stars stay on the main sequence at a luminosity and lifetime given by the prescription given above, and then vanish. *Even if the luminosity function is given, the initial mass function implied depends upon the stellar evolutionary theory used in the fitting procedure.* The initial mass function is taken to have the form $\xi(m) = \xi_0 m^{-\alpha}$; the fit shown used $\tau_0 = 9$ Gy, $\xi_0 = 1.5 \times 10^{-2}$ pc^{-3}, and $\alpha = 1.33$, which, considering the changes in bolometric corrections and stellar evolution, is in good agreement with Salpeter's original values of $\tau_0 = 6$ Gy, $\xi_0 = 3 \times 10^{-2}$, and $\alpha = 1.35$.

If the present rate of star formation were much different from the average value over the history of the solar neighborhood, then there would not be a simple change in slope at stars with main sequence ages near τ_0. Thus figure 14.1 is consistent with the rate of star formation in the solar neighborhood having been constant (on average) over the last $\tau_0 = 9$ Gy.

Stellar masses are indicated in figure 14.1 by short vertical lines at their average main sequence luminosity. The fit is best for stars in the range $0.7 \leq m \leq 2$. It is this range that determines the value of τ_0, the age of the solar neighborhood, and presumably the galactic disk. For lower masses, these original data are suspect: the surveys were done in visual magnitudes, while these stars radiate more to the red (in the K band). Consequently they were subject to undercounting; the data do lie below the estimated values. To compensate for this failing, we have added data

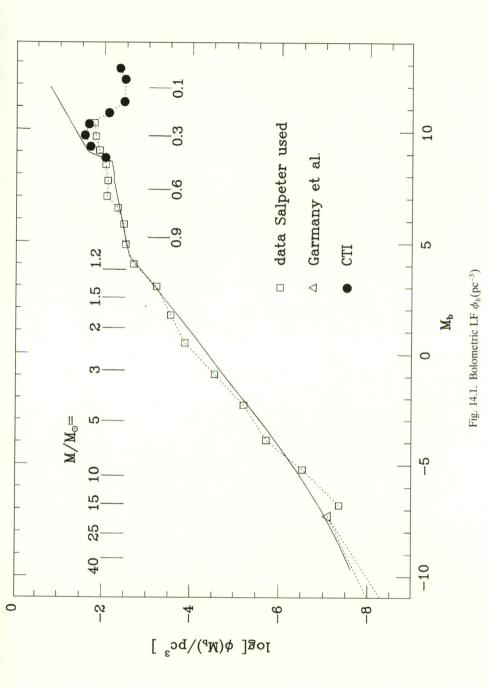

Fig. 14.1. Bolometric LF $\phi_b(\mathrm{pc}^{-3})$

from the CCD Transit Instrument (CTI [356]; see also [621]), which are shown as filled circles for luminosity $M_b \geq 9$. Notice that the change in the mass-luminosity relation makes the predicted curve rise at $m \approx 0.45$, as the CTI data does. Below a mass of $0.3M_\odot$, the IMF falls below the $m^{-4/3}$ power law.

For the larger-mass stars, $m > 2$, the curve is relatively insensitive to the mass-luminosity relation: changes in that also change the lifetime, and move the stars on a trajectory almost parallel to the power law shown. An increased age for these stars, keeping the luminosity fixed, would raise the theoretical curve. At the largest masses the theoretical IMF flattens; inclusion of the effects of mass loss will steepen it again.

The distribution of massive stars is nonuniform in space [426], and probably in time. Garmany, Conti, and Chiosi [260] find that within 2.5 kpc of the Sun, the counts of O stars are complete, with a density (projected on the Galactic plane) of about 22 kpc^{-2}. To continue our analysis we must convert this to a density per unit volume (however, see below). Taking a scale height [260] of 45 pc, this implies an equivalent volume density of 2.4×10^{-7} pc^{-2}. This may be directly related to the bolometric luminosity function. The number of O stars is

$$\Delta N = \int_{-\infty}^{M_1} \tau(m)\Lambda\xi(m) \, d\ln m, \qquad (14.35)$$

where M_1 is the bolometric magnitude of the intrinsically dimmest O star. With a little algebra this becomes

$$\Delta N = \xi_1\tau_1\Lambda/(\chi + \gamma - 1), \qquad (14.36)$$

where ξ_1 is the IMF value for the least massive O star, τ_1 its lifetime, and γ the exponent for the IMF above that mass. This represents the birthrate function only over the last $\tau_1 \leq 2 \times 10^7$ y. Given the spectacularly splotchy nature of the O-star distribution in spiral galaxies, this birthrate might well deviate from its average value. Using Eq. 14.32, this implies

$$\phi_b(M_1) = \left(\ln(10)/2.5\chi\right)\Delta N(\chi + \gamma - 1). \qquad (14.37)$$

The least-luminous stars in the sample have $M_b \approx -7.3$. The open triangle in figure 14.1 represents this data; the analysis of Garmany et al. suggests a slope of about 4/3 to 5/3, depending upon the stellar evolution assumed. The dotted lines indicate this range. The older data are significantly lower, possibly due to incompleteness.

Areal IMF

Consideration of the massive stars exposes a problem with this formulation. In analogy to the kinetic theory of gases, we have considered a number density per unit volume as a primary statistical entity. The stars in a galactic disk are unlike molecules in a gas in that their net angular momentum is far from zero. This gives a pronounced anisotropy to the problem. The distributions of stars perpendicular to the disk fall off at different rates, depending upon the stellar mass; formally this would make our volumetric IMF a function of height above the galactic plane. This effect is mirrored in the velocity distribution [439], as is shown in figure 14.2. The solid squares represent the mean observed velocities perpendicular to the galactic plane, for main sequence stars of different spectral types; the masses are inferred from stellar models. There is a flatter portion at $m < 1$ and another at $m > 5$, with a ramp in between. The stars of highest mass have the lowest velocity, comparable to that of the interstellar gas. The error bars are taken from a suggestion of Jahreiss and Wielen [337]. The lower-mass stars, having higher velocities, also have higher average heights above the galactic plane.

If the stars formed a Maxwell-Boltzmann gas at a constant "temperature," then since $\frac{1}{2}mu^2 = \frac{1}{2}kT$ for the single (vertical) degree of freedom

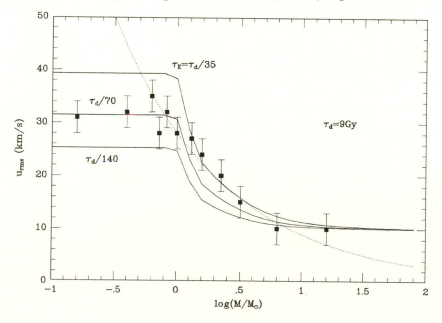

Fig. 14.2. Stellar Velocities u_z

which we consider, the root mean square velocity u_{rms} should be inversely proportional to the square root of the stellar mass. This is shown as a dotted line, normalized at $m = 1$. It fails badly for low masses. As these stars live longest, they would have had the most time, on average, to "thermalize," so this is a fatal flaw. The higher-mass stars also deviate from the curve.

Spitzer and Schwarzschild [585] explained this velocity distribution in terms of energy exchange. Stars are formed with the velocity distribution of the interstellar gas, and gain energy by dynamical encounters with large masses ($m \approx 5 \times 10^5$, typical of molecular cloud complexes). They found

$$u_{rms}(t) \approx u_{rms}(0)\left[1 + (t/\tau_E)\right]^{1/3}, \qquad (14.38)$$

where $\tau_E \approx 2 \times 10^8$ y. This particular form is derived from a specific mechanism for the energy exchange, but there is a range of possibilities; see [439, 682] for a clear discussion. If we assume a constant rate of star formation, and integrate this over either the age of the disk or the age of stars of that mass, whichever is least, then we can derive the average value of u_{rms} which should be observed. This is shown in figure 14.2 as three solid lines. The middle one corresponds to $\tau_E = \tau_d/70 \approx 1.3 \times 10^8$ y; the upper and lower curves are for twice and half this value, respectively. The relevant parameter is τ_E/τ_d, the number of heating time-scales available in the age of the disk. The low-mass stars all have similar velocities because they have been accelerated for the same time, τ_d, by dynamical encounters with much larger masses.

To account for these complications, we follow Oort [468] and consider a new reference volume: a cylinder centered near the Sun and oriented perpendicular to the Galactic disk. We may hope, statistically at least, that the oldest stars do not elude us, and the youngest are properly represented. Stars of a given spectral type are observed to be distributed above (and below) the disk with an approximately exponential distribution [66, 439] ($N_{sp} \propto e^{-|z|/\ell}$). If the velocity distribution is well mixed [468] (so that we may assume that this distribution is not transient), the problem reduces to determining the scale heights ℓ for each set of stars. For an exponential distribution symmetric about the plane of the disk, the surface density σ is related to the volumetric density ρ by

$$\sigma = 2\ell\rho(0), \qquad (14.39)$$

where ℓ is the scale height and $\rho(0)$ is the volumetric density evaluated at the midplane. For more general distributions of density, this relation may be regarded as a definition of the scale height ℓ. We will convert a volumetric IMF to an areal one by multiplying by 2ℓ.

The density of a set of stars is connected to their velocity dispersion by dynamics. We will consider a typical star and derive a self-consistent relation between its average velocity perpendicular to the Galactic plane, and its typical height (which we will identify with its scale height). Strictly speaking, we should consider an ensemble of stars having the properties of this typical star as the ensemble average. Because of the time-dependent increase in perpendicular velocity, the time dependences implied by stellar lifetimes, galactic orbits, and star-formation history must be examined.

For an axisymmetric disk, Poisson's equation is [468]

$$\partial K_\varpi/\partial \varpi + K_\varpi/\varpi + \partial K_z/\partial z = -4\pi G\rho, \tag{14.40}$$

where K_z is the force per unit mass in the perpendicular direction, z is the perpendicular coordinate, and ϖ is the (cylindrical) radial dimension. Close to the plane (small z), this approaches the plane-parallel case, so that

$$\partial K_z/\partial z = -4\pi G\rho. \tag{14.41}$$

We approximate Oort's estimate of K_z by

$$K_z \approx 2.6 \times 10^{-8}x/(1 + 1.6|x| + 0.4x^2), \tag{14.42}$$

where x is the z coordinate expressed in kiloparsecs, and K_z is in cm s^{-2}. For heights larger than one kiloparsec, the planar approximation begins to break down. At the midpoint of the plane, $z = 0$, so

$$\partial K_z/\partial z \approx 2.6 \times 10^{-8} \text{ cm s}^{-2} \text{ Kpc}^{-1}, \tag{14.43}$$

which implies a density at the midplane of

$$\rho(0) \approx 1.0 \times 10^{-23} \text{ g cm}^{-3}, \tag{14.44}$$

or about $0.148 M_\odot$ pc^{-3}. Estimates range from 0.10 [368] to 0.185 [64].

Suppose we place a test particle (star) at some height z with zero velocity and examine its oscillations above and below the plane. For the moment we ignore the Spitzer-Schwarzschild effect. This is equivalent to a nonlinear oscillator with no damping or heating. Table 14.2 gives the results for a variety of initial heights, z_{max}; see also [367], table 6. The rms distance from the plane is $z_{rms} \equiv (< x^2 >)^{1/2}$, the time to

TABLE 14.2

Height and Speed

z_{max} (pc)	$<z>$ (pc)	τ (My)	w_{max} (km/s)	$<w>$ (km/s)
10	7	69.4	0.89	0.63
50	35	71.1	4.30	3.06
75	53	72.2	6.34	4.52
100	71	73.3	8.29	5.94
150	106	75.5	12.01	8.66
200	142	77.7	15.48	11.24
300	214	82.1	21.80	15.99
400	285	86.3	27.42	20.31
500	358	90.6	32.46	24.26
600	430	94.7	37.02	27.89
700	503	98.8	41.16	31.24
800	576	102.9	44.95	34.35
900	648	106.9	48.44	37.24
1000	721	110.9	51.67	39.95
1250	904	120.8	58.75	46.03
1500	1088	130.4	64.73	51.29
1750	1271	140.0	69.86	55.90

execute one complete period of oscillation is τ, the maximum velocity in the perpendicular direction is w_{max}, and the average (rms) velocity is w_{rms}. Notice that the period of oscillation $\tau \approx 60$ to 100 My, which is only slightly less than the energy exchange time-scale of Spitzer and Schwarzschild ($\tau_E \approx 130$ My). A better model would be a stochastically driven aharmonic oscillator with moderately large amplitude kicks; however, we will use table 14.2 as representative of ensemble average values. Notice that this period is the main sequence lifetime for a star of about $4M_\odot$.

Even if the gravitational potential and the energy exchange rate are fixed, the scale height of stars depends upon their age, their lifetimes, their birthplace, and their birth velocity. Young stellar objects are correlated with molecular clouds [439]; presumably they form from these

TABLE 14.3

Height and Age

log(Age(y))	w_{rms} (km s^{-1})	z_{rms} (pc)	w_{rms} (km s^{-1})	z_{rms} (pc)
7.1	8.1	99	10.2	127
8.1	9.1	112	11.4	144
8.5	10.3	129	12.9	167
8.7	11.3	142	14.1	185
8.9	12.5	161	15.6	208
9.1	14.0	184	17.5	239
9.3	15.9	212	19.8	278
9.5	18.2	250	22.7	329
9.7	20.9	296	26.1	395
9.9	24.1	356	30.2	479

clouds, inheriting their positions and velocities. Possibly the process of star formation further modifies both. We assume that all stars are born near the plane (for simplicity), and consider two values of initial velocity: $w(0)_{rms} = 8$ and 10 km s^{-1}. The energy exchange is that used above. The corresponding rms velocities and heights are given as a function of logarithm of the stellar age (in years) in table 14.3. If we take the lower-mass stars as representative of most of the mass of the disk, this would give an average height of $\ell_{disk} \approx 350$ pc (see [66, 364]), for a velocity dispersion corresponding to $< w >= 25$ km s^{-1}, which is similar to observed values [337] for low-mass stars. In this simple picture, the corresponding surface density would be $\sigma \approx 100$ M$_\odot$ pc^{-2}, with about half of this within 200pc of the plane. This errs because the density falls as a gaussian ($e^{-(z/z_0)^2}$) near the plane,[7] so that the linearity of the force law—which depends upon constant mass density—fails when we consider stars which make up much of the mass. The surface density, and hence the disk scale height, are probably less well determined than the density at the midplane [64]. Bahcall finds $\sigma = 67 \pm 5$ M$_\odot$ pc^{-2}; Kuijken and Gilmore [367] estimate $\sigma = 46 \pm 9$ M$_\odot$ pc^{-2}.

---

[7]Statistical mechanics tells us that an isothermal gas will be distributed in height according to the exponential of the potential. Near the plane, the force law is linear, so the potential is quadratic.

Notice that the IMF depends upon galactic dynamics as well as stellar evolution. It is important to use self-consistent models across these different aspects of the analysis.

Figure 14.3 shows the resulting IMF, with a format similar to that used previously. The solid line is a fit corresponding to $\xi_0 = 15$ pc^{-2}, $\tau_0 = 9$ Gy, and $\alpha = 5/3$. Also shown, as a dashed line, is the case for $\alpha = 4/3$, the Salpeter value. The lower-mass stars have the same scale height, so that this part of the curve is unchanged. The $\alpha = 5/3$ curve does not fit quite as well at $m < 0.3$, but these data and stellar models are still being improved. For the stars of $m > 1$, the scale height decreases, systematically lowering the curve for increasing mass ($1 < m < 5$). For $m > 5$, the curve simply is shifted down by a factor of $\ell_{highm}/\ell_{lowm} \approx 1/4$. The vertical dotted line denotes the mass above which the stars do not live long enough on the main sequence to complete even a single period of oscillation about the midplane of the disk. Such stars will not have time to develop well-mixed distributions along their orbits, but will retain more information about their origins. The older data for high-mass stars may be incomplete, but the density scale heights may also have been different. The newer data [260] are already in areal form and fit nicely onto the $\alpha = 5/3$ power law for these models; it is independent of these complications of scale height.

The major difference between the latest IMFs for the solar neighborhood and Salpeter's original result is due to the change from a volumetric to an areal representation. Scalo [440, 542] has critically examined many sources of uncertainty. Basu and Rana [81] consider an areal IMF and find that, above $m = 6.5$, a slope of 1.67 rather than 1.33 is appropriate, which is consistent with our values; see their paper for further discussion.

While use of the Garmany, Conti, and Chiosi [260] data gives a better average over space than older work, it represents an *instantaneous* value in Galactic terms and does not necessarily provide the proper average over time. What we need are several sets of such data, scattered over time through several Galactic rotations. Because such a direct time average seems hardly feasible, we examine an ensemble average of similar systems. Studies of other disk galaxies [561, 385, 345], and of young open clusters in our own galaxy [494], suggest that the upper part of the IMF is better represented by a slope of 4/3, the original Salpeter value, extended to 100M$_\odot$ and above. The observed properties of late-type spiral galaxies in the *UBVR* bands and in Hα are not reproduced with IMFs having a steeper power law in the range $1 \leq m \leq 50$, while an IMF with a slope of 4/3 works well. This result may extend as well to the Type Sa spirals [137], and be appropriate for the disks of all spirals.

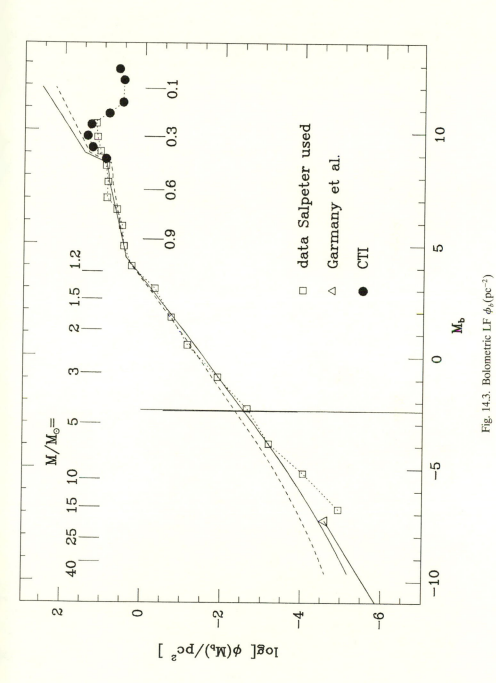

Fig. 14.3. Bolometric LF $\phi_b(pc^{-2})$

For properties relating to the current status of "nearby" massive stars in the Galactic disk, the steeper slope may be preferable, but for properties depending upon broader averages over time and space, the original Salpeter slope seems to be a viable alternative at present.

Implications

Having found a simple form for the IMF, it is easy to determine some of its properties, and to compare it to other representations. The total *mass* in stars formed with stellar masses from m_1 to m_2 is

$$\mathcal{M}(m_1, m_2) = \int_{m_1}^{m_2} \xi(m)\, dm, \tag{14.45}$$

and the total *number* of such stars is

$$\mathcal{N}(m_1, m_2) = \int_{m_1}^{m_2} (\xi(m)/m)\, dm. \tag{14.46}$$

There are some blunders in the literature in which these have been confused, so we quote them explicitly. Using $\xi(m) = \xi_0 m^{-\alpha}$, we have

$$\mathcal{M}(m_1, m_2) = \xi_0(m_1^{-\alpha+1} - m_2^{-\alpha+1})/(\alpha - 1), \tag{14.47}$$

and

$$\mathcal{N}(m_1, m_2) = \xi_0(m_1^{-\alpha} - m_2^{-\alpha})/\alpha. \tag{14.48}$$

Integrating from $m_1 = 0.3$ to $m_2 = 10$, the total mass of stars formed in this range is $\mathcal{M}(0.3, 10) = 3.09\xi_0$ M$_\odot$, and the total *number* (0.3,10) of these stars is $\mathcal{N} = 3.70\xi_0$. The average stellar mass in this range is therefore $< m > = 0.835$. Notice that specifying ξ_0 here allows a normalization to any galactic mass scale desired.

Table 14.4 gives numbers and masses contained in several interesting ranges of stellar mass, m_1 and m_2. Below $m = 1$ the IMF exponent α is taken to be 4/3; for $m > 1$, both $\alpha = 4/3$ and 5/3 are presented. The two cases in which the range overlaps $m = 1$, so that both exponents are used, are enclosed in brackets.

The range $0.1 < m < 0.3$ is difficult [440, 542, 81, 288, 520, 356]. The work of Tinney et al. [622] suggests that for $0.1 \leq m \leq 0.3$, $\xi(m) \approx \xi_0(0.3^{-4/3})$, that is, the IMF is constant, with a value near the Salpeter one for mass $m = 0.3$. Then

$$\mathcal{M}(0.1, 0.3) \geq \xi_0(0.3)^{-4/3} \int_{0.1}^{0.3} dm = 0.996\xi_0. \tag{14.49}$$

TABLE 14.4

IMF

Quantity	m_1	m_2	$\alpha = 4/3$	$\alpha = 5/3$
$\mathcal{M}/\xi_0$	0.3	1	1.481	
	0.3	100	3.835	[2.911]
	1	8	1.500	1.125
	1	100	2.354	1.430
	8	15	0.284	0.128
	8	100	0.854	0.305
	15	30	0.251	0.091
	30	100	0.319	0.086
$\mathcal{N}/\xi_0$	0.3	1	2.985	
	0.3	100	3.733	[3.584]
	1	8	0.703	0.581
	1	100	0.748	0.600
	8	15	0.0266	0.0122
	8	100	0.0453	0.0185
	15	30	0.0122	0.0045
	30	100	0.0063	0.0018

If this constancy extends to lower masses as well, the total mass in these stars $(0 \leq m \leq 0.1)$ is

$$\mathcal{M}(0, 0.1) = \xi_0 (0.3)^{-4/3} \int_0^{0.1} dm = 0.498 \xi_0. \qquad (14.50)$$

Such small-mass stars are called *brown dwarfs*; the minimum mass for hydrogen burning is $m \approx 0.08$. This contribution is small, but the estimate is merely an extrapolation to stellar masses $m < 0.1$. The results of Kirkpatrick et al. [356], which we used in figures 14.1 and 14.3, suggest less mass in the range $0.1 \leq m \leq 0.3$. For simplicity, we will ignore $\mathcal{M}(0, 0.3)$ in what follows, but further observational tests are needed. A novel possible method is the use of gravitational microlensing to detect such faint stars [477, 274].

At the time of writing, the evolution in the HR diagram of stars above $10M_\odot$ is uncertain, and consequently so is the IMF for $m > 10$. However, these stars are the dominant sites for nucleosynthesis.[8] The amount of mass $\mathcal{M}_{up} = \mathcal{M}(8, m_3)$ is given in table 14.4. The mass contained in $8 < m < 100$ is 0.22 and 0.10 of the total, for the two choices of IMF. In a stellar generation, only this fraction of the mass is processed by massive stars. The mass contained in $1 < m < 8$ is 0.391 and 0.386 of the total; this corresponds roughly to the stars that can die in the age of the universe. About 40 to 50 percent of the mass is tied up in low-mass, long-lived stars in each epoch of star formation.

Suppose that above a mass $m_3 = 30$, the stars became black holes (with no mass loss); this would swallow $\mathcal{M}(30, 100)/\xi_0$ of 0.319 and 0.0857 respectively, or 8.3 and 2.9 percent of the total. Supernovae 1987A and 1993J suggest that m_3 is well above twenty solar masses, so that this simple model implies an interesting constraint.

The stars in the range $8 \leq m \leq 100$ are thought to produce neutron stars and black holes. Their number is $\mathcal{N}_{cr}$; the subscript denotes *condensed remnant*. Suppose that most of these condensed remnants are neutron stars, which must be the case if stars above $m_3 = 30$ are the black hole progenitors. The fraction by number of all stars which form neutron stars is roughly $\mathcal{N}(8, 100)/\mathcal{N}(0.3, 100)$, which is 1.2 and 0.5 percent respectively. There is relatively little difference in these ratios for a smaller upper mass, because most of the contribution comes from stars of mass nearer the lower limit. More neutron stars would be produced in the range $8 < m < 15$ than in $15 < n < 100$.

If we take the mass range for the progenitors of white dwarfs to be $1 < m < 8$, the number of white dwarfs is larger than the number of neutron stars by $\mathcal{N}(1, 8)/\mathcal{N}(8, 100)$, which gives 15.5 and 31.5. The typical mass of a star in the range $8 < m < 100$ is $\mathcal{M}_{cr}(8, 100)/\mathcal{N}_{cr}(8, 100)$, or 18.8 and 16.5 $M_\odot$ respectively, which—in this picture—is the typical progenitor of a neutron star, and the typical mass of the progenitor of a core-collapse supernova. This is somewhat smaller than the typical mass for the production of fresh nucleosynthesis products, as we shall see.

Using this simple analytic form for the IMF, other possibilities may be quickly explored.

14.3   ONE-ZONE MODELS

A *one-zone model* describes the evolution of a single system of gas and stars with properties that are homogeneous in space. Because the com-

---

[8] Emphasizing the bolometric luminosity rather than effective temperature may be advisable, as this is a better measure of the rate of nuclear burning.

position of the interstellar gas is described by a single composition at any instant of time, the volume must be assumed to be well mixed. As pointed out above, this is only realistic for small volumes, in which case fluxes through the surface *cannot* be ignored. The one-zone models have been extensively discussed and used, but it is seldom noted that, because of this mixing problem, they may be inconsistent with the physical systems they are used to describe. The early history of one-zone models is given in the classic review by Audouze and Tinsley [60].

The Salpeter-Schmidt model

The fundamental equations are

$$d\mathscr{G}_i/dt = -X_i \int_{min}^{max} \mathscr{B}(m, t)\, dm + \int_{min}^{max} \mathscr{B}(m, t - \tau_m)\, R_i(m)\, dm,$$
(14.51)

where $\mathscr{G}_i = \mathscr{G}X_i$ is the fraction of nucleons in the form of gas of species i, $\mathscr{B}(m, t) = \Lambda(t)\xi(m)$ is the birthrate of stars in the stellar mass interval m to $m + dm$ at time t, $R_i(m)$ is the fraction of its mass which the star returns to the gas in the form of species i by the time it dies, and τ_m is the lifetime of a star of mass m. The limits of integration are: *min* is the mass for which τ_0, the stellar lifetime, equals the lifetime of the "galaxy," and *max* is the mass of the most massive star formed. Integrals of this type are insensitive to *max* if the integrand falls rapidly with increasing mass. The initial mass function is $\xi(m)$. With the instantaneous recycling approximation this simplifies to

$$d\mathscr{G}_i/dt \approx -X_i\, \Lambda(t) + \Lambda(t)\, f_i$$
(14.52)

where

$$f_i = \int_{min}^{max} \xi(m)\, R_i(m)\, dm$$
(14.53)

is the fraction of returned gas in the form of species i.

To obtain quantities referring to all the matter, not just one component i, we simply sum over the i species. Thus, as $\mathscr{G} = \sum_i \mathscr{G}_i$, $f = \sum_i f_i$, and $\sum_i X_i = 1$, we have

$$d\mathscr{G}/dt = -\Lambda(t)(1 - f).$$
(14.54)

Several choices have been suggested for the form of the star-formation rate $\Lambda(t)$. The Salpeter [537] choice is

$$\Lambda(t) = \lambda\mathscr{G},$$
(14.55)

while Schmidt [550] also considered

$$\Lambda(t) = \lambda \mathscr{G}^2/\mathscr{G}(0). \tag{14.56}$$

The corresponding solutions for the evolution of the gas density are

$$\mathscr{G}(t) = \mathscr{G}(0) \exp[-\lambda(1-f)t], \tag{14.57}$$

and

$$\mathscr{G}(t) = \mathscr{G}(0)/(1 + \lambda \mathscr{G}(0)(1-f)t), \tag{14.58}$$

which have qualitatively similar behavior.

Using equations 14.52 and 14.54, and the identity

$$d\mathscr{G}_i/dt = d(\mathscr{G}X_i)/dt,$$
$$= X_i d\mathscr{G}/dt + \mathscr{G}dX_i/dt, \tag{14.59}$$

we have

$$dX_i/dt = \left(\frac{\Lambda(t)}{\mathscr{G}}\right)(f_i - fX_i). \tag{14.60}$$

We note that

$$\frac{\Lambda(t)}{\mathscr{G}} = -d\ln\mathscr{G}/dt/(1-f) = \frac{1}{1-f}dy/dt, \tag{14.61}$$

where $y = \ln[\mathscr{G}(0)/\mathscr{G}(t)]$. The history of the model is specified by the one relationship between y and t given by Eq. 14.61. For the evolution of abundances, we then get

$$dX_i/dy + X_i f/(1-f) = f_i/(1-f). \tag{14.62}$$

The quantity $f_i/(1-f)$ is called the *yield* of species i; in this model the abundances typically approach values near these yields.[9] A specific relation between y and t is necessary only if one is discussing (a) abundances as a function of time, and (b) cosmochronology. For the Salpeter birthrate function,

$$X_i(t) = X_i(0)e^{-\lambda ft} + \frac{f_i}{f}(1 - e^{-\lambda ft}), \tag{14.63}$$

---

[9]The yield defined here is more precise than the often-used quantity of Searle and Sargent [562], which includes time-delayed effects as older generations of stars eject matter. Consequently, it combines effects due to the star-formation rates with the different effects of the IMF, and of stellar evolution. The yield defined above is for a single generation, and thus isolates the latter two effects from the former.

and we see that the asymptotic abundance for long times is f_i/f, the abundance in the ejecta.

In this model, the system is closed, so $\mathscr{G}(0) = \mathscr{G}(t) + \mathscr{S}(t)$, where $\mathscr{S}(t)$ is the mass in the form of stars at time t. For a general choice of the form of the birthrate,

$$X_i(t) = X_i(0)\Gamma^{f/(1-f)} + \frac{f_i}{f}(1 - \Gamma^{f/(1-f)}), \qquad (14.64)$$

where $\Gamma = \mathscr{G}(t)/\mathscr{G}(0)$.

The Larsen model

An important modification of the one-zone models is the addition of flow of matter into a homogeneous "box"; this is the *infall* model of Larsen [384, 404, 175]. We illustrate this class of models by the simplest case: a constant rate of infall I. Here we can also think of the infall as a crude approximation to the results of finite gradients in a heterogeneous system. Then,

$$d\mathscr{G}_i/dt \approx -X_i\Lambda(t) + \Lambda(t)f_i + I\,X_i^{in}, \qquad (14.65)$$

where X_i^{in} is the abundance of species i in the infalling gas. Similarly,

$$d\mathscr{G}/dt = -\Lambda(t)(1 - f) + I. \qquad (14.66)$$

The solution, using the Salpeter birthrate function, is

$$\mathscr{G}(t) = \mathscr{G}(0)\,e^{-\lambda(1-f)t} + \frac{I}{\lambda(1 - f)}(1 - e^{-\lambda(1-f)t}). \qquad (14.67)$$

The abundance equation is

$$\mathscr{G}(t)\,dX_i(t)/dt = \Lambda f_i - X_i(\Lambda f + I) + IX_i^{in}. \qquad (14.68)$$

Both the gas and the abundances tend toward a steady-state value:

$$\mathscr{G} \to I/\lambda(1 - f) = \Lambda/\lambda, \qquad (14.69)$$

and

$$X_i \to f_i + (1 - f)X_i^{in}. \qquad (14.70)$$

Figure 14.4 shows the different behavior of the closed box and the infall models. The upper panel illustrates the change in abundance versus time. Both models have a final gas fraction of about 12 percent. The infall model abundances rise quicker, and then stay nearly constant, while the closed box model shows an almost linear increase in abundance with time.[10] The lower panel shows the change in abundance versus the fraction of total mass in the form of stars. This indicates the distribution of metallicity in existing stars. This infall model has few stars with less than half the average abundance (about 8 percent), while the corresponding number for the closed box model is about 35 percent. The solar neighborhood has few stars which are metal poor; this is the classical "G-K dwarf problem" (see [646, 551, 60]). Notice that the most metal-rich stars in the closed box model are about twice as rich as those for the infall model (lower gas fractions give a larger difference). The same amount of nucleosynthesis occurs—it is distributed differently.

The Hartwick model

If we allow matter to flow into our box, why not out? Suppose that gas outflow is driven by heating by supernovae or OB stars. Because these stars evolve so quickly, the outflow would occur at a rate approximately proportional to the rate of star formation, or $dD/dt = c\Lambda$, where D is the amount of mass that escapes. Then,

$$d\mathcal{G}/dt = -(1-f)\Lambda - c\Lambda, \tag{14.71}$$

and the net rate at which mass is converted into stars is

$$d\mathcal{S}/dt = (1-f)\Lambda. \tag{14.72}$$

This is the Hartwick [284] model for the Galactic halo. Now if we choose initial conditions so that we start with only gas, $\mathcal{D}(0) = \mathcal{S}(0) = 0$, and normalize so that $\mathcal{G} = 1$, we have $\mathcal{D} + \mathcal{S} + \mathcal{G} = 1$,

$$\mathcal{D} = c\mathcal{S}/(1-f), \tag{14.73}$$

and

$$\mathcal{G} = 1 - \frac{\alpha + c}{\alpha}\mathcal{S}, \tag{14.74}$$

---

[10]Relaxation of the instantaneous recycling approximation allows relatively unprocessed matter to be stored in moderate mass stars and later returned to the interstellar medium. This is a bit like infall and modifies the closed box model accordingly.

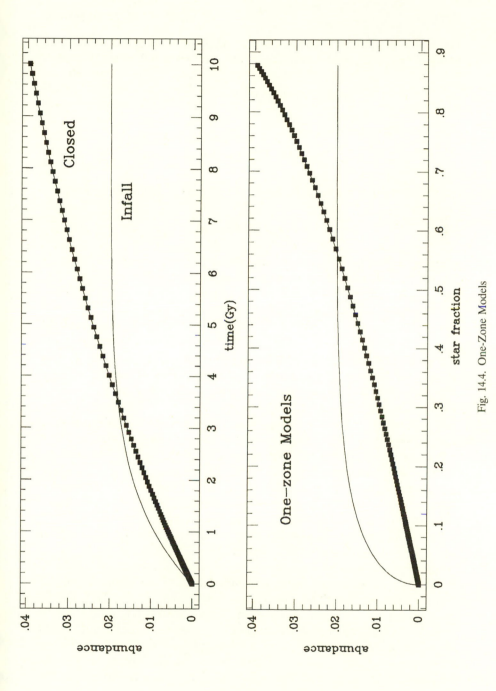

Fig. 14.4. One-Zone Models

where $\alpha = 1 - f$. The equation for dX_i/dt is the same as before; the gas removed has the same abundance as the gas that is left. Then, because

$$\Lambda/\mathcal{G} = (\frac{1}{\alpha} d\mathcal{S}/dt)/(1 - \frac{\alpha+c}{\alpha}\mathcal{S}), \tag{14.75}$$

the abundance equation,

$$dX_i/dt = (f_i - X_i f)\Lambda/\mathcal{G}, \tag{14.76}$$

may be written in terms of the variables X_i and $\mathcal{S}$, with the solution

$$\mathcal{S} = \frac{\alpha+c}{\alpha}(1 - (1 - \chi)^{\frac{\alpha+c}{f}}), \tag{14.77}$$

where $\chi = fX_i/f_i$. This gives the distribution of mass in stars of different levels of abundance, if we take X_i to denote the total metallicity. The differential distribution is

$$d\mathcal{S}/d\ln X_i = \frac{\alpha}{f}\chi(1 - \chi)^{\frac{\alpha+c}{f}-1}. \tag{14.78}$$

Figure 14.5 shows this distribution for several choices of c. The value $c = 0$ recovers the closed box model; $c = 10$ was chosen by Hartwick [284] to represent the halo of the Galaxy. For large c, most of the mass is lost: $\mathcal{D} = c\mathcal{S}/(1 - f) \approx c\mathcal{S}$, for $\alpha = 1 - f \approx 0.7$. As the gas disappears, $\mathcal{S} \rightarrow \alpha/(\alpha + c)$. Note that the distribution function is scaled upward by the factor $(\alpha + c)/\alpha$ to keep it visible in the figure.

For the $c = 10$ model, the average metallicity is $0.068z_\odot$. The fraction of stars with metallicity less than some value $\log_{10}(z/z_\odot) \leq -1.5$ is approximately $F = 100(z/z_\odot)^2$. This is relevant to searches for metal-poor stars. It implies that one percent of the halo stars would have a metallicity less than 0.01 solar, and 10^{-4} would have a metallicity less than 10^{-3} solar. A million halo stars would have, on average, only one star with a metallicity of 10^{-4} solar in this model!

Many possibilities can be examined by weaving together these three classes of models. For example, the initial stage of the formation of the Galaxy might be described by a Hartwick halo model, and the gas shed might dissipate, and sink into a disk. Then a closed box model would begin, with this halo epoch as an initial ("prompt") burst of nucleosynthesis, or more elegantly, a Larsen infall model could begin with halo abundances in the infalling gas. See the work of Clayton [175, 177], Matteucci [415, 414], Pagel [480], Tosi [626], and their collaborators, for example.

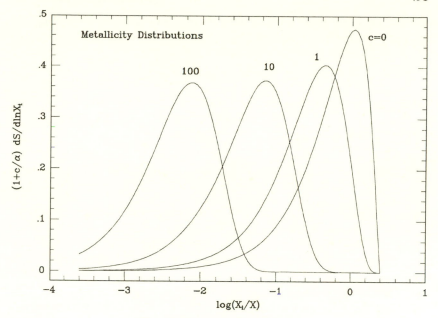

Fig. 14.5. Metallicity Distribution

14.4 ABSOLUTE YIELDS

The first step toward building a theory of nucleosynthesis in galaxies is to consider what is contributed by each type of star. One of the successes of stellar evolution theory is the recognition that the primary determining parameter for the behavior of a star is its mass. Initial abundance and interaction with a companion also are important.

By *bulk yields* we mean the production of those nuclei which comprise most of the mass of newly synthesized material. These might be determined by general features that even an immature theory could correctly represent.

The mass exchange between each star and the interstellar gas may be described by an integral of the mass flux through the surface σ which surrounds the star. It is thought that most of the infall occurs during the star-formation process (interacting binary systems are more complicated), and that for the rest of the lifetime of the star, mass is lost by the star. In such a case, the infall mass is

$$m_{star} = \int_{form} \int \rho \mathbf{v_{in}} \cdot d\sigma \, dt, \qquad (14.79)$$

where the time integration is over the duration of the star-formation process. For each separate nuclear species i, its infall mass is

$$m_{in}^i = m_{in} X_i^0, \tag{14.80}$$

where X_i^0 is its nucleon fraction in the gas out of which the star was formed.

The ejected mass is

$$m_{ej} = \int_{life} \int \rho \mathbf{v}_{out} \cdot d\sigma \, dt, \tag{14.81}$$

where the time integration is over the duration of the stellar lifetime, including its demise. If the time-scale for ejecting matter is short compared to galactic time-scales (the usual case), it is useful to think of the process as a mapping of initial abundances onto a set of final abundances. Mathematically this corresponds to defining a *production matrix* [606, 607]. For stars that lose mass over long time-scales, it may be necessary to consider the time dependence explicitly. The nucleon fraction of species i in the ejected matter is $\tilde{X}_i = m_{ej}^i / m_{ej}$, where

$$m_{ej}^i = \int_{life} \int X_i \rho \mathbf{v}_{out} \cdot d\sigma \, dt. \tag{14.82}$$

The production matrix Q_{mij} is defined by

$$m_{ej}^i(t_{death}) = \sum_j Q_{mij} X_j(t_{birth}) m. \tag{14.83}$$

The remnant mass is $m_{rem} = m_{star} - m_{ej}$. The net effect of star formation and mass loss upon the mass in the interstellar medium is

$$\Delta m^i = \tilde{X}_i m_{ej}^i - m_{star} X_i^0 \tag{14.84}$$

$$= m_{ej}^i (\tilde{X}_i - X_i^0) - m_{rem} X_i^0. \tag{14.85}$$

Here $\tilde{X}_i$ is the abundance in the ejected matter. For a net increase in a species i, its enhancement in the ejecta must be large enough to make up for loss due to remnant formation.

A complete evolutionary history of the star, including its possibly violent end, is needed in order to estimate Δm^i. Then, to get the total yield, all stars that make a significant contribution must be considered.

It is prudent to consider separate contributions first, and then attempt a synthesis later. Let us focus our attention on massive stars, which do most of the nucleosynthesis and which undergo core collapse.

Two complications immediately arise: massive stars often occur with binary companions [2], and massive stars are known to suffer extensive mass loss during their evolution [168]. The evolution of single massive stars with mass loss is sufficiently difficult; the binary situation is clouded by our lack of understanding of the interaction of binary mass transfer, mass and angular momentum loss from the system, and the evolution of both stars [572, 649]. Even the simplest case, conservative mass transfer, is complicated [487, 649], and this case is not appropriate for massive binaries because of the extensive mass loss from these systems. We note that a hydrodynamic investigation of a related but simpler problem, Bondi-Hoyle accretion onto a single gravitating object, revealed chaotic behavior [530]. Clearly there remains much to learn about the interacting binary problem!

Fortunately there are mitigating effects. Not all binaries are so close that they interact strongly. Most of the lifetime of a star is spent burning hydrogen, so that a likely possibility is the loss of the hydrogen envelope. We saw in chapter 8 that, while there were differences in detail, the yields were similar from helium cores and from stars without mass loss having cores of the same mass. A possible exception would be the yield of the products of hydrogen burning itself (e.g., ^{14}N and ^4He). Also, if there is extensive mass loss during helium burning, the yields of ^4He and the products of helium burning could be affected. However, because neutrino cooling accelerates the later burning stages, it is much less likely that they will be significantly modified. It appears that models of interacting systems and models of single stars with mass loss tend to give similar results [206]; roughly, the effective helium core mass M_α is decreased relative to the single-star, no mass loss case.

Table 14.5 gives abundances (in units of $M_\odot$) of several important elements prior to core collapse, for a variety of values for M_α. The most dramatic variation with core mass is for oxygen, which is richly produced in the most massive objects listed. This is simply a consequence of the increased yield of oxygen from hydrostatic helium burning at increased radiation entropy (chapter 8). Helium also shows an increase in absolute yield per star with stellar mass, but its fractional production actually decreases. Core convergence caused by neutrino cooling insures that the remaining yields vary less rapidly with stellar mass. Numerically integrating such results over an initial mass function gives an estimate for the absolute yields from the hydrostatic burning in layers that are ejected but not reprocessed by the explosion [607, 608, 35].

TABLE 14.5
Estimated Yields before Explosion (M_α)

M_α	He	C	O	Ne	Mg	SiCa	"Fe"	M_{Y_e}
3.52	1.83	0.066	0.107	0.033	0.025	0.333	1.13	1.42
4.02	2.06	0.112	0.190	0.056	0.045	0.156	1.41	1.50
5.02	2.46	0.218	0.430	0.141	0.087	0.254	1.44	1.57
6.03	2.81	0.339	0.758	0.240	0.129	0.123	1.63	1.63
7.04	3.09	0.349	1.056	0.358	0.225	0.460	1.51	1.72
8.05	3.42	0.338	1.626	0.302	0.291	0.480	1.59	1.62
9.05	3.64	0.653	2.019	0.597	0.197	0.314	1.63	1.64
10.06	3.86	0.816	2.498	0.614	0.197	0.422	1.65	1.56
12.07	4.37	0.986	3.338	0.594	0.261	0.330	2.19	1.65
16.09	5.22	1.293	5.774	1.069	0.349	0.302	2.09	1.61
20.13	5.84	1.877	7.874	0.422	0.453	1.396	2.26	1.82
24.14	6.50	2.138	10.163	0.548	0.584	1.781	2.43	2.03

As we have seen, in the ejected matter only the processes of oxygen burning and silicon burning are explosive; for these layers a model of the explosion is necessary. Also note the last two columns in table 14.5, which are simple estimates of the mass of the condensed remnant. Ultimately such estimates should come from a quantitative theory of the collapse and explosion. The first estimate is the mass of silicon-burning ashes (the iron-peak elements). The second (M_{Y_e}) is defined as matter which has had sufficient electron capture to ruin its chances for making a solar system distribution of isotopes. This corresponds to matter which has completely burned silicon and oxygen hydrostatically; it is deep in the star's gravitational potential well and has little energy to release if heated. Such matter does tend to become part of the condensed remnant in simulations of core collapse and explosion; it provides an interesting test of explosion mechanisms. The two estimates coincide well except for the highest and the lowest masses. These are "nucleon masses"; if they collapse to neutron stars, the mass must be reduced by the gravitational binding energy of the neutron star, which is roughly a factor of 0.9. For many of the cores, the estimates imply $M_{ns} \approx 0.9(1.6) = 1.44$, which is the Chandrasekhar mass limit for white dwarfs. Notice that the larger

TABLE 14.6
Estimated Yields before Explosion ($M_\odot$)

M	H	He	C	O	Ne	Mg	SiCa	"Fe"
15.1	8.6	4.77	0.141	0.142	0.028	0.019	0.059	1.342
20.1	10.5	6.91	0.346	0.579	0.010	0.143	0.173	1.451
25.1	12.2	9.29	0.541	0.991	0.350	0.163	0.117	1.545
30.2	13.9	11.40	0.695	1.640	0.230	0.248	0.383	1.742
40.2	17.0	15.10	1.310	3.490	0.337	0.248	0.771	1.984
60.3	22.8	22.90	2.190	8.120	0.393	0.398	1.280	2.276
90.5	30.9	32.90	2.890	16.200	0.845	0.842	3.140	2.8916

cores have values which may not be stable as neutron stars, but would collapse to black holes directly, or upon cooling and loss of leptons.

The yields from helium cores resemble the yields from hydrogen stars which develop helium cores of similar size. As seen in chapters 7 and 8, there are differences in detail. Table 14.6 gives the yields from stars of initial mass from 15.1 to $90.5 M_\odot$. Comparing this with the previous table allows us to construct a crude equivalence table between total mass and equivalent helium core mass; this is only a convenient approximation, not an exact correspondence. The results are shown in table 14.7.

Unfortunately, these details, while tantilizing, do depend upon algorithms used for convection in the late stages of evolution. As we have

TABLE 14.7
Mass Relations

M	M_α	M_{Y_e}
15.1	3.90	1.39
20.1	5.97	1.45
25.1	8.02	1.56
30.2	10.30	1.81
40.2	14.70	1.92
60.3	24.02	1.80
90.5	38.04	2.46

seen, the general trends in the yields depend upon quasi-hydrostatic and thermal balance, and seem robust. The precise quantitative details do not seem so (see chapter 10). This justifies using an approximate but simpler approach to illustrate the estimation of yields by integration over the IMF. We approximate the yields per stellar mass by linear functions of the mass, which gives fair accuracy and makes the integrations trivial. The two most fundamental are the relation between the total mass and the helium core mass,

$$m = 5.71 + 2.32m_\alpha, \tag{14.86}$$

and that between the helium core mass and the carbon-oxygen core,

$$m_\alpha = 0.92 + 1.54m_{CO}, \tag{14.87}$$

which we have constructed using tables 14.5, 14.6, and 14.7. These approximations are intended for the range $13 \leq m \leq 70$ and $3.5 \leq m_\alpha \leq 12$. The upper ranges are relatively well behaved, but at the lower ranges new sorts of evolution occur [598, 460]. We will only examine the production above these problem masses, which are not now thought to contribute much to the total yield. The masses in these formulae refer to mass coordinates; for example, the mass processed by hydrogen burning only would be $m_\alpha - m_{CO}$.

The mass processed at least through helium burning (ashes with nuclei having $Z > 2$) is $m_{CO} - m_{rem} \approx m_{CO} - 1.45$. This is positive only for $M > 13$, which means that for smaller masses, the CO core is smaller than the Chandrasekhar mass. Such stars must have an extensive shell-burning stage, during which the core grows to this size [598, 460]. The amount of such mass ejected from a generation of stars is

$$\mathcal{M}_{COej} = \int_{13}^{100} ((m_{CO} - 1.45)/m)\xi \, dm. \tag{14.88}$$

This gives

$$\mathcal{M}_{COej} = 0.280\mathcal{M}(13, 100) - 3.65\mathcal{N}(13, 100) \tag{14.89}$$

$$= 0.0926\xi_0 \tag{14.90}$$

for $\alpha = 4/3$, and

$$\mathcal{M}_{COej}/\xi_0 = 0.0270\xi_0 \tag{14.91}$$

for $\alpha = 5/3$.

TABLE 14.8

Bulk Yields

α	f	f_z	$f_z/(1-f)$	f_z/f
4/3	0.481	0.0241	0.0464	0.0501
5/3	0.346	0.00928	0.0142	0.0268

We may now estimate the absolute values of the parameters which appeared in the discussion of one-zone models. The return fraction is $f = \mathcal{M}_{ej}/\mathcal{M}$, where

$$\mathcal{M}_{ej} \approx \int_{1.21}^{100} (1 - 0.56/m)\xi \, dm, \tag{14.92}$$

and the lower-limit corresponds to the least-massive stars now making white dwarfs.[11] We use an average white dwarf mass of $0.56M_\odot$. Similarly, the return fraction of "metals" is $f_z \approx \mathcal{M}_{COej}/\mathcal{M}(0.3, 100)$. These values are given in table 14.8. In the closed box model, the abundance approaches $f_z/(1 - f)$, while in the infall model it approaches f_z. In the solar neighborhood, the typical abundance is about that of the Sun, so that the $\alpha = 4/3$ case is too large by a factor of two to three for the closed box model, but only slightly high for the infall model. For the $\alpha = 5/3$ case, the value for the closed box model is essentially solar, but that for the infall model is low by a factor of two. These estimates make no allowance for mass loss or binary effects, which must eventually be included. It is encouraging that the stellar evolutionary estimates of the yields, taken with the plausible range of IMF values, does give the range of metallicity seen in Population I stars.

14.5   THE GALACTIC DISK

The best-observed part of all galactic systems is the solar neighborhood, which is dominated by disk stars. Throughout this book we have used solar system abundances as our basis for comparison; now we extend our scope and attempt to describe the nature and composition of the Galactic disk, as typified by the solar neighborhood.

[11]See §14.5 below. For Population II, the lower abundances give less opacity, making the stars brighter. Thus, they burn faster, so that the corresponding mass would be lower.

Suppose white dwarfs are formed by all stars which have a mass below some value m_u but are still massive enough to evolve during the time since the disk formed. We found the age of the solar neighborhood to be $\tau_0 \approx 9$ Gy. This age is consistent with the cooling ages of white dwarfs [688, 325] To the main sequence lifetime ($\tau(m) = 10.8m^{-3}$ Gy), we must add the time spent in later evolution. Most disk stars have abundances similar to the solar system; most of the extra time is spent in the shell-narrowing stage [518]. We take $\tau = 16m^{-3}$ Gy for Population I stars. Equating this to 9 Gy gives $m_w = 1.21$. Then the number of white dwarfs in the solar neighborhood is

$$\mathcal{N}_{wd} = \frac{\xi_0}{\alpha}\big[m_w^{-\alpha} - m_u^{-\alpha}\big], \tag{14.93}$$

where $\alpha = 4/3$. For a value [526] of $m_u = 8$, $\mathcal{N}_{wd} = 8.0 \times 10^{-3}$ pc^{-3}, which is identical to the estimate of Bahcall and Soneira [66]. For $\alpha = 5/3$ for masses above $m = 1$, $\mathcal{N}_{wd} = 6.2 \times 10^{-3}$ pc^{-3}. Using Population II lifetimes would increase the density because more stars would evolve to be white dwarfs. These values are consistent with Salpeter [537]. The white dwarf density of the sample of Liebert, Dahn, and Monet [395], which they do not expect to be complete, gives 3×10^{-3} pc^{-3}. In particular, a proper motion survey undercounts objects with apparently small velocities (see [395] for further discussion). Weidemann [673] attempts to correct for incompleteness and arrives at a density of 5×10^{-3} pc^{-3}. If we take the average mass of these white dwarfs to be 0.56 M$_\odot$, as in [90], we find a mass density of 0.0045 M$_\odot$ pc^{-3} and 0.0035 M$_\odot$ pc^{-3}. This is similar to the estimate of Bahcall and Soneira [66], which is encouraging. However, note that these values are significantly smaller than the value of 0.02 M$_\odot$ pc^{-3} quoted in Mihalas and Binney [439], or the 0.028 M$_\odot$ pc^{-3} of Binney and Tremaine [99].

If we assume that stars above $m = 8$ up to $m = 25$ make neutron stars of mass 1.45 M$_\odot$, we find a mass density of $0.056\xi_0 = 8.4 \times 10^{-4}$ M$_\odot$ pc^{-3}. For an upper mass limit of $m = 100$, we have $0.068\xi_0 = 1.02 \times 10^{-3}$ M$_\odot$ pc^{-3}, which is little different. These values are about twice those of Shapiro and Teukolsky [567]; they assumed the progenitors were in the range $4 < m < 10$. Following our previous discussion of black hole production, the number density of black holes in the solar neighborhood is $\mathcal{N}_{bh} \leq \mathcal{N}(30, 100) = 0.0063\xi_0$ to $0.0018\xi_0$, or 9.4 and 2.7×10^{-5} pc^{-3}. These upper limits are a factor of ten lower than the estimates of Shapiro and Teukolsky.

Using our fitted value of $\xi_0 = 1.5 \times 10^{-2}$ pc^{-3}, we can estimate the mass in stars in the solar neighborhood. First, consider the stars that live too long to have reached the white dwarf stage within the lifetime of the disk, $m \leq m_w = 1.21$. We will assume that integrating the Salpeter IMF down to $m = 0.2$ will give an adequate estimate; see figure 14.1. Notice that this result is sensitive to the mass tied up in these very-low-mass stars. Extending the range from 0.3 to 0.2 adds a mass of $0.65\xi_0$. Then

$$\mathcal{M}_{MS} = \mathcal{M}(0.2, 1.21) \tag{14.94}$$

gives $\mathcal{M}_{MS} = 2.315\xi_0 = 0.035$ $M_\odot$ pc^{-3}, compared to the 0.040 $M_\odot$ pc^{-3} of Bahcall and Soneira. We now account for stars of mass $m \geq 1.21$, which can have died since the disk formed. In this case, the integral becomes

$$\mathcal{M}_{evol} = \int_{m_w}^{m_2} \Lambda t \xi(m)\, dm, \tag{14.95}$$

where $t = 16m^{-3}$ Gy and (for our IMF fit) $\Lambda^{-1} = 9$ Gy. Because of the added mass dependence, the integral is insensitive to the upper mass limit, and we have $\mathcal{M}_{evol} = 0.281\xi_0$, which is a small correction. Thus the mass in visible stars is

$$\mathcal{M}_{vis} = 2.60\xi_0 = 0.0389 \text{ M}_\odot \text{ pc}^{-3}, \tag{14.96}$$

in rough agreement with the value of 0.044 $M_\odot$ pc^{-3} quoted in Mihalas and Binney [439].

In the solar neighborhood, the surface density of visible stars plus stellar remnants is $\sigma \approx 44.6$ $M_\odot$pc^{-2} ([98]). Setting $\mathcal{M} = \sigma$ and using $\xi_0 \approx 15$pc^{-2} gives 57.5 and 43.7pc^{-2}, which is consistent. For $\tau_0 = 9$Gyr^{-1}, this corresponds to an average rate of gas consumption by star formation of approximately 5 $M_\odot$ pc^{-2} Gyr^{-1}. This would be partially compensated for by gas return (in the instantaneous recycling approximation, this reduction would be $1 - f \approx 0.7$ to 0.5). For a surface density for total mass of 75 $M_\odot$pc^{-2} and a total disk mass of 6×10^{10} $M_\odot$ ([98]), the area of the galactic disk must be $A_{disk} \approx 8 \times 10^8$ pc^2. This rate of star formation, averaged over the disk, would correspond to a mass consumption of 3.9 $M_\odot$ yr^{-1}. A decreasing rate of star formation would reduce these values.

From the motion of stars perpendicular to the plane, the mass density at the plane was found to be $0.148 M_\odot$ pc^{-3}. Our IMF fit gives $\mathcal{M} \approx 3\xi_0 = 0.04$ $M_\odot$ pc^{-3}. Interstellar matter is estimated [584] to give 0.045 $M_\odot$ pc^{-3}. This would leave us short by 0.05 $M_\odot$ pc^{-3}, which is thought to be made up by dark matter, but the uncertainty in the estimates of the gravitational mass seems uncomfortably similar in size to the shortfall. Suppose the "other matter" were contributed by stars of mass $m \leq 0.3$. Our largest estimate from §14.2 was $1.5\xi_0 \approx 0.022$ M$_\odot$ pc^{-3}, which is about half the deficit. The estimate is quite uncertain, but at least the difficulty with dark matter is not restricted to our problem [628].

The rate of formation of stars in the mass range $m_1 \leq m \leq m_2$ is

$$d\mathcal{N}/dt = \int_{m_1}^{m_2} \Lambda\big(\xi(m)/m\big)\, dm, \tag{14.97}$$

for a constant formation rate Λ. The rate at which gas is consumed in star formation for $m_1 \leq m \leq m_2$ is

$$d\mathcal{M}_{gas}/dt = \int_{m_1}^{m_2} \Lambda\xi(m)\, dm. \tag{14.98}$$

In some domain of mass $\mathcal{M}$, the rate of formation of stars in the range $m_1 < m < m_2$ is

$$d\mathcal{N}/dt = F\Lambda\mathcal{M}\mathcal{N}(m_1, m_2)/\mathcal{M}(m_1, m_2). \tag{14.99}$$

Here F is the fraction of gravitating matter in visible form (gas plus stars); probably it lies in the range $0.5 \leq F \leq 1$. Using $\Lambda = 1/9$Gy and taking the mass of the disk [66] to be $5.6 \times 10^{10} M_\odot$, the rate of stars dying by core collapse (and of Type II supernovae) is $d\mathcal{N}_{snii}/dt = F\Lambda\mathcal{N}(8, 100) = 0.0734F\ y^{-1}$ for $\alpha = 4/3$, and $d\mathcal{N}_{snii}/dt = F\Lambda\mathcal{N}(8, 100) = 0.0394F\ y^{-1}$ for $\alpha = 5/3$, giving one per $13.6F^{-1}$ y and $25.3F^{-1}$ y. This procedure is based on the assumption that the solar neighborhood is a perfect average of the disk. If supernova units (SU) are defined as the number of supernovae per century per 10^{10} L$_B(\odot)$, and the Galaxy is 1.7 times brighter than this [66], these rates correspond to $4.3F$ and $2.3F$ SU. For comparison, Tammann [610] estimated 5.5 ± 0.6 SU, while Evans, van den Bergh, and McClure [225] got 1.9 ± 0.6 SU. It appears that Tammann's rates may need to be revised downward, due to an overestimate of the inclination effect [648]. The factor F relating to dark matter comes in here because we take the scale of the Galaxy from gravitational information, while the stellar content (and rate of star-formation) involves the visible matter.[12] If the

---

[12] We could avoid this dependence on F by using the B band luminosity for the solar neighborhood as defined in the IMF discussion, and scale the rate to the B luminosity of the galaxies in which supernovae are seen. This approach would not give a rate for the Galaxy, however, without additional information.

star-formation rate decreased in time by a factor of 0.83 (0.44) from its average value, the agreement with Evans et al. would be exact for the $\alpha = 5/3$ ($\alpha = 4/3$) case, with $F = 1$. Taking a decreasing star-formation rate (by 0.7), and a modest estimate for dark matter ($F = 0.7$), will put the $\alpha = 4/3$ case within the Evans et al. error limits. The assumptions about birthrate and disk averaging also introduce significant uncertainty. These estimates are only for core collapse events; thermonuclear events must be added. Probably these would correspond to SNIa's. Taking these estimates at face value, the rate of thermonuclear events might have to be significantly smaller, as seems to be the case [225]. Given the uncertainties, the agreement is encouraging.

14.6 PRIMORDIAL STELLAR YIELDS

Can we go beyond these bulk yield estimates? It should be clear that we can use the results from previous chapters, adding contributions from core collapse and from thermonuclear supernovae, seasoned with some shell nucleosynthesis from intermediate-mass stars, to postdict the solar system abundances. The mathematicians would say that we have established existence of such a solution, but not its uniqueness. Too many possibilites lie within the range of reasonable uncertainty. For example, do the VMOs of chapter 11 have a role? Are the Type Ia supernovae due to delayed detonation, helium detonation, both, or something else? Are there significant effects from binary interaction and from mass loss? We need to test the predictions from the potential sources separately, to avoid glossing over inconsistencies by adjustment of poorly constrained parameters. We begin by focusing attention on the detailed yield from massive stars, both as a step toward a more complete understanding and as an illustration of the method.

Despite the unsatisfactory state of the details of mixing during the late stages of massive star evolution, it is of interest to examine detailed quantitative predictions of nucleosynthesis. We attempt a strategy that differs from simple summation of the results of evolutionary sequences of several stellar masses. We concentrate on the "average supernova for nucleosynthesis," which means the average effect of those events that contribute the most to the yield integrals, or $m \approx 25 \pm 5$. We use existing evolutionary sequences to define typical conditions for the convective burning regions (such as radiation entropy), and weighting factors for each contribution. We then use our previous methods (chapters 7 through 10) to evolve the abundances with a network of 300 nuclei, from hydrogen burning through explosion. This is essentially an integration by sampling of the yield integrals, in which we concentrate upon the most representative regions. In order to build up a complete picture, we begin with only massive stars; contributions from intermediate mass stars and from thermonuclear explosions will need to be added. The extent to which this restricted procedure works will give an estimate of the importance of the omitted contributions. The limitation of the network to 300 nuclei means that Kr (Z=36) is the most massive element considered. This implies that we ignore the r-process and the canonical s-process, but include the weak exposure s-process nucleosynthesis.

Although we know that massive stars with Population II abundances, or even primordial Big Bang abundances, will evolve differently in detail from Population I massive stars, as a first step we ignore this subtlety. This is a crude attempt to simulate nucleosynthesis in the primordial generation of stars, using the most cautious hypothesis. Comparison with observation will reveal the flaws which may provide clues as to what is missing. We begin with stars having primordial abundances, derived from Big Bang nucleosynthesis.

For the explosive stages, the conventional explosive approximations are used (see [30] and chapter 9 of the present volume). This then gives estimates of the total yield of those nuclei included in the network. The weights given for explosive burning to ^{56}Ni were taken to be roughly consistent with SN1987A: a typical core collapse was assumed to produce $0.075M_\odot$ of ^{56}Ni. This range can be easily attained in hydrodynamic modeling of the explosion.

This sampling approach to the integration over the IMF has the virtue of simplicity and speed. Because of the uncertainty in explosion mechanism, and in presupernova evolution, it is not clear that more detail is warranted. The sampling approach has the corresponding difficulty that the choice of trajectories must be carefully made (chapters 7 through 10), and tested in a way independent of the abundance patterns to be predicted. We examine a skeleton model, with few trajectories, which shows the key features of a complex problem.

The typical trajectories are only allowed to vary due to changes in composition; this is equivalent to forcing both the IMF and the relative masses in different stellar zones to be constant in time. This constraint gives an appropriate limiting case, but it would be surprising if it were correct in detail. How does this procedure work? The weight factors for the trajectories are fixed, but for example, hydrogen burning may occur at higher temperature if CNO catalysts are rare. Different stellar evolution trajectories could be implemented for each different metallicity, but the problem of the uncertain IMF at low metallicity would remain.

Table 14.9 gives the key features of those trajectories which were hydrostatic rather than explosive. Some matter was processed not only in

TABLE 14.9

Hydrostatic Trajectories

id	$M_\odot$ /SN	S_γ /$\mathfrak{R}$	From	To	Fuel	F_x	F_t	$T_9(f)$ (K)	$\rho(f)$ (g/cc)
l	12.0	1	BB	0.770	—	1	40	1.01(−2)	1.24(−1)
k	1.0	3	BB	0.675	H	1	40	1.02(−1)	4.26(1)
j	1.0	3	k	0.550	H	1	40	1.00(−1)	4.10(1)
i	2.9	1	l	1(−7)	H	1	40	1.28(−1)	2.55(2)
g	0.16	3	i	0.35	He	3	40	2.04(−1)	3.40(2)
f	0.25	1	i	1(−4)	He	1	40	2.77(−1)	2.59(3)
e	0.75	1.5	f	0.02	C	1	40	1.308	1.81(5)
d	0.4	1	f	0.02	Ne	1	40	1.756	6.58(5)
c	0.1	1	d	0.02	O	1	40	2.286	1.45(6)
w	0.0	0.25	c	0.05	O	1	100	2.366	6.43(6)

this way but then modified further by the explosion, as we shall see below. The column labeled $M_\odot/SN$ corresponds to the mass that was processed only to this level and no further. Therefore, by adding the entries in this column, and the corresponding one for explosive processing in table 14.10 below, we find this "typical supernova" has a mass of $18.8M_\odot$ (see §14.2 and 14.4). The first line corresponds to matter that was ejected but not thermonuclearly processed at all. It is not really true that two-thirds of the matter is unprocessed entirely; we simply are ignoring some processes which are both complex and peripheral to our main concern here. The elements LiBeB were not followed carefully, so that their variation was not reliably predicted. Thus, Big Bang abundances simply mean nucleon fractions for ^{1}H and ^{4}He of 0.23 and 0.77 respectively, in this approximation. Also note that our representation of very incomplete hydrogen burning underestimates the modification of the rare isotopes of C and O, as well as ^{14}N, ^{15}N, and ^{19}F. Accurate predictions for these nuclei require an understanding of mixing, mass loss, and binary interaction in massive stars that the author does not have.

The k trajectory corresponds to incomplete hydrogen burning (the fuel is H), in which the abundance drops from its initial value of 0.77 (from the Big Bang, hence BB) down to 0.675. During this burning, the radiation entropy is $S_\gamma = 3\mathfrak{R}$; this represents the contribution of shell

TABLE 14.10

Explosive Trajectories

id	$M_\odot$ /SN	S_γ /$\Re$	From	Burning	$T_9(i)$ (K)	$\rho(i)$ (g/cc)
y	0.05	9.7	d	Ne	3.13	3.82(5)
b	0.1	9.7	c	O	3.60	5.84(5)
z	0.04	16.8	c	α-rich	5.50	1.2(6)
a	0.03	9.7	w	Si	4.80	1.38(6)

burning, and of core burning in very massive stars. The convective region is forty times more massive than the flame zone; this is indicated by $F_t = 40$. The evolution is similar to that described in detail in chapter 7. Because there are no CNO catalysts initially, a brief period of high temperature burning occurs in which they are made in trace amounts by helium burning. Then expansion and cooling occur, and hydrogen burning proceeds by the CNO cycle. The final temperature (T_9) and density (ρ) are given.

The j trajectory represents a more extended stage of incomplete hydrogen burning; it is a continuation of the k trajectory. The i trajectory corresponds to complete burning of hydrogen in the stellar core (note the lower value of $S_\gamma = 1\Re$); it begins from the initial condition l. This branching of paths reflects some of the complexity of the problem.

With trajectory g, helium burning occurs in a shell, and depletes the ^{4}He to 0.35. Core helium burning is represented by sequence f. These stages were described in chapter 8. The factor F_x is discussed there, and corrects the relative rates over the convective core of the 3α and the ^{12}C(α, γ) reactions.

Shell carbon burning is represented by e; it goes almost to completion ($X(^{12}$C $\to$ 0.02). Neon shell burning corresponds to d, and oxygen shell burning by c. Silicon shell burning is w; notice that none of this matter escapes without further processing by the explosion. All these stages were discussed in chapter 10.

The detailed yields from explosive burning may be significantly modified by the nonspherical behavior observed in recent numerical simulations [85]. We take a simple approach, which categorizes the major processes. Table 14.10 defines the explosive trajectories; see chapter 9 for further detail. The mass ejected, the radiation entropy, the previous trajectory being continued, and the initial (postshock) temperature and density are given.

The outermost layer which might have some explosive processing is y, which is labeled "Ne" burning. This corresponds to 0.043 of the mass ejected as hydrostatic carbon and neon burning, which is roughly consistent with the constraint in §9.6 (from the observed ^{70}Zn/^{68}Zn ratio in the solar system). Explosive oxygen-burning is b, explosive silicon burning is a, and the alpha-rich freezeout trajectory is z. The entries are in order of increasing density. The densest layer, which is explosive silicon burning, or essentially just the e-process, will have the highest neutron excess because it was hydrostatically processed by almost complete oxygen burning at low entropy. Explosive oxygen burning to alpha-rich freezeout is the most vigorous process and may increase the entropy significantly (as is assumed here).

The r-process contribution, although thought to be related to the core collapse of massive stars and expected to be important at low metallicities, was not included because of uncertainty concerning the mechanism of explosion (chapter 12). Note that r-process isotopes of Ge, Se, and Kr comprise about 0.1 of the solar system abundance pattern. If such an r-process accompanied the production of Fe in these supernovae, an elemental ratio to oxygen of order 0.1 of the solar system value might be expected.

These trajectories are incomplete because the yields from other types of stars are ignored. In particular, the rate of nucleosynthesis from thermonuclear supernovae may be slower because of longer times needed to evolve to explosion. Because of an increasing awareness of the complexity of the issue, the effects of thermonuclear supernovae were not included (see chapter 11, however). Consequently, the Population II abundances, which are presumably not contaminated by those supernovae having slower evolving progenitors, attain a new importance.

Figure 14.6 shows the resulting yields if the initial abundance distribution was that produced by the Big Bang. The abundances are presented in a conventional astronomical notation, as ratios to Fe relative to the same ratio in solar system abundances [13],

$$[X_i/X_{Fe}] \equiv \log\left((X_i/X_{Fe})/(X_i/X_{Fe})_\odot\right), \qquad (14.100)$$

where the X_i are nucleon fractions ("mass" fractions). A perfect reproduction of the solar system abundances from a generation of stars would be represented by each point lying on a horizontal line which intersects the vertical axis at zero. The dotted horizontal lines show abundances,

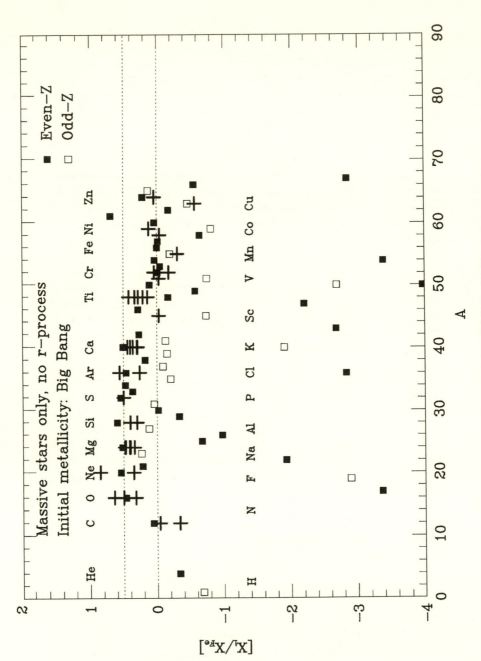

Fig. 14.6. Primordial Stellar Yields

relative to iron, of solar value and three times solar. All isotopes are shown, with unstable nuclei decayed into their stable daughters (except for those of exceedingly long lifetimes, in this case ^{40}K, which are plotted as their long-lived unstable form). The most abundant isotope of each element is labeled by the symbol for the element. Even-Z nuclei are represented by solid squares, and odd-Z nuclei by open squares. The corresponding abundances observed in some Population II stars are shown as large crosses (see [283, 406, 710, 270, 455, 456]).

Except for nitrogen and fluorine, which are strongly underproduced, the nuclei up to Zn are surprisingly solar in abundance. The yields of N and F are very low, which suggests a flawed choice of trajectories representing the weakly burned matter in these stars, and may also reflect the neglect[13] of nucleosynthesis from intermediate-mass stars [517]. Observations by [154, 370] suggest a relatively high nitrogen abundance. If these low-metallicity stars formed quickly, then one attractive possibility is that intermediate-mass stars do not evolve fast enough to contribute; if so, then this is evidence for primary nitrogen production in massive stars.

Beyond Zn, there is essentially no production of these nuclei in the nucleosynthesis trajectories represented. However, yields of these nuclei might be dominated by the neglected r-process from other layers of the ejecta. The high abundance of O, Ne, Mg, Si, S, Ar, and Ca relative to Fe is the long-known "alpha effect" [272]. The enhancement is a factor of three; assuming a different production of ^{56}Ni per core collapse would affect this directly. The consistency of this enhancement should not be taken too seriously, as variations of 0.2 in the base ten logarithm are easily made. Here carbon is not enhanced relative to iron; the precise value of C is a function of the convective mixing algorithm. A reasonable measure of the a priori uncertainty in this value for C, relative to O, is at least 50 percent. The odd-Z nuclei are underabundant relative to the even-Z neighbors, but the effect is much less than predicted by simple arguments of linear increase of neutron excess with metallicity. This is due to electron capture and positron decay during hydrostatic carbon burning and beyond, a possibility worried about before [23]. If we had assumed that the radiation entropy of the trajectories for the first stars was higher (lower), the electron capture effects would be lower (higher), and therefore the odd-Z/even-Z ratios would be lower (higher).

---

[13]The itermediate-mass stars return about three times as much matter to the interstellar medium as do the massive stars (see §14.4), but with much less processing. The processing that is done is likely to reflect hydrogen burning, possibly some helium burning, and the s-process.

Several of the features seen here agree with observations of extreme Population II stars. The C/Fe ratio is essentially solar, while the plateau from O to Ca corresponds to an increase of a factor of three or so relative to iron. The predicted Sc/Fe is low relative to solar, much lower than the observational value. The Sc yield is sensitive to the neutron excess in the matter undergoing explosive oxygen burning; presumably it therefore reflects the conditions in the oxygen-burning shell of the presupernova star. The predicted ratio of ^{48}Ti/Fe is too low; but ^{46}Ti, which makes up about 0.1 of solar system Ti, is enhanced, giving an elemental value for Ti/Fe that is at the bottom of the observed range. The prediction for vanadium is discrepant, being lower than observations of Population II stars by a factor of almost ten. This is a flaw in the simple model.

The iron group, Cr, Fe, and Ni, are solar. However, Mn is predicted to be low, and is observed to be just this low. From the nuclear point of view, this prediction is more reliable than Sc or V; it involves the production of a much less rare nucleus, and is therefore less sensitive to small errors of omission. The Mn abundance seems to be a direct measure of the neutron excess in the matter undergoing explosive silicon burning (e-process). This matter seems to have had a lower entropy and higher neutron excess than that undergoing alpha-rich freezeout (see below). The predicted yield of Co is lower than the observations by a factor of six (another flaw).

Finally, the predicted Zn/Fe and Cu/Fe are close to those observed. However, this is sensitive to the choice of parameters; this choice illustrates a dependence of abundance upon explosion model. The Zn and Cu were produced in an alpha-rich freezeout, and their abundances are sensitive to the initial neutron excess and the entropy of the explosive trajectory. This matter presumably came from the inner part of the oxygen-burning shell of the presupernova, and its high entropy related to explosive oxygen burning after the supernova shock arrived. These abundances represent the higher end of the range of entropy seen in hydrodynamic simulations of the explosive burning of the oxygen shell.

Except for these three flaws, at Sc, V and Co, the predictions of the trends agree well in this preliminary comparison with the pattern observed in Population II stars.

Population II stars have a metallicity of about 0.01 solar; what might their supernovae produce? Except for a weak s-process production of elements beyond Zn, at a level of 0.01 relative to Fe, the yields change little. The elements beyond Zn are expected to be produced at a higher level, as a group, by the r-process. Thus figure 14.6 is a reasonable representation for these supernovae as well, to the extent that changes in the IMF and in the stellar evolution do not change our trajectories.

At what level do metallicity effects—from purely nuclear causes—become important? Figure 14.7 shows an estimate; the initial abundances were taken to be 0.1 of the yields from the previous figure, mixed with primordial matter. This is a metallicity of 0.1 solar for O through Ca, and about 0.03 for Fe. Most of the abundance features are changed little. Although still woefully underproduced (see above), N and F have moved upward in the graph. Above Zn another change has occurred: the small seed abundance of iron-peak nuclei has undergone s-processing during hydrostatic He, C, and Ne burning. This produces an enhancement of nuclei, from Ga to Kr, up to 0.1 of the solar ratio. This is small, supporting the idea of Truran [680] that such nuclei were produced predominantly by an r-process at early epochs.

For conventional models of galactic evolution, higher metallicities are affected significantly by the evolution of lower-mass stars and by Type Ia supernovae. The comparison of observations of abundance and theoretical yields becomes more complex, involving new uncertainties. However, as an aid to insight, it is interesting to consider the yields if the initial abundances reflect 0.3 of the yield; this is roughly the metallicity of the Large Magellanic Cloud. Here, for purposes of analysis, we add no contribution from Type Ia supernovae or from intermediate-mass stars; we do not expect this to be realistic. We should see flaws related to the neglect of these and other sources of nucleosynthesis. The results are shown in Figure 14.8. Now N is still underproduced, but not so badly as before; F/Fe is also low. Na and Al have almost solar ratios relative to Ne and Mg. Cl and K are mildly underproduced relative to Si, S, Ar, and Ca. Sc and Ti are still low. The iron-peak is changed little. The biggest change is for Zn and beyond, where an enhancement is seen. The Ge, Se, and Kr are approaching solar relative to the "Fe produced only by core collapse supernovae." Both s- and p-process isotopes are produced.

In fact, if the metallicity is increased by another factor of three, to solar for O-Ca and one-third solar for Fe, there is a pronounced depression in the yield curve between A=40 and 70. It is this gap that the yields from Type Ia supernovae should fill. For these parameters, 0.33 of the solar iron would come from core collapse; this could equally well be varied however, since the integrated Ni yield from all masses of these objects is not well determined observationally, nor plausibly computed theoretically. A better quantitative limit on this important parameter is needed; at present it is simply a parameter which may be adjusted within this range, for nucleosynthetic and not hydrodynamic reasons.

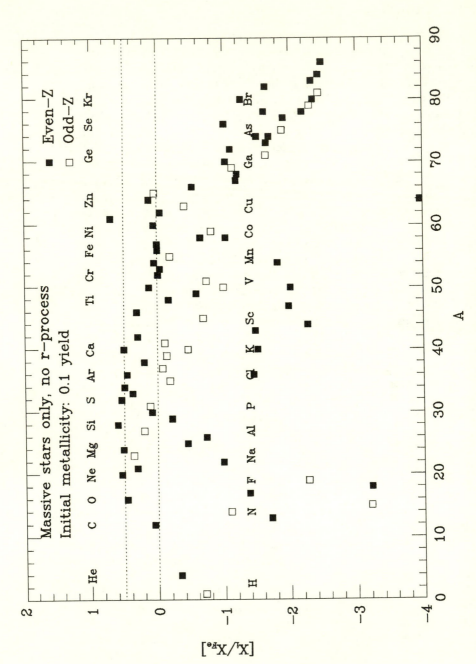

Fig. 14.7. Stellar Yields ($z/z_\odot = 0.1$)

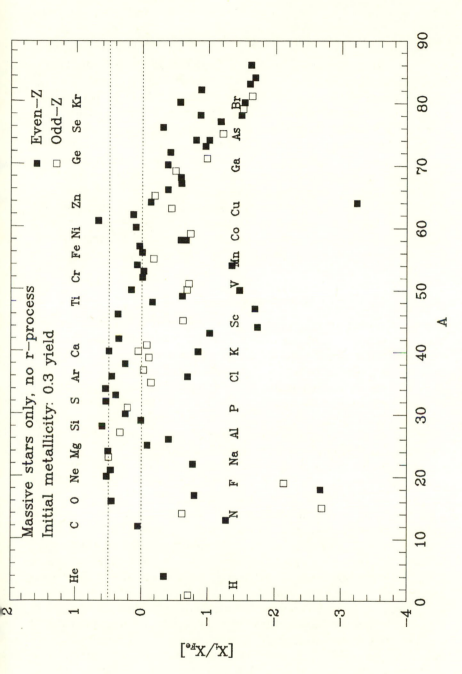

Fig. 14.8. Stellar Yields ($z/z_\odot = 0.3$)

14.7   CRITICAL UNCERTAINTIES

Clearly, serious attempts to predict nucleosynthesis yields from massive stars give attractive results. By adding additional sources, and adjusting computational treatments, excellent representations of the solar abundance distribution can be constructed which agree with the available data to within reasonable estimates of the observational and interpretational error. Comparison to elemental abundances in low-metallicity stars, and in supernovae, give encouragement. We now have the beginnings of a quantitative understanding of how the abundances came to be, and a set of powerful tools to extend that understanding.

But despite this flurry of success, we must not forget that our ultimate goal should be to get it right! Therefore even the most beautiful result must be examined with skepticism; any possible flaw may provide a chance for improvement. Fortunately there are many such chances.

Thermonuclear explosions

The solar system abundance data is still the most demanding because of its accuracy for isotopic abundances. As we have seen, connection of theory to this data set requires that we know the yields from Type Ia supernovae. Differences between predictions and solar data may be due to lack of proper understanding of these Type Ia yields, even if our predictions for massive stars were perfect! This is not an idle complaint; for the same O-Ca metallicity, using solar system abundances as the initial values, gives yields that differ in nontrivial ways. The subject of thermonuclear explosion of electron-degenerate stellar objects is becoming increasingly complex, as we saw in chapter 11. Yield estimates have been presented for the popular W7 model [463, 619], but see [9, 53, 353] for new views on this approach, and [700] and [400] on detonations of sub-Chandrasekhar mass white dwarfs. This suggests that a high priority be put on continuing and expanding the effort to obtain high-quality abundances for metal-poor stars, which presumably probe the time before the Type Ia supernovae became important. Similarly, analysis of spectra of Type Iabc supernovae should indicate more precisely what these objects are, and what they synthesize.

An interesting issue concerns the yields (and brightness) of Type Ia supernovae and the "alpha effect" seen in Population II abundances. If the Type Ia supernovae produce as much Si and Ca, for example, as Ni (which decays to Fe), then they cannot produce the "alpha effect." Detonations produce Si-Ca at lower densities, and ^{56}Ni at higher; so the way to make dim supernovae, with little Ni, is to have the explosion occur at such low densities, whatever the detonation/deflagration mechanism. Thus the explanation of the "alpha effect" in terms of Type Ia supernovae implies a minimum luminosity for them, at least in a statistical sense (due to the integration of yields over a mass function). As shown from examination of figures 14.6, 14.7, and 14.8, this corresponds to an average Type Ia supernova which produces a fair amount of ^{56}Ni, and so is fairly bright.

Oxygen shell burning

The practical limitation on numerical computations of compressible hydrodynamics is the time that sound takes to travel across a grid element: the Courant time. The time required to consume a nuclear fuel is the ratio of the available nuclear energy to the burning rate: the burning time. Cooling by neutrino-antineutrino emission forces oxygen burning to such a high rate that the burning time becomes a computationally feasible number of Courant times [45, 84, 85]. This allows us to avoid use of the mixing-length theory of convection, which was designed for conditions such as those in the solar photosphere (a factor of about 10^{14} lower in density). This is relevant to the yield problem because as yet only algorithms have been used which did not resolve the flame zone for oxygen burning. In fact, all the burning stages dominated by neutrino cooling suffer from this flaw.

The direct simulations of convective oxygen burning are qualitatively different from the 1D evolutionary computations which must use mixing-length theory. Therefore our predictions of yields are uncertain to this extent. The 2D simulations smoothly diverge from the initial 1D structure, and approach a new solution. Some significant differences are: (1) Y_e has significant variation through the convection zone, (2) density and temperature show significant variation with coordinate angle for a fixed radius, (3) burning occurs in a very sporadic way, in which hot spots appear and disappear (and have peak heating rates that are 100 times the average value), and (4) abundances mix more (for example, ^{12}C wanders down into the oxygen convective zone!). Extensive testing of zoning and boundary conditions indicates that these results are robust.

The general idea of a thermal balance over the convective zone between heating and neutrino cooling seems still to be valid (§10.1), but the behavior of the boundaries of the convective region is different. The flame zone is no longer entirely convective. It is the long-term evolution of this region that determines the core mass for collapse, and the yields for the different burning stages. Computations for longer times, and computations in 3D, will be necessary to understand these long-term implications.

A Shocked O-Shell

The two-dimensional models of the oxygen-burning shell present an opportunity. It is this region in which explosive burning actually makes the ^{56}Ni seen in core collapse supernovae. In SN1987A we have considerable information about this matter, and in particular about its mixing and its velocity structure. Simulations involving Richtmeyer-Meshkov and Rayleigh-Taylor instabilities at the H-He interface [248, 447], and the hot nickel bubble [290, 44], have been inadequate to explain the observations quantitatively. A supernova shock will drive instabilities as it moves through the oxygen shell, and the initial perturbations are no longer arbitrary but determined by the previous hydrostatic evolution. If the promise of preliminary simulations is fulfilled, SN1987A will provide fresh constraints on the nature of the oxygen shell, its explosion, and its nucleosynthetic yield.

Old, neglected worries

These efforts represent significant progress, but—of course—problems remain: mass loss, binary interaction, and the initial mass function for primordial and Population II stars. Silicon burning produces nuclei which, having higher proton number, have lower thresholds to electron capture. Electron capture can be driven by the electron fermi energy, but these nuclei can then be carried by convection to lower density where decay occurs. This Urca process couples evolution to convection in a way that has not yet been simulated properly because of its complexity (see [61] and chapters 11 and 12 of the present volume). It would happen during the stages of core and shell silicon burning, in which the core collapse is triggered. Reliable models of the explosion mechanisms still elude us, although both thermonuclear and core collapse simulations are now promising.

The data set of abundances for testing nucleosynthesis theory has improved enormously, but is still limited. Questions of conversion of observations into abundances remain. Perhaps these results will help observers by providing the necessary quantitative targets for disproof!

Having presented a list of possible flaws, a more positive statement is appropriate. The preliminary comparison with Population II abundances is favorable. But the most startling implication is that little else, besides ordinary massive stars in a Salpeter-like initial mass function, is required to produce a reasonable approximation to the abundances of the elements seen in the oldest stars. Will this hold true to even lower metallicities, and with better analyses, or will there be surprises? Reasonable estimates of the yields from thermonuclear models of Type Ia supernovae (see chapter 11), and from asymptotic giant branch stars (see [116, 309, 668]), seem capable of filling in the gaps as needed to transform the Population II abundance pattern into that of the solar system.

APPENDIXES

Appendix A

Solar System Abundances

*After the natal season of the world, the birthday of sea
and lands and the uprising of the sun, many atoms have
been added from without, many seeds contributed on
every side by bombardment from the universe at large.*
—Lucretius (98?–55 B.C.),
On the Nature of Things

This appendix summarizes the features of solar system (or "cosmic")
abundances. To the extent that this pattern is ubiquitous in the cosmos
(and it seems to be amazingly common), it indicates the existence of
an ongoing, universal set of processes. Here we mean "universal" in its
literal sense: processes that occur throughout the Universe.

As a step toward understanding the nature of these processes, the
naturally occurring nuclei are placed in categories, according to the
nuclear fire by which they are thought to be produced. We would like to
know the *sites* of these processes, the particular astronomical situations
in which they happen. These sites are not unique; the same process
can occur in different astronomical objects. For example, their light
curves show that both Type I supernovae and Type II supernovae produce ^{56}Ni, which is synthesized in explosive silicon burning. At present
our knowledge of processes is more secure than our knowledge of
sites.

A.1 TABLES

The following tables give the abundance data of Anders and Grevesse, as
presented in [13], and an attempt to categorize each nucleus according to
how it is made. For historical perspective, the tables may be compared
to the appendix in Burbidge, Burbidge, Fowler, and Hoyle [135], the
abundance tables of Suess and Urey [566] and of Cameron [140, 146, 147,
148], and the review of Trimble [627]. The entries are the element symbol
(such as H, He, O, C, etc.), the proton number Z, the nucleon number
A, the nucleon fraction X_i (sometimes called the "mass" fraction), the
log of the number (normalized so that the base 10 logarithm of the

element silicon is 6), and the probable source of the nucleus. The notes that follow the tables give further information concerning the attribution to a particular source.

The entries up to proton number $Z = 36$, the element krypton, reflect the calculations described in the text; they emphasize the yield from massive stars, and the Big Bang. The yields from intermediate and low-mass stars seem to be important for hydrogen burning (rare CNO isotopes) and the s-process. The attribution of heavier nuclei to the s, r, and p processes follows Cameron [148]. There seem to be no major gaps. Plausible scenarios have been constructed for synthesis of nuclei not produced in massive stars or cosmologically, but are not as yet uniquely constrained.

TABLE A.1
H to S

Element	Z	A	Nucleon Fraction	$\log y_i$ (Si=6)	Source(s) (see notes)
H	1	1	7.057E-01	10.446	BB
	1	2	2.317E-05	5.661	BB
He	2	3	3.453E-05	5.658	BB, H
	2	4	2.752E-01	9.435	BB, H
Li	3	6	6.435E-10	0.627	x
	3	7	9.367E-09	1.723	BB, x, H?
Be	4	9	1.662E-10	−0.137	x
B	5	10	1.052E-09	0.619	x
	5	11	4.730E-09	1.230	x
C	6	12	3.032E-03	7.000	He, C
	6	13	3.683E-05	5.049	H?
N	7	14	1.105E-03	6.494	H
	7	15	4.363E-06	4.061	H?
O	8	16	9.592E-03	7.375	He, C, Ne
	8	17	3.827E-06	3.949	H?
	8	18	2.208E-05	4.686	He?
F	9	19	4.052E-07	2.926	H?
Ne	10	20	1.548E-03	6.486	C, Ne
	10	21	4.935E-06	3.968	C, He
	10	22	2.076E-04	5.572	He
Na	11	23	3.339E-05	4.759	C, H?
Mg	12	24	5.130E-04	5.927	Ne
	12	25	6.893E-05	5.037	C, He, Ne
	12	26	7.892E-05	5.079	C, He, Ne
Al	13	27	5.798E-05	4.929	C, Ne
Si	14	28	6.530E-04	5.965	O
	14	29	3.448E-05	4.672	Ne
	14	30	2.345E-05	4.490	Ne
P	15	31	8.155E-06	4.017	Ne
S	16	32	3.958E-04	5.689	O
	16	33	3.264E-06	3.592	O
	16	34	1.866E-05	4.336	O
	16	36	6.374E-08	1.845	C, Ne

TABLE A.2
Cl to Ni

Element	Z	A	Nucleon Fraction	$\log y_i$ (Si=6)	Source(s) (see notes)
Cl	17	35	3.506E-06	3.598	O
	17	37	1.198E-06	3.107	O
Ar	18	36	7.740E-05	4.929	O
	18	38	1.538E-05	4.204	O
	18	40	1.750E-08	1.238	C, Ne, He
K	19	39	3.463E-06	3.545	O
	19	40	5.221E-09	0.713	C, Ne, He
	19	41	2.686E-07	2.413	O
Ca	20	40	5.990E-05	4.772	O, Si
	20	42	4.154E-07	2.592	O
	20	43	9.637E-08	1.947	
	20	44	1.402E-06	3.100	α-f
	20	46	2.350E-09	0.305	C, Ne
	20	48	1.372E-07	2.053	
Sc	21	45	3.893E-08	1.534	C
Ti	22	46	2.211E-07	2.279	O
	22	47	2.081E-07	2.243	
	22	48	2.149E-06	3.248	Si, α-f
	22	49	1.636E-07	2.121	Si
	22	50	1.619E-07	2.107	
V	23	50	8.891E-10	−0.153	
	23	51	3.767E-07	2.465	Si
Cr	24	50	7.361E-07	2.765	O
	24	52	1.486E-05	4.053	Si
	24	53	1.729E-06	3.111	Si
	24	54	4.385E-07	2.507	
Mn	25	55	1.329E-05	3.980	Si
Fe	26	54	7.158E-05	4.719	Si
	26	56	1.169E-03	5.916	Si, α-f
	26	57	2.840E-05	4.294	Si, α-f
	26	58	4.357E-06	3.473	He, Ne
Co	27	59	3.358E-06	3.352	α-f, He
Ni	28	58	4.915E-05	4.525	Si, α-f
	28	60	1.958E-05	4.111	α-f

TABLE A.3
Ni to Kr

Element	Z	A	Nucleon Fraction	$\log y_i$ (Si=6)	Source(s) (see notes)
Ni	28	61	9.057E-07	2.769	C, Ne, α-f, He
	28	62	2.823E-06	3.255	C, Ne, α-f, He
	28	64	8.612E-07	2.726	C, Ne
Cu	29	63	5.753E-07	2.558	C, Ne, He
	29	65	2.647E-07	2.207	C, Ne, He
Zn	30	64	9.972E-07	2.790	He
	30	66	5.843E-07	2.544	C, Ne, He
	30	67	8.779E-08	1.714	C, Ne, He
	30	68	4.025E-07	2.369	C, Ne, He
	30	70	1.383E-08	0.893	Ne
Ga	31	69	3.979E-08	1.358	C, Ne, He
	31	71	2.694E-08	1.176	C, Ne, He
Ge	32	70	4.320E-08	1.387	C, Ne, He
	32	72	5.937E-08	1.513	C, Ne, He
	32	73	1.704E-08	0.965	C, Ne, He
	32	74	8.143E-08	1.638	Ne, He
	32	76	1.774E-08	0.965	Ne
As	33	75	1.245E-08	0.817	s, r, Ne, He
Se	34	74	1.011E-09	−0.268	p
	34	76	1.077E-08	0.748	s, C, Ne, He
	34	77	9.174E-09	0.673	s, r, C, Ne, He
	34	78	2.881E-08	1.164	s, r, C, Ne, He
	34	80	6.253E-08	1.490	s, r, C, Ne
	34	82	1.184E-08	0.757	r, Ne
Br	35	79	1.191E-08	0.775	s, r, Ne, He
	35	81	1.197E-08	0.766	s, r, Ne, He
Kr	36	78	3.137E-10	−0.799	p
	36	80	2.064E-09	0.009	s, p, He
	36	82	1.079E-08	0.716	s, Ne, C, He
	36	83	1.092E-08	0.716	s, r, Ne, C, He
	36	84	5.439E-08	1.408	s, r, Ne, C, He
	36	86	1.701E-08	0.893	r, Ne, C, He

A.2   NOTES: H to Kr

The following notes refer to the symbols in the preceding tables; they are a code for the burning shell which contributes, or the process which produces, a solar yield of the nucleus.

BB   Big Bang, cosmological nucleosynthesis (chapter 5)

H   Hydrostatic hydrogen burning by CNO cycle in massive stars (chapter 7)

H?   Hydrogen burning *not* by CNO cycle in massive stars

x   Spallation by cosmic-ray interactions with the interstellar medium

He   Hydrostatic helium burning in massive stars (chapter 8)

C   Carbon burning (in a hydrostatic shell, chapter 10)

Ne   Neon burning (in a hydrostatic shell, chapter 10)

O   Oxygen burning, probably explosive (chapter 9.3)

Si   Explosive silicon burning or *e*-process (chapter 9.4)

α-f   Alpha-rich freezeout, explosive (chapter 9.4)

s   Neutron capture on a slow time-scale, or *s*-process; hydrostatic

r   Neutron capture on a fast time-scale, or *r*-process; explosive

p   *p*-process, neon and oxygen shell burning (chapter 10.4).

TABLE A.4
Rb to Pd

Element	Z	A	Nucleon Fraction	$\log y_i$ (Si=6)	Source(s) (see notes)
Rb	37	85	1.101E-08	0.709	s, r, C, Ne, He
	37	87	4.555E-09	0.316	r, He
Sr	38	84	2.805E-10	−0.879	p
	38	86	5.047E-09	0.365	s
	38	87	3.411E-09	0.190	s
	38	88	4.318E-08	1.288	s, r
Y	39	89	1.045E-08	0.667	s, r
Zr	40	90	1.336E-08	0.769	s, r
	40	91	2.946E-09	0.107	s, r
	40	92	4.538E-09	0.290	s, r
	40	94	4.708E-09	0.297	s, r
	40	96	7.746E-10	−0.496	r
Nb	41	93	1.642E-09	−0.156	s, r
Mo	42	92	9.402E-10	−0.394	p
	42	94	5.493E-10	−0.636	p
	42	95	9.636E-10	−0.397	s, r
	42	96	1.025E-09	−0.375	s
	42	97	5.913E-10	−0.618	s, r
	42	98	1.502E-09	−0.218	s, r
	42	100	6.223E-10	−0.609	r
Ru	44	96	2.477E-10	−0.991	p
	44	98	8.627E-11	−1.458	p
	44	99	5.935E-10	−0.625	s, r
	44	100	5.944E-10	−0.629	s
	44	101	8.124E-10	−0.498	s, r
	44	102	1.517E-09	−0.231	s, r
	44	104	9.102E-10	−0.461	r
Rh	45	103	8.963E-10	−0.463	s, r
Pd	46	102	3.432E-11	−1.876	p
Pd	46	104	3.999E-10	−0.818	s
	46	105	8.207E-10	−0.510	s, r
	46	106	1.019E-09	−0.420	s, r
	46	108	1.014E-09	−0.431	s, r
	46	110	4.563E-10	−0.785	r

TABLE A.5
Ag to Te

Element	Z	A	Nucleon Fraction	$\log y_i$ (Si=6)	Source(s) (see notes)
Ag	47	107	6.766E-10	−0.602	s, r
	47	109	6.507E-10	−0.627	s, r
Cd	48	106	5.255E-11	−1.708	p
	48	108	3.852E-11	−1.851	p
	48	110	5.537E-10	−0.701	s
	48	111	5.756E-10	−0.688	s, r
	48	112	1.099E-09	−0.411	s, r
	48	113	5.631E-10	−0.706	s, r
	48	114	1.341E-09	−0.333	s, r
	48	116	3.580E-10	−0.914	r
In	49	113	2.252E-11	−2.103	s, p
	49	115	5.120E-10	−0.754	s, r
Sn	50	112	1.040E-10	−1.435	p
	50	114	7.267E-11	−1.599	p
	50	115	3.898E-11	−1.873	p, s, r
	50	116	1.602E-09	−0.263	s
	50	117	8.612E-10	−0.536	s, r
Sn	50	118	2.740E-09	−0.037	s, r
	50	119	9.873E-10	−0.484	s, r
	50	120	3.794E-09	0.097	s, r
	50	122	5.555E-10	−0.745	r
	50	124	7.120E-10	−0.644	r
Sb	51	121	5.417E-10	−0.752	p
	51	123	4.107E-10	−0.879	s
Te	52	120	1.299E-11	−2.369	p
	52	122	3.641E-10	−0.928	s
	52	123	1.301E-10	−1.379	s
	52	124	6.963E-10	−0.654	s
	52	125	1.062E-09	−0.474	s, r
	52	126	2.868E-09	−0.046	s, r
	52	128	4.954E-09	0.185	r
	52	130	5.459E-09	0.220	r

TABLE A.6
I to Nd

Element	Z	A	Nucleon Fraction	$\log y_i$ (Si=6)	Source(s) (see notes)
I	53	127	2.891E-09	−0.046	s, r
Xe	54	124	1.857E-11	−2.228	p
	54	126	1.721E-11	−2.268	p
	54	128	3.303E-10	−0.991	s
	54	129	4.209E-09	0.111	s, r
Xe	54	130	6.577E-10	−0.699	s
	54	131	3.347E-09	0.004	s, r
	54	132	4.074E-09	0.086	s, r
	54	134	1.620E-09	−0.321	r
	54	136	1.355E-09	−0.405	r
Cs	55	133	1.252E-09	−0.429	s, r
Ba	56	130	1.490E-11	−2.344	p
	56	132	1.456E-11	−2.361	p
	56	134	3.695E-10	−0.963	s
	56	135	1.011E-09	−0.529	s, r
	56	136	1.207E-09	−0.455	s
	56	137	1.760E-09	−0.294	s, r
	56	138	1.124E-08	0.508	s, r
La	57	138	1.435E-12	−3.386	p
	57	139	1.568E-09	−0.351	s, r
Ce	58	136	7.534E-12	−2.660	p
	58	138	9.914E-12	−2.547	p
	58	140	3.577E-09	0.004	s, r
	58	142	4.526E-10	−0.900	r
Pr	59	141	5.956E-10	−0.777	s, r
Nd	60	142	8.046E-10	−0.650	s
	60	143	3.653E-10	−0.996	s, r
	60	144	7.176E-10	−0.706	s, r
	60	145	2.520E-10	−1.163	s, r
	60	146	5.281E-10	−0.845	s, r
	60	148	1.775E-10	−1.324	r
	60	150	1.764E-10	−1.333	r

TABLE A.7
Sm to Er

Element	Z	A	Nucleon Fraction	log y_i (Si=6)	Source(s) (see notes)
Sm	62	144	2.907E-11	−2.098	p
	62	147	1.476E-10	−1.401	s, r
	62	148	1.086E-10	−1.538	s
	62	149	1.346E-10	−1.447	s, r
	62	150	7.285E-11	−1.717	s
	62	152	2.653E-10	−1.161	r
	62	154	2.283E-10	−1.232	r
Eu	63	151	1.776E-10	−1.333	s, r
	63	153	1.966E-10	−1.294	s, r
Gd	64	152	2.538E-12	−3.180	p
	64	154	2.766E-11	−2.149	s
	64	155	1.905E-10	−1.313	s, r
	64	156	2.668E-10	−1.170	s, r
	64	157	2.053E-10	−1.287	s, r
	64	158	3.281E-10	−1.086	s, r
	64	160	2.926E-10	−1.141	r
Tb	65	159	2.425E-10	−1.220	s, r
Dy	66	156	8.168E-13	−3.684	p
	66	158	1.423E-12	−3.449	p
	66	160	3.659E-11	−2.044	s
	66	161	3.030E-10	−1.128	s, r
	66	162	4.139E-10	−0.996	s, r
	66	163	4.057E-10	−1.007	s, r
	66	164	4.605E-10	−0.955	s, r
Ho	67	165	3.710E-10	−1.051	s, r
Er	68	162	1.397E-12	−3.467	p
	68	164	1.622E-11	−2.408	s, p
Er	68	166	3.519E-10	−1.077	s, r
	68	167	2.429E-10	−1.240	s, r
	68	168	2.885E-10	−1.168	s, r
	68	170	1.604E-10	−1.428	r

TABLE A.8
Tm to Os

Element	Z	A	Nucleon Fraction	$\log y_i$ (Si=6)	Source(s) (see notes)
Tm	69	169	1.616E-10	−1.423	s, r
Yb	70	168	1.424E-12	−3.475	p
	70	170	3.229E-11	−2.124	s
	70	171	1.536E-10	−1.450	s, r
	70	172	2.354E-10	−1.267	s, r
	70	173	1.750E-10	−1.398	s, r
	70	174	3.473E-10	−1.103	s, r
	70	176	1.407E-10	−1.500	r
Lu	71	175	1.576E-10	−1.449	s, r
	71	176	4.897E-12	−2.959	s
Hf	72	174	1.219E-12	−3.558	p
	72	176	3.566E-11	−2.096	s
	72	177	1.276E-10	−1.545	s, r
	72	178	1.882E-10	−1.379	s, r
	72	179	9.599E-11	−1.674	s, r
	72	180	2.472E-10	−1.265	s, r
Ta	73	180	1.161E-14	−5.593	p
	73	181	9.477E-11	−1.684	s, r
W	74	180	8.196E-13	−3.745	p
	74	182	1.616E-10	−1.455	s, r
	74	183	8.888E-11	−1.717	s, r
	74	184	1.899E-10	−1.389	s, r
	74	186	1.778E-10	−1.423	r
Re	75	185	8.985E-11	−1.717	s, r
	75	187	1.585E-10	−1.475	s, r
Os	76	184	5.632E-13	−3.917	p
	76	186	4.098E-11	−2.060	s
	76	187	3.765E-11	−2.099	s
	76	188	4.270E-10	−1.047	s, r
	76	189	5.211E-10	−0.963	s, r
	76	190	8.555E-10	−0.750	s, r
	76	192	1.345E-09	−0.558	r

TABLE A.9
Ir to U

Element	Z	A	Nucleon Fraction	$\log y_i$ (Si=6)	Source(s) (see notes)
Ir	77	191	1.193E-09	−0.607	s, r
	77	193	2.021E-09	−0.383	s, r
Pt	78	190	8.170E-13	−3.770	p
	78	192	5.099E-11	−1.979	s
	78	194	2.164E-09	−0.356	s, r
	78	195	2.234E-09	−0.344	s, r
	78	196	1.681E-09	−0.470	s, r
	78	198	4.838E-10	−1.015	r
Au	79	197	9.318E-10	−0.728	s, r
Hg	80	196	2.459E-12	−3.305	p
	80	198	1.738E-10	−1.460	s
	80	199	2.884E-10	−1.242	s, r
	80	200	3.976E-10	−1.105	s, r
	80	201	2.283E-10	−1.348	s, r
	80	202	5.161E-10	−0.996	s, r
	80	204	1.202E-10	−1.633	r
Tl	81	203	2.788E-10	−1.265	s, r
	81	205	6.741E-10	−0.886	s, r
Pb	82	204	3.204E-10	−1.207	s
	82	206	3.090E-09	−0.227	s, r
	82	207	3.398E-09	−0.188	s, r
	82	208	9.681E-09	0.265	s, r
Bi	83	209	7.613E-10	−0.842	s, r
Th	90	232	1.966E-10	−1.475	r
U	92	235	1.284E-11	−2.666	r
	92	238	4.118E-11	−2.165	r

A.3   NOTES: Rb to U

The following notes refer to the symbols in the preceding tables; they are a code for the burning shell which contributes, or the process which produces, a solar yield of the nucleus.

He   Hydrostatic helium burning in massive stars (chapter 8)

C   Carbon burning (in a hydrostatic shell, chapter 10)

Ne   Neon burning (in a hydrostatic shell, chapter 10)

s   Neutron capture on a slow time-scale, or s-process; hydrostatic

r   Neutron capture on a fast time-scale, or r-process; explosive

p p-process, neon and oxygen shell burning (chapter 10.4)

Appendix B

Equations of State

... the atoms ... have been rushing everlastingly throughout
all space in their myriads, undergoing a myriad changes under
the disturbing impact of collisions, till they have fallen into the
particular pattern by which this world of ours is constituted.
—Lucretius (98?–55 B.C.),
On the Nature of Things

The relationship between thermodynamic properties of a parcel of matter is called its *equation of state*. For example, with a given composition, if we know temperature and density, the equation of state will give pressure, energy density, entropy, or ... The equation of state is a determining factor in the onset of electron degeneracy, the size of the Chandrasekhar mass, the entropy gradient (and hence convective instability), the onset of electron-positron pair formation, the ionization and excitation states, and many other important aspects of stellar physics.

The chemical potential is particularly important for nucleosynthesis theory because of its key role in equilibrium with respect to reaction rates. Unfortunately its nature is not widely understood or appreciated. In determining the most probable distribution, the chemical potential is introduced as the undetermined Lagrange multiplier that is used to constrain the distribution to have the correct number of particles. It should not be surprising that the chemical potential has a deep connection with reaction processes—which change the number of particles, and (through them) with conservation laws. The temperature, which has a more intuitively obvious meaning, is introduced as the undetermined Lagrange multiplier to constrain the distribution to have the correct total energy, and in this sense is a similar quantity.

B.1 COULOMB INTERACTIONS

The coulomb interactions in an ionized gas are expected to have little effect on the equation of state if the kinetic energy of a particle is large compared to its potential energy due to electrostatic forces (Clayton [173], §2-3). The interparticle spacing for ions is of order $r \approx$

$(3/4\pi N_i)^{1/3} = (3/4\pi\rho a Y_i)^{1/3}$. For a pure gas, $X_i = 1$ and $Y_i = 1/A_i \approx 1/2Z_i$. The coulomb energy is of order $(Ze)^2/r$, so we require $kT \gg (Ze)^2/r \approx 5.99 Z^{5/3}\rho^{1/3}$ eV for ρ in g/cm^3, or

$$T \gg 6.95 \times 10^4 Z^{5/3}\rho^{1/3} \text{ K}. \tag{B.1}$$

This is usually (but not always) satisfied in stars. Clayton discusses the corrections necessary when the Debye-Huckel method can be used; Landau and Lifshitz ([381], ch. 7) discuss this and more general methods.

A convenient aspect of noninteracting gases is that Dalton's law of partial pressures holds: the pressure is simply the sum of the pressures expected from each component separately.

B.2 INTEGRALS

This discussion follows Landau and Lifshitz [381]. The number density of "elementary" noninteracting particles is

$$N = \frac{4\pi g}{h^3}\int_0^\infty \frac{p^2\,dp}{e^{(w-\mu)/kT}\pm 1} = \int_0^\infty \frac{dN}{dp}\,dp \tag{B.2}$$

where $g = 2s + 1$ (s is the spin), p the momentum and $w = \sqrt{(pc)^2 + (mc^2)^2}$ the total energy of a particle. The upper sign is for Fermi-Dirac, the lower Bose-Einstein statistics. Here μ is an undetermined Lagrange multiplier for number conservation, and for a gas in "chemical" equilibrium (i.e., with regard to reactions that give composition change, see chapter 4); μ is the chemical potential. Clayton ([173], p. 86) uses a quantity α which is related to μ by

$$\alpha = -\frac{(\mu - mc^2)}{kT}. \tag{B.3}$$

The energy density is given by

$$\rho E = \int_0^\infty \epsilon\,\frac{dN}{dp}\,dp \tag{B.4}$$

where $\epsilon = w - mc^2$; the rest mass energy is subtracted out so that the usual result is obtained in the nondegenerate and nonrelativistic limit. The pressure may be obtained from

$$P = \int_0^\infty \left(\frac{1}{3}vp\right)\frac{dN}{dp}\,dp \tag{B.5}$$

where the velocity is $v = pc^2/w$ and the factor in parentheses is for the net momentum flux through a plane for a perfect gas (Clayton [173], p. 79).

The chemical potential μ enters in the combination $w - \mu = \epsilon + mc^2 - \mu = \epsilon - u$ if $u \equiv \mu - mc^2$. In the nonrelativistic limit this subtraction of the rest mass energy removes a large constant value from both w and μ, giving the numerically more convenient quantities ϵ and u.

The entropy can be calculated directly from the Boltzmann expression or from

$$S = (E + VNmc^2 + PV - VN\mu)/T,$$
$$= (E + PV - VNu)/T, \tag{B.6}$$

where $V = 1/\rho$. Therefore,

$$S = \frac{V}{T} \int_0^\infty \left(\epsilon - u + \frac{vp}{3}\right) \frac{dN}{dp}\, dp. \tag{B.7}$$

The free energy is $F = E - TS$ and the enthalpy is $W = E + PV$.

In general these expressions must be evaluated by numerical integration, but simpler expressions can be found in limiting cases (see below).

B.3   PHOTON GAS

Blackbody radiation can be treated as a gas of zero-mass bosons ($m = 0$). The number of photons is not conserved, but is determined by the conditions of thermal equilibrium. There is no separate constraint on number, so $\mu = 0$. Because there are two linearly independent states of polarization for electromagnetic waves, $g = 2$, then the integrals simplify to

$$N = 8\pi \left(\frac{kT}{\hbar c}\right)^3 \int_0^\infty \frac{x^2\, dx}{e^x - 1}, \tag{B.8}$$

$$P = \frac{\rho E}{3} = \frac{\rho ST}{4} \tag{B.9}$$

$$= \frac{8\pi}{3} \left(\frac{kT}{hc}\right)^3 kT \int_0^\infty \frac{x^3\, dx}{e^x - 1}. \tag{B.10}$$

The dimensionless integrals are $\Gamma(3)\zeta(3) = 2.404$ and $\pi^4/15$ respectively (Landau and Lifshitz [381], §60). Using the Stefan-Boltzmann constant,

$$\sigma = ac/4$$
$$= \pi^2 k^4/60\hbar^3 c^2,$$

we find

$$P = \frac{aT^4}{3}, \tag{B.11}$$

$$\rho E = aT^4, \tag{B.12}$$

and

$$S = \frac{4aT^3}{3\rho}. \tag{B.13}$$

B.4 FERMI-DIRAC GAS

The integrands of the general expressions for N, E, P, and S, are functions of three energies: kT, u, and mc^2. There are two independent ratios that can be formed; the integrals therefore have four limiting cases. The relativistic limit occurs for $(kT/mc^2$ and $u/mc^2) \gg 1$; the opposite (nonrelativistic) case occurs for large mc^2. The nondegenerate limit occurs if

$$\exp\left[\frac{\epsilon - u}{kT}\right] \gg 1. \tag{B.14}$$

The opposite (degenerate) case occurs for high density and low temperature.

In the nonrelativistic limit, $\epsilon = p^2/2m$, so

$$N = \frac{2\pi g}{h^3}(2m)^{3/2} \int_0^\infty \frac{\epsilon^{1/2}\, d\epsilon}{1 + e^{(\epsilon - u)/kT}} \tag{B.15}$$

$$\begin{aligned} P &= \frac{2}{3}\rho E \\ &= \frac{4\pi g}{3h^3}(2m)^{3/2} \int_0^\infty \frac{\epsilon^{3/2}\, d\epsilon}{1 + e^{(\epsilon - u)/kT}}, \end{aligned} \tag{B.16}$$

and the entropy S can be obtained from

$$\rho TS = \frac{5}{2}P - Nu = \frac{5}{2}P + \alpha NkT \tag{B.17}$$

where α is defined above. These can be expressed as

$$N = \frac{2\pi g}{h^3}(2mkT)^{3/2}F_{1/2}(\alpha), \tag{B.18}$$

where the integrals $F_n(\alpha)$ are defined as

$$F_n(\alpha) \equiv \int_0^\infty \frac{x^n \, dx}{1 + \exp(\alpha + x)}. \tag{B.19}$$

Clayton [173] tabulates $F_{1/2}(\alpha)$ and $F_{3/2}(\alpha)$. Also,

$$P = \frac{4\pi g}{3h^3}(2mkT)^{3/2}kTF_{3/2}(\alpha). \tag{B.20}$$

It is useful to consider the nondegenerate, nonrelativistic limit in more detail. Then, using B.14 and $\epsilon = p^2/2m$, we have

$$N = \frac{2\pi g}{h^3}(2mkT)^{3/2}e^{u/kT}\int_0^\infty e^{-x}x^{1/2}\, dx, \tag{B.21}$$

where the integral is $\sqrt{\pi}/2$, and

$$P = \frac{4\pi g}{3h^3}(2mkT)^{3/2}kTe^{u/kT}\int_0^\infty e^{-x}x^{3/2}\, dx, \tag{B.22}$$

where the integral is $3\sqrt{\pi}/4$. Now N can be inverted to give an expression for the chemical potential in this limit:

$$u = kT \ln\left[\frac{N}{g}\left(\frac{2\pi\hbar^2}{mkT}\right)^{3/2}\right]$$

$$= \mu - mc^2. \tag{B.23}$$

Here,

$$u/kT = \ln\left[Y\theta/gA^{3/2}\right] \tag{B.24}$$

where

$$\theta = 0.1013\rho_9/T_9^{3/2}. \tag{B.25}$$

This expression will prove extremely useful. Eliminating u between N and P gives $P = NkT$ and a convenient check on our algebra. Then,

$$\rho TS = N\left(\frac{5}{2}kT - u\right) \tag{B.26}$$

gives a simplified expression for the entropy.

In the extreme relativistic limit, $\epsilon = pc$, so

$$N = \frac{4\pi g}{(hc)^3} \int_0^\infty \frac{\epsilon^2 \, d\epsilon}{1 + \exp\left((\epsilon - u)/kT\right)}$$

$$= 4\pi g \left(\frac{kT}{hc}\right)^3 F_2(\alpha) \tag{B.27}$$

and

$$P = \frac{1}{3}\rho E$$

$$= \frac{4\pi g}{3(hc)^3} \int_0^\infty \frac{\epsilon^3 \, d\epsilon}{1 + e^{(\epsilon - u)/kT}}$$

$$= \frac{4\pi g}{3} \left(\frac{kT}{hc}\right)^3 kTF_3(\alpha), \tag{B.28}$$

where the integrals $F_n(\alpha)$ are defined above. Bludman and Van Riper [106] give fitting formulae for several of these integrals.

This discussion of the degenerate limit follows Chandrasekhar ([162], ch. 10). In this limit, $kT \to 0$. Thus for $\epsilon < u$, the denominator $1 + e^{(\epsilon-u)/kT}$ is unity, and for $\epsilon > u$ it approaches $e^{(\epsilon-u)/kT}$, which goes to infinity as $kT \to 0$. Thus

$$N = \frac{4\pi g}{3} \left(\frac{p_f}{h}\right)^3 \tag{B.29}$$

where p_f is the fermi momentum and $u = \epsilon_f$ is the fermi energy. If we use $N = \rho N_A Y$ (where N_A is Avogadro's number) and define

$$x = \frac{p_f}{mc} \tag{B.30}$$

this becomes

$$\rho Y = \frac{4\pi g}{3N_A} \left(\frac{mc}{h}\right)^3 x^3. \tag{B.31}$$

For electrons,

$$\rho Y_e = \left(0.9735 \times 10^6 \, \text{g cm}^{-3}\right) x^3. \tag{B.32}$$

Similarly,

$$P = \frac{4\pi g}{3} \left(\frac{mc}{h}\right)^3 mc^2 \frac{f(x)}{8}$$

$$= 6.003 \times 10^{22} f(x) \, \text{dynes/cm}^2, \tag{B.33}$$

where

$$f(x) = x(2x^2 - 3)(x^2 + 1)^{1/2} + 3\sinh^{-1}x, \tag{B.34}$$

and

$$\rho E = \frac{4\pi g}{3} \left(\frac{mc}{h}\right)^3 \frac{mc^2}{8} g(x) \tag{B.35}$$

where

$$g(x) = 8x^3 \left(\sqrt{x^2 + 1} - 1\right) - f(x). \tag{B.36}$$

For $x \ll 1$ (low density),

$$f(x) = \frac{8}{5}x^5 - \frac{4}{7}x^7 + \dots \tag{B.37}$$

and

$$g(x) = \frac{12}{5}x^5 - \frac{3}{7}x^7 + \frac{1}{6}x^9 - \dots, \tag{B.38}$$

while for $x \gg 1$ (high density),

$$f(x) = 2x^4 - 2x^2 + \dots \tag{B.39}$$

and

$$g(x) = 6x^4 - 8x^3 + 7x^2 - \dots \tag{B.40}$$

In the limit $x \ll 1$,

$$P_e = K_{5/3}(\rho Y_e)^{5/3} = \frac{2}{3}\rho E_e, \tag{B.41}$$

where $K_{5/3} = 1.004 \times 10^{13}$ dynes/cm^2. In the limit $x \gg 1$,

$$P_e = K_{4/3}(\rho Y_e)^{4/3} = \frac{1}{3}\rho E_e, \tag{B.42}$$

where $K_{4/3} = 1.244 \times 10^{15}$ dynes/cm^2.

Using $TS = E + PV - VNu$ with any set of these $T = 0$ expressions gives $TS = 0$. Corrections for finite temperature can be found in Chandrasekhar [162], or derived directly by expanding the integrals in the appropriate limit. The finite temperature corrections are of order T^2 or higher, so $S \to 0$ as $T \to 0$.

Finally, the chemical potential equals the fermi energy, so

$$u = \epsilon_f = mc^2(\sqrt{1 + x^2} - 1). \tag{B.43}$$

B.5 EQUILIBRIA

A reaction occurring in a mixture of reacting substances ultimately leads to the establishment of a state in which the concentration (abundance) of each reactant no longer changes. For historical reasons this type of thermal equilibrium is called "chemical" equilibrium although the nature of the reactions is not specified. The state of chemical equilibria is not dependent upon the nature of the reactions by which the equilibria was established (merely that it did happen), but does depend upon the conditions under which mixture of reactants exists in equilibrium. For a reaction mixture, the first law of thermodynamics may be written as

$$dE + PdV = dQ = TdS + \mu_i dN_i \tag{B.44}$$

where summation over i reactants is implied. Here dQ is the energy gained by actual flow into the volume element. In equilibrium a steady state must obtain, so

$$\frac{dE}{dt} = \frac{dS}{dt} = \frac{dV}{dt} = 0, \tag{B.45}$$

and thus

$$\sum_i \mu_i dN_i = 0 \tag{B.46}$$

for the reactants. Symbolically a reaction can be expressed as

$$\sum_i \nu_i S_i = 0 \tag{B.47}$$

where S_i is the symbol for the ith reactant. For example, the reaction

$$2\,^{12}\mathrm{C} \rightarrow\,^{20}\mathrm{Ne} +\,^{4}\mathrm{He} = 0 \tag{B.48}$$

can be written as

$$2\,^{12}\mathrm{C} -\,^{20}\mathrm{Ne} -\,^{4}\mathrm{He} = 0 \tag{B.49}$$

so

$$\nu(^{12}\mathrm{C}) = 2, \quad \nu(^{20}\mathrm{Ne}) = \nu(^{4}\mathrm{He}) = -1. \tag{B.50}$$

In particular,

$$dN(^{12}\mathrm{C}) = \frac{\nu(^{20}\mathrm{Ne})}{\nu(^{12}\mathrm{C})}\, dN(^{20}\mathrm{Ne}) \tag{B.51}$$

and so on, so that the ν_i are constraints on the dN_i in this example to insure nucleon conservation. Pick a reactant k. Then

$$\sum_i \mu_i \nu_i \frac{dN_k}{\nu_k} = \left(\frac{dN_k}{\nu_k}\right) \sum_i \mu_i \nu_i = 0, \tag{B.52}$$

but the choice of k (and therefore dN_k) was arbitrary, so

$$\sum_i \mu_i \nu_i = 0, \tag{B.53}$$

which is the required condition for chemical equilibrium. This can also be written as

$$\sum_i (u_i \nu_i) = - \sum_i (\nu_i m_i c^2) \tag{B.54}$$

where the RHS is the Q-value for the reaction. The chemical potentials u_i can be simply expressed in some cases (see above), and are expressed implicitly in terms of number density and temperature by the integral defining N.

B.6   URCA RATES

In the discussion in chapter 11 of the combined electron capture and beta-decay at high density which is associated with the Urca process, dimensionless integrals were encountered for the rate of capture,

$$\mathcal{W}_c = \int_q^\infty \frac{(y-q)^2 x^2}{1 + e^{\beta(y-u)}} dx, \tag{B.55}$$

for the rate of decay,

$$\mathcal{W}_d = \int_0^q \frac{(q-y)^2 x^2}{1 + e^{-\beta(y-u)}} dx, \tag{B.56}$$

for the rate of emission of energy by neutrinos from electron capture,

$$\mathcal{E}_c = \int_q^\infty \frac{(y-q)^3 x^2}{1 + e^{\beta(y-u)}} dx, \tag{B.57}$$

and for the rate of emission of energy by antineutrinos from decay,

$$\mathcal{E}_d = \int_0^q \frac{(q-y)^3 x^2}{1 + e^{-\beta(y-u)}} dx, \tag{B.58}$$

where $y = \sqrt{x^2 + 1} - 1$, $q = Q/m_e c^2$, Q is the end-point energy of beta-decay, and the other symbols have been previously defined. In many cases of interest the electrons are relativistic, so the electron momentum is simply related to the energy by $x = y+1$, and $x^2 dx \to (y+1)^2 dy$. Because these integrals emphasize the higher energies, this approximation has a wider range of validity than might be guessed. Notice that the sign of the exponential in the denominator changes between capture and decay; this is because capture is enhanced by *filled* electron states, while decay is repressed. All of these rates are per target nucleus.

In the limit of complete electron degeneracy,

$$\mathcal{W}_c = W(u, q) - W(q, q), \tag{B.59}$$

where

$$W(a, b) \equiv a^5/5 - a^4(b - 1)/2 + a^3(b^2 - 4b + 1)/3 \tag{B.60}$$
$$+ a^2 b(b - 1) + ab^2. \tag{B.61}$$

For decay,

$$\mathcal{W}_d = W(q, q) - W(u, q). \tag{B.62}$$

Similarly, for neutrino emission,

$$\mathcal{E}_c = E(u, q) - E(q, q), \tag{B.63}$$

where

$$E(a, b) \equiv a^6/6 - a^5(2b - 3)/5 + a^4(3b^2 - 6b + 1)/4 \tag{B.64}$$
$$+ a^3 b(-b^2 + 6b - 3)/3 + a^2 b^2(-2b + 3) - ab^3, \tag{B.65}$$

and for antineutrino emission,

$$\mathcal{E}_d = E(q, u) - E(q, q). \tag{B.66}$$

Clearly these rates are zero at $u = q$. If temperature is not zero, but still small, so that $|u - q| \gg kT$, then we may calculate thermal corrections for these expressions. Following Landau and Lifshitz [381], an integral of the form

$$I = \int_0^\infty \frac{f(\epsilon)}{1 + e^{(\epsilon - u)/kT}} d\epsilon \tag{B.67}$$

may be approximated in this limit by

$$I = \int_0^u f(\epsilon)d\epsilon + \frac{\pi^2}{6}(kT)^2 f'(u) + \frac{7\pi^4}{360}(kT)^4 f'''(u) + \cdots \qquad \text{(B.68)}$$

Rewriting the range of integration

$$\mathcal{W}_c = \int_0^\infty \frac{(y-q)^2 x^2}{1 + e^{\beta(y-u)}} dx - \int_0^q \frac{(y-q)^2 x^2}{1 + e^{\beta(y-u)}} dx, \qquad \text{(B.69)}$$

and using $u > q + kT/m_e c^2$,

$$\mathcal{W}_c \approx \int_0^\infty \frac{(y-q)^2 x^2}{1 + e^{\beta(y-u)}} dx - \int_0^q (y-q)^2 x^2 dx. \qquad \text{(B.70)}$$

This procedure will be incorrect as $u \to q$; see below. Thus, for $u > q$,

$$\mathcal{W}_c \approx W(u, q) - W(q, q) + \frac{\pi^2}{6} W_1(u, q)/\beta^2 + \frac{7\pi^4}{30} W_2(u, q)/\beta^4, \quad \text{(B.71)}$$

where

$$W_1(u, q) = 4u^3 - 6u^2(q - 1) + 2u(q^2 - 4q + 1) + 2q(q - 1) \quad \text{(B.72)}$$

and

$$W_2(u, q) = 2u - q + 1. \qquad \text{(B.73)}$$

The particular form of $f(\epsilon)$ insures that W_1 has a zero at $u = q$ while W_2 does not.

The corresponding rate of emission of neutrinos is, for $u > q$,

$$\mathcal{E}_c \approx E(u, q) - E(q, q) + \frac{\pi^2}{6} E_1(u, q)/\beta^2 + \frac{7\pi^4}{30} E_2(u, q)/\beta^4 \quad \text{(B.74)}$$

where

$$E_1(u, q) = 5u^4 + 4u^3(-3q + 2) + 3u^2(3q^2 - 6q + 1) \qquad \text{(B.75)}$$
$$+ 2u(-q^3 + 6q^2 - 3q) + (-2q^3 + 3q^2), \qquad \text{(B.76)}$$

and

$$E_2(u, q) = 60u^2 + 24u(-3q + 2) + 6(3q^2 - 6q + 1). \qquad \text{(B.77)}$$

Again, E_1 has a zero at $u = q$ while E_2 does not.

For decay, we rewrite the range of integration

$$\mathcal{W}_d = \int_0^q (q-y)^2 x^2 dx - \int_0^q \frac{(q-y)^2 x^2}{1 + e^{\beta(y-u)}} dx. \tag{B.78}$$

This time we consider the case of $u < q + kT/m_e c^2$, so

$$\mathcal{W}_d \approx \int_0^q (y-q)^2 x^2 dx - \int_0^\infty \frac{(y-q)^2 x^2}{1 + e^{\beta(y-u)}} dx. \tag{B.79}$$

Notice that the integrands are the same as for capture, so for $u < q$,

$$\mathcal{W}_d \approx W(q,q) - W(u,q) - \frac{\pi^2}{6} W_1(u,q)/\beta^2 - \frac{7\pi^4}{30} W_2(u,q)/\beta^4. \tag{B.80}$$

For antineutrino emission, again rewriting the range of integration,

$$\mathcal{E}_d = \int_0^q (q-y)^3 x^2 dx - \int_0^q \frac{(q-y)^3 x^2}{1 + e^{\beta(y-u)}} dx. \tag{B.81}$$

For $u < q + kT/m_e c^2$,

$$\mathcal{E}_d \approx \int_0^q (y-q)^3 x^2 dx - \int_0^\infty \frac{(y-q)^3 x^2}{1 + e^{\beta(y-u)}} dx. \tag{B.82}$$

Notice that the integrands are the negative of those for capture, so for $u < q$,

$$\mathcal{E}_d \approx E(q,q) - E(u,q) - \frac{\pi^2}{6} E_1(u,q)/\beta^2 - \frac{7\pi^4}{30} E_2(u,q)/\beta^4. \tag{B.83}$$

These approximations only give us half of the solution we need, and break down as $u = q$ is approached. If we set $u = q$ and let $z = \beta(y-u)$,

$$\mathcal{W}_c = \beta^{-5} \int_0^\infty \frac{z^2(z + \beta(q+1))^2 dz}{1 + e^z}. \tag{B.84}$$

This may be expanded in powers of z.

Integrals of this type may be expressed in terms of the gamma function and the Riemann zeta function,

$$\int_0^\infty \frac{z^{x-1} dz}{1 + e^z} = (1 - 2^{1-x})\Gamma(x)\zeta(x), \tag{B.85}$$

where $(\Gamma(n) = (n-1)!$ for integer n, $\Gamma(3/2) = \pi^{1/2}/2$, $\Gamma(5/2) = 3\pi^{1/2}/2$, $\zeta(3/2) = 2.612$, $\zeta(5/2) = 1.341$, $\zeta(3) = 1.202$, and $\zeta(5) = 1.037$; see [1] for further values). For odd powers, the zeta function may be expressed in terms of the Bernoulli numbers,

$$\int_0^\infty \frac{z^{2n-1}dz}{1+e^z} = (2^{2n-1} - 1)\pi^{2n}|B_{2n}|/2n, \qquad (B.86)$$

where $B_2 = 1/6$, $B_4 = -1/30$, $B_6 = 1/42$, and $B_8 = -1/30$; see [1] for further values.[1]

This gives

$$W_c(0) = (45\zeta(5)/2 + 7\pi^4\beta(q + 1)/60 + 3\zeta(3)\beta^2(q + 1)^2/2)/\beta^5. \quad (B.87)$$

For $q \geq u$,

$$W_c \approx e^{\beta(u-q)}W_c(0). \qquad (B.88)$$

For decay, the same procedure gives

$$W_d(0) = (45\zeta(5)/2 - 7\pi^4\beta(q + 1)/60 + 3\zeta(3)\beta^2(q + 1)^2/2)/\beta^5. \quad (B.89)$$

For $q \leq u$,

$$W_d \approx e^{-\beta(q-u)}W_d(0). \qquad (B.90)$$

A similar procedure gives the emission of neutrinos as

$$\mathcal{E}_c(0) = (31\pi^6/252 + 45\zeta(5)\beta(q + 1) + 7\pi^4\beta^2(q + 1)^2/120)/\beta^6. \quad (B.91)$$

For $q \geq u$,

$$\mathcal{E}_c \approx e^{\beta(u-q)}\mathcal{E}_c(0). \qquad (B.92)$$

For emission of antineutrinos,

$$\mathcal{E}_d(0) = (31\pi^6/252 - 45\zeta(5)\beta(q + 1) + 7\pi^4\beta^2(q + 1)^2/120)/\beta^6, \quad (B.93)$$

and for $q \leq u$,

$$\mathcal{E}_d \approx e^{-\beta(q-u)}\mathcal{E}_d(0). \qquad (B.94)$$

[1]Note that this convention for the Bernoulli numbers is consistent with [1] rather than [381].

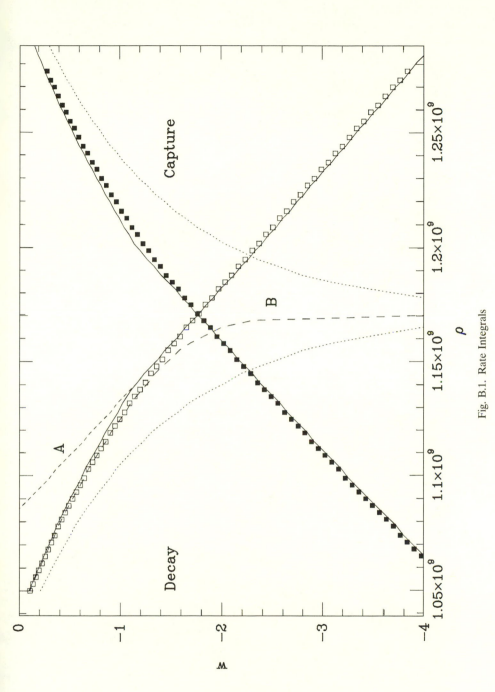

Fig. B.1. Rate Integrals

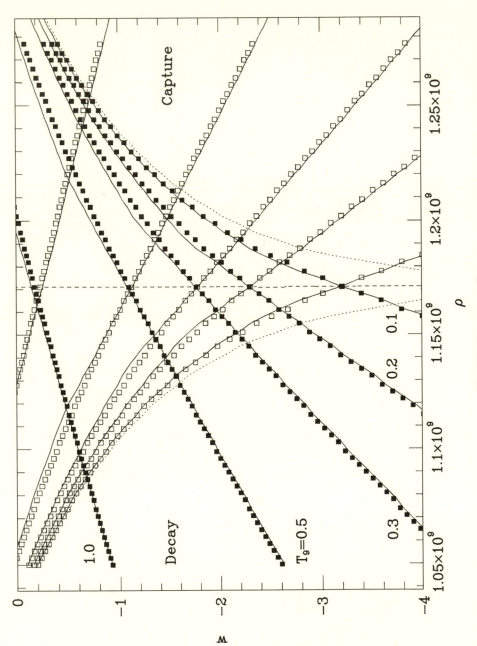

Fig. B.2. Rates versus T

A reasonably good approximation may be developed from a simple matching of these solutions. Consider capture first. The approximation for $u \approx q$, equation B.88, is quite good for $u \leq q$. The degenerate expansion is good for $u > q$, but the zeroth and first terms go to zero as $u \to q$, although the third term does not. The third term is small in our region of interest, so we use the first two terms of the degenerate expansion plus the value at $u = q$, namely $\mathcal{W}_c(0)$. The minimum of this and equation B.88 gives an exact match at $u = q$, and a moderately good approximation for all u.

For decay the roles reverse. The approximation for $u \approx q$, equation B.90, is quite good for $u \geq q$. The degenerate expansion is good for $u < q$, but breaks down at $u \to q$. Taking the first two terms, which go to zero there, and adding the value at $u = q$, namely $\mathcal{W}_d(0)$, gives the trial value. The minimum of this and equation B.90 gives a reasonable approximation. This is shown in figure B.1. The open squares are a direct numerical integration of $\mathcal{W}_d$; the filled squares, $\mathcal{W}_c$. The solid lines are the approximations just suggested. The temperature was $T_9 = 0.3$ and the endpoint energy was 3.833 MeV (the ^{25}Mg-^{25}Na pair). The curves intersect at u slightly below q. The dotted lines represent the decay and capture rates, respectively, at zero temperature. The dashed curve labeled A corresponds to $\mathcal{W}_d \approx e^{\beta(q-u)}\mathcal{W}_d(0)$. The other dashed curve, labeled B, corresponds to the first two terms of the degenerate expansion. Notice how it dives toward zero as $u \to q$, so that simply adding the value at $u = q$ is roughly correct. This gives a slight excess at $\rho \approx 1.135 \times 10^9$ g cm^{-3}; a more sophisticated matching procedure would give an even better approximation. The capture integrals are very similar; the combined graphs are very nearly symmetrical about $u = q$.

How well do these approximations work for different temperatures? This is indicated in figure B.2. The notation is as before, but the curves for decay are now labeled with their temperature; the temperatures for the capture curves may be inferred by the symmetry. The vertical dashed line is at $u = q$; notice the slight skewing of the intersections, which correspond to steady states, as the temperature increases. Even at higher temperature ($T_9 = 1.0$), the electron gas is still highly degenerate, and the approximations are still adequate. This matching of asymptotic values has an accuracy of about 10 percent (better at high degeneracy) in intermediate regions, and eventually breaks down (in the manner one would expect) as the electron gas becomes nondegenerate.

A precisely analogous procedure may be used for the integrals for neutrino and antineutrino emission.

Appendix C

Stellar Structure

To some extent a small thing may afford an
illustration and an imperfect image of great things.
—Lucretius (98?–55 B.C.), *On the Nature of Things*

The equations governing stellar evolution were examined in chapter 6, and some approximate solutions were shown. While identifying an astrophysicist's understanding with the ability to compute may be too narrow a conception, the inability to be quantitative certainly suggests a lack of understanding. Accurate computation often involves great complexity. Simplicity is a goal of this discussion.

A compromise is available. By judicious use of approximate solutions we may avoid much of the real complexity involved in a "state-of-the-art" calculation. In the earlier chapters, crude but robust approximations are freely used. To obtain more accuracy with little extra effort, we may use the more elegant theory of polytropes, in which a particular (power law) relation between pressure and density is assumed to hold throughout the star. This allows us to solve the equations of mass and momentum conservation to obtain hydrostatic configurations. In this approach we do not delve into the details of the thermal balance which sets up such a relation.

This appendix summarizes the features of polytropes which are used in the text, and contains a sketch of their derivation. For further information, see the pioneering books by Eddington [218] and Chandrasekhar [162].

At first it is ironic that with the impressively expanding power of computer technology, we revert to precomputer methods. Actually it is a natural consequence of that progress. Mechanical calculators were a labor-saving aid to the scientist; modern computers are something more. Computers allow us to attempt problems and use methods that would take an unaided human a time longer than a normal life expectancy, much less the more relevant interval of a reasonable attention span. How can we validate computations of such complexity? Comparison with approximate analytic solutions may help. Perhaps we should regard reading the classical work as of more than historical interest.

C.1 POLYTROPES

Let us assume that the pressure and density vary through a star as

$$P = K\rho^{1+1/n} \tag{C.1}$$

where K and n are constants. Also let

$$\rho = \lambda\theta^n \tag{C.2}$$

whre λ is a constant scaling factor. If we choose $\theta = 1$ at the center then $\lambda = \rho_c$, the central density. Combining the equations for hydrostatic equilibrium and mass conservation we have

$$d\left(\frac{r^2}{\rho}\frac{dP}{dr}\right) = -4\pi\rho Gr^2. \tag{C.3}$$

In order to get a dimensionless version of this equation let

$$\xi = \frac{r}{a}, \tag{C.4}$$

and choose

$$a = \sqrt{(n+1)K\lambda^{\frac{1-n}{n}}/4\pi G}. \tag{C.5}$$

Then,

$$\frac{d\left(\xi^2 d\theta/d\xi\right)}{d\xi} = -\xi^2\theta^n \tag{C.6}$$

or

$$\frac{d^2\theta}{d\xi^2} = -\left(\frac{2}{\xi}\right)\frac{d\theta}{d\xi} - \theta^n \tag{C.7}$$

which are the usual forms of the Lane-Emden equation.

What are the boundary conditions? At the origin ($\xi = 0$) we have already specified that $\theta = 1$. Noting that

$$\frac{dP}{dr} = -\frac{GM\rho}{r^2}$$

$$\simeq -\frac{4\pi G\rho_c^2 r}{3}$$

for small r, gives us

$$\theta^n \frac{d\theta}{d\xi} = -\frac{\xi}{3}, \tag{C.8}$$

for $\xi \ll 1$, and

$$\frac{d\theta}{d\xi} = 0 \quad \text{at} \quad \xi = 0. \tag{C.9}$$

At the surface (ξ is some value ξ_1), pressure and density go to zero, or

$$\theta = 0 \quad \text{at} \quad \xi_1. \tag{C.10}$$

At $\xi = 0$, C.7 has a singularity. Some care must be used to begin the solution. We expand θ around $\xi = 0$ in a Taylor series,

$$\theta = 1 - \frac{\xi^2}{6} + n\frac{\xi^4}{120} - \dots \tag{C.11}$$

which is valid for $\xi \ll 1$. Having begun the solution we can continue by using standard techniques of numerical integration.

The mass $M(r)$ contained inside a radius r is

$$M(r) = \int_0^r 4\pi x^2 \rho \, dx$$

$$= 4\pi \left(\frac{(n+1)K}{4\pi G} \right)^{3/2} \lambda^{\frac{3-n}{2n}} \left(\frac{-\xi^2 d\theta}{d\xi} \right). \tag{C.12}$$

The total mass may be obtained by taking $\xi = \xi_1$. If the average density of a star of radius R is

$$\langle \rho \rangle = \frac{3M}{4\pi R^3}, \tag{C.13}$$

then the ratio of average to central density is

$$\frac{\langle \rho \rangle}{\rho_c} = -\left[\left(\frac{3}{\xi} \right) \frac{d\theta}{d\xi} \right]_{\xi_1}. \tag{C.14}$$

Using $R = a\xi_1$, we can construct

$$P_c = \frac{GM^2/R^4}{4\pi(n+1)(d\theta/d\xi)^2}. \tag{C.15}$$

Then,

$$\frac{P_c}{\rho_c} = C_n \frac{GM}{R}, \tag{C.16}$$

where

$$C_n = \frac{-1}{((n+1)\xi \, d\theta/d\xi)_{\xi_1}}. \tag{C.17}$$

Another useful form may be obtained by using $<\rho>$ to eliminate R,

$$\frac{P_c}{\rho_c^{4/3}} = D_n \, GM^{2/3}, \tag{C.18}$$

where

$$D_n = C_n \left(\frac{4\pi \langle\rho\rangle}{3\rho_c}\right)^{1/3}. \tag{C.19}$$

If we define

$$\beta = \frac{\Re YT\rho}{P}, \tag{C.20}$$

then

$$\frac{\Re YT_c}{\beta} = C_n \frac{GM}{R}$$
$$= D_n \, GM^{2/3} \rho_c^{1/3}. \tag{C.21}$$

Using solar values for the mass, $M_\odot = 1.987 \times 10^{33}$ g, and the radius, $R_\odot = 0.6965 \times 10^{11}$ cm, we have $G M_\odot/\Re R_\odot = 2.2886 \times 10^7$ K and alternatively, $G M_\odot^{2/3}/\Re = 1.2679 \times 10^7$ K cm/g$^{1/3}$. For a polytrope of index $n = 3$,

$$Y \frac{T_c}{\beta} = 1.955 \times 10^7 \text{ K} \left(\frac{M}{M_\odot}\right)\left(\frac{R_\odot}{R}\right)$$
$$= 4.6145 \times 10^6 \text{ K} \left(\frac{M}{M_\odot}\right)^{2/3} \rho_c^{1/3}, \tag{C.22}$$

where the density is in units of g/cc.

Table C.1 summarizes some of these quantities for various values of the polytropic index n.

TABLE C.1

Polytropic Constants

n	ξ	$\xi^2(d\theta/d\xi)_1$	$\rho_c/\langle\rho\rangle$	C_n	D_n
0	2.44949	4.89898	1.00000	0.50000	0.80600
1	3.14159	3.14159	3.28986	0.50000	0.54193
1.5	3.65375	2.71406	5.99071	0.53849	0.47795
3	6.89685	2.01824	54.1825	0.85431	0.36394
3.25	8.01894	1.94980	88.1530	0.96769	0.35050
4	14.9715	1.79723	622.408	1.66607	0.31456
5	∞	1.73205	∞	∞	0.26867

For $n = 1$, 2, and 3, analytic solutions exist (see Chandrasekhar [162], ch. 4). They are, for $n = 0$,

$$\theta = 1 - \xi^2/6, \tag{C.23}$$

for $n = 1$,

$$\theta = \frac{\sin \xi}{\xi}, \tag{C.24}$$

and for $n = 5$,

$$\theta = \frac{1}{(1 + \frac{1}{3}\xi^2)}. \tag{C.25}$$

The last has infinite radius, but finite mass.

Table C.2 gives the structure of a polytrope of index $n = 3$. This polytrope is of special importance because it is used in Eddington's Standard Model, and it is the limiting case for a massive white dwarf. The variables are in dimensionless form. The first column is the radius ξ, the second the mass coordinate $-\xi^2 d\theta/d\xi$, the third the temperature θ, the fourth the density θ^3, and the last is the pressure θ^4.

C.2 GRAVITATIONAL BINDING

The internal energy of a star is

$$U = \int_0^M E \, dm \tag{C.26}$$

TABLE C.2

Polytropic Structure ($n = 3$)

ξ	$-\xi^2 d\theta/d\xi$	θ	θ^3	θ^4
0.00000	0.00000	1.000E-00	1.000E-00	1.000E-00
0.02000	0.00000	9.999E-01	9.998E-01	9.997E-01
0.10000	0.00033	9.983E-01	9.950E-01	9.934E-01
0.50000	0.03871	9.598E-01	8.843E-01	8.488E-01
1.00000	0.25213	8.551E-01	6.252E-01	5.345E-01
1.20000	0.38979	8.026E-01	5.170E-01	4.149E-01
1.40000	0.54688	7.475E-01	4.176E-01	3.121E-01
1.60000	0.71413	6.915E-01	3.307E-01	2.287E-01
1.80000	0.88286	6.363E-01	2.576E-01	1.639E-01
2.00000	1.04596	5.829E-01	1.980E-01	1.154E-01
2.20000	1.19827	5.319E-01	1.505E-01	8.005E-02
2.40000	1.33652	4.839E-01	1.133E-01	5.484E-02
2.60000	1.45907	4.391E-01	8.468E-02	3.719E-02
2.80000	1.56555	3.976E-01	6.285E-02	2.499E-02
3.00000	1.65645	3.592E-01	4.636E-02	1.665E-02
3.50000	1.82336	2.763E-01	2.108E-02	5.825E-03
4.00000	1.92271	2.093E-01	9.166E-03	1.918E-03
4.50000	1.97678	1.551E-01	3.729E-03	5.782E-04
5.00000	2.00315	1.108E-01	1.361E-03	1.508E-04
5.50000	2.01411	7.429E-02	4.099E-04	3.045E-05
6.00000	2.01758	4.374E-02	8.367E-05	3.660E-06
6.50000	2.01821	1.787E-02	5.703E-06	1.019E-07
6.80000	2.01824	4.168E-03	7.239E-08	3.017E-10
6.87000	2.01824	1.143E-03	1.495E-09	1.709E-12
6.89000	2.01824	2.907E-04	2.456E-11	7.139E-15
6.89650	2.01824	1.460E-05	3.111E-15	4.541E-20
6.89684	2.01824	6.453E-09	2.687E-25	1.734E-33

where E is the internal energy per unit mass. The gravitational potential energy is just the total work done by the gravitational field as each element of mass is added to the star:

$$\Omega = \int_0^M \int_r^\infty \left(-GM' \frac{dM'}{r^2} \right) dr$$
$$= \int_0^M \left(\frac{GM'}{r} \right) dM'.$$

The gravitational binding energy B (that is, the negative of the total energy) is

$$B = \Omega - U. \tag{C.27}$$

With this sign convention Ω is intrinsically positive, hence the explicit sign difference between U and Ω. Using

$$\frac{GM}{r} = -\frac{r}{\rho} \frac{dP}{dr} \tag{C.28}$$

from the condition of hydrostatic equilibrium and integrating by parts gives

$$\Omega = \int_0^M \frac{3P}{\rho} \, dm - 4\pi r^3 P \Big|_0^R . \tag{C.29}$$

Thus, for a hydrostatic star,

$$B = \int_0^M \left(\frac{3P}{\rho} - E \right) dm - 4\pi r^3 P \Big|_0^R . \tag{C.30}$$

The polytropic assumption implies

$$\frac{1}{\rho} dP = (n+1) \, d \left(\frac{P}{\rho} \right). \tag{C.31}$$

Integration by parts gives

$$\Omega = - \int 3M \, d \left(\frac{P}{\rho} \right). \tag{C.32}$$

But using the condition of hydrostatic equilibrium gives

$$\Omega = \frac{3}{n+1} \int_0^R \frac{GM^2}{r^2} \, dr. \tag{C.33}$$

Integrating this by parts gives

$$\Omega = \left[\frac{3}{n+1}\right]\left[2\int_0^M \frac{Gm}{r}\,dm - \frac{GM^2}{R}\right],$$

$$= \left(\frac{3}{n+1}\right)\left(2\Omega - \frac{GM^2}{R}\right),$$

or finally,

$$\Omega = \left(\frac{3}{5-n}\right)\left(\frac{GM^2}{R}\right). \tag{C.34}$$

If $E = PV/(\gamma - 1)$ where γ is a constant, then,

$$\Omega = 3(\gamma - 1)U \tag{C.35}$$

which for $\gamma = 5/3$ is a simple form of the virial theorem. Also,

$$B = \left(\gamma - \frac{4}{3}\right)\left(\frac{\Omega}{\gamma - 1}\right)$$

$$= 3\left(\gamma - \frac{4}{3}\right)\left(\frac{GM^2/R}{(5-n)(\gamma-1)}\right). \tag{C.36}$$

For a mixture of ideal gas and radiation this reduces to

$$B = \left(\frac{\beta}{2}\right)\Omega. \tag{C.37}$$

For $\gamma = 4/3$, or $\beta = 0$, the binding energy is zero, so that the star is in a state of marginal stability.

C.3   ENERGY GENERATION

Following Fowler and Hoyle [238] we can use the polytropic approximation to relate the average energy generation (or loss) rate to that at the center of the star. This allows us to explicitly integrate over the spatial structure of the star, so that we then need only consider the evolution of conditions at one characteristic point—the center. Define the average energy generation rate as

$$\langle \epsilon \rangle = \int_0^m \epsilon \frac{dm}{M}. \tag{C.38}$$

Approximate the energy generation rate per gram, ϵ, by

$$\frac{\epsilon}{\epsilon_c} \approx \left(\frac{\rho}{\rho_c}\right)^{u-1} \left(\frac{T}{T_c}\right)^s \tag{C.39}$$

where the c denotes central values. For our purposes u and s will be constant or at most slowly varying.

To the extent that β/Y is constant through the star, for a polytrope

$$\frac{\rho}{\rho_c} = \left(\frac{T}{T_c}\right)^n, \tag{C.40}$$

and $T/T_c = \theta$. Thus

$$M\langle\epsilon\rangle = 4\pi\epsilon_c\rho_c a^3 \int_0^{\xi_1} \theta^{nu+s} \xi^2 \, d\xi. \tag{C.41}$$

If the exponent $(nu + s)$ is large, the integrand is large only for small ξ. We note that

$$\exp(-\xi^2/6) = 1 - \frac{\xi^2}{6} + \frac{\xi^4}{72} - \cdots \tag{C.42}$$

agrees with C.11 up to the fourth-order term, so we use

$$\theta \approx \exp\left(-\frac{\xi^2}{6}\right) \tag{C.43}$$

for $\xi \ll 1$. Thus

$$\int_0^{\xi_1} \theta^{nu+s} \xi^2 \, d\xi \approx \int_0^{\infty} \exp\left(-(nu+s)\frac{\xi^2}{6}\right) \xi^2 \, d\xi,$$

$$= \frac{\sqrt{27\pi/2}}{(nu+s)^{3/2}}. \tag{C.44}$$

From C.5 and C.12 we have

$$M = 4\pi\rho_c a^3 \left(\frac{-\xi^2 \, d\theta}{d\xi}\right)_{\xi_1}. \tag{C.45}$$

If we take $n = 3$, then

$$\frac{\langle\epsilon\rangle}{\epsilon_c} = \frac{3.23}{(3u+s)^{3/2}}, \tag{C.46}$$

and for $n = 1.5$ it becomes

$$= \frac{2.40}{(1.5u + s)^{3/2}}. \tag{C.47}$$

Having a similar approximation for burning shells will also be useful. For a sufficiently thin shell, the run of density and temperature may be approximated by a power law in radius. Suppose that $T/T_s = x^{-\alpha}$, where T_s is the temperature at the shell, and $x = r/r_s$ is the radius measured in units of the shell radius r_s. Obviously this implies that $\rho/\rho_s = x^{-n\alpha}$, where n is the local value of the polytropic index. Using the same approach as before,

$$\langle \epsilon \rangle = \frac{1}{(m_{cc} - m_s)} \int_{m_s}^{m_{cc}} \epsilon \, dm, \tag{C.48}$$

where m_s is the Lagrangian mass coordinate of the inner edge of the burning shell, and m_{cc} for the outer edge of the associated convective region. This becomes

$$\langle \epsilon \rangle / \epsilon_s = \frac{4\pi \rho_s r_s^3}{(m_{cc} - m_s)} \int_1^{r_{cc}/r_s} x^{-\alpha(nu+s)+2} \, dx, \tag{C.49}$$

where r_s is the radius of the inner edge of the burning shell, and r_{cc} for the outer edge of the associated convective region. Here it is assumed that $\alpha(nu+s)$ is large enough to avoid the logarithmic solution, as should ususally be the case. Then,

$$\langle \epsilon \rangle / \epsilon_s = \frac{4\pi \rho_s r_s^3}{(m_{cc} - m_s)} \frac{[1 - (r_s/r_{cc})^{\alpha(nu+s)-3}]}{\alpha(nu + s) - 3}. \tag{C.50}$$

If the exponent is large (as it often is) or the ratio r_s/r_{cc} is small, this simplifies to

$$\langle \epsilon \rangle / \epsilon_s = 4\pi \rho_s r_s^3 / (m_{cc} - m_s)(\alpha(nu + s) - 3). \tag{C.51}$$

Now, the mass integral is

$$(m_{cc} - m_s)/4\pi \rho_s r_s^3 = \int_1^{r_{cc}/r_s} x^{2-n\alpha} \, dx, \tag{C.52}$$

$$= [(r_{cc}/r_s)^{n(\alpha-1)} - 1]/[n(\alpha - 1)] \text{ for } (n\alpha \neq 3),$$

$$= \ln(r_{cc}/r_s) \text{ for } (n\alpha = 3).$$

For the second stage of shell carbon burning, typical values are $\alpha \approx 0.8$, $n \approx 2.5$, and $r_{cc}/r_s \approx 5$.

Appendix D

Supernova Light Curves

No other substance is so rigidly held together by
the intertanglement of its elemental atoms as
cold iron, that stubborn and benumbing metal.
—Lucretius (98?–55 B.C.),
On the Nature of Things

Most of the spectacular optical display that is a supernova explosion
can be interpreted in terms of two concepts: (1) a shock wave running
through a stellar envelope of large radius, and (2) the radioactive decay
of newly synthesized ^{56}Ni to its daughter ^{56}Co, and then to ^{56}Fe, the fifth
most abundant nucleus in the Universe.

The nature of explosions is the expansion away from the region of
energy release. The pushing and shoving of matter to gain elbow room
gives a characteristic pattern of overall velocity. The matter at the outer
edge gets shoved the most, and has the highest velocity. The matter near
the center is pushed and pulled, backwards and forwards, so that the
net effect is a smaller velocity. The explosion becomes more spherically
symmetric with time, and tends toward a small-scale Hubble flow. In this
sense it is a Big Bang in miniature.

Until it expands enormously, the star remains opaque to its thermal
radiation. It shines because particles of light diffuse to its surface and
escape. It is cooled by this loss of photons, and by its own expansion. It is
heated by the shock wave which blows it apart, and by radioactive decay.
This fortunate situation allows the application of classical mathematical methods: the equations may be transformed into a time-dependent
diffusion equation, which has known solutions. While the assumptions
necessary for this simplification are only satisfied approximately, the resulting solutions are a useful tool for creating an understanding of the
real events.

D.1 SIMPLE COOLING

The simplest light curves are those resulting from the diffusive cooling of an envelope heated by the supernova shock associated with the

explosion, with no later heating mechanisms in action. This discussion follows [37].

Propagation of a supernova shock through the envelope of the pre-supernova star can be complex [228, 48, 221]. The strong shock leaves behind a radiation-dominated gas, with energy about equally divided between kinetic and internal. Except for acceleration from PdV work, further expansion is essentially free coasting. Except for a small amount of mass affected by shock steepening near the stellar surface, the expansion is more or less homologous. For a given temperature, the densities will be less than usually considered in stellar evolution; we will assume that Thomson scattering dominates the opacity. The effects of expansion on line opacity [344] must be considered for quantitative precision. Spherical symmetry is assumed.

Reduction to ODEs

The partial differential equations governing light curves may be reduced to ordinary differential equations (ODEs) in a useful special case. The thermal state of the expanding envelope evolves in time according to the first law of thermodynamics,

$$dE/dt + PdV/dt = -\partial L/\partial m, \tag{D.1}$$

where E is the internal energy per unit mass, P the pressure, $V = 1/\rho$ is the specific volume, L the radiative luminosity, and m the spherical mass coordinate. For a radiation gas, in which the matter contributes little to the pressure and the thermal energy (see ch. 13), $E = aT^4V$ and $P = aT^4/3$. In the diffusion approximation the luminosity L is

$$L/4\pi r^2 = -(\lambda c/3)\partial aT^4/\partial r, \tag{D.2}$$

where the mean free path is $\lambda = 1/\rho\kappa$. Therefore,

$$4T^4(\dot{T}/T + \dot{V}/3V) = \frac{1}{r^2}\frac{\partial}{\partial r}\left(\frac{c}{3\kappa\rho}r^2\frac{\partial T^4}{\partial r}\right). \tag{D.3}$$

Here $R(t)$ is the "surface" of the expanding envelope (to be defined more carefully below).

Only a small fraction of the explosion energy (a few percent) escapes as radiation; the rest remains as kinetic energy of the ejecta. A strictly adiabatic solution would give $T \propto R(t)^{-1}$, which suggests that we change to a variable which factors out this dependence. Further, we wish to separate dependencies on space (r) and time (t), so

$$T(r, t)^4 \equiv \Psi(x)\phi(t)T(0, 0)^4 R(0)^4/R(t)^4. \tag{D.4}$$

Further,

$$\dot{V}/V = 3\dot{R}/R = 3v_{sc}/R(t), \tag{D.5}$$

for homologous expansion; to the extent acceleration from PdV work is ignored, v_{sc} is a constant that sets the velocity scale. Let $x \equiv r/R(t)$ be a dimensionless radial coordinate that follows the expansion of the matter. Then,

$$3R(t)^2\dot{\phi}/c\phi = \frac{1}{\Psi x^2}\frac{\partial}{\partial x}\left(\frac{x^2}{\rho\kappa}\frac{\partial\Psi}{\partial x}\right). \tag{D.6}$$

Now, let

$$\rho(r, t) = \rho(0, 0)\eta(x)[R(0)/R(t)]^3, \tag{D.7}$$

where $\eta(x)$ defines the shape of the density distribution (assumed constant in time, or slowly varying). If the opacity κ is a function of x, then this variation may be added here as a function multiplying $\eta(x)$. The opacity is a function of temperature, density, and composition; in general this implies that it is a function of time as well as space. The question of variation in composition is of the utmost importance, but for the moment we will ignore such effects for simplicity. This is not absurd because Thomson scattering often dominates, in which case the opacity is almost constant. At sufficiently low temperatures recombination causes the opacity to drop. If the drop in opacity is steep at recombination, the recombination wave can be taken as a moving outer boundary, beyond which the matter is transparent. If the matter has a large velocity gradient across a mean free path, resonance lines become smeared over a wider range of frequency, given an enhanced opacity ([344]); this effect is time dependent and is important for detailed understanding of the early stages of Type I supernovae.

If $\kappa = \kappa(x)$, then we may separate variables:

$$\alpha = -\frac{1}{x^2\Psi(x)}\frac{\partial}{\partial x}\left(\frac{x^2}{\eta(x)}\frac{\kappa(0)}{\kappa(x)}\frac{\partial\Psi}{\partial x}\right), \tag{D.8}$$

and

$$\frac{\partial\phi}{\partial t} = -R(t)\phi(t)/R(0)\tau_0, \tag{D.9}$$

where

$$\tau_0 \equiv 3R(0)^2\rho(0,0)\kappa(0)/\alpha c \qquad (D.10)$$

is the diffusion time-scale (at time $t = 0$) and α is the eigenvalue. Note that the constants for temperature $T(0,0)$, and specific volume $V(0,0)$, are taken at the center ($x = 0$) and at the fiducial time ($t = 0$); with radius $R(0)$ taken at ($t = 0$), this sets the scale for the problem.

Equation D.9 may be written as

$$\ln\left[\phi(t)/\phi(0)\right] = -\int_0^t \frac{R(t)}{\tau_0 R(0)} dt. \qquad (D.11)$$

With no loss in generality we may take $\phi(0) = 1$. For a coasting phase,

$$R(t) \approx R(0) + v_{sc}t, \qquad (D.12)$$

so

$$\phi(t) = \exp\left[-\frac{t}{\tau_0}(1 + t/2\tau_h)\right], \qquad (D.13)$$

where $\tau_h \equiv R(0)/v_{sc}$. Soon after the shock hits the surface, we have $t > 2R(0)/v_{sc}$; therefore, for later times, $\phi(t)$ falls in time as a gaussian rather than an exponential. To evaluate τ_0 (or α) we will solve the eigenvalue problem (equation D.8) subject to appropriate boundary conditions.

Boundary conditions

At the center we impose the symmetry constraint $d\Psi/dx = 0$. At the surface we use the Eddington surface boundary condition,

$$T^4 = \frac{3}{4}T_e^4(\tau + q), \qquad (D.14)$$

where we take $q \approx \frac{2}{3}$, and define the usual optical depth as

$$\tau = \int_r^{R(t)} \rho\kappa dr. \qquad (D.15)$$

The radius of the photosphere is found by taking the mean free path λ to be slowly varying, so that

$$R_{ph}(t) = R(t) - \frac{2}{3}\lambda. \qquad (D.16)$$

The choice of $q \approx \frac{2}{3}$ is appropriate for a plane-parallel atmosphere, and should be modified when λ becomes significant with respect to $R_{ph}(t)$. With separation of variables,

$$\Psi(x) = \frac{3}{4}\Psi_e(\tau + q) \tag{D.17}$$

and

$$\frac{\partial \Psi(x)}{\partial x} = -\frac{3}{4}\Psi_e \frac{R(t)}{\lambda(x)}, \tag{D.18}$$

so that at the outer boundary, $x = a$,

$$\Psi(a) = -\frac{2}{3} \frac{\partial \Psi}{\partial x}\bigg|_a \frac{\lambda(a)}{R(t)}. \tag{D.19}$$

For a dense object, $\lambda(a) \ll R(t)$, so $\Psi(a) \to 0$. However, as transparency begins, $\Psi(a)$ increases (see below).

We must also consider conditions at the boundary in time: the initial spatial state. As the shock emerges at the surface ($r = R(0)$) at time $t = 0$, the stellar ejecta have been heated and accelerated. For a photon-dominated gas, $\gamma = 4/3$; behind a strong shock the energy is divided in thermal and kinetic modes (E_{Th} and E_K). We take this to be approximately true for the whole of the ejected matter at $t = 0$. Adiabatic, homologous expansion would give

$$E_{Th}(t) = E_{Th}(0)R(0)/R(t), \tag{D.20}$$

and the total energy of the supernova explosion E_{SN} is

$$E_{SN} = E_{Th}(0) + E_K(0) = E_{Th}(0)/f \tag{D.21}$$

where $f \approx 0.5$ for a strong radiation-dominated shock (see [709], §28 and §29). Thus if radiation diffusion losses are small,

$$E_{Th}(t) = gE_{SN} = gE_K(t)/(1 - g), \tag{D.22}$$

where

$$g \equiv fR(0)/R(t). \tag{D.23}$$

The thermal energy can also be expressed as

$$E_{Th}(t) = \int_0^{R(t)} aT^4 4\pi r^2 \, dr \tag{D.24}$$

for a radiation-dominated gas, and the kinetic energy as

$$E_K(t) = \frac{1}{2} \int_0^{R(t)} \rho v^2 4\pi r^2 dr. \tag{D.25}$$

These are time dependent. At late times, $R(t) \gg R(0)$, so $g \to 0$ and $E_K \to E_{SN}$; all the thermal energy is converted into kinetic energy in this approximation. Thus $\dot{R}(t) = v_{sc}(t) \to \sqrt{2}v_{sc}(0)$; this effect gives a 40 percent increase in velocity during expansion. Most of this occurs in the first doubling of $R(t)$, which is roughly the first day (or less) of the event. This suggests that a constant velocity is a useful approximation for later epochs.

Equation D.24 can be written as

$$E_{Th}(t) = [4\pi R(0)^3 aT(0,0)^4] I_{Th}\phi(t)R(0)/R(t), \tag{D.26}$$

where

$$I_{Th} \equiv \int_0^1 \Psi(x)x^2 dx \tag{D.27}$$

is a dimensionless form factor for the distribution of the energy of the radiation-gas. The first factor in equation D.26 has units of energy, and sets that scale. The second factor has the time dependence: $R(t)^{-1}$ from expansion, and $\phi(t)$ from radiative diffusion.

For homologous expansion,

$$v = \frac{\partial}{\partial t}[xR(t)] = xv_{sc}. \tag{D.28}$$

Using this, we have

$$E_K(t) = [2\pi R(0)^3 \rho(0,0)]v_{sc}(t)^2 I_K, \tag{D.29}$$

where

$$I_K \equiv \int_0^1 \eta(x)x^4 dx \tag{D.30}$$

is the dimensionless form factor for the kinetic energy distribution. The first factor has units of mass; the second factor sets the expansion velocity scale. The mean squared velocity $< v^2 >= 2E_K/M$ is

$$< v^2 >= (I_K/I_M)v_{sc}^2, \tag{D.31}$$

where

$$I_M \equiv \int_0^1 \eta(x)x^2 dx. \tag{D.32}$$

For uniform density,

$$< v^2 > = \frac{3}{5}v_{sc}^2, \tag{D.33}$$

or $v_{sc} \approx 1.3 v_{rms}$. More centrally condensed density structure gives a larger factor (see below). Shock steepening near the surface has the same tendency. The actual maximum velocity observed should exceed our scaling velocity v_{sc}.

Spatial part

The nature of the spatial solutions will be illustrated for $\eta(x)$ of the form

$$\eta(x) = \exp(1.723 A x), \tag{D.34}$$

with $A = 0, \pm 1, \pm 2, \pm 3$. The first case, $A = 0$, represents uniform density, and the other values range from centrally condensed to shell-like structure. The shock wave modifies any *initial* density distribution, so that these forms correspond to the *postshock* density, not the preexplosion density structures. Values of $A \approx 0$ resemble density structures found in many numerical simulations. The factor 1.723 is a historical artifact; for $A = -1$, $\eta(x)$ resembles a *preexplosion* red supergiant structure of Paczynski (private communication).

For uniform density,

$$\Psi(x) = \frac{\sin(\alpha^{\frac{1}{2}} x)}{\alpha^{\frac{1}{2}} x} \tag{D.35}$$

where we have used $\Psi(0) = 1$. For the Eddington outer boundary condition,

$$\frac{\sin(\sqrt{\alpha})}{\sqrt{\alpha}} = \frac{2}{3} \frac{\lambda(1)}{R(t)}. \tag{D.36}$$

For λ/R small, this gives

$$\sqrt{\alpha} \approx \pi \left[1 - \frac{2}{3} \frac{\lambda(1)}{R(t)}\right]. \tag{D.37}$$

TABLE D.1

Eigenfunctions

α	$\Psi(1)$	$-\Psi(1)'$	$R(t)/\lambda(1)$
2	0.6985	0.5425	0.518
4	0.4547	0.8708	1.277
7	0.1798	1.059	3.928
8	0.1089	1.060	6.490
9.849	1.054(-3)	1.001	633.1
π^2	0.	1.	∞

Note that $R(t)/\lambda(1)$ is the optical depth. The "radiative zero" solution occurs for $\alpha = \pi^2$. For larger α the function $\Psi(x)$ has nonphysical (negative) values. The solutions $\alpha < \pi^2$ correspond to decreasing optical depth, as table D.1 demonstrates.

Note that the total optical depth must become quite small ($R/\lambda < 10$) before Ψ deviates much from the $\alpha = \pi^2$ solution. The eigenfunctions for nonuniform density give similar results. Only the "radiative-zero" results will be discussed for various choices of $\eta(x)$; these are given in table D.2, but for small optical depths, use caution. The variable $\eta(1)$ is the ratio of surface to central density; it varies by a factor of roughly

TABLE D.2

"Radiative-Zero" Solutions

A	+2	+1	0	-1	-2	-3
α	0.60978	2.5655	9.8696	32.931	89.007	189.72
$-\Psi(1)'$	3.49923	2.1038	1.000	0.3303	7.05(-2)	1.11(-2)
I_K	3.897	0.8635	0.2	0.0492	0.0131	3.83(-3)
I_M	5.305	1.2765	0.3333	0.0973	0.0327	0.0129
v_{sc}/v_{rms}	1.167	1.216	1.291	1.407	1.582	1.834
I_T	5.830(-3)	2.614(-2)	0.1013	0.3148	0.7791	1.806
$\eta(1)$	3.137(1)	5.601	1.0	0.1785	0.319	5.690(-3)
αI_M	3.235	3.275	3.290	3.204	2.911	2.442

10^4 on the range of $A = +2$ to -3. This corresponds to variation from a shell to a centrally condensed cloud. The important variable combination αI_M is very insensitive to this large change in structure, and the velocity ratio v_{sc}/v_{rms} does not vary strongly either. The values for the $A = 0$ case are given as a measure of the accuracy of the numerical integration as well as for completeness.

Solution in time

The surface luminosity is

$$L(1, t) = \frac{4\pi}{3}\left[\frac{R(0)c}{\kappa(0)}\right]\frac{E_{Th}(0)}{M}\alpha I_M\phi(t), \qquad (D.38)$$

where the identity

$$I_T\alpha\eta(1) = -\left.\frac{d\Psi}{dx}\right|_1 \qquad (D.39)$$

has been used. Note that αI_M in Table D.2 is insensitive to the choice of density profile. This may be written as

$$L(1, t) = L(1, 0)\phi(t), \qquad (D.40)$$

so that $\phi(t)$ contains all the time dependence. The luminosity scale, $L(1, 0)$, may be written as

$$L(1, 0) \approx \left[\frac{2\pi}{3}c\alpha I_K f\right]R(0)v_{sc}^2/\kappa(0). \qquad (D.41)$$

The factor in square brackets is insensitive to the details of the model. Thus, in this approximation, the luminosity scale depends only on the initial radius of the star, $R(0)$, the opacity $\kappa(0)$, and the *energy per unit ejected mass* (or v_{sc}^2).

The behavior of the bolometric luminosity gives a distance-independent characteristic of the supernova through the function $\phi(t)$. If the quadratic term dominates,

$$\phi(t) \approx \exp(-t^2/2\tau_0\tau_h). \qquad (D.42)$$

Some n e-folds occur after a time

$$t_n = (2n\tau_0\tau_h)^{\frac{1}{2}} = [M\kappa(0)/v_{sc}]^{\frac{1}{2}}/[2\pi c\alpha I_M/3n]^{\frac{1}{2}}. \qquad (D.43)$$

The denominator is insensitive to the density structure, so that t_n is a measure of the "diffusion mass" M of the matter ejected by the supernova explosion. Notice that it depends upon the value of the opacity and upon its constancy in time. Recombination would reduce the effective value of the opacity. Less obvious is the problem that it depends upon the assumption of spherical symmetry—a clumpy medium could have significant "holes" for radiation to leak through more quickly that a smooth model would predict. Both these errors are in the sense that the mass is at least as large as inferred.

D.2   HEATING

The light curves of the previous section may be generalized to include the effects of heating. Most important is heating due to radioactive decay of ^{56}Ni and ^{56}Co. The solutions are "analytic" in the sense that they are expressed in terms of integrals that are easy to do numerically, and may be tabulated as functions, for example.

A simple case which includes heating, and is of astrophysical importance, is: small initial radii and the energy source simply exponential decay of ^{56}Ni. This may be extended to include the effects of larger initial radii; we recover the previous light curves with no heating as a limiting case. The method may be further generalized to include ^{56}Co decay and increasing transparency to γ-rays and to x-rays. The effect of an inward-moving recombination wave can be included as a changing outer boundary.

The simplifications are: (1) homologous expansion in spherical symmetry, (2) radiation pressure dominant, (3) ^{56}Ni present in the ejected matter, and (4) ^{56}Ni distribution only mildly peaked toward the center of the ejected mass.[1] The first law of thermodynamics becomes

$$dE/dt + PdV/dt = -\partial L/\partial m + \epsilon, \qquad (D.44)$$

where ϵ is the rate of heating per unit mass. Let

$$\epsilon = \epsilon^0 \xi(x)\nu(t), \qquad (D.45)$$

where ϵ^0 is a constant, $\xi(x)$ gives the spatial distribution of the abundance of that matter which causes the heating, and $\nu(t)$ gives the time dependence of the heating. For example, for simple decay of ^{56}Ni alone, $\epsilon^0 = 4.78 \times 10^{10}$ erg g^{-1} s^{-1}, and $\nu(t) = e^{-t/\tau_{Ni}}$, where $\tau_{Ni} = 7.605 \times 10^5$ s.

[1] The success of these analytic light curves was one of the clues that the Ni in NS1987A was mixed outward beyond the predictions of one-dimensional models.

Proceeding as before we obtain the same spatial equation if we define the eigenvalue α as previously done (see D.8), but the time equation becomes

$$\frac{1}{\phi}\frac{\partial \phi}{\partial t} = -R(t)/R(0)\tau_0 + \epsilon/aT^4V. \tag{D.46}$$

In order for this to be a function of time alone, the spatial dependence in the last term, ϵ/aT^4V, must factor out. Actually, this is not unreasonable. We saw previously that the spatial distribution of temperature was peaked toward the center. The hotter central regions might be expected to be richer in products of fresh nucleosynthesis, and the heating function might then be peaked toward the center too. Hydrodynamic instabilities would smooth an initial spike into a more gently peaked distribution. As the heating becomes more important, it adds to the local heat content, and so tends toward the required condition. Finally, as the optical depth decreases, we found that the temperature distribution became flatter. Similarly, as the depth to γ and x-rays decreases, the deposition of energy becomes less localized, and flatter. Therefore,

$$\epsilon/aT^4V = \left[\frac{\epsilon^0}{aT(0,0)V(0,0)}\right]\left[\frac{\xi(x)\eta(x)}{\Psi(x)}\right]v(t), \tag{D.47}$$

where we take $b \equiv [\xi(x)\eta(x)/\Psi(x)]$ to be constant. Thus we select a set of solutions which, while not appropriate to all possiblities, seem to be approximately correct for many interesting cases.

Small radii

We now proceed with the solution for the special case of simple decay of ^{56}Ni alone, and small initial radius. Then

$$\dot\phi + \phi R(t)/R(0)\tau_0 = \tilde\epsilon R(t)e^{-t/\tau_{Ni}}/R(0), \tag{D.48}$$

where the combination

$$\tilde\epsilon \equiv b\epsilon_{Ni}^0/aT(0,0)^4V(0,0) \tag{D.49}$$

is presumed to be constant. Let $\dot u \equiv R(t)/R(0)\tau_0$, so from Eq. D.12,

$$u(t) = t/\tau_0 + (t/\tau_m)^2, \tag{D.50}$$

where $\tau_m^2 \equiv 2R(0)\tau_0/v_{sc} = 2\tau_0\tau_h$, and $\tau_h \equiv R(0)/v_{sc}$. This time-scale τ_m determined the behavior of the light curves for simple cooling discussed above, and will be a fundamental parameter here as well. Now,

$$R(t)\tilde{\epsilon}e^{-t/\tau_{Ni}}/R(0) = \dot{\phi} + \phi\dot{u} = e^{-u}\frac{d}{dt}(\phi e^u). \tag{D.51}$$

For simplicity we will consider the case of $R(0)$ small; the more general case is given below. Before we take the limit $R(0) \to 0$, we need to consider the diffusion time τ_0 more carefully. Denote

$$I_M \equiv \int_0^1 \eta(x)x^2 dx = V(0,0)M/4\pi R(0)^3, \tag{D.52}$$

where M is the total mass ejected, and use this to rewrite Eq. D.10 as

$$\tau_0 = \kappa M/\beta c R(0), \tag{D.53}$$

where $\beta \equiv 4\pi(\alpha I_M/3)$. From table D.2 we see that $\beta \approx 13.8$ for a variety of density distributions. Now, $\tau_0 R(0)$ is constant for any $R(0)$, so as $R(0)$ goes to zero,

$$u(t) \to (t/\tau_m)^2. \tag{D.54}$$

To evaluate $\tilde{\epsilon}$, we need to use some identities. The mass of ^{56}Ni initially present must be

$$M_{Ni}^0 = 4\pi R(0)^3 V(0,0)^{-1} \int_0^1 \xi(x)\eta(x)x^2 dx. \tag{D.55}$$

The total thermal energy content E_{Th} is obtained by integrating aT^4 over the volume, which gives

$$E_{Th} = 4\pi R(0)^3 aT(0,0)^4 I_{Th}\phi(t)R(0)/R(t), \tag{D.56}$$

where

$$I_{Th} \equiv \int_0^1 \Psi(x)x^2 dx. \tag{D.57}$$

Combining these results,

$$b = \left(\frac{M_{Ni}^0}{M}\right)\frac{I_M}{I_{Th}}. \tag{D.58}$$

Eliminating $T(0, 0)$ and $V(0, 0)$ from the definition of $\tilde{\epsilon}$ gives

$$\tilde{\epsilon} = \epsilon_{Ni}^0 M_{Ni}^0 \phi(0)/E_{Th}(0), \tag{D.59}$$

which is finite as $R(0) \to 0$ if $E_{Th}(0) > 0$.

Finally,

$$\phi(t) = \frac{\epsilon_{Ni}^0 M_{Ni}^0 \tau_0}{E_{Th}(0)} \Lambda(x, y) + e^{-x^2}, \tag{D.60}$$

and

$$\Lambda(x, y) \equiv e^{-x^2} \int_0^1 e^{-2zy+z^2} 2z\,dz, \tag{D.61}$$

where $x \equiv t/\tau_m$ and $y \equiv \tau_m/2\tau_{Ni}$. Here the extra scale factor $\phi(0)$ can be set to unity.

The function $\Lambda(x, y)$ is a one-parameter (y) family of curves. The parameter $y = (2\kappa M/\beta c v_{sc})^{1/2}/2\tau_{Ni}$ is small for small $\kappa M/v_{sc}$. In the limit $y \ll x$,

$$\Lambda(x, y) \approx e^{-t/\tau_{Ni}} - e^{-(t/2y\tau_{Ni})^2}. \tag{D.62}$$

For short "effective diffusion times" τ_m, the luminosity follows the radioactive decay (after a rapid rise).

For $y \gg x$,

$$\Lambda(x, y) \approx [1 - e^{2xy}(1 + 2xy)]/2. \tag{D.63}$$

Recall that $t/\tau_{Ni} = 2xy$. This is a sort of saturation curve which rises to a maximum luminosity in a few decay times. Such an approximation is appropriate to the rising part of the light curve, when diffusion times are long. For small xy,

$$\Lambda(x, y) \approx x^2, \tag{D.64}$$

for all y.

For constant opacity, there is an interesting and simple result: by direct differentiation of $\Lambda(x, y)$ with respect to x (time), the maximum of the light curve occurs at

$$\Lambda(x, y) = e^{-2xy}. \tag{D.65}$$

At maximum light the diffusion loss equals the radioactive input. The Ni decay curve intersects each luminosity curve at its maximum value. This means that the luminosity at that time depends upon the opacity only indirectly through y, and y is given independently by the shape of the light curve.

Arbitrary radii

If we relax the restriction that $R(0)$ is small, we obtain

$$\phi(t) = \phi(0)\left[e^{-u(x)} + \{\epsilon_{Ni}^0 M_{Ni}^0 \tau_0 / E_{Th}(0)\}\Omega(x, y, w)\right], \qquad \text{(D.66)}$$

where $u(x) \equiv wx + x^2$, $w \equiv (2\tau_h/\tau_0)^{1/2}$, and

$$\Omega(x, y, w) \equiv e^{-u(x)} \int_0^x (w + 2z)e^{-2yz+u(z)}dz. \qquad \text{(D.67)}$$

Again we may choose $\phi(0) = 1$ with no loss of generality. If we let $w \to 0$, then $\Omega(x, y, w) = \Omega(x, y, 0) \to \Lambda(x, y)$, and our previous solution is recovered. If we set $\epsilon_{Ni}^0 M_{Ni}^0$ to zero, we find $\phi(t) = e^{-u(x)}$, which is the result found above for simple cooling.

References

[1] Abramowitz, M., and Stegun, I. 1968. *Handbook of Mathematical Functions*. New York: Dover.

[2] Abt, H. A. 1983. *Annual Review of Astronomy and Astrophysics* (hereafter, *Ann. Rev. A. Ap.*) **21**, 343.

[3] Alastuey, A., and Jancovici, B. 1978. *Astrophysical Journal* (hereafter, *Ap. J.*) **226**, 1034.

[4] Alekseev, E. N., Alekseev, L. N., and Krisvosheina, I. V. 1987. *JETP Lett.* **45**, 589.

[5] Allen, C. W. 1955. *Astrophysical Quantities*. 2d ed. London: Athlone Press.

[6] Aller, L. H. 1960. In *Stellar Atmospheres*, ed. J. L. Greenstein (Chicago: University of Chicago Press), p. 156.

[7] Aller, L. H. 1963. *Astrophysics* (New York: Roland Press), especially chapters 1 and 8.

[8] Aller, L. H. 1989. In *Cosmic Abundances of Matter*, ed. C. J. Waddington (New York: American Institute of Physics), p. 224.

[9] Aldering, G., Humphreys, R. M., and Richmond, M. W. 1994. *Astronomical Journal* (hereafter, *A. J.*) **107**, 662.

[10] Alpher, R. A., and Herman, R. C. 1953. *Annual Review of Nuclear Science* (hereafter, *Ann. Rev. Nucl. Sci.*) **2**, 1.

[11] Amari, S., Lewis, R. S., and Anders, E. 1990. *Nature* **345**, 238.

[12] Anders, E., and Ebihara, M. 1982. *Geochim. Cosmochin. Acta* **46**, 2363.

[13] Anders, E., and Grevesse, M. 1989. *Geochim. Cosmochin. Acta* **53**, 197.

[14] Anselmann, P., et al. 1992. *Physics Letters* (hereafter, *Phys. Lett.*) B, in press.

[15] Appenzeller, I. 1970. *Astronomy and Astrophysics* (hereafter, *A. and A.*) **9**, 216.

[16] Arnett, D. 1966. *Can. J. Phys.* **44**, 2553.

[17] Arnett, D. 1967. *Can. J. Phys.* **45**, 1621.

[18] Arnett, D. 1968a. *Nature* **219**, 1344.

[19] Arnett, D. 1968b. *Ap. J.* **153**, 341.

[20] Arnett, D. 1969. *Ap. J.* **157**, 1369.

[21] Arnett, D. 1969. *Astrophysics and Space Science* (hereafter, *Ap. Space Sci.*) **5**, 180.

[22] Arnett, D. 1969. In *Supernovae and Their Remnants*, ed. P. Brancazio and A. G. W. Cameron. (New York: Gordon and Breach), 89.

[23] Arnett, D. 1971a. *Ap. J.* **166**, 153.

[24] Arnett, D. 1971b. *Ap. J.* **169**, 113.

[25] Arnett, D. 1971c. *Ap. J.* **170**, L45.

[26] Arnett, D. 1972a. *Ap. J.* **173**, 393.

[27] Arnett, D. 1972b. *Ap. J.* **176**, 681.

[28] Arnett, D. 1972c. *Ap. J.* **176**, 699.

[29] Arnett, D. 1973a. In *Explosive Nucleosynthesis*, ed. W. D. Arnett and D. N. Schramm (Austin: University of Texas).

[30] Arnett, D. 1973b. *Ann. Rev. A. Ap.* **11**, 73.

[31] Arnett, D. 1974. *Ap. J.* **191**, 727.

[32] Arnett, D. 1975. *Ap. J.* **195**, 727.

[33] Arnett, D. 1977a. *Ann. N.Y. Acad. Sci.* **302**, 90 (8th Texas Conference on Relativistic Astrophysics).

[34] Arnett, D. 1977b. *Ap. J.* **218**, 815.

[35] Arnett, D. 1978. *Ap. J.* **219**, 1008.

[36] Arnett, D. 1979. *Ap. J.* **230**, L37.

[37] Arnett, D. 1980. *Ap. J.* **237**, 541.

[38] Arnett, D. 1980. *Ann. N.Y. Acad. Sci.* **336**, 379 (9th Texas Conference on Relativistic Astrophysics).

[39] Arnett, D. 1982. *Ap. J.* **253**, 785.

[40] Arnett, D. 1986. In *The Origin and Evolution of Neutron Stars*, ed. D. J. Helfand and J.-H. Huang (Dordrecht: D. Riedel), p. 273.

[41] Arnett, D. 1987. *Ap. J.* **319**, 136.

[42] Arnett, D. 1988. *Ap. J.* **331**, 377.

[43] Arnett, D. 1991a. *Ap. J.* **383**, 295.

[44] Arnett, D. 1991b. In *High-Energy Astrophysics: American and Soviet Perspectives*, ed. W. H. G. Lewin, G. W. Clark, R. A. Sunyaev, with K. K. Trivers and D. M. Abramson, (Washington: National Academy Press), p. 1

[45] Arnett, D. 1994. *Ap. J.* **427**, 932.

[46] Arnett, D. 1995. *Ann. Rev. A. Ap.*, **33**, 115.

[47] Arnett, D., Bahcall, J. N., Kirshner, R. P., and Woosley, S. E. 1989. *Ann. Rev. A. Ap.* **27**, 629.

[48] Arnett, D., Fryxell, B., and Müller, E. 1989. *Ap. J.* **341**, L63.

[49] Arnett, D., Truran, J. W., and Woosley, S. E. 1971. *Ap. J.* **165**, 87.

[50] Arnett, D., and Falk, S. W. 1976. *Ap. J.* **210**, 733.

[51] Arnett, D., and Fu, A. 1989. *Ap. J.* **340**, 396.

[52] Arnett, D., and Livne, E. 1994a. *Ap. J.* **427**, 315.

[53] Arnett, D., and Livne, E. 1994b. *Ap. J.* **427**, 330.

[54] Arnett, D., and Truran, J. W. 1969. *Ap. J.* **157**, 1369.

[55] Arnett, D., and Truran, J. W. 1969. *Ap. J.* **157**, 339.

[56] Arnould, M. 1976. *A. and Ap.* **46**, 117.

[57] Arnould, M., and Prantzos, N. 1986. *Nucleosynthesis and Its Implications on Nuclear and Particle Physics*, ed. J. Audouze and N. Mathieu (Dordrecht: D. Reidel), p. 363.

[58] Arnould, M., and Takahashi, K. 1990. In *Astrophysical Ages and Dating Methods*, ed. E. Vangioni-Flam, M. Cassé, J. Audouze and J. Tran Thanh Van (Gif sur Yvette Cedex: Editions Frontieres), p. 325.

[59] Audouze, J., and Reeves, H. 1982. In *Essays in Nuclear Astrophysics*, ed. C. A. Barnes, D. D. Clayton, and D. N. Schramm (Cambridge: Cambridge University Press), p. 355.

[60] Audouze, J., and Tinsley, B. M. 1976. *Ann. Rev. A. Ap.* **14**, 43.

[61] Aufderheide, M. B. 1993. *Ap. J.* **411**, 813.

[62] Axford, W. I. 1981. *Proc. 17th. International Cosmic Ray Conf.* (Paris) **12**, 155.

[63] Baade, W., and Zwicky, F. 1934. *Proc. Nat. Acad. of Sciences* **20**, 254.

[64] Bahcall, J. N. 1983. *The Nearby Stars and the Stellar Luminosity Function*, I. A. U. Colloquium no. 76 (Schenectady, N.Y.: L. Davis Press), p. 271.

[65] Bahcall, J. N. 1989. *Neutrino Astrophysics*. Cambridge: Cambridge University Press.

[66] Bahcall, J. N., and Soneira, R. M. 1980. *Ap. J. Suppl.* **44**, 73.

[67] Barbaro, G., Bertelli, G., Chiosi, C., and Nasi, E. 1973. *A. and A.* **29**, 185.

[68] Barbon, R., Ciatti, F., and Rosino, L. 1979. *A. and A.* **72**, 287.

[69] Barbuy, B., Spite, F., and Spite, M. 1987. *A. and A.* **178**, 199.

[70] Barkat, Z., Buchler, J.-R., and Wheeler, J. C. 1971. *Ap. Letters* **8**, 21.

[71] Barkat, Z., Rakavy, G., and Sack, N. 1967. *Physical Review Letters* (hereafter, *Phys. Rev. Lett.*) **18**, 379.

[72] Barkat, Z., and Wheeler, J. C. 1990. *Ap. J.* **355**, 602.

[73] Barnes, C. A. In *Essays in Nuclear Astrophysics*, ed. C. A. Barnes, D. D. Clayton, and D. N. Schramm (Cambridge: Cambridge University Press), p. 193.

[74] Baron, E., Cooperstein, J., and Kahana, S. 1985a. *Nucl. Phys.* **A440**, 744.

[75] Baron, E., Cooperstein, J., and Kahana, S. 1985b. *Phys. Rev. Lett.* **55**, 126.

[76] Baron, E. 1988. *Physics Reports* **163**, 37.

[77] Baron, E., Hauschildt, P. H., Branch, D., Wagner, R. M., Austin, S. J., Filipenko, A. V., and Matheson, T. 1993. *Ap. J.* **416**, L21.

[78] Baron, E., et al. 1995. *Ap. J.*, in press.

[79] Bartunov, O. S., Blinnikov, S. I., Pavlyuk, N. N., and Tsvetkov, D. Yu. 1994. *A. and A.* **281**, L53.

[80] Baschek, B. 1962. *Z. Astrophys.* **56**, 207.

[81] Basu, S., and Rana, N. C. 1992. *Ap. J.* **393**, 373.

[82] Batten, A. H. 1973. *Binary and Multiple Systems of Stars*. Oxford: Pergamon.

[83] Baym, G., Bethe, H. A., and Pethick, C. J. 1971. *Nucl. Phys.* **A175**, 225.

[84] Bazan, G., and Arnett, D. 1994. *Ap. J.* **433**, L41.

[85] Bazan, G., and Arnett, D. 1995. *Ap. J.*, submitted.

[86] Beaudet, G., Petrosian, V., and Salpeter, E. E. 1967. *Ap. J.* **150**, 979.

[87] Beer, H. 1988. In *Origin and Distribution of the Elements*, ed. G. J. Mathews (Singapore: World Scientific), p. 505.

[88] Bell, R. A., and Branch, D. 1970. *Astrophys. Lett.* **5**, 203.

[89] Benz, W., Colgate, S. A., and Herant, M. 1994. *Physica D* **77**, 305.

[90] Bergeron, P., Saffer, R. A., and Liebert, J. 1992. *Ap. J.* **394**, 228.

[91] Bernatowicz, T., Fraundorf, G., Tang, M., Anders, E., Wopenka, B., Zinner, E., and Fraundorf, P. 1987. *Nature* **330**, 728.

[92] Bessel, M. S. 1991. *A. J.* **101**, 662.

[93] Bethe, H. A. 1936. *Rev. Mod. Phys.* **8**, 82.

[94] Bethe, H. A. 1990. *Rev. Mod. Phys.* **62**, 801.

[95] Bethe, H. A., and Johnson, M. 1974. *Nucl. Phys.* **A230**, 1.

[96] Bethe, H. A., Brown, G. E., Applegate, J., and Lattimer, J. M. 1979. *Nucl. Phys.* **A372**, 496.

[97] Bethe, H. A., and Wilson, J. R. 1985. *Ap. J.* **295**, 14.

[98] Binney, J., and Tremaine, S. 1987. *Galactic Dynamics*. Princeton: Princeton University Press.

[99] Bionta, R. M., et al. 1987. *Phys. Rev. Lett.* **58**, 1494.

[100] Blaauw, A. 1963. In *Basic Astronomical Data*, ed. K. Aa. Strand (Chicago: University of Chicago Press), p. 383.

[101] Black, D. C., and Pepin, R. O. 1969. *Earth Planet. Sci. Lett.* **6**, 395.

[102] Blanco, V. M., et al. 1987. *Ap. J.* **320**, 589.

[103] Blandford, R. D., and Ostriker, R. P. 1980. *Ap. J.* **237**, 793.

[104] Blatt, J. M., and Weisskopf, V. F. 1952. *Theoretical Nuclear Physics*. New York: John Wiley.

[105] Blondin, J., and Lundquist, P. 1993. *Ap. J.* **405**, 337.

[106] Bludman, S., and Van Riper, K. A. 1977. *Ap. J.* **212**, 859.

[107] Bludman, S., and Van Riper, K. A. 1978. *Ap. J.* **224**, 631.

[108] Bludman, S. 1988. *Physics Reports* **163**, 47.

[109] Bodansky, D., Clayton, D. D., and Fowler, W. A. 1968. *Ap. J. Suppl.* **16**, 299.

[110] Bodenheimer, P., and Woosley, S. E. 1983. *Ap. J.* **269**, 381.

[111] Boesgaard, A. M. 1968. *Ap. J.* **154**, 185.

[112] Bohr, A., and Mottleson, B. R. 1969. *Nuclear Structure*, Vol. 1, *Single Particle Motions*. New York: W. A. Benjamin.

[113] Bond, J. R., Arnett, W. D., and Carr, B. J. 1982. In *Supernovae*, ed. M. J. Rees and R. J. Stoneham. Dordrecht: Reidel.

[114] Bond, J. R., Arnett, W. D., and Carr, B. J. 1984. *Ap. J.* **280**, 825.

[115] Bondi, H., and Hoyle, F. 1944. *Monthly Notices of the Royal Astronomical Society* (hereafter, *M. N. R. A. S.* **104**, 273.

[116] Boothroyd, A. I., Sackmann, I.-J., and Wasserburg, G. J. 1994. *Ap. J.* **430**, L77.

[117] Bouchet, P., Moneti, A., Slezak, E., Le Bertre, T., and Manfroid, J. 1989. *A. and A.* **224**, 367.

[118] Bouchet, P., Phillips, M. M., Suntzeff, N. B., Gouiffes, C., Hanuschik, R. W., and Wooden, D. H. 1991. *A. and A.* **245**, 490.

[119] Bowers, R. L., and Arnett, D. 1977. *Ap. J. Suppl.* **33**, 415.

[120] Bowers, R. L., and Wilson, J. R. 1982. *Ap. J. Suppl.* **50**, 115.

[121] Bowers, R. L., and Wilson, J. R. 1991. *Numerical Modeling in Applied Physics and Astrophysics*. Boston: Jones and Bartlett.

[122] Branch, D., Drucker, W., and Jeffery, D. J. 1988. *Ap. J.* **330**, L117.

[123] Branch, D., Lacy, C. H., McCall, M. L., Sutherland, P. G., Uomoto, A., Wheeler, J. C., and Wills, B. J. 1983. *Ap. J.* **270**, 123.

[124] Brown, G. E., Bethe, H. A., and Baym, G. 1982. *Nucl. Phys. A* **375**, 481.

[125] Brown, G. E. 1988. *Physics Reports* **163**, 3 and 167.

[126] Bruenn, S. W. 1972. *Ap. J. Suppl.* **24**, 283.

[127] Bruenn, S. W. 1973. *Ap. J.* **183**, L125.

[128] Bruenn, S. W., Buchler, J. R., and Livio, M. 1979. *Ap. J.* **234**, L183.

[129] Bruenn, S. W. 1985. *Ap. J. Suppl.* **58**, 771.

[130] Bruenn, S. W. 1989a. *Ap. J.* **340**, 955.

[131] Bruenn, S. W. 1989b. *Ap. J.* **341**, 385.

[132] Bruenn, S. W. 1992. In *First Symposium of Nuclear Physics in the Universe*. Oak Ridge: ORNL.

[133] Buchler, J.-R., Wheeler, J. C., and Barkat, Z. 1971. *Ap. J.* **167**, 465.

[134] Buchmann, L., et al. 1993. *Phys. Rev. Lett.* **70**, 726.

[135] Burbidge, E. M., Burbidge, G., Fowler, W. A., and Hoyle, F. 1957. *Rev. Mod. Phys.* **29**, 547.

[136] Burrows, A., and Fryxell, B. A. 1993. *Ap. J.* **418**, L33.

[137] Caldwell, N., Kennicutt, R., Phillips, A. C., and Schommer, R. A. 1991. *Ap. J.* **370**, 526.

[138] Cameron, A. G. W. 1955. *Ap. J.* **121**, 144.

[139] Cameron, A. G. W. 1957. *Atomic Energy of Canada, Ltd.*, **CRL-41**.

[140] Cameron, A. G. W. 1959a. *Ap. J.* **129**, 676.

[141] Cameron, A. G. W. 1959b. *Ap. J.* **130**, 429.

[142] Cameron, A. G. W. 1959c. *Ap. J.* **130**, 884.

[143] Cameron, A. G. W. 1959d. *Ap. J.* **130**, 895.

[144] Cameron, A. G. W. 1960. *A. J.* **65**, 485.

[145] Cameron, A. G. W. 1962. *Icarus* **1**, 13.

[146] Cameron, A. G. W. 1968. In *Origin and Distribution of the Elements*, ed. L. H. Ahrens. Oxford: Pergamon Press.

[147] Cameron, A. G. W. 1973. *Space Sci. Rev.* **15**, 121.

[148] Cameron, A. G. W. 1982. In *Essays in Nuclear Astrophysics*, ed. C. A. Barnes, D. D. Clayton, and D. N. Schramm (Cambridge: Cambridge University Press), p. 23.

[149] Cameron, A. G. W. 1988. *Ann. Rev. A. Ap.* **26**, 441.

[150] Cameron, A. G. W. 1993. In *Protostars and Planets III*, ed. E. H. Levy and J. I. Lunine (Tucson: University of Arizona Press), p. 47.

[151] Cameron, A. G. W., and Elkin, R. M. 1965. *Can. J. Phys.* **43**, 1288.

[152] Canal, R., Isern, J., and Labay, J. 1990. *Ann. Rev. A. Ap.* **28**, 183.

[153] Carbon, D. F., Barbuy, B., Kraft, R. P., Friel, E. D., and Suntzeff, N. B. 1987. *PASP* **99**, 335.

[154] Catchpole, R. M., et al. 1987. *M. N. R. A. S.* **231**; *M. N. R. A. S.* **231**, 75P; *M. N. R. A. S.* **237**, 55P.

[155] Caughlan, G. R., Fowler, W. A., Harris, M. J., and Zimmerman, B. A. 1985. ATOMIC DATA AND NUCLEAR DATA TABLES **32**, 197.

[156] Caughlan, G. R., and Fowler, W. A. 1988. ATOMIC DATA AND NUCLEAR DATA TABLES **40**, 283.

[157] Cernohorsky, J. 1994. *Ap. J.* **433**, 247.

[158] Cernohorsky, J., and Bludman, S. A. 1994. *Ap. J.* **433**, 250.

[159] Cesarsky, C. J. 1987. *Proc. 20th. International Cosmic Ray Conf.* (Moscow) **8**, 87.

[160] Champagne, A. E., and Wiescher, M. 1992. *Ann. Rev. Nucl. Part. Sci.* **43**, 39.

[161] Chandrasekhar, S. 1931. *Ap. J.* **74**, 81.

[162] Chandrasekhar, S. 1939. *An Introduction to the Study of Stellar Structure.* Chicago: University of Chicago Press.

[163] Chandrasekhar, S. 1950. *Radiative Transfer.* Oxford: Oxford University Press.

[164] Chevalier, R. A. 1977. *Ann. N. Y. Acad. Sci.* **302**, 106.

[165] Chevalier, R. A. 1990. In *Supernovae*, ed. A. G. Petschek (Berlin: Springer-Verlag), p. 91.

[166] Chevalier, R. A., and Kirshner, R. P. 1978. *Ap. J.* **219**, 931.

[167] Chevalier, R. A., and Kirshner, R. P. 1979. *Ap. J.* **233**, 154.

[168] Chiosi, C., and Maeder, A. 1986. *Ann. Rev. A. Ap.* **24**, 329.

[169] Chiosi, C., and Summa, C. 1970. *Ap. Space Sci.* **8**, 478.

[170] Chiu, H-Y. 1966. *Ann. Rev. Nucl. Sci.* **16**, 591.

[171] Chiu, H-Y. 1968. *Stellar Physics.* Waltham, Mass.: Blaisdell.

[172] Chiu. H-Y., and Stabler, R. 1961. *Phys. Rev.* **122**, 1317.

[173] Clayton, D. D. 1968. *Principles of Stellar Evolution and Nucleosynthesis.* New York: McGraw-Hill.

[174] Clayton, D. D. 1982. *Q. Jl. R. Astr. Soc.* **23**, 174.

[175] Clayton, D. D. 1985. In *Nucleosynthesis: Challenges and New Developments*, ed. W. D. Arnett and J. W. Truran (Chicago: University of Chicago Press), 65.

[176] Clayton, D. D., Colgate, S. A., and Fishman, G. J. 1969. *Ap. J.* **155**, 75.

[177] Clayton, D. D., and Pantelaki, I. 1986. *Ap. J.* **307**, 441.

[178] Clayton, D. D., and Woosley, S. E. 1969. *Ap. J.* **157**, 1381.

[179] Clayton, R. N. 1978. *Ann. Rev. Nucl. Part. Sci.* **28**.

[180] Cohen, J. G. 1978. *Ap. J.* **223**, 487.

[181] Colgate, S. A., and Johnson, M. H. 1960. *Phys. Rev. Lett.* **5**, 235.

[182] Colgate, S. A. 1971. *Ap. J.* **163**, 221.

[183] Colgate, S. A., Grassberger, W. H., and White, R. H., 1961. Lawrence Radiation Laboratory UCRL-6471.

[184] Colgate, S. A., and White, R. H. 1966. *Ap. J.* **143**, 626.

[185] Colgate, S. A., and Petschek, A. G. 1980. *Ap. J.* **236**, L115.

[186] Conti, P. S. 1988. In *O Stars and Wolf-Rayet Stars*, NASA SP-497, ed. P. S. Conti, and A. B. Underhill. Washington: NASA.

[187] Contrell, P. L., and Da Costa, G. S. 1981. *Ap. J.* **245**, L79.

[188] Cook, C. W., Fowler, W. A., Lauritsen, C. C., and Lauritsen, T. 1957. *Phys. Rev.* **107**, 508.

[189] Cook, W. R., Palmer, D. M., Prince, T. A., Schindler, S. M., Starr, C. H., et al. 1988. *Ap. J.* **334**, L87.

[190] Cooperstein, J., and Wambach, J. 1984. *Nucl. Phys.* **A420**, 591.

[191] Cooperstein, J. 1988. *Physics Reports* **163**, 95.

[192] Couch, R. G., Schmiedekamp, A. B., and Arnett, W. D. 1974. *Ap. J.* **190**, 95.

[193] Couch, R. G., and Arnett, W. D. 1972. *Ap. J.* **178**, 711.

[194] Couch, R. G., and Arnett, W. D. 1974. *Ap. J.* **194**, 537.

[195] Couch, R. G., and Arnett, W. D. 1975. *Ap. J.* **196**, 791.

[196] Courant, R., Fredrichs, K. O., and Lewy, H. 1928. *Math. Ann.* **100**, 32.

[197] Courant, R., and Friedrichs, K. O. 1948. *Supersonic Flow and Shock Waves*. New York: Interscience Publishers.

[198] Cowan, J. J., Thielemann, F. K., and Truran, J. W. 1986. In *Advances in Nuclear Astrophysics*, ed. E. Vangioni-Flam, J. Audouze, M. Cassé, J. P. Chieze, and J. Tran Thanh Van (Gif Sur Yvette: Editions Frontiéres), p. 477.

[199] Cox, A. N. 1965. In *Stellar Structure*, ed. L. H. Aller and D. B. McLaughlin. Chicago: University of Chicago Press.

[200] Cox, J. P. 1980. *Theory of Stellar Pulsation*. Princeton: Princeton University Press.

[201] Cox, J. P., and Guili, R. T. 1968. *Principles of Stellar Structure*. Vols. 1 and 2. New York: Gordon and Breach.

[202] Crotts, A. P. S., Kunkel, W. E., and Heathcote, S. R. 1994. *Ap. J.*, in press.

[203] Cvitanović, P. 1989. *Universality in Chaos* (Bristol: Adam Hilger), p. 3.

[204] D'Antona, F., and Mazzitelli, I. 1985. *Ap. J.* **296**, 502.

[205] Dayras, R., Switkowski, Z. E., and Woosley, S. E. 1977. *Nucl. Phys. A* **279**, 70.

[206] De Greve, J. P. 1993. *A. and A.* **277**, 475.

[207] De Groot, S. R., and Mazur, P. 1984. *Non-Equilibrium Thermodynamics* (New York: Dover), p. 197.

[208] Deinzer, W., and Salpeter, E. E. 1964. *Ap. J.* **140**, 499.

[209] Denisenkov, P. A., and Denisenkova, S. N. 1990. *Soviet Astron. Lett.* **16**, 275.

[210] Dicus, D. 1972. *Phys. Rev.* **D6**, 941.

[211] Diehl, R., et al. 1995. *A. and A.*, in press.

[212] Doering, W. 1943. *Ann. Phys.* **43**, 421.

[213] Dotani, T., Hayashida, K., Inoue, H., Itoh, J., Koyama., K., et al. 1987. *Nature* **330**, 230.

[214] Dupree, A. K. 1986. *Ann. Rev. A. and Ap.* **24**, 377.

[215] Dyer, P., and Barnes, C. A. 1974. *Nucl. Phys.* **A233**, 495.

[216] Eastman, R. G., and Kirshner, R. P. 1989. *Ap. J.* **347**, 771.

[217] Eastman, R. G., and Pinto, P. 1995. in preparation.

[218] Eddington, A. S. 1926. *The Internal Constitution of Stars*. Cambridge University Press: Cambridge.

[219] El Eid, M. F., Fricke, K. J., and Ober, W. W. 1983. *A. and A.* **119**, 54.

[220] Elmegreen, B. G. 1983. *The Nearby Stars and the Stellar Luminosity Function*, IAU Colloquium 76, ed. A. G. Davis Philip and A. R. Upgren (Schenectady, N.Y.: L. Davis Press), 235.

[221] Ensman, L., and Burrows, A. 1992. *Ap. J.* **393**, 742.

[222] Epstein, R. I. 1979. *M. N. R. A. S.* **188**, 305.

[223] Eriguchi, Y., and Müller, E. 1985. *A. and A.* **147**, 161.

[224] Evans, R. D. 1955. *The Atomic Nucleus*. New York: McGraw-Hill.

[225] Evans, R., van den Bergh, S., and McClure, R. D. 1989. *Ap. J.* **345**, 752.

[226] Falk, S. W. 1973. *Ap. J.* **180**, L65.

[227] Falk, S. W. 1977. *Ap. J. Suppl.* **33**, 515.

[228] Falk, S. W. 1978. *Ap. J.* **226**, L113.

[229] Fermi, E. 1934. *Zeit. Physik* **88**, 161.

[230] Fermi, E. 1949a. *Nuclear Physics*. Chicago: University of Chicago Press.

[231] Fermi, E. 1949b. *Phys. Rev.* **75**, 1169.

[232] Festa, G. C., and Ruderman, M. A. 1969. *Phys. Rev.* **180**, 1227.

[233] Feynman, R. P., and Gell-Mann, M. 1958. *Phys. Rev.* **109**, 193.

[234] Fickett, W., and Davis, W. C. 1979. *Detonation*. Berkeley: University of California Press.

[235] Forestini, M., Paulus, G., and Arnould, M. 1991. *A. and A.* **252**, 597.

[236] Fowler, W. A., Caughlan, G. R., and Zimmerman, B. A. 1967. *Ann. Rev. A. Ap.* **5**, 525.

[237] Fowler, W. A., Caughlan, G. R., and Zimmerman, B. A. 1975. *Ann. Rev. A. Ap.* **13**, 69.

[238] Fowler, W. A., and Hoyle, F. 1964. *Ap. J. Suppl.* **9**, 201. See also Fowler, W. A., and Hoyle, F., 1964, *Nucleosynthesis in Massive Stars and Supernovae* (Chicago: University of Chicago Press).

[239] Fowler, W. A., and Meisl, C. C. 1986. *Cosmogonical Processes*, ed. W. D. Arnett, C. Hansen, J. Truran, and S. Tsuruta (Utrecht: VNU Science Press), p. 83.

[240] Fraley, G. S. 1968. *Ap. Space Sci.* **2**, 96.

[241] Fransson, C., and Lundqvist, P. 1989. *Ap. J.* **341**, L59.

[242] Freedman, D. Z. 1974. *Phys. Rev. D* **9**, 1389.

[243] Freedman, D. Z., Schramm, D. N., and Tubbs, D. L. 1977. *Ann. Rev. Nucl. Sci.* **27**, 167.

[244] Freedman, W. L., et al. 1994. *Ap. J.* **427**, 628.

[245] Fricke, K. J. 1973. *Ap. J.* **183**, 941.

[246] Fricke, K. J. 1974. *Ap. J.* **189**, 535.

[247] Fryxell, B. A., and Woosley, S. E. 1982. *Ap. J.* **261**, 332.

[248] Fryxell, B. A., Müller, E., and Arnett, D. 1991. *Ap. J.* **367**, 619.

[249] Fu, A., and Arnett, D. In *Supernovae as Distance Indicators*, ed. N. Bartel, (Berlin-Heidelberg: Springer-Verlag), 209.

[250] Fuller, G. M. 1982. *Ap. J.* **252**, 741.

[251] Fuller, G. M., Fowler, W. A., and Newman, M. J. 1982a. *Ap. J. Suppl.* **42**, 447.

[252] Fuller, G. M., Fowler, W. A., and Newman, M. J. 1982b. *Ap. J. Suppl.* **48**, 279.

[253] Fuller, G. M., Fowler, W. A., and Newman, M. J. 1982c. *Ap. J.* **252**, 741.

[254] Fuller, G. M., Fowler, W. A., and Newman, M. J. 1985. *Ap. J.* **293**, 1.

[255] Fuller, G. M., Mayle, R. W., Wilson, J. R., and Schramm, D. N. 1987. *Ap. J.* **322**, 795.

[256] Fuller, G. M., Mayle, R. W., Meyer, B. S., and Wilson, J. R. 1992. *Ap. J.* **389**, 517.

[257] Gallino, R., Busso, M., Picchio, G., Raiteri, C. M., and Renzini, A. 1988. *Ap. J.* **334**, L45.

[258] Gamow, G. 1937. *Structure of Atomic Nuclei and Nuclear Transformations*. Oxford: Clarendon Press.

[259] Gamow, G., and Schönberg, M. 1941. *Phys. Rev.* **59**, 539.

[260] Garmany, C. D., Conti, P., and Chiosi, C. 1982. *Ap. J.* **263**, 777.

[261] Garvey, G. T., Gerace, W. J., Jaffe, R. L., and Talmi, I. 1969. *Rev. Mod. Phys.* **41**, S1.

[262] Garz, T., Kock, M., Richter, J., Baschek, B., Holweger, H., and Unsöld, A. 1969. *Nature* **223**, 1254.

[263] Gehrels, N., Leventhal, M., and MacCallum, C. J. 1988. In *Nuclear Spectroscopy of Astrophysical Sources, AIP Conf Proc. No. 170*, ed. N. Gehrels and G. Share (New York: AIP), p. 87.

[264] Gilbert, A., and Cameron, A. G. W. 1965. *Can. J. Phys.* **43**, 1446.

[265] Ginsburg, V. L., and Syrovatskii, S. I. 1964. *The Origin of Cosmic Rays*. Oxford: Pergamon Press.

[266] Glashow, S. 1961. *Nucl. Phys.* **22**, 579.

[267] Glendenning, N. K. 1986. *Phys. Rev. Lett.* **57**, 1120.

[268] Goldreich, P., and Weber, S. V. 1980. *Ap. J.* **238**, 991.

[269] Grasberg, E. K., Imshennik, V. S., and Nadyoshin, D. K. 1971. *Ap. Space Sci.* **10**, 28.

[270] Gratton, R. G., and Sneden, C. 1991. *A. and A.* **241**, 501.

[271] Green, A. E. S. 1954. *Phys. Rev.* **95**, 1006.

[272] Greenstein, J. L. 1970. *Comments on Astrophys. Space Sci.* **2**, 85.

[273] Grevesse, N., and Swings, J. P. 1969. *A. and A.* **2**, 28.

[274] Griest, K. et al. 1991. *Ap. J.* **372**, L79.

[275] Grossman, A. S., Hays, D., and Graboske, Jr., H. C. 1974. *A. and A.* **30**, 95.

[276] Gunn, J. E., and Ostriker, J. P. 1971. *Ap. J.* **160**, 979.

[277] Hainebach, K. L., Clayton, D. D., Arnett, W. D., and Woosley, S. E. 1974. *Ap. J.* **193**, 157.

[278] Hamuy, M., Suntzeff, N. B., Gonzalez, R., and Martin, G. 1988. *Astron. J.* **95**, 63.

[279] Hamuy, M., Phillips, M. M., and Maza, J. 1994. *Ap. J.* **108**, 2226.

[280] Hansen, C. J. 1971. *Ap. and Space Sci.* **14**, 389.

[281] Harris III, D. L., Strand, K. Aa., and Worley, C. E. 1963. *Basic Astronomical Data*, ed. K. Aa. Strand (Chicago: University of Chicago Press), p. 273.

[282] Harris, M. J., Fowler, W. A., Caughlan, G. R. and Zimmerman, B. A. 1983. *Ann. Rev. A. Ap.* **21**, 165.

[283] Hartmann, K., and Gehren, T. 1988. *A. and A.* **199**, 269.

[284] Hartwick, F. D. A. 1976. *Ap. J.* **209**, 418.

[285] Hashimoto, M., Nomoto, K., Arai, K., and Kaminishi, K. 1986. *Ap. J.* **307**, 687.

[286] Hayashi, C., Nishida, M., Ohyama, N., and Tsuda, H. 1959. *Prog. Theo. Phys.* **22**, 101.

[287] Heitler, W. 1960. *The Quantum Theory of Radiation*. 3d ed. Oxford: Clarendon Press.

[288] Henry, T. J., and McCarthy, D. W. 1990. *Ap. J.* **350**, 344.

[289] Henry, T. J., and McCarthy, D. W. 1993. *A. J.* **106**, 773.

[290] Herant, M., and Benz, W. 1992., *Ap. J.* **387**, 294.

[291] Herant, M., Benz, W., and Colgate, S. A. 1992. *Ap. J.* **395**, 642.

[292] Hershkowitz, S., Linder, E., and Wogoner, R. 1986. *Ap. J.* **303**, 800.

[293] High, M. D., and Cujek, B. 1977. *Nucl. Phys.* **A282**, 181.

[294] Hillebrandt, W. 1987. In *High Energy Phenomena around Collapsed Stars*, ed. F. Pacini (Drodrecht: Ridel), p. 73.

[295] Hirata, K. S., Kajita, T., Koshiba, M., et al. 1987. *Phys. Rev. Lett.* **58**, 1490.

[296] Hirata, K. S., et al. 1990. *Phys. Rev. Lett.* **65**, 1297.

[297] Höflich, P., Khokhlov, A., and Müller, E. 1991. *A. and A.* **248**, L7.

[298] Höflich, P., Khokhlov, A., and Müller, E. 1992. *A. and A.* **259**, 549.

[299] Hollowell, D., and Iben, Jr., Icko. 1988. *Ap. J.* **333**, L25.

[300] Holmes, J. A., Woosley, S. E., Fowler, W. A., and Zimmerman, B. A. 1976. Atomic Data and Nuclear Data Tables **18**, 305.

[301] Howard, W. M., Arnett, W. D., Clayton, D. D., and Woosley, S. E. 1972. *Ap. J.* **175**, 201.

[302] Hoyle, F. 1946. *M. N. R. A. S.* **106**, 343.

[303] Hoyle, F. 1954. *Ap. J. Suppl.* **1**, 121.

[304] Hoyle, F. 1994. *Home Is Where the Wind Blows*. Mill Valley, Calif.: University Science Books.

[305] Hoyle, F., and Fowler, W. A. 1960. *Ap. J.* **132**, 565.

[306] Hulke, G., Rolfs, C., and Trautvetter, H. P. 1980. *Zeit. Physik A* **297**, 161.

[307] Huss, G. R., Hutcheon, I. D., Wasserburg, G. J., and Stone, J. 1992. *Lunar Planet. Sci.* **23**, 563.

[308] Huss, G. R., Hutcheon, I. D., Fahey, A. J., and Wasserburg, G. J. 1993. *Meteoritics* **28**, 369.

[309] Huss, G. R., Fahey, A. J., Galino, R., and Wasserburg, G. J. 1994. *Ap. J.* **430**, L81.

[310] Hutcheon, I. D., Huss, G. R., Fahey, A. J., and Wasserburg, G. J. 1994. *Ap. J.* **425**, L27.

[311] Iben, I., Jr. 1965a. *Ap. J.* **141**, 993.

[312] Iben, I., Jr. 1965b. *Ap. J.* **142**, 1447.

[313] Iben, I., Jr. 1966a. *Ap. J.* **143**, 483.

[314] Iben, I., Jr. 1966b. *Ap. J.* **143**, 505.

[315] Iben, I., Jr. 1966c. *Ap. J.* **143**, 516.

[316] Iben, I., Jr. 1967. *Ann. Rev. A. Ap.* **5**, 571.

[317] Iben, I., Jr. 1969. *Ann. Physics* **54**, 164.

[318] Iben, I., Jr. 1974. *Ann. Rev. A. Ap.* **12**, 215.

[319] Iben, I., Jr. 1977. *Ap. J.* **196**, 549.

[320] Iben, I., Jr. 1978a. *Ap. J.* **219**, 213.

[321] Iben, I., Jr. 1978b. *Ap. J.* **226**, 996.

[322] Iben, I., Jr. 1982. *Ap. J.* **253**, 248.

[323] Iben, I., Jr. 1991. *Ap. J. Suppl.* **76**, 55.

[324] Iben, I., Jr. 1993. *Ap. J.* **415**, 767.

[325] Iben, I., Jr., and Laughlin, G. P. 1989. *Ap. J.* **341**, 312.

[326] Iben, I., Jr., and Rood, R. 1970. *Ap. J.* **161**, 587.

[327] Iben, I., Jr., and Tutakov, A. V. 1984. *Ap. J. Suppl.* **54**, 335.

[328] Iben, I., Jr., and Tutakov, A. V. 1991. *Ap. J.* **370**, 615.

[329] Imshennik, V. S., and Nadyozhin, D. K. 1972. *Zh. Eksp. Teor. Fiz.* **63**, 1548; *Soviet Phys.-JETP* **36**, 821.

[330] Imshennik, V. S., and Nadyozhin, D. K. 1978. *Ap. Space Sci.* **62**, 309.

[331] Imshennik, V. S., and Khokhlov, A. M. 1984. *Soviet Astron. Lett.* **10**, L262.

[332] Itoh, N., Adachi, T., Nakagawa, M., Kohyama, Y., and Munakata, H. 1989. *Ap. J.* **339**, 354.

[333] Itoh, N., Kuwashima, F., and Munakata, H. 1990. *Ap. J.* **362**, 620.

[334] Itoh, N., Totsuji, H., and Ichimaru, S. 1977. *Ap. J.* **218**, 477.

[335] Jackson, J. D. 1962. *Classical Electrodynamics*. New York: John Wiley.

[336] Jacoby, G. H., et al. 1992. *Publ. Astron. Soc. Pacific* **104**, 599.

[337] Jahreiss, H., and Wielen, R. 1983. *The Nearby Stars and the Stellar Luminosity Function*, I. A. U. Colloquium No. 76 (Schenectady, N.Y.: L. Davis Press), 277.

[338] Jakobsen, P., et al. 1991. *Ap. J.* **369**, L63.

[339] Janka, H.-T., and Hillebrandt, W. 1989. *A. and A. Suppl.* **78**, 375.

[340] Janka, H.-T., and Müller, E. 1989. In *Frontiers of Neutrino Astrophysics*. Tokyo: Universal Academy Press.

[341] Kahler, J. B. 1989. *Stars And Their Spectra*. Cambridge: Cambridge University Press.

[342] Kamper, K., and van den Bergh, S. 1976. *Ap. J. Suppl.* **32**, 351.

[343] Kar, K., Ray, A., and Sarkar, S. 1994. *Ap. J.* **434**, 662.

[344] Karp, A. H., Chan, K. L., Lasher, G. J., and Salpeter, E. E. 1977. *Ap. J.* **214**, 161.

[345] Kennicutt, R. C., Jr. 1983. *Ap. J.* **272**, 54.

[346] Kenyon, S. J., Livio, M., Mikolajewska, J., and Tout, C. A. 1993. *Ap. J.* **407**, L91.

[347] Kettner, K. U., Lorenz-Wirzba, H., Rolfs, C., and Winkler, H. 1977. *Phys. Rev. Lett.* **38**, 337.

[348] Kettner, K. U., Lorenz-Wirzba, H., and Rolfs, C. 1980. *Zeit Physik* **A298**, 65.

[349] Khokhlov, A. M. 1989. *M. N. R. A. S.* **239**, 785.

[350] Khokhlov, A. M. 1991a. *A. and A.* **245**, 114.

[351] Khokhlov, A. M. 1991b. *A. and A.* **246**, 246.

[352] Khokhlov, A. M. 1992. Private communication.

[353] Khokhlov, A. 1993. *Ap. J.* **419**, L77.

[354] Khokhlov, A. M., Müller, E., and Höflich, P. 1992. *A. and A.* **253**, L9.

[355] Kippenhahn, R., and Weigert, A. 1990. *Stellar Structure and Evolution.* Berlin: Springer-Verlag.

[356] Kirkpatrick, J. D., McGraw, J. T., Hess, T. R., Dahn, C. C., Harris, H. C., Liebert, J., and McCarthy, D. W., Jr. 1994. *Ap. J.*, in press.

[357] Kirshner, R. P., and Blair, W. P. 1980. *Ap. J.* **236**, 135.

[358] Kodira, K., Greenstein, J. L., and Oke, J. B. 1970. *Ap. J.* **159**, 485.

[359] Kolb, E. W., and Turner, M. S. 1988. *The Early Universe: Reprints.* Reading, Mass.: Addison-Wesley.

[360] Kolb, E. W., and Turner, M. S. 1990. *The Early Universe.* Reading, Mass.: Addison-Wesley.

[361] Koslovsky, B.-Z. 1970. *Ap. and Space Sci.* **8**, 114.

[362] Kraft, R. P. 1993. *A. J.* **106**, 1490.

[363] Kremer, R. M., et al. 1988. *Phys. Rev. Lett.* **60**, 1475.

[364] Kroupa, P., Tout, C. A., and Gilmore, G. 1993. *M. N. R. A. S.* **262**, 545.

[365] Kudryashov, A. D., and Tutakov, A. V. 1988. *Astron. Tsirk.* **no. 1525**, 11.

[366] Kuijken, K., and Gilmore, G. 1989a. *M. N. R. A. S.* **239**, 571.

[367] Kuijken, K., and Gilmore, G. 1989b. *M. N. R. A. S.* **239**, 605.

[368] Kuijken, K., and Gilmore, G. 1989c. *M. N. R. A. S.* **239**, 651.

[369] Käppeler, F., Beer, H., and Wisshak, K. 1989. *Rep. Prog. Theo. Phys.* **52**, 945.

[370] Laird, J. B. 1985. *Ap. J.* **289**, 556.

[371] Lamb, D. Q., Lattimer, J. M., Pethick, C. J., and Ravenhall, D. G. 1978. *Phys. Rev. Lett.* **41**, 1623.

[372] Lamb, D. Q., Lattimer, J. M., Pethick, C. J., and Ravenhall, D. G. 1983. *Nucl. Phys.* **A 411**, 449.

[373] Lamb, F. K. 1991. In *Frontiers of Stellar Evolution*, ed. D. L. Lambert (San Francisco: Astronomical Society of the Pacific), p. 299.

[374] Lamb, S. A., Howard, W. M., Truran, J. W., and Iben, I., Jr. 1977. *Ap. J.* **217**, 213.

[375] Lamb, S. A., Iben, Icko, Jr., and Howard, W. M. 1976. *Ap. J.* **207**, 209.

[376] Lambert, D. L. 1989. In *Cosmic Abundances of Matter*, ed. C. J. Waddington, (New York: American Institute of Physics), p. 168.

[377] Lambert, D. L. 1992. *A. and Ap. Rev.* **3**, 201.

[378] Landau, L. D., and Lifshitz, E. M. 1959. *Fluid Mechanics.* New York: Addison-Wesley.

[379] Landau, L. D., and Lifshitz, E. M. 1962. *Classical Theory of Fields.* Oxford: Pergamon Press.

[380] Landau, L. D., and Lifshitz, E. M. 1965. *Quantum Mechanics.* New York: Addison-Wesley.

[381] Landau, L. D., and Lifshitz, E. M. 1969. *Statistical Physics.* New York: Addison-Wesley.

[382] Langer, G. E., Hoffman, R., and Sneden, C. 1993. *Publ. Astron. Soc. Pacific* **105**, 301.

[383] Langer, N. 1991. In *Supernovae*, ed. S. E. Woosley (Berlin: Springer-Verlag), p. 549.

[384] Larsen, R. 1972. *Nature Phys. Sci.* **236**, 7.

[385] Larsen, R., and Tinsley, B. M. 1978. *Ap. J.* **219**, 46.

[386] Larson, R., and Starrfield, S. 1971. *A. and A.* **13**, 190.

[387] LeBlanc, J. M., and Wilson, J. R. 1970. *Ap. J.* **161**, 541.

[388] Lederer, C. M., and Shirley, V. S. 1978. Ed. *Table of Isotopes*. 7th ed. New York: John Wiley.

[389] Ledoux, P. 1941. *Ap. J.* **94**, 537.

[390] Ledoux, P. 1947. *Ap. J.* **105**, 305.

[391] Lequeux, J. 1979. *A. and A.* **80**, 35.

[392] Lewis, J. M., et al. 1994. *M. N. R. A. S.* **266**, L27.

[393] Lewis, R. S., Tang, M., Wacker, J. F., Anders, E., and Steel, E. 1987. *Nature* **326**, 160.

[394] Liebert, J. 1980. *Ann. Rev. A. Ap.* **18**, 363.

[395] Liebert, J., Dahn, C. C., and Monet, D. G. 1988. *Ap. J.* **332**, 891.

[396] Liebert, J., and Probst, R. G. 1987. *Ann. Rev. A. Ap.* **25**, 473.

[397] Livio, M., Buchler, J. R., and Colgate, S. A. 1980. *Ap. J.* **238**, L139.

[398] Livne, E. 1990. *Ap. J.* **354**, L53.

[399] Livne, E., and Arnett, W. D. 1993. *Ap. J.* **415**, L107.

[400] Livne, E., and Arnett, W. D. 1995. *Ap. J.*, submitted.

[401] Livne, E., and Glasner, A. S. 1991. *Ap. J.* **370**, 272.

[402] Lund, N. 1989. *Cosmic Abundances of Matter*, ed. J. Waddington (New York: American Institute of Physics), p. 111.

[403] Luo, D., and McCray, R. 1991. *Ap. J.* **379**. 659.

[404] Lynden-Bell, D. 1975. *Vistas in Astronomy* **19**, 299.

[405] MacCallum, C. J., Huters, A. F., Stang, P. D., and Leventhal, M. 1987. *Ap. J.* **317**, 877.

[406] Magain, P. 1989. *A. and A.* **209**, 211.

[407] Mahoney, W. A., Ling, J. C., Jacobson, A. S., and Lingenfelter, R. E. 1982. *Ap. J.* **262**, 742.

[408] Mahoney, W. A., Ling, J. C., Wheaton, W. A., and Jacobson, A. S. 1984. *Ap. J.* **286**, 578.

[409] Mahoney, W. A., et al. 1988. *Ap. J.* **334**, L81.

[410] Malet, I., et al. 1991. In *Gamma-Ray Line Astrophysics*, ed. P. Durouchoux and N. Prantzos (New York: American Institute of Physics), p. 123.

[411] Marshak, R., and Sudarshan, E. 1958. *Phys. Rev.* **109**, 1860.

[412] Martin, C., and Arnett, D. 1995. *Ap. J.*, submitted.

[413] Martinez, M., King, R. B., and Whaling, W. 1969. *Bull. Amer. Phys. Soc.* **13**, 1674.

[414] Matteucci, F. 1991. In *Frontiers of Stellar Evolution*, ed. D. Lambert (San Francisco: Astronomical Soc. Pacific), p. 539.

[415] Matteucci, F., Ferrini, F., Pardi, C., and Penco, U. 1990. In *Chemical and Dynamical Evolution of Galaxies*, ed. F. Ferrini, J. Frano, and F. Matteucci (Pisa: Ets Editrice Pisa), p. 586.

[416] Matz, S. M., Share, G. H., Leising, M. D., Chupp, E. L., Vestrand, W. T., et al. 1988. *Nature* **331**, 416.

[417] May, M. M., and White, R. H. 1967. In *Meth. in Comp. Physics* **7**, 219.

[418] May, R. M. 1976. *Nature* **261**, 459.

[419] Maza, J., Hamuy, M., Phillips, M. M., Suntzeff, N. B., and Aviles, R. 1994. *Ap. J.* **424**, L107.

[420] Mazarakis, M., and Stephens, W. 1973. *Phys. Rev.* **C7**, 1280.

[421] Mazarakis, M., and Stephens, W. 1972. *Ap. J.* **171**, L97.

[422] Mazurek, T. J. 1975. *Ap. Space Sci.* **35**, 117.

[423] Mazurek, T. J., Meier, D. L., and Wheeler, J. C. 1977. *Ap. J.* **213**, 518.

[424] McCray, R. 1993. *Ann. Rev. A. Ap.* **31**, 175.

[425] McCrea, W. H., and Milne, E. A. 1934. *A. J. Math.* Oxford Series, **5**, 73.

[426] McCuskey, S. W. 1965., In *Galactic Structure*, ed. A. Blaauw and M. Schmidt (Chicago: University of Chicago Press), 1.

[427] Mengel, J. G., and Sweigart, A. V. 1981. In *Astrophysical Parameters for Globular Cluster Stars*, ed. A. G. D. Davis Philip (Dordrecht: Reidel), p. 277.

[428] Merrill, P. W. 1952. *Ap. J.* **116**, 21.

[429] Mewaldt, R. A. 1989. In *Cosmic Abundances of Matter*, ed. C. J. Waddington, (New York: American Institute of Physics), p. 124.

[430] Meyer, J-P. 1989. In *Cosmic Abundances of Matter*, ed. C. J. Waddington, (New York: American Institute of Physics), p. 245.

[431] Mezzacappa, A., and Bruenn, S. W. 1993a. *Ap. J.* **405**, 637.

[432] Mezzacappa, A., and Bruenn, S. W. 1993b. *Ap. J.* **405**, 699.

[433] Mezzacappa, A., and Bruenn, S. W. 1993c. *Ap. J.* **410**, 740.

[434] Michaud, G. 1972. *Ap. J.* **175**, 751.

[435] Michaud, G., and Fowler, W. A. 1970. *Phys. Rev.* **C2**, 2041.

[436] Michaud, G., and Vogt, E. 1972. *Phys. Rev. Lett.* **C5**, 350.

[437] Michel, F. C. 1991. *Theory of Neutron Star Magnetospheres*. Chicago: University of Chicago Press.

[438] Mihalas, D. 1978. *Stellar Atmospheres*, San Francisco: W. H. Freeman.

[439] Mihalas, D., and Binney, J. J. 1981. *Galactic Astronomy*. 2d. ed. San Francisco: W. H. Freeman.

[440] Miller, G. E., and Scalo, J. M. 1979. *Ap. J. Suppl.* **41**, 513.

[441] Mönchmeyer, R., and Müller, E. 1989. *A. and A.* **271**, 351.

[442] Müller, E., Różyczka, M., and Hillebrandt, W. 1980. *A. and A.* **81**, 288.

[443] Müller, E. 1981. *A. and A.* **103**, 358.

[444] Müller, E. 1986. *A. and A.* **162**, 103.

[445] Müller, E., and Arnett, W. D. 1982. *Ap. J.* **261**, L109.

[446] Müller, E., and Arnett, W. D. 1986. *Ap. J.* **307**, 619.

[447] Müller, E., Fryxell, B. A., and Arnett, D. 1991. *A. and A.* **251**, 505.

[448] Müller, E., Höflich, P., and Khokhlov, A. 1991. *A. and A.* **249**, L1.

[449] Munari, U., and Renzini, A. 1992. *Ap. J.* **397**, L87.

[450] Myra, E. S., Bludman, S., Hoffman, Y., Lichtenstadt, I., Sack, N., and Van Riper, K. 1987. *Ap. J.* **318**, 744.

[451] Myra, E. S. 1988. *Physics Reports* **163**, 127.

[452] Myra, E. S., and Bludman, S. 1989. *Ap. J.* **340**, 384.

[453] Nadyoshin, D. K. 1993. private communication.

[454] Narayan, R. 1987. *Ap. J.* **319**, 162.

[455] Nissen, P. E. 1988. *A. and A.* **199**, 146.

[456] Nissen, P. E., Gustafsson, B., Edvardsson, B., and Gilmore, B. 1994. *A. and A.* **285**, 440.

[457] Nomoto, K. 1982. *Ap. J.* **253**, 798.

[458] Nomoto, K., and Hashimoto, M. 1988. *Physics Reports* **163**, 13.

[459] Nomoto, K., Susuki, T., Shigeyama, T., Kumagai, S., Yamaoka, H., and Saio, H. 1993. *Nature* **364**, 507.

[460] Nomoto, K., Sparks, W., Fesen, R., Gull, T., Miyaji, S., and Sugimoto, D. 1982. *Nature* **299**, 803.

[461] Nomoto, K., and Sugimoto, D. 1977. *Publ. Astron. Soc. Japan* **29**, 765.

[462] Nomoto, K., Sugimoto, D., and Neo, S. 1976. *Ap. Space Sci.* **39**, L37.

[463] Nomoto, K., Thielemann, F. K., and Yokoi, Y. 1984. *Ap. J.* **286**, 644.

[464] Nozakura, T., Ikeuchi, S., and Fujimoto, M. Y. 1984. *Ap. J.* **286**, 221.

[465] Ober, W. W., El Eid, M. F., and Fricke, K. J. 1982. In *Supernovae*, ed. M. J. Rees and R. J. Stoneham. Dordrecht: Reidel.

[466] Ober, W. W., El Eid, M. F., and Fricke, K. J. 1983. *A. and A.* **119**, 61.

[467] Oort, J. H. 1958. *La Structure et l'evolution de l'universe*, 11th Solvay Conf. (Brussels: Éditions Stoops), p. 163.

[468] Oort, J. H. 1965. In *Galactic Structure*, ed. A. Blaauw and M. Schmidt (Chicago: University of Chicago Press), p. 455.

[469] Oppenheimer, J. R., and Snyder, H. 1939. *Phys. Rev.* **56**, 455.

[470] Oppenheimer, J. R., and Volkoff, G. M. 1938. *Phys. Rev.* **55**, 374.

[471] Ostriker, J. P., Richstone, D. O., and Thuan, T. X. 1974. *Ap. J.* **188**, L87.

[472] Ouellet, J. M. L., et al. 1992. *Phys. Rev. Lett.* **69**, 1896.

[473] Paczynski, B. 1970a. *Acta Astron.* **20**, 47.

[474] Paczynski, B. 1970b. *Acta Astron.* **20**, 47.

[475] Paczynski, B. 1971. *Ann. Rev. A. Ap.* **9**, 183.

[476] Paczynski, B. 1972. *Ap. Letters* **11**, 53.

[477] Paczynski, B. 1991. *Ap. J.* **371**, L63.

[478] Pagel, B. E. J. 1972. *Nature* **236**, 9.

[479] Pagel, B. E. J. 1985. In ESO workshop proceedings: *Production and Distribution of C, N, O Elements* (Garching, Germany: ESO), p. 155.

[480] Pagel, B. E. J. 1989. *Rev. Mex. Astr. Astrofis.* **18**, 161.

[481] Pagel, B. E. J., and Edmunds, M. G. 1981. *Ann. Rev. A. Ap.* **19**, 77.

[482] Panagia, N. 1994. *Nature* **369**, 364.

[483] Pankey, T., Jr. 1962. "Possible Thermonuclear Activities in Natural Terrestrial Materials." Ph.D. diss., Howard University (Ann Arbor: University Microfilms).

[484] Papaloizou, J. C. B. 1973. *M. N. R. A. S.* **162**, 169.

[485] Pardo, R. C., Couch, R. G., and Arnett, D. 1974. *Ap. J.* **191**, 711.

[486] Parker, E. N. 1979. *Cosmical Magnetic Fields*. Oxford: Clarendon Press.

[487] Patterson, J. R., Nagorka, B. N., Symons, G. D., and Zuk, W. M. 1971. *Nucl. Phys.* **A165**, 545.

[488] Patterson, J. R., Winkler, H., and Zaidins, C. 1969. *Ap. J.* **157**, 367.

[489] Peebles, P. J. E. 1966. *Ap. J.* **146**, 542.

[490] Peebles, P. J. E. 1971. *Physical Cosmology.* Princeton: Princeton University Press.

[491] Peimbert, M., and Torres-Peimbert, S. 1974. *Ap. J.* **193**, 327.

[492] Peters, J. G. 1968. *Ap. J.* **154**, 225.

[493] Peterson, R. C. 1980. *Ap. J.* **237**, L87.

[494] Phelps, R. L., and Janes, K. A. 1993. *A. J.* **106**, 1870.

[495] Phillips, M. M. 1993. *Ap. J.* **413**, L105.

[496] Pierce, M. J. 1994. *Ap. J.* **430**, 53.

[497] Podosek, F. A., and Swindle, T. D. 1988. *Meteorites and the Early Solar System* (Tucson: University of Arizona Press), p. 1093.

[498] Podsiadlowski, P. H., Fabian, A. C., and Stevens, I. R. 1991. *Nature* **354**, 43.

[499] Podsiadlowski, P. H., Hsu, J. J. L., Joss, P. C., and Ross, R. R. 1993. *Nature* **364**, 509.

[500] Pontecorvo, B. 1959. *Zh. Ekspr. Teor. Fiz.* **36**, 615; 1960, *Soviet Physics. JETP*, **9**, 1148.

[501] Popper, D. M. 1980. *Ann. Rev. A. Ap.* **18**, 115.

[502] Potter, D. M. 1973. *Computational Physics.* New York: John Wiley.

[503] Prantzos, N. 1991. *Gamma-Ray Line Astrophysics*, ed. P. Durouchoux and N. Prantzos (New York: American Institute of Physics), p. 129.

[504] Prantzos, N., Arnould, M., Arcoragi, J. P., and Cassè, M. 1985. *Proc. 19th International Cosmic Ray Conf.* (La Jolla) **3**, 167.

[505] Press, W. H., Teukolsky, S. A., Vetterling, W. T., and Flannery, B. P. 1992. *Numerical Recipes.* 2d ed. (Cambridge: Cambridge University Press), ch. 2.

[506] Rakavy, G., Shaviv, G., and Zinamon, Z. 1967. *Ap. J.* **150**, 131.

[507] Ray, A., Singh, K. P., and Sutaria, F. K. 1993. *Journal of Ap. and Astron.* **14**, 53.

[508] Rast, M. P., Nordlund, A., Stein, R. F., and Toomre, J. 1993. *Ap. J.* **408**, L53.

[509] Redder, A., et al. 1987. *Nucl. Phys.* **A462**, 385.

[510] Reeves, H. 1965. In *Stellar Structure*, ed. L. H. Aller and D. B. McLaughlin. Chicago: University of Chicago Press.

[511] Reeves, H. 1966. *Ap. J.* **146**, 477.

[512] Reeves, H., and Salpeter, E. E. 1959. *Phys. Rev.* **116**, 1505.

[513] Reinhard, P. G., Friedrich, J., Goeke, K., Gümmer, F., and Gross, D. H. E. 1984. *Phys. Rev. C* **30**, 878.

[514] Renzini, A. 1993. In *Supernovae and Supernova Remnants*, ed. R. McCray (Cambridge: Cambridge University Press), in press.

[515] Renzini, A., and Fusi Pecci, F. 1988. *Ann. Rev. A. and Ap.* 26, 199.

[516] Renzini, A., Greggio, L., Ritossa, C., and Ferrario, L. 1992. *Ap. J.* **400**, 280.

[517] Renzini, A., and Voli, M. 1981. *A. and A.* **94**, 175.

[518] Rester, A. C., et al. 1988. *Ap. J.* **342**, L71.

[519] Reynolds, J. H., and Turner, G. 1964. *J. Geophys. Res.* **69**, 3263.

[520] Richer, H. B., Fahlman, G. G., Buonanno, R., and Pecci, F. F. 1990. *Ap. J.* **359**, L11.

[521] Richmond, M. W., et al. 1994. *Ap. J.* **107**, 1022.

[522] Robertson, J. W. 1972. *Ap. J.* **177**, 473.

[523] Rolfs, C. E., and Rodney, W. S. 1974. *Ap. J.* **194**, L63.

[524] Rolfs, C. E., and Rodney, W. S. 1982. private communication.

[525] Rolfs, C. E., and Rodney, W. S. 1988. *Cauldrons in the Cosmos*. Chicago: University of Chicago Press.

[526] Romanishin, W., and Angel, R. 1980. *Ap. J.* **235**, 992.

[527] Rood, R. T. 1973. *Ap. J.* **184**, 815.

[528] Ruderman, M. 1965. *Rept. Prog. Phys.* **28**, 411.

[529] Ruelle, D. 1980. *Math. Intelligencer* **2**, 126.

[530] Ruffert, M., and Arnett, D. 1994. *Ap. J.* **427**, 351.

[531] Rybiki, G. B., and Lightman, A. P. 1979. *Radiative Processes in Astrophysics*. New York: John Wiley.

[532] Sakashita, S., and Hayashi, C. 1961. *Prog. Theo. Phys.* **26**, 942.

[533] Sakharov, A. D. 1967. *Zh. Eksp. Teor. Fiz. Pis'ma* **5**, 32 [(1967) *JETP Letters* **5**, 24].

[534] Salam, A. 1968. In *Elementary Particle Theory: Relativistic Groups and Analyticity, Proceedings of the Eighth Nobel Symposium*, ed. N. Svartholm (Stockholm: Almqvist and Wiksell), p. 367.

[535] Salpeter, E. E. 1952. *Ap. J.* **115**, 326.

[536] Salpeter, E. E. 1955. *Ap. J.* **121**, 161.

[537] Salpeter, E. E. 1959. *Ap. J.* **129**, 608.

[538] Salpeter, E. E., and Van Horn, H. 1969. *Ap. J.* **155**, 183.

[539] Sandie, W. G., et al. *Ap. J.* **342**, L91.

[540] Sargent, A. I., and Welch, W. J. 1993. *Ann. Rev. A. Ap.* **31**, 297.

[541] Sato, K. 1975. *Prog. Theo. Phys.* **54**, 1325.

[542] Scalo, J. M. 1986. *Fund. Cosmic Phys.* **11**, 1.

[543] Scalo, J. M., and Ulrich, R. K. 1973. *Ap. J.* **183**, 149.

[544] Schatzman, E. 1958. *White Dwarfs*. Amsterdam: North-Holland.

[545] Scheffler, H., and Elsässer, H. 1988. *Physics of the Galaxy and Interstellar Medium*. Berlin: Springer-Verlag.

[546] Schmidt, B., et al. 1993. *Nature* **364**, 600.

[547] Schmidt, B., et al. 1994. *Ap. J.* **432**, 42.

[548] Schmidt, B., Eastman, R., and Kirshner, R. P. 1995. In preparation.

[549] Schmidt, B. 1996. In preparation.

[550] Schmidt, M. 1959. *Ap. J.* **129**, 243.

[551] Schmidt, M. 1963. *Ap. J.* **137**, 758.

[552] Schramm, D. N. 1990. *Astrophysical Ages and Dating Methods*, ed. E. Vangioni-Flam, M. Cassé, J. Audouze, and J. Tran Thanh Van (Gif sur Yvette Cedex: Editions Frontieres), p. 365.

[553] Schramm, D. N., and Wagoner, R. V. 1977. *Ann. Rev. Nucl. Sci.* **27**, 37.

[554] Schwartz, R. A. 1967. *Ann. Phys.* **43**, 42.

[555] Schwarzschild, M. 1958. *The Structure and Evolution of Stars* (Princeton: Princeton University Press), p. 216.

[556] Schwarzschild, M., and Härm, R. 1958. *Ap. J.* **128**, 348.

[557] Schwarzschild, M., and Härm, R. 1959. *Ap. J.* **129**, 637.

[558] Schwarzschild, M., and Härm, R. 1962. *Ap. J.* **136**, 158.

[559] Schwarzschild, M., and Härm, R., 1965. *Ap. J.* **142**, 855.

[560] Schönberg, M., and Chandrasekhar, S. 1942. *Ap. J.* **96**, 161.

[561] Searle, L., Sargent, W. L. W., and Bagnuolo, W. G. 1973. *Ap. J.* **179**, 427.

[562] Searle, L., and Sargent, W. L. W. 1972. *Ap. J.* **173**, 25.

[563] Seeger, P. A. 1965. *U. S. Atomic Energy Commission Report* **LA-3380-MS**.

[564] Seeger, P. A., Clayton, D. D., and Fowler, W. A. 1965. *Ap. J. Suppl.* **11**, 121.

[565] Segre, E. 1964. *Nuclei and Particles*. New York: Benjamin.

[566] Seuss, H. E., and Urey, H. C. 1956. *Rev. Mod. Phys.* **28**, 53.

[567] Shapiro, S. L., and Teukolsky, S. A. 1983. *Black Holes, White Dwarfs, and Neutron Stars*. New York: John Wiley.

[568] Share, G. H., Kinzer, R. L., Kurfess, J. D., Forrest, D. J., Chupp, E. L., and Rieger, D. 1985. *Ap. J.* **292**, L61.

[569] Shen, B. S. P. 1967. *High-Energy Nuclear Reactions in Astrophysics*. New York: Benjamin.

[570] Shigeyama, T., Nomoto, K., and Hashimoto, M. 1988. *A. and A.* **196**, 141.

[571] Shu, F. H., Adams, F. C., and Lizano, S. 1987. *Ann. Rev. A. Ap.* **25**, 23.

[572] Shu, F. H., and Lubow, S. H. 1981. *Ann. Rev. A. Ap.* **19**, 277.

[573] Simpson, J. A. 1983. *Ann. Rev. Nucl. and Part. Sci.*, **33**, 323.

[574] Smarr, L., Wilson, J. R., Barton, R. T., and Bowers, R. L. 1981. *Ap. J.* **246**, 515.

[575] Smith, L. F., Biermann, P., and Metzger, P. G. 1978. *A. and A.* **66**, 65.

[576] Smith, V. V. 1989. In *Cosmic Abundances of Matter*, ed. C. J. Waddington, (New York: American Institute of Physics), p. 200.

[577] Sneden, C., Kraft, R. P., Prosser, C. F., and Langer, G. E. 1991. *A. J.* **102**, 2001.

[578] Snedon, C. 1992. private communication.

[579] Spiegel, E. 1971. *Ann. Rev. A. Ap.* **9**, 323.

[580] Spinka, H. 1970. Ph.D. diss., California Institute of Technology.

[581] Spinka, H., and Winkler, H. 1972. *Ap. J.* **174**, 455.

[582] Spinka, H., and Winkler, H. 1974. *Nucl. Phys.* **A233**, 456.

[583] Spite, F., and Spite, M. 1982. *A. and A.* **115**, 357.

[584] Spitzer, L. 1978. *Physical Processes in the Interstellar Medium*. New York: John Wiley.

[585] Spitzer, L., and Schwarzschild, M. 1953. *Ap. J.* **118**, 106.

[586] Spruit, H. C. 1992. *A. and A.* **253**, 131.

[587] Starrfield, S., Sparks, W. B., and Truran, J. W. 1974. *Ap. J. Suppl.* **28**, 247.

[588] Stein, R. F. 1966. *Stellar Evolution*, ed. R. F. Stein and A. G. W. Cameron (New York: Plenum Press), p. 3.

[589] Steinmetz, M., Müller, E., and Hillebrandt, W. 1992. *A. and A.* **254**, 177.

[590] Stephenson, D. J. 1991. *Ann. Rev. A. Ap.* **29**, 163.

[591] Stephenson, G. J., Jr. 1966. *Ap. J.* **129**, 243.

[592] Stothers, R. 1970. *M. N. R. A. S.* **151**, 65.

[593] Stothers, R., and Simon, N. R. 1970. *Ap. J.* **160**, 1019.

[594] Struve, O., and Elvey, C. T. 1934. *Ap. J.* **79**, 409.

[595] Suess, H. E. 1947. *Z. Naturforsh.* **2a**, 311 and 604.

[596] Suess, H. E. 1968. *Nucleosynthesis*, ed. W. D. Arnett, C. J. Hansen, J. W. Truran, and A. G. W. Cameron, (New York: Gordon and Breach), p. 21.

[597] Suess, H. E., and Urey, H. C. 1956. *Rev. Mod. Phys.* **28**, 53.

[598] Sugimoto, D., and Nomoto, K. 1980. *Space Sci. Rev.* **25**, 155.

[599] Suntzeff, N. B. 1994. In *Supernovae and Supernovae Remnants*, ed. R. McCray, IAU Colloquium 145 (Cambridge: Cambridge University Press), in press.

[600] Sweigart, A. V., and Mengel, J. G. 1979. *Ap. J.* **229**, 624.

[601] Symbalisty, E. 1984. *Ap. J.* **285**, 729.

[602] Taam, R. E. 1980. *Ap. J.* **237**, 142.

[603] Takarada, K., Sato, H., and Hayashi, C. 1966. *Prog. Theo. Phys.* **36**, 504.

[604] Talbot, R. J., Jr. 1971. *Ap. J.* **165**, 121.

[605] Talbot, R. J., Jr. 1980. *Ap. J.* **235**, 821.

[606] Talbot, R. J., Jr., and Arnett, D. 1971a. *Ap. J.* **170**, 409.

[607] Talbot, R. J., Jr., and Arnett, D. 1973. *Ap. J.* **186**, 51.

[608] Talbot, R. J., Jr., and Arnett, D. 1974. *Ap. J.* **190**, 605.

[609] Talbot, R. J., Jr., and Arnett, D. 1971b. *Nature* **229**, 150.

[610] Tammann, G. A. 1982. In *Supernovae: A Survey of Current Research*, ed. M. J. Rees and R. J. Stoneham (Dordrecht: Reidel), p. 371.

[611] Taylor, R. J. 1954. *Ap. J.* **120**, 332.

[612] Teegarden, B. J., Barthelmy, S. D., Gehrels, N., Tueller, J., Leventhal, M., and MacCallum, C. 1991. In *Gamma-Ray Line Astrophysics*, ed. P. Durouchoux and N. Prantzos (New York: American Institute of Physics), p. 116.

[613] Teegarden, B. J., et al. 1989. *Nature* **339**, 122.

[614] Tennekes, H., and Lumley, J. L. 1972. *A First Course in Turbulence*. Cambridge: MIT Press.

[615] Thielemann, F. K., Arnould, M., and Truran, J. W. 1986. *Advances in Nuclear Astrophysics*, ed. E. Vangioni-Flam, J. Audouze, M. Cassé, J. P. Chieze, and J. Tran Thanh Van (Gif sur Yvette: Editions Frontières), p. 525.

[616] Thielemann, F. K., Nomoto, K., and Yokoi, K. 1986. *A. and A.* **158**, 17.

[617] Thielemann, F. K., and Arnett, W. D. 1985. *Ap. J.* **295**, 604. (*See also* [667].)

[618] Thomas, H.-C. 1967. *Zeit. Ap.* **67**, 420.

[619] Thomas, J., Chen, Y. T., Hinds, S., Meredith, D., and Olson, M. 1986. *Phys. Rev. C* **33**, 1679.

[620] Timmes, F. X., and Woosley, S. E. 1992. *Ap. J.* **396**, 649.

[621] Tinney, C. G. 1993. *Ap. J.* **414**, 279.

[622] Tinney, C. G., Mould, J. R., and Reid, I. N. 1992. *Ap. J.* **396**, 173.

[623] Toline, J. E., Schombert, J. M., and Boss, A. P. 1980. *Space Sci. Rev.* **27**, 555.

[624] Tolman, R. C. 1934. *Relativity, Thermodynamics, and Cosmology*. Oxford: Clarendon Press.

[625] Tomkin, J., and Lambert, D. L. 1980. *Ap. J.* **235**, 925.

[626] Tosi, M. 1990. In *Chemical and Dynamical Evolution of Galaxies*, ed. F. Ferrini, J. Frano, and F. Matteucci (Pisa: Ets Editrice Pisa), p. 586.

[627] Trimble, V. 1975. *Rev. Mod. Phys.* **47**, 877.

[628] Trimble, V. 1987. *Ann. Rev. A. Ap.* **25**, 425.

[629] Truran, J. W. 1972. *Ap. J.* **157**, 339.

[630] Truran, J. W., Arnett, D., and Cameron, A. G. W. 1967. *Can. J. Phys.* **45**, 2315.

[631] Truran, J. W., Cameron, A. G. W., and Gilbert, A. 1966. *Can. J. Phys.* **44**, 563.

[632] Truran, J. W., Hansen, C. J., Cameron, A. G. W., and Gilbert, A. 1966. *Can. J. Phys.* **44**, 151.

[633] Truran, J. W., and Arnett, D. 1970. *Ap. J.* **160**, 181.

[634] Truran, J. W., and Arnett, D. 1970. *Ap. Space Sci.* **18**, 306.

[635] Truran, J. W., and Arnett, D. 1971. *Ap. Space Sci.* **11**, 430.

[636] Truran, J. W., and Iben, Icko, Jr. 1977. *Ap. J.* **216**, 797.

[637] Tsuda, H. 1963. *Prog. Theo. Phys.* **146**, 437.

[638] Tsuruta, S., and Cameron, A. G. W. 1966. *Can. J. Phys.* **44**, 1895.

[639] Tsuruta, S., and Cameron, A. G. W. 1970. *Ap. Space Sci.* **7**, 374.

[640] Tueller, J., Barthelmy, S., Gehrels, N., Teegarden, B. J., Leventhal, M., and MacCallum, C. J. 1990. *Ap. J.* **351**, L41.

[641] Ulrich, R. K. 1973. In *Explosive Nucleosynthesis*, ed. D. N. Schramm and D. Arnett (Austin: University of Texas Press), p. 139.

[642] Ulrich, R. K. 1982. In *Essays in Nuclear Astrophysics*, ed. C. A. Barnes, D. D. Clayton, and D. N. Schramm (Cambridge: Cambridge University Press), p. 355.

[643] Ulrich, R. K., and Scalo, J. M. 1972. *Ap. J.* **176**, L37.

[644] Unsöld, A. 1969. *Science* **163**, 1015.

[645] Utrobin, V. 1994. *A. and A.* **281**, L89.

[646] Van den Bergh, S. 1962. *A. J.* **67**, 486.

[647] Van den Bergh, S., McClure, R. D., and Evans, R. 1987. *Ap. J.* **323**, 44.

[648] Van den Bergh, S., and McClure, R. D. 1990. *Ap. J.* **359**, 277.

[649] Van den Heuvel, E. P. J. 1977. *Ann. New York Acad. Sci.* **302**, 14.

[650] VandenBerg, D. A., Hartwick, F. D. A., Dawson, P., Alexander, D. R. 1983. *Ap. J.* **266**, 747.

[651] Van Riper, K. A. 1978. *Ap. J.* **221**, 304.

[652] Van Riper, K. A., and Arnett, D. 1979. *Ap. J.* **225**, L129.

[653] Vogt, E., MacPherson, D., Kuehner, J., and Almqvist, E. 1964. *Phys. Rev.* **136**, B99.

[654] Von Ballmoos, P., Diehl, R., and Schönfelder, V. 1987. *Ap. J.* **318**, 654.

[655] Von Neumann, J. 1942. In *John von Neumann, Collected Works*, vol **6**, ed. A. J. Taub (New York: Macmillan), p. 203.

[656] Von Weizsäcker, C. F. 1935. *Zeit. Physik.* **77**, 1.

[657] Wagoner, R. V. 1969. *Ap. J. Suppl.* **18**, 247.

[658] Wagoner, R. V. 1973. *Ap. J.* **179**, 343.

[659] Wagoner, R. V., Fowler, W. A., and Hoyle, F. 1967. *Ap. J.* **148**, 3.

[660] Wagoner, R. V. 1981. *Ap. J.* **250**, L65.

[661] Wallace, R. K., and Woosley, S. E. 1981. *Ap. J. Suppl.* **45**, 389.

[662] Wallerstein, G. 1968. *Science* **162**, 625.

[663] Walter, G., Beer, H., Käppeler, F., Reffo, G., and Fabbri, F. 1986. *A. and Ap.* **167**, 186.

[664] Wampler, E. J., et al. 1990. *Ap. J.* **362**, L13.

[665] Wang, L., and Mazzali, P. A. 1992. *Nature* **355**, 58.

[666] Wang, L., and Wampler, E. J. 1992. *A. and A.* **262**, L9.

[667] Wang, R. P., and Thielemann, F.-K. 1992, private communication.

[668] Wasserburg, G. J., Busso, M., Gallino, R., and Raiteri, C. M. 1994. *Ap. J.* **424**, 412.

[669] Weaver, T. A., and Woosley, S. E. 1980. *Ann. New York Acad. Sci.* **336**, 335.

[670] Weaver, T. A., and Woosley, S. E. 1993. *Physics. Reports* **227**, 65.

[671] Wehrse, R. 1990. *Accuracy of Element Abundances from Stellar Atmospheres.* Berlin: Springer-Verlag.

[672] Weidemann, V. 1990. *Ann. Rev. A. Ap.* **28**, 103.

[673] Weidemann, V. 1991. In *White Dwarfs*, ed. G. Vauclair and E. Sion (Dordrecht: Kulwer Academic), 67.

[674] Weinberg, A. M., and Wigner, E. P. 1958. *The Physical Theory of Neutron Chain Reactors* (Chicago: University of Chicago Press), p. 110.

[675] Weinberg, S. 1967. *Phys. Rev. Lett.* **19**, 1264.

[676] Weinberg, S. 1972. *Gravitation and Cosmology.* New York: John Wiley.

[677] Weisberg, J. M., and Taylor, J. H. *Phys. Rev. Lett.* **52**, 1348.

[678] Wells, L. A., et al. 1994. *A. J.* **108**, 2233

[679] Wheeler, J. C. 1977. *Ap. Space Sci.* **50**, 125.

[680] Wheeler, J. C., Sneden, C., and Truran, J. W. 1989. *Ann. Rev. A. Ap.* **27**, 279.

[681] Whitelock, P., et al. *M. N. R. A. S.* **234**, 5P.

[682] Wielen, R. 1977. *A. and A.* **60**, 263.

[683] Williams, F. A. 1985. *Combustion Theory.* Menlo Park, Calif.: Benjamin/Cummings.

[684] Wilson, J. R. 1971. *Ap. J.* **163**, 209.

[685] Wilson, J. R. 1979. In *Sources of Gravitational Radiation*, ed. L. Smarr (Cambridge: Cambridge University Press), pp. 335, 423.

[686] Wilson, J. R. 1985. In *Numerical Astrophysics*, ed. J. M. Centrella, J. M. LeBlanc, and R. L. Bowers (Boston: Jones and Bartlett), p. 422.

[687] Wilson, J. R., Mayle, R. W., Woosley, S. E., and Weaver, T. 1986. *Ann. New York Acad. Sci.* **470**, 267.

[688] Winget, D. E., Hansen, C. J., Liebert, J., Van Horn, H. M., Fontaine, G., Nether, R. E., Kepler, S. O., and Lamb, D. Q. 1987. *Ap. J.* **315**, L77.

[689] Woltjer, L. 1958. *Bull. Astr. Inst. Netherlands* **14**, 39.

[690] Woosley, S. E. 1986. In *Nucleosynthesis and Chemical Evolution*, ed. B. Hauck, A. Maeder, and G. Meynet. Geneva: Swiss Society of Astrophysics and Astronomy, Geneva Observatory.

[691] Woosley, S. E. 1988. *Ap. J.* **330**, 218.

[692] Woosley, S. E., Arnett, W. D., and Clayton, D. D. 1973. *Ap. J. Suppl.* **26**, 231.

[693] Woosley, S. E., Eastman, R. G., Weaver, T. A., and Pinto, P. A. 1994. *Ap. J.* **429**, 300.

[694] Woosley, S. E., Fowler, W. A., Holmes, J. A., and Zimmermann, B. A. 1978. ATOMIC DATA AND NUCLEAR DATA TABLES **22**, 266.

[695] Woosley, S. E., Taam, R. E., and Weaver, T. A. 1986. *Ap. J.* **301**, 601.

[696] Woosley, S. E., and Howard, W. M. 1978. *Ap. J. Suppl.* **36**, 285.

[697] Woosley, S. E., and Weaver, T. 1981. *Ap. J.* **243**, 561.

[698] Woosley, S. E., and Weaver, T. A. 1982. In *Supernovae*, ed. M. J. Rees and R. J. Stoneham. Dordrecht: Reidel.

[699] Woosley, S. E., and Weaver, T. A. 1988. *Physics Reports* **163**, 79.

[700] Woosley, S. E., and Weaver, T. A. 1994. *Ap. J.* **423**, 371.

[701] Worrall, G., and Wilson, A. M. 1972. *Nature* **236**, 15.

[702] Wu, A., and Barnes, C. A. 1984. *Nucl. Phys.* **A422**, 373.

[703] Yahil, A. 1983. *Ap. J.* **265**, 1047.

[704] Yahil, A., and Lattimer, J. M. 1982. In *Supernovae*, ed. M. J. Rees and R. J. Stoneham. Dordrecht: Reidel.

[705] Yang, J., Turner, M. J., Steigman, G., Schramm, D. N., and Olive, K. A. 1984. *Ap. J.* **281**, 493.

[706] Zel'dovich, Ya. B. 1940. *Zh. Eksp. Teor. Fiz.* **10**, 542 (English translation: NACA TM 1261, 1960).

[707] Zel'dovich, Ya. B., Barenblatt, G. I., Librovich, V. B., and Makhviladze, G. M. 1985. *The Mathematical Theory of Combustion and Explosions* (New York: Plenum), p. 487.

[708] Zel'dovich, Ya. B., and Novikov, I. D. 1971. *Relativistic Astrophysics*. Chicago: University of Chicago Press.

[709] Zel'dovich, Ya. B., and Razier, Yu. P. 1966. *Physics of Shock Waves and High Temperature Hydrodynamic Phenomena*, ed. W. D. Hayes and R. F. Probstein. New York: Academic Press.

[710] Zhao, G., and Magain, P. 1990. *A. and A.* **238**, 242.

[711] Zhao, Z., et al. 1993. *Phys. Rev. Lett.* **70**, 2066.

[712] Ziebarth, K. 1970. *Ap. J.* **162**, 947.

Index

About the Author

David Arnett is a Regents Professor at the University of Arizona and
an astrophysicist at the Steward Observatory. He has been an active
contributor to the development of a quantitative understanding of the
birth of the elements and the death of stars, with particular interests
in the use of computers in science, and in the interface
between physics and astronomy.